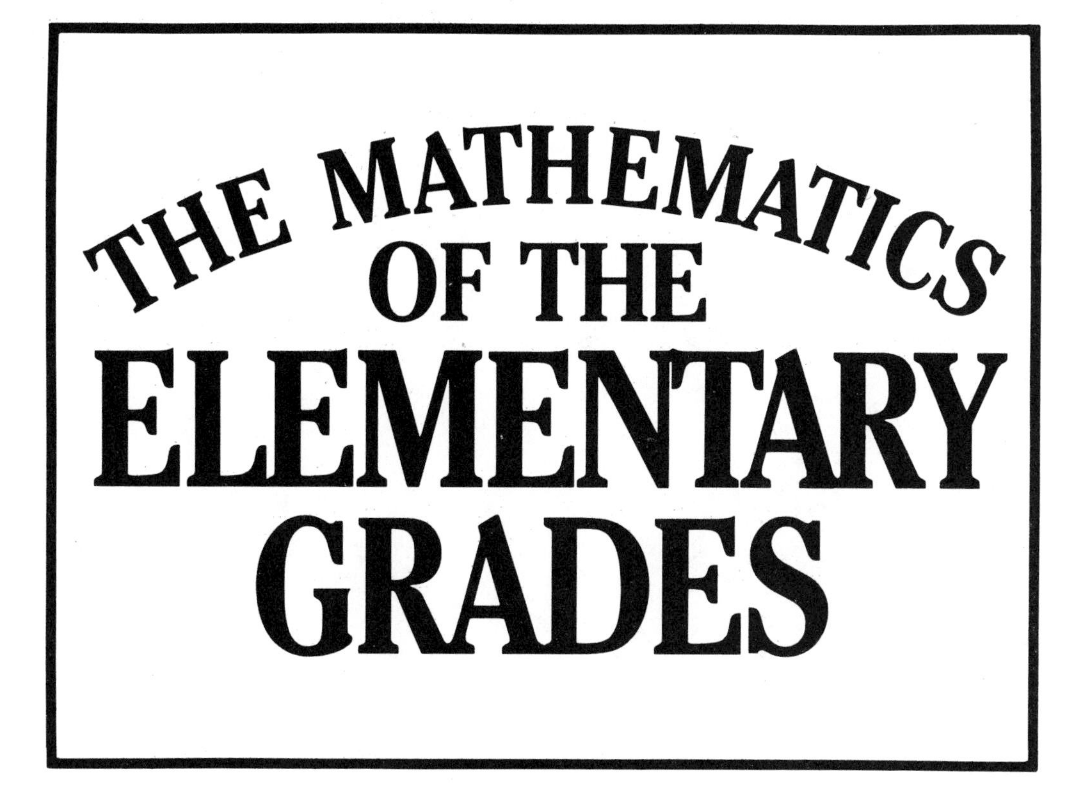
THE MATHEMATICS
OF THE
ELEMENTARY
GRADES

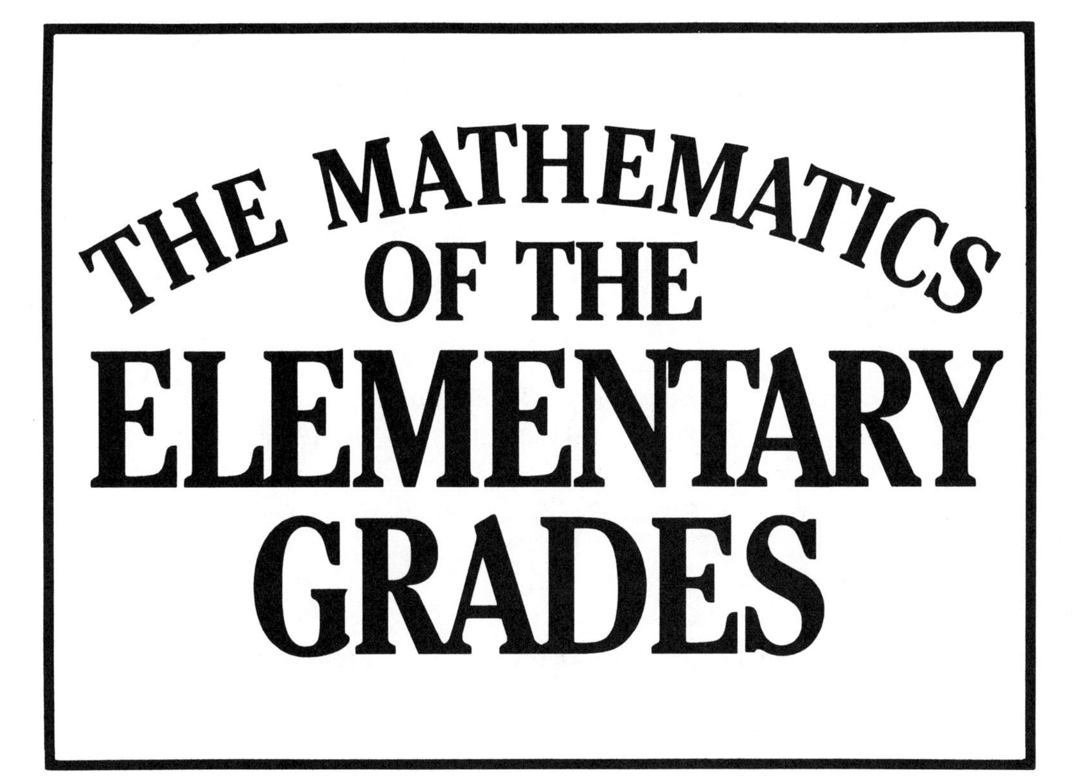

THE MATHEMATICS OF THE ELEMENTARY GRADES

William P. Berlinghoff
COLBY COLLEGE

&

Robert M. Washburn
SOUTHERN CONNECTICUT
STATE UNIVERSITY

ARDSLEY HOUSE PUBLISHERS, INC., NEW YORK

Address orders and editorial
correspondence to:
Ardsley House, Publishers, Inc.
320 Central Park West
New York, NY 10025

Other Books by the Same Authors

William P. Berlinghoff:

- *Mathematics, The Art of Reason*. Boston: D. C. Heath & Co., 1968.
- *A Mathematical Panorama*, with B. R. Gulati and K. E. Grant. Lexington, MA: D. C. Heath & Co., 1980.
- *A Mathematics Sampler*, with K. E. Grant. New York: Ardsley House, Publishers, Inc., 1987.

Robert M. Washburn:

- *Geometry For Teachers*, with Robert A. Nowlan. New York: Harper & Row, 1975.

ISBN: 0-912675-80-2

Printed in the United States of America

10 9 8 7 6 5 4 3 2 1

To Dorothy Schrader,
mathematician,
educator,
colleague,
mentor,
friend.

Preface

Recent trends in mathematics education underscore the fact that teachers of the elementary grades need to know far more about the roots of arithmetic and other mathematical topics than the skills and ideas that they actually teach to their students. As they prepare their students for a future shaped by rapid advances in science, mathematics, and computer technology, teachers must be able to deal intelligently with varied curricular approaches, teaching methods, and software packages. Moreover, they are often faced with bright, learning-disabled children, whose special needs require their teachers to circumvent problems of perception or communication by explaining mathematical ideas in many different ways. These demands of diversity and flexibility require that a first college mathematics course for prospective elementary-school teachers—which is often the *last* mathematics course they take—explore a variety of mathematical topics in considerable depth. This book aims at setting a new trend in the preparation of future teachers by providing the scope and richness of material necessary for them to understand not only the content, but the spirit of modern mathematical ideas, and to convey both content and spirit to their students.

In 1989, the National Council of Teachers of Mathematics (NCTM) published *Curriculum and Evaluation Standards for School Mathematics*, a lengthy document that proposes specific standards for elementary-school mathematics and describes in detail the recommended implementation of these standards at various grade levels. This book has been designed to complement the NCTM's recommendations by providing future teachers with the depth of mathematical experience needed to meet these standards in their classrooms. Our coverage of the following NCTM-recommended topical areas provides what future teachers need in ways often ignored by current texts.

- **Mathematics as Problem Solving.** Chapter 1 describes a dozen specific problem-solving techniques and illustrates how they are used. In the rest of the book, "Problem-Solving Comments" point out how these techniques relate to the ideas being presented and serve as a reminder that good problem-solving habits are important everywhere.

- **Mathematics as Reasoning.** An entire chapter (Chapter 2) is devoted to explaining logical connectives, quantified statements, and valid arguments. The rest of the book illustrates these principles by presenting the logical justification that supports the algorithms and ideas of elementary arithmetic and the basic concepts of geometry. This process is aided in several key places by the appearance of a fictional, but typical, student who thinks he/she is terrible at solving problems, but who succeeds in showing how reasoning and problem-solving skills can be used to handle difficult questions.

- **Mathematical Connections.** The conceptual strands of sets, equivalence relations, and functions introduced in Chapter 3 tie together the development of all the layers of the number systems, as well as unify the treatment of geometry and probability. An important theme is the use of analogy in developing parallel ideas, especially in the structure of the number systems. In particular, the wording in Chapter 9 deliberately mimics that of Chapter 8 in order emphasize how the construction of the rational-number system is analogous to that of the integers.

- **Number Sense and Numeration.** The concept of *number* is a main theme of this book. Eight of its fourteen chapters (4–10 and 13) explore numbers, numeration systems, and algorithms. The question "What is a number?" is posed in Chapter 1, and then is answered in detail for each number system—whole numbers, integers, rationals, reals, and complex numbers. Each type of number is *constructed* set-theoretically at a level suited for the intended audience, with the constructions tied to questions that arise naturally in elementary arithmetic. Numeration systems of ancient civilizations are examined as a source of comparisons and contrasts to our familiar Hindu-Arabic system. Chapter 5 also discusses numeration systems with bases other than ten, with special emphasis on the bases two, eight, and sixteen, and their importance in relation to computers.

- **Estimation.** Considerable attention is given to the importance of estimation and approximation skills. These ideas appear first in Chapter 1 as part of problem solving, and then frequently throughout the text as these problem-solving tactics are put to work. A careful discussion of the ideas of approximation, rounding, digits of precision, and error analysis appears in Section 10.7.

- **Measurement.** Most of Chapter 12 is related to the ideas of measurement in one form or another. In particular, this chapter contains a detailed explanation of the metric system and its relationship to other measurement systems. Angular measure is introduced in a unique manner that makes clear the connection between degrees and radians.

- **Geometry and Spatial Sense.** Chapters 11 and 12 blend axiomatics and intuition to explore this area. Spatial sense is developed through the informal discussion of various kinds of geometric shapes and their relationships to one another. Some of the classical postulates and theorems of Euclidean geometry are explained, leading historically and axiomatically to the story of the non-Euclidean geometries.

- **Probability and Statistics.** Chapter 14 presents the fundamental ideas of probability theory, drawing heavily on the language and algebra of sets covered in Chapter 2. Sections 14.4 and 14.5 provide an informal, but careful, introduction to both descriptive and inferential statistics. Considerable attention is paid to the underlying assumptions of probability and statistics, so that students may clearly see the limitations as well as the strengths of the techniques discussed.

In addition to these specific topical areas, the NCTM *Standards* also state, as the first of five general goals for all students:

> Students should have numerous and varied experiences related to the cultural, historical, and scientific evolution of mathematics so that they can appreciate the role of mathematics in the development of our contemporary society and explore the relationships among mathematics and the disciplines it serves.[1]

To this end, each chapter contains at least one major **Historical Note** to provide the basis for a historical exploration of one of the topics under discussion in that chapter.

We would be remiss in our discussion of the NCTM *Standards* if we were to ignore the fact that these proposals are curricular goals for elementary-school *students*, not for their teachers. As we noted at the outset, prospective teachers should be able to understand and communicate ideas at a more mature level than that at which they actually teach. Only then will these teachers be equipped to make sound pedagogical decisions about the best way to teach something to a particular group of students at a particular level. To accomplish this objective, we have blended rigorous, abstract presentations of fundamental concepts with a variety of concrete examples and intuitive, common-sense descriptions. To assist students in distinguishing between the adult, college-level treatment of a topic and a treatment that is appropriate for an elementary classroom, we have provided numerous ex-

---

[1]Page 5.

cerpts from current elementary-school texts to illustrate how the topics are typically handled in the elementary grades. By striking this balance between rigor and intuition, by blending abstraction with common sense, we believe that this book provides a vehicle for meeting a fundamental NCTM standard:

> Prospective teachers must be taught in a manner similar to how they are to teach—by exploring, conjecturing, communicating, reasoning, and so forth. Thus, colleges of education and mathematical sciences departments should reconsider their teacher preparation programs in light of these curriculum and evaluation criteria.[2]

The chapters of this book can be used in a variety of combinations to form one-, two-, or three-semester courses, either at the undergraduate level or in post-baccalaureate teacher preparation programs:

- Chapters 1–4, Sections 8.1–8.5, Chapter 9, and Chapter 13 can be used for a one-semester course that emphasizes the formal development of the real-number system. This choice of sections minimizes the treatment of numeration and algorithms.
- An alternative one-semester course, focusing on numeration and algorithms while minimizing the formal theory of the number systems, can be taught from Chapters 1–7 (treating Sections 3.4 through 4.3 lightly), Sections 8.6–8.8, and Chapter 10.
- For a two-semester development of the structure of the number systems, along with numeration and algorithms, we suggest Chapters 1–6 in the first semester, and Chapters 7–10, Sections 12.1–12.4, and Chapter 13 in the second semester. Sections 5.3 and 7.5 can be considered optional.
- A two-semester course that includes more geometry topics than the previous option might use Chapters 1–7 in the first semester (treating Section 5.3 as optional), and cover Chapters 8–12 in the second semester.
- Chapter 14, Probability and Statistics, is independent of Chapters 11–13 and can be used in place of (or in addition to) them.
- The entire book can be covered in a three-semester (or possibly in a two-semester, 8-credit) course. Such a course would provide a comprehensive mathematical background for future elementary teachers.

---

[2] Page 253.

All the material in this book has been thoroughly class-tested. It has been taught by us and our colleagues to more than a thousand students at Southern Connecticut State University during the past several years. As a result, this book has benefited from the comments of students and faculty members far too numerous to cite individually, but to whom we are grateful. We would like to express our special thanks to Professors Helen Bass, Henry Gates, Elizabeth Johnston, Leo Kuczynski, Karen Latil, Dorothy Schrader, and J. Philip Smith, all of Southern Connecticut State University, for their many valuable suggestions, and to Professors J. C. Ferrar of Ohio State University, Maita Levine of the University of Cincinnati, W. Gary Martin of Northern Illinois University, Jean M. Prendergast of Bridgewater State College, and Hubert Voltz of Slippery Rock University for their helpful reviews of the preliminary edition. We are also grateful to Scott, Foresman and Company and to Houghton Mifflin Company for permitting us to use excerpts from their elementary mathematics text series, and to Ms. Georgia Tobin for the special numeral symbols that appear in Chapter 5.

W. P. B.
R. M. W.

Acknowledgements

The following pages contain excerpts from *Invitation to Mathematics*, by L. Carey Bolster, et al. Copyright © 1985, 1987 by Scott, Foresman & Co. Reprinted by permission.

58, 78, 88, 95, 101, 104, 107, 113, 116, 122, 135, 136, 145, 154, 181, 186, 226, 256, 270, 277, 280, 309, 336, 347, 351, 354, 357, 377, 413, 423, 424, 459, 465, 484, 518.

The following pages contain excerpts from *Houghton Mifflin Mathematics*, by W. G. Quast, William L. Cole, Thelma Sparks, Mary Ann Haubner, and Charles E. Allen. Copyright © 1987 by Houghton Mifflin Company. Used by permission.

5, 27, 44, 197, 217, 233, 263, 294, 536, 541, 552, 569.

Contents

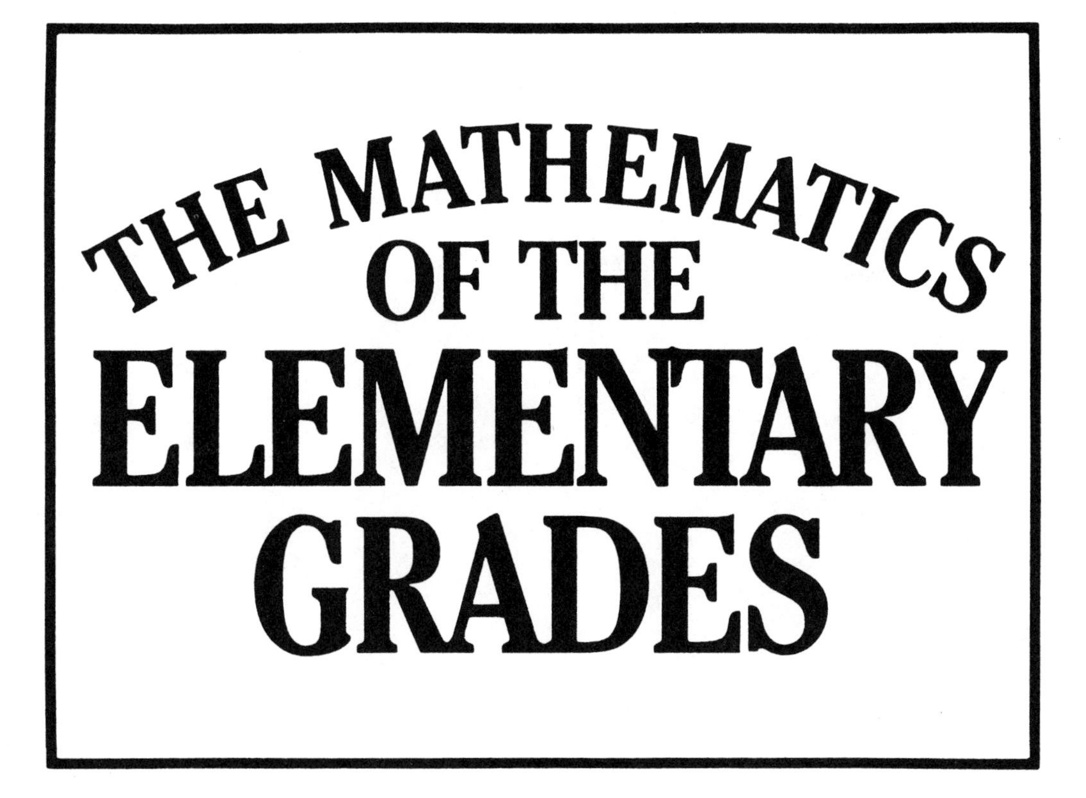
THE MATHEMATICS
OF THE
ELEMENTARY
GRADES

1.1 How To Solve A Problem

> Problem solving is the process of applying previously acquired knowledge to new and unfamiliar situations .... Problem-solving strategies involve posing questions, analyzing situations, translating results, illustrating results, drawing diagrams, and using trial and error. Students should ... be able to identify and extend patterns and use experiences and observations to make conjectures (tentative conclusions). They should ... learn to use models, known facts, and logical arguments to validate a conjecture.[1]

The bewildering variety of topics called "mathematics" is unified by a common theme—problem solving. Of course, solving problems is not limited to mathematics; nor, strictly speaking, is mathematics limited to solving problems. Nevertheless, the problem-solving process, which is part of almost every human activity, is most clearly described in the simple, abstract world of mathematics. It is at the heart of all mathematical endeavors, from primitive tallying and measuring to modern abstract speculations. All the mathematical topics that you will see later in this book, or in your teaching

[1]"Essential Mathematics for the Twenty-first Century; The Position of the National Council of Supervisors of Mathematics," *Arithmetic Teacher* 36:9 (May 1989), p. 28.

career, or in other parts of your life are united by that one common bond—a way of thinking, a way of using reason to approach and solve problems of any sort. In this section we examine the problem-solving process in general terms and describe the main features that apply to a wide variety of situations. In the next section we show in detail how this process may be used to solve a specific problem.

The study of how things are discovered or invented, of how problems are solved and how proofs are found, is called **heuristics**. Writings on this subject date back almost two thousand years. Pappus of Alexandria, a 4th-century geometer, devoted considerable attention to problem-solving methods. In the 17th century, both Descartes and Leibniz tried to formulate systematic treatments of heuristic processes. However, in the 18th and 19th centuries, as intellectual activity became more specialized, this odd mixture of logic, philosophy, and psychology was largely neglected.

Polya's Approach

Recently, there has been renewed interest in problem solving, especially among mathematicians and mathematics educators, including elementary-school teachers faced with the challenge of teaching their students to be successful problem solvers. Much of this interest is due to the remarkable work of George Polya (1888–1985), a Stanford University geometer who spent much of his long professional life teaching and writing about heuristic methods. The first of his several books on the subject, a small but insightful volume entitled *How To Solve It*, appeared in 1945. Much of the material in this section is based on that book.

Polya begins with a four-part overview of how to solve *any* problem:

- Understand the problem.
- Devise a plan.
- Carry out the plan.
- Look back at the completed solution.

Taken all by itself, this list is of little use, except to convey the important message: "Slow down! A worthwhile problem is not likely to be solved at first glance." But if we examine each of the four parts, expanding on them a bit, there emerges a truly valuable scheme to help any problem solver in any situation. Let us look at each part in turn.

Understand the Problem

Exactly what are you trying to find or prove? What facts do you have at your disposal? What parts of the context of the problem might be relevant to what you are trying to find or prove? To answer these questions you will have to *check the definitions* of all the words used in the problem to make sure you understand them thoroughly. In doing this, it is useful to *restate the problem* in as many different forms as you can think of. Sometimes it is helpful to *draw a diagram* to give your visual intuition an image of the problem as a whole. You might also want to *introduce appropriate notation* to symbolize essential pieces of the data or the question for ease in remembering and working with them.

Devise a Plan

This is the hard part, but also the part that's the most fun. The main trouble here is having to cope with the uncertainty of not knowing what to do first, or how long the job will take, or even if you will be able to solve the problem at all. In this aspect of problem solving, the main difference between students (even college students) and professional mathematicians is not that the mathematicians are necessarily brighter or quicker, but rather that, through experience, they have developed more patience. Polya provides some comforting advice for all of us:

> The first rule of discovery is to have brains and good luck. The second rule of discovery is to sit tight and wait till you get a bright idea.[2]

Of course, while you "sit tight," there are many things you can do to help you get a bright idea. You can *look for a pattern* in the data you have. If you don't have much data, you might *construct examples* to generate some. You can look for a similar problem that has already been solved and try to *argue by analogy.* You might *first solve a simpler problem* by assuming additional useful information; then try to *generalize that solution.* If the problem asks you to find something, you might *approximate the answer,* then check to see how far off you are and why. (Some approaches of this kind are called "trial-and-error methods.") If the problem asks you to prove something, it is often useful to *reason backwards from the desired conclusion.* Finally, you might check to *see if you used all the data.* Remember—none of these techniques is guaranteed to produce a plan for solving the problem, but some combination of them often works *if* you have the patience and perseverance to work at them until you get the right "bright idea."

Carry Out the Plan

Once you have devised a plan to solve the problem, you must carry it out step by step, checking each one carefully to see that you have not made a reasoning error or introduced some unjustified assumption. This part of the problem-solving process is not nearly as much fun as the exciting moment

---

[2]George Polya, *How To Solve It* (2nd ed.), Princeton, NJ: Princeton University Press, 1975, p. 172.

when you first see your plan, but it must be done to insure that you actually have a correct solution. (It's like having to split the firewood after cutting down the tree.) Sometimes there are unpleasant surprises in store for you at this stage; sometimes the plan turns out to be less complete or accurate than you first thought. On the other hand, good plans often yield dividends when you carry them out; the step-by-step logical development may turn up unexpected connections with other ideas, other problems. In the event that your plan is wrong in some fundamental way, learning to cope with the disappointment and frustration as you force yourself to search for another plan is itself a valuable lesson (in mathematics and in living).

Look Back at the Completed Solution

This is a critical part of the process for any researcher or serious problem solver. Once the problem is solved, it sometimes is possible to find an easier way to comprehend the whole situation. Often you can check your solution in that way. You might also check it by finding a simple approximation of the answer. Looking back also provides a chance to see if you can *generalize the solution* to get a "stronger" result (one with fewer restrictions), and to look for any new questions suggested by your solution.

For convenient reference, we provide here a summary list of the dozen specific problem-solving techniques discussed in this section. You might find it helpful to refer to these suggestions from time to time as you wrestle with the questions and exercises in the text.

- Check the definitions.
- Restate the problem.
- Draw a diagram.
- Introduce appropriate notation.
- Look for a pattern.
- Construct examples.
- Argue by analogy.
- First solve a simpler problem.
- Approximate the answer.
- Reason backwards from the desired conclusion.
- See if you used all the data.
- Generalize the solution.

Figure 1.1 shows how Polya's four-part approach is presented in the elementary grades. The words used to describe the steps have been simplified,

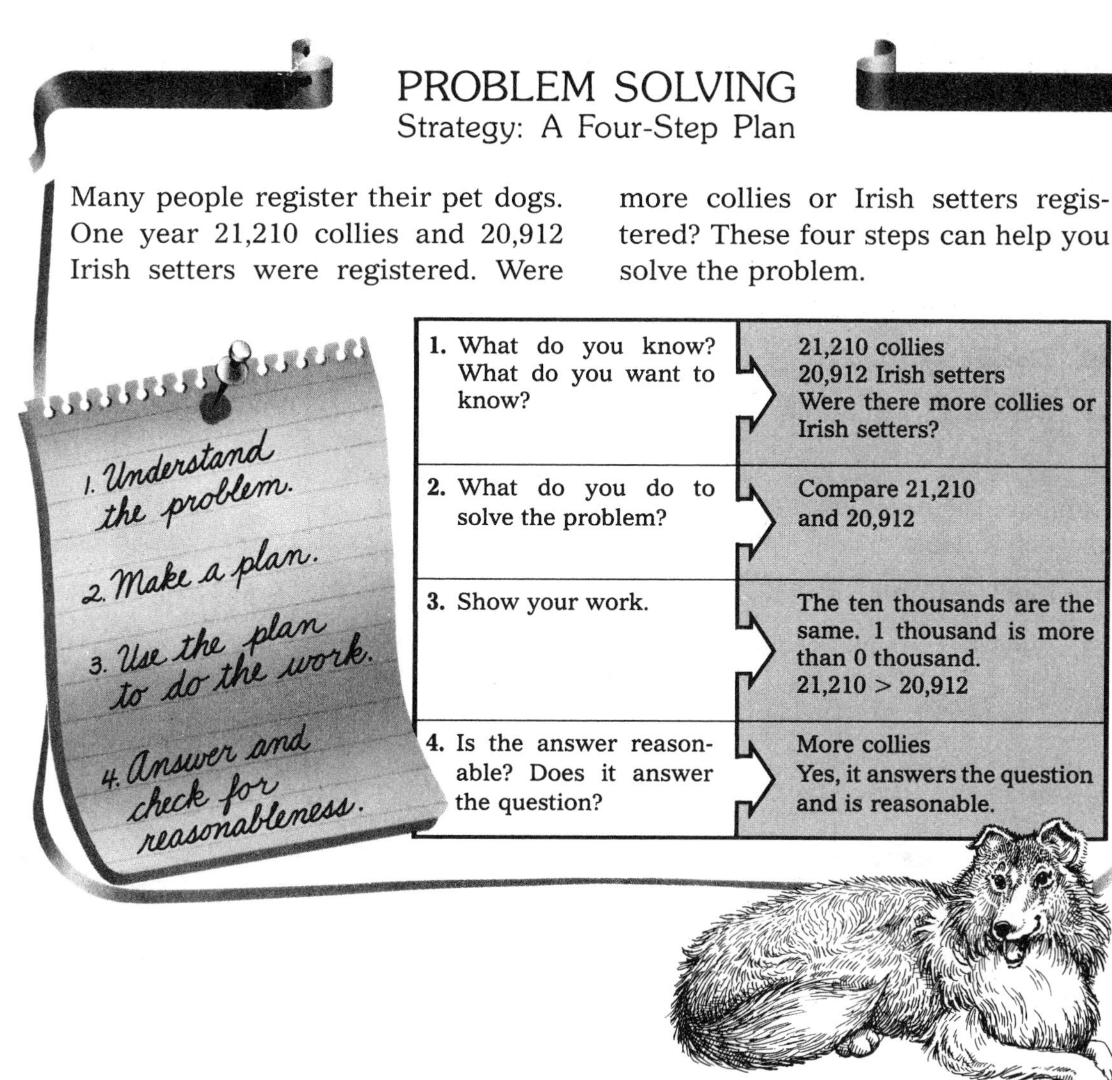

PROBLEM SOLVING

Strategy: A Four-Step Plan

Many people register their pet dogs. One year 21,210 collies and 20,912 Irish setters were registered. Were more collies or Irish setters registered? These four steps can help you solve the problem.

1. Understand the problem.
2. Make a plan.
3. Use the plan to do the work.
4. Answer and check for reasonableness.

1. What do you know? What do you want to know?	21,210 collies 20,912 Irish setters Were there more collies or Irish setters?
2. What do you do to solve the problem?	Compare 21,210 and 20,912
3. Show your work.	The ten thousands are the same. 1 thousand is more than 0 thousand. 21,210 > 20,912
4. Is the answer reasonable? Does it answer the question?	More collies Yes, it answers the question and is reasonable.

Figure 1.1 HOUGHTON MIFFLIN MATHEMATICS, Grade 5, p. 12

but the essential ideas are all there. Of course, the problem given in this example is too simple to need many of the specific techniques, but the main approach is appropriate even in this case.

There are no exercises for this section, but let us leave you with some questions to ponder: Suppose that, instead of just presenting the four major parts of Polya's scheme to a fifth-grade class, you wanted to teach them some of the dozen specific techniques. Mimicking the approach shown in Figure 1.1, can you design a problem, suitable for fifth-graders, that illustrates most of these twelve techniques? How many of them would be listed on the notebook sheet that is "pinned up" at the left of the figure? Which ones might be difficult to present? Which ones do you think would be the most helpful? How would you simplify the college-level language we have used in describing them without losing the essential ideas? Do you think this approach will help older elementary-school students to understand the ideas behind the mathematical skills they are taught?

1.2 The Problem-Solving Process At Work

When students are having trouble with mathematics, you will often hear them say, "I can always solve problems in class with the teacher, but when I'm home alone I just can't do them." (You might have felt that way yourself, sometimes.) Have you ever wondered why that is? If you think about it for a minute, it's not as surprising as it sounds. In class a teacher often guides students to the solution of a problem by asking a series of pertinent questions, causing the students to think of using some of the problem-solving techniques discussed in the previous section. When the student is at home, there may be no one else to ask the important questions; each student must learn to *ask himself or herself* these same questions.

Students who think they "can't do mathematics" or "can't solve word problems" often have that opinion simply because they have never been taught to frame the proper questions before looking for answers. They spend lots of frustrating time searching for things that won't be recognized even when found. (It's a little like looking for a matching pair of socks in a dark bedroom without taking time to turn on the light.) A good problem solver is a person who discovers what questions need to be asked before trying to answer them. The significance of Polya's work is that he, perhaps more than anyone else, made mathematics educators aware that solving problems by asking a series of helpful questions is a skill that can be taught and learned. His approach has a general form, as we saw in the previous section. That form can be applied to a wide range of problems (both within and outside of mathematics), with very little change.

In this section we illustrate this questioning approach using a single problem. Let us meet Terry, a fictional, but typical, student who (mistakenly) thinks he/she is *terrible* at solving problems. We shall ask Terry about each of the problem-solving techniques in turn, and observe that Terry's natural response to each one gives Terry (and us) important information leading ultimately to the solution of the original problem. *Before you read the answer to each question, try to answer it yourself.* Then try to solve the problem. If you can't, see if you can decide what question should be asked next. Then move on to the next question and repeat the process, until you have either solved the problem or read through to the solution provided.

Using the Problem-Solving Techniques

THE PROBLEM: In Figure 1.2, the letters of MATHEMATICS are used to create a triangular array. Determine the total number of ways MATHEMATICS can be spelled within that array by using a sequence of adjacent letters that are above, below, to the right, or to the left of each other.

```
                    M
                  M A M
                M A T A M
              M A T H T A M
            M A T H E H T A M
          M A T H E M E H T A M
        M A T H E M A M E H T A M
      M A T H E M A T A M E H T A M
    M A T H E M A T I T A M E H T A M
  M A T H E M A T I C I T A M E H T A M
M A T H E M A T I C S C I T A M E H T A M
```

Figure 1.2 The triangular array.

Question 1: Can you *restate the problem* in your own words?

Terry: I'm supposed to count how many ways to spell MATHEMATICS. That doesn't seem too hard; I only see a few.

Question 2: Did you *check the definitions*? What are the key words in the problem? What do they mean?

Terry: One key phrase is "adjacent letters that are above, below, to the right, or to the left of each other." Let's see ... "adjacent letters" means letters that are next to each other ... and they can be above or below, to the right or to the left. Hmmmmm! That must mean I have to look for MATHEMATICS spelled forwards, backwards, or up and down, or maybe even in a zigzag pattern!

Question 3: Can you *draw a diagram*?

Terry: Well, I could start by tracing a few of the possible "routes" in the triangle. (See Figure 1.3.)

As I start to trace out these paths, I can see that there are lots of possible ways to spell out MATHEMATICS in the triangle, but this picture doesn't seem to help much in telling me how to find *all* of them.

Question 4: Can you *approximate the answer*?

Terry: There seem to be lots of ways—hundreds; maybe even thousands!

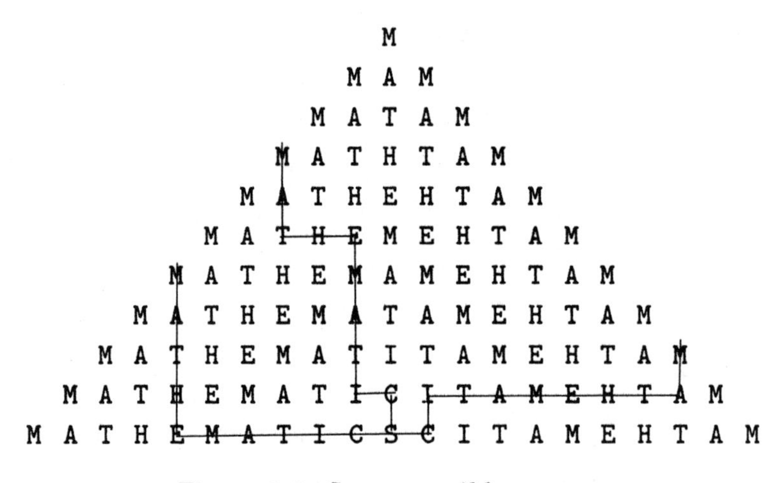

Figure 1.3 Some possible routes.

Question 5: How about trying to *solve a simpler problem* that's like this one? Can you think of one?

Terry: Maybe I could try a smaller triangle, using a shorter word.

Question 6: A good idea! Can you *construct examples* of that kind?

Terry: Sure. I could use a 2-letter word, or 3-, 4-, or 5-letter words, like NO, NOT, NOTE, NOTES. These words give me triangles like this:

```
  N
N O N

    N
  N O N
N O T O N

      N
    N O N
  N O T O N
N O T E T O N

        N
      N O N
    N O T O N
  N O T E T O N
N O T E S E T O N
```

Figure 1.4 Some simpler cases.

Question 7: Now can you *approximate the answers* to these simpler problems?

Terry: I can do better than that. I can get an exact answer to the first couple of them, and maybe to all of them, if you give me time to work at them for awhile.

– In the first triangle, there are exactly 3 different ways to spell NO—down, from the left, and from the right:

```
    N
NO  O  ON
```

– In the second triangle, I can find NOT spelled in 7 ways:

```
             N
     N   NO  O  ON    N
NOT  OT   T  T  T    TO  TON
```

– ...(*long pause*)...In the third triangle, I can find NOTE spelled in 15 ways.

– ......(*much longer pause, with lots of searching; questioner takes coffee break*)......And, in the fourth triangle I can find NOTES spelled in 31 different ways!

Question 8: That's very good! Now, can you *find a pattern* in this information?

Terry (*temporarily exhausted*): No.

Question 9: You answered that much too quickly. Try again. For instance, you might at least say something general, like: "The number of ways increases as the size of the triangle gets bigger." Try to improve on this statement, perhaps by making a table of your results.

Terry: OK. What I've found so far is this:

Number of Letters	Number of Ways
2	3
3	7
4	15
5	31

Table 1.1 Terry's first table.

That doesn't seem to help much; I still can't see a pattern.

Question 10: How about trying to *solve a simpler problem* again? What happens if you cut your triangles in half?

Terry: Let's see .... If I do what I did before, I get:

	Letters	Ways
N NO	2	2
N NO NOT	3	4
N NO NOT NOTE	4	8
N NO NOT NOTE NOTES	5	16

Figure 1.5 The half-triangle ways.

Question 11: Now can you *find a pattern*?

Terry: I see that the number of ways the word appears in each case is a power of 2; let me add that to the table:

Letters	*Ways*
2	$2 = 2^1$
3	$4 = 2^2$
4	$8 = 2^3$
5	$16 = 2^4$

Table 1.2 Terry's second table.

Question 12: Can you *generalize the solutions* of these simpler problems?

Terry: Well, ...in each case, the exponent of 2 is one less than the number of letters in the word used. I suppose that's a sort of generalization because it suggests that if there were 6 letters in the word, the number of ways should be 32, which is 2^5; if there were seven letters, the number of ways should be 64; and so on.

Question 13: Can you *introduce appropriate notation* to express this generalized conclusion in a convenient formula?

Terry: If we let n be the number of letters in the word and let s be the number of ways it is spelled in the triangle, then

$$s = 2^{n-1}$$

Question 14: Terrific! Now how about using this to *argue by analogy* about your examples of the "double triangles"?

Terry: Sure! If the triangle includes a backwards spelling on each line along with the forward spelling, then the total number of ways the word appears should be twice what we got using just the forward spellings, like this:

Letters	*Ways*
2	4
3	8
4	16
5	32

Table 1.3 Terry's third table.

And the general formula would be $s = 2 \cdot 2^{n-1} = 2^n$. Right?

Question 15: Are you sure? Did you check to *see if you used all the data*?

Terry: I'll check the table I made earlier to see if the entries are twice what I got in this table .... Something's wrong; they don't agree. But I rechecked all my counting, and the data seems to be right. What did I do wrong?

Question 16: Have you tried to *reason backwards from the desired conclusion*?

Terry: The conclusions I want—that is, the results in the first table—all seem to be one less than I expected them to be from generalizing the simpler results of the second table.

Question 17: But that expectation came from arguing by analogy from the simpler results to the more complicated problems. Did you *use all the data* when you set up that argument? If not, the analogy might be misleading.

Terry: Well, I said that the number of spellings in the whole triangle should be twice the number of forward spellings, but it seems to come out one less than that number each time. Why? .... Oh, I see! The backwards part of the triangle also contains the vertical word down the center, which I already counted in the forward part. So when I doubled the number of forward spellings I counted that one twice. That means I have to *subtract one* after I double the number of ways the word is spelled in the forward half. So, the formula should be

$$s = 2^n - 1$$

Question 18: It looks like you've got it! Now, can you use this *generalized solution* to solve the original problem?

Terry: It's easy now. MATHEMATICS has 11 letters, so the total number of ways it's spelled in that triangle is $2^{11} - 1$, which is 2047, according to my calculator. I was right; there *are* thousands of ways to do it!

With a little perseverance, Terry (and you and your students) can get used to asking questions like these to help you solve problems, and the encouraging comments along the way soon become superfluous. All it takes is some time, a little determination, and a touch of faith that it really *will* work out for you. The exercises at the end of this section give you a chance to try out some of that faith and determination.

HISTORICAL NOTE: THE DELIAN PROBLEM

Mathematical history contains many famous problems, but only a few whose renown stems from the consequences of an *incorrect* solution. One of the best known of these is the so-called "Delian Problem," which dates back to the 5th Century B.C. Tradition tells its story this way:

In 430 B.C., a terrible plague was ravaging Athens. The desperate Athenians, seeking a way to stop the plague, appealed to the oracle of Apollo, which was on the island of Delos. The oracle told them to double the size of Apollo's cubical altar. They constructed a new altar, a cube with *each side* twice as long as the side of the old altar! Of course, this made the new altar eight times the size of the old one, and, instead of stopping, the plague got even worse. In Polya's terms, they had *misunderstood the problem*, so they *devised the wrong plan* for its solution. Realizing their error, the Athenians appealed to Plato for help. After telling them that the oracle gave them this problem "to reproach the Greeks for their neglect of mathematics and their contempt of geometry," [3] Plato set about finding the proper side length for a new altar whose volume ("size") was exactly twice that of the old one.

Accounts differ as to whether or not Plato himself found a solution, but the problem was eventually solved by other Greek mathematicians, notably Eratosthenes (276–194 B.C.). Long after the plague of Athens had run its course, mathematicians of many countries were still devising new methods for "duplicating the cube." However, all these efforts shared one common failing—none of them conformed to the traditional "compass-and-straightedge construction" rules required by classical Greek geometry. This annoying imperfection remained for more than 2000 years as one of the most tantalizing unsolved problems of mathematics, and attempts to solve it produced some major advances in mathematical theory. Finally, by *restating the problem* in terms of a field known as abstract algebra, mathematicians of the 19th century were able to prove that duplication of the cube by compass-and-straightedge methods is impossible! Thus ends the saga of The Delian Problem, one of the most productive irritants in the history of problem-solving. ◇

[3] David M. Burton, *The History of Mathematics,* Boston: Allyn and Bacon, Inc., 1985, p. 134.

Exercises 1.2

For Exercises 1 and 2, apply the method used to solve the problem that was explained in this section.

1. (a) How many ways can the word NOT be spelled in the diamond-shaped array in Figure 1.6?

(b) How many ways can the word MATHEMATICS be spelled if it is arranged in a diamond-shaped array like that of Figure 1.6?

Figure 1.6 The NOT diamond.

2. How many ways are there to spell the word MATHEMATICS in each of the rectangular arrays given in Figure 1.7?

```
M A M
A T A        M A T A M
T H T        A T H T A
H E H        T H E H T
E M E        H E M E H
M A M        E M A M E
A T A        M A T A M
T I T        A T I T A
I C I        T I C I T
C S C        I C S C I

 (a)            (b)
```

Figure 1.7
Two rectangular arrays of MATHEMATICS.

3. Figure 1.8 shows the first six rows of a famous triangular array of numbers known as **Pascal's Triangle**.

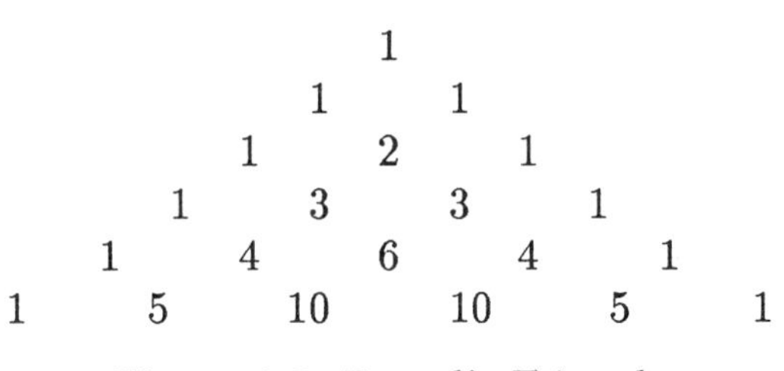

Figure 1.8 Pascal's Triangle

(a) What do you think the next three rows are?

(b) How many numbers are in the first 100 rows of Pascal's Triangle?

(c) What is the sum of the numbers of the 100th row?

(d) What is the sum of all the numbers in the first 100 rows?

(e) What is the third number of the 100th row?

(f) What is the sum of all the numbers in the diagonal farthest to the left cutting across the first 100 rows?

(g) What is the sum of all the numbers in the next diagonal cutting across the first 100 rows?

(h) What is the sum of all the numbers in the third such diagonal cutting across the first 100 rows?

4. Figure 1.9 shows the first five rows of a triangular array of numbers.

(a) What is the general pattern for this array?

(b) What are the next three rows?

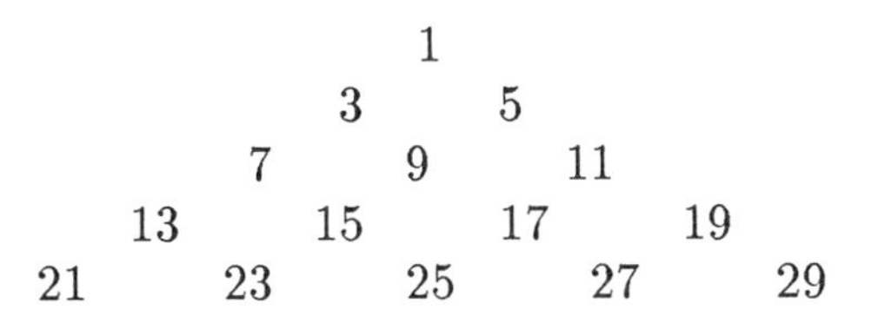

Figure 1.9 A triangular array of numbers.

(c) What is the sum of the numbers of the 100th row?
(d) What is the sum of all the numbers in the first 100 rows?
(e) What is the first number of the 100th row?

5. One hundred points are marked on a circle, and all possible chords are drawn with these points as their endpoints.

(a) How many chords are there?
(b) How many chords join nonconsecutive points?

6. A club has 100 members. Its by-laws contain the following rules for the formation of committees:

i. A committee must have at least two members.
ii. A committee cannot contain all members of the club.

How many different committees can be formed? (*Note*: Two committees are regarded as different if there is at least one club member who is on one committee and not on the other.)

7. A bear, starting from point P, walked one mile due south. Then he changed direction and walked one mile due east. Then he turned again to the left and walked one mile due north, arriving exactly at the point P he started from. What was the color of the bear?

8. Lisa has 12 boxes and 64 marbles. She wants to store all the marbles in the boxes so that each box contains a different number of marbles. Can she do this? Justify your answer.

9. Find the value of

$$\frac{1}{2!}+\frac{2}{3!}+\frac{3}{4!}+\ \dots\ +\frac{99}{100!}$$

(*Note*: $n! = 1\cdot 2\cdot 3\cdot 4\cdot\ \dots\ \cdot(n-1)\cdot n$)

1.3 Estimation

One of the questions to be asked in trying to devise a problem-solving plan is: Can you approximate an answer? Recently, mathematics educators have been placing renewed emphasis on developing this skill in students of all ages. The National Council of Teachers of Mathematics has presented the case for starting the development of approximation and estimation skills in the early grades:

> In grades K–4, the curriculum should include estimation so students can—
>
> ◇ explore estimation strategies;
> ◇ recognize when an estimate is appropriate;
> ◇ determine the reasonableness of results;
> ◇ apply estimation in working with quantities, measurement, computation, and problem solving.[4]

[4]NCTM, *Curriculum and Evaluation Standards for School Mathematics*, March 1989, p. 36.

For these reasons we emphasize the tactic of estimation repeatedly throughout this book. Whenever appropriate, the exercise sets include problems that involve estimation and approximation.

Using Estimation

To illustrate how approximation might have played a role in solving the problem described in Section 1.2, let us insert some additional dialogue:

Question 18: It looks like you've got it! Now, can you use the *generalized solution* to solve the original problem?

Terry: It's easy now. MATHEMATICS has 11 letters, so the total number of ways it's spelled in that triangle is $2^{11} - 1$, and ... according to my calculator, that's 1024.

Question 19: Could you estimate the answer? That is, could you find it approximately, without using your calculator?

Terry: I guess so, if I had to.

Question 20: How?

Terry: Well, I could start with 2 and just keep doubling.

Question 21: But that's *computing* the answer, not estimating it; besides, all that doubling can get pretty tedious after a while. How about using the laws of exponents to break 2^{11} into more convenient pieces?

Terry: I don't understand. I know that $2^{11} = 2 \cdot 2^{10}$, but that doesn't seem very useful; it just describes the doubling process backwards.

Question 22: Can you break 2^{10} into smaller powers of 2?

Terry: Well, 2^{10} is $(2^2)^5$ or $(2^5)^2$. I get it! $2^5 = 32$, which is a little more than 30, so $(2^5)^2$ must be larger than 900, maybe near 1000. So $2^{11} - 1$ is about twice that, somewhere around 2000.

Question 23: In the light of that estimate, does the answer you got on your calculator seem reasonable?

Terry: No; it's way off. I must have done something wrong. Let me try it again. Oh, I see. I should have used the power button *before* I subtracted the 1. If I do that, I get 2047, which fits my estimate pretty well.

You might find it useful to practice this kind of estimation before going on. In Exercise 1, which follows, we suggest that you look back at some of the exercises for Section 1.2 that are similar to this example and try to approximate the answers in ways analogous to the method just described. Compare your estimates with the answers you found previously, and explain any discrepancies.

Exercises 1.3

1. Check your answers to Exercises 1, 2, 5, and 9 of Section 1.2 by estimating the answers.

In Exercises 2–6, estimate the value of the given exponential expression.

2. 2^{12}
3. 2^{20}
4. $4^{10} - 1$
5. 11^3
6. 29^3

In Exercises 7–16, estimate the value of each expression by rounding the given numbers to ones that are easier to handle. Then check the accuracy of your estimates with a calculator.

7. 21×59
8. 30.71×4.995
9. $108 \div 5.21$
10. $3.1 \times 78.31 \times 121 \div 61.2$
11. $\dfrac{\sqrt{99.2} \times 1002 \times 3.87 \times .492}{4.0997}$
12. $\dfrac{31 \times 29}{14.9}$
13. $\dfrac{12.07 \times 98.5}{21 \times 196}$
14. $\dfrac{48.7 \times 5.8}{5.01 \times 6.2}$
15. $\dfrac{\sqrt{80} \times 7.3}{\sqrt{24.9 \times 1.01}}$
16. $\dfrac{\sqrt{41 \times 295 \div 30.1}}{19.9 \times 49 \div 24.9}$

1.4 Defining Numbers and Other Things

Part of the first phase of understanding any problem is *checking the definitions.* This is a critically important task. Often a problem seems far more difficult than it really is simply because the would-be solver does not have a precise understanding of the terms being used, or does not accurately define key concepts that are needed for the solution. Here we discuss properties that good definitions ought to have—properties that make them conceptually clear and logically usable. As an introduction to these properties, we focus on a question that is fundamental to every elementary-school mathematics program:

What is a number?

"That's easy," we hear you say. "I know what a number is; it's just a symbol for a quantity. Let's get on with the arithmetic review to refresh my memory about that other stuff. After all, I've already learned those elementary things; with a little brushing up I should be ready to teach them really well."

We hope so. But understanding elementary mathematics well enough to teach it competently involves much more than being able to do arithmetic. For example, in order to deal with conflicting educational theories (such as those of Jean Piaget and Jerome Bruner), their experimental bases, and their implications for teaching mathematics in the early grades, you need an intelligent acquaintance with such mathematical topics as logic, the theory of order, elementary properties of figures in space, and modern algebra. But thoughts like these may seem far removed from your immediate needs at this stage of your college career, so let us defer this discussion for a while and deal, instead, with what you think you know.

The Definition of "Number"

You said that a number is "a symbol for a quantity," and at least one respectable dictionary agrees with you.[5] But a symbol, according to this same dictionary, is "a written or printed mark, letter, abbreviation, etc. . . . ," implying that two symbols are the same only when they look the same and are different whenever they look different. Let us consider the number *three*, which you claim is a symbol. What symbol is it—

3 ?,   III ?, | | | ?

These symbols (and others) are used to represent the number *three*, but they are different marks and thus are different symbols. So, if the number *three* is a symbol, it can be at most one of these and not the others.

"O. K.," you say, "so it's not a symbol. It's really the idea represented by those symbols. It's the quantity itself."

That sounds reasonable. Now all we need to know is what a *quantity* is. Back to the dictionary:

A quantity is "an amount . . . or number."

The second part is no help, because it's the word we are trying to define; let's look up *amount*. The same dictionary says that an amount is

"a quantity," or "the sum of two or more quantities."

Now what? We have three words, each defined using the others, so there is no way of knowing what *any* of them really means.

"Oh, come on!" you say, "We all know what *quantity* and *amount* are. There's no need to be so picky about writing down formal definitions for them."

Perhaps you do know what they are. Perhaps your intuitive idea of their meanings is clear enough for you to see how they apply to 3 and 29 and 56^{148}, to $1/3$ and $-2/5$, to $\sqrt{2}$ and $\sqrt{-17}$, to .33 and .333..., to *lb.* and *cm*, to x and $2y + 5$. Can you say which of these symbols represent specific

[5] The first definition of *number* in Webster's *New World Dictionary*, Second College Edition, is: "a symbol or word, or a group of either of these, showing how many or which one in a series . . . ."

quantities or amounts? Can you then distinguish between the ones that represent numbers and the ones that do not? Before you go about tackling the much harder job of trying to teach young minds about numbers, you owe it to them to have the underlying concepts absolutely clear in your own mind!

Definitions In General

The problem is this: In choosing a place to begin our development of the number systems, we need to find some initial concepts that are simple enough to be *assumed* as obvious and reliable. Any attempt to define every word we use would simply result in a "circular" set of statements, each dependent on another, and hence all meaningless. (Think of trying to learn Japanese by using an all-Japanese dictionary.) Similarly, not every statement can be proved from previous ones. We must have some logical starting point, some words that are assumed to be meaningful without formal definition, and some statements that are assumed to be true without formal proof. Once we have chosen these basic concepts, we must build everything else from them by careful definitions and logical arguments so that all the mathematical systems we derive will be unambiguous and error-free.

The preceding discussion indicates that *number* (or *quantity* or *amount*) is not a simple enough idea to be assumed as a starting point that everyone understands clearly in the same way; we must look for something more basic. Once we have chosen the starting point, every other concept we introduce must be defined in a way that is based logically on that beginning. A vital part of this development, then, is the formation of good definitions.

A definition states a condition that something must satisfy in order to be labeled with the word being defined. A good definition must be:

characteristic

i. Given any object whatsoever, we can determine whether or not that object satisfies the condition, and

ii. the word being defined is used to label every object that satisfies the condition and is not used to label anything else;

noncircular

The condition is not described by using the word we are defining or by using words that are in turn defined by the word we are defining.

The proposed definition of a number as a symbol is faulty because it is not characteristic; the symbols themselves should not be called numbers. Moreover, its revision in terms of *quantity* or *amount* is circular because these terms are defined using *number.* To be useful, any definition must avoid both of these traps.

When you write a definition of anything (in any subject), keep in mind that it must be characteristic and noncircular. You might also find it helpful to remember another hint about forming good definitions:

> Every definition of a noun (a thing) should have two parts—a *generic* part and a *specific* part. The **generic** part indicates the *general* kind of thing you are defining, and the **specific** part tells what makes this *particular* thing distinctive.

If you want to define something called a "gzink," for instance, your definition should have a form like this:

"A **gzink** is a ______[generic part] such that ______[specific part]."

For example:

> "A **square** is a rectangle [generic part] such that all four sides are equal [specific part]."

The phrase "such that" may not be there explicitly, but there usually is some grammatical separation between the generic and specific parts of a definition. Keeping this generic/specific structure in mind will help you avoid definition errors like:

"A **square** is when all four sides are equal."

There are many things wrong with this definition. To begin with, there is no generic part. The proposed definition doesn't say what *kind* of thing a square is. The phrase "is when" seems to imply that a square is a *time*, which is nonsense. (*Remember*: Do not use "when" in a definition unless you are speaking about time, and *never* use the grammatical barbarism "is when.") One might guess that the person who wrote this definition really means to say a square is some kind of planar figure, but the definition does not supply that generic information. The generic part is present if the definition begins

"A **square** is a rectangle . . . ,"

assuming, of course, that *rectangle* has already been defined. The specific part of the definition must distinguish the rectangles we want to call "squares" from all the rest of the rectangles, so if we add

". . . with four equal sides."

the resulting definition of "square" is characteristic.

A good definition of an adjective must also be characteristic and noncircular. However, its form usually differs from that of a noun. An adjective is a label for a property of a thing, so its definition should specify

i. what kind of thing has the property, and

ii. exactly how to decide whether or not such a thing has this property.

The first part tells you that (this meaning of) the adjective can only be applied to things of a certain kind, and the second part specifies the property that the word being defined stands for. Thus, when we state

"A triangle is **isosceles** if two of its sides are equal in length."

we are saying that

i. the word "isosceles" applies to triangles, and

ii. to see if it applies to a particular triangle we must check to see if two sides of that triangle are of equal length.

That is, the adjective is just an abbreviation for the property.

(*A study hint*: When you are learning a formal definition of a thing, separate it into its generic and specific parts. When you are learning a formal definition of a property, separate it into the description of the property itself and the statement of the kind of things it modifies. These extra steps may seem a little artificial at first, but with a little practice they will become easy and useful habits.)

Our early attempts at defining "number" have failed generically

—it's not a symbol—

and specifically

—if it's a quantity of some sort, we haven't distinguished it from other quantities.

We need to look for another starting point. But even if we find a satisfactory place to begin, we shall need to know how to proceed from there without making errors. Thus, before we go back to the question "What is a number?" again, it seems wise to take a careful look at the rules of the game of reason, otherwise known as *basic logic*. That is the subject of the next chapter.

PROBLEM-SOLVING COMMENT

One common reason why people have difficulty solving problems is that they do not always understand precisely enough all the terms being used. When that happens, it is usually because either

(a) the definitions were not considered carefully enough, or

(b) the definitions given were not well designed, in the sense that they do not have the essential properties described above.

As a problem solver, you can't do much about (b), but that's a relatively rare occurrence in mathematics courses. To guard against running afoul of (a), it is a good idea to begin solving any problem by *checking the definitions*. Ask yourself: Do I thoroughly understand all the terms being used? Can I write down an accurate definition of each one? Make a habit of writing out a list of all the technical terms that are relevant to the problem; then see if you can state a formal definition of each one. Don't be satisfied with "the general idea." If you have any doubt about what a term means, look it up and write out its definition so that you have it right in front of you while you're working on the problem. ◇

Exercises 1.4

1. Find dictionary definitions of any three objects and discuss whether they are characteristic and/or noncircular.

2. Find dictionary definitions of any three objects and separate them into their generic and specific parts, if possible. If that is not possible for one or more of the definitions you find, indicate how the definition might be improved (without changing its meaning) to make it fit that form.

In Exercises 3–10, separate each definition into a generic part and a specific part. You may have to rephrase some of them. (These definitions are from *Webster's New World Dictionary*, Second College Edition.)

3. A **prototype** is a person or thing that serves as a model for one of a later period.

4. An **average** is the numerical result obtained by dividing the sum of two or more quantities by the number of quantities.

5. A **canon** is an established or basic rule or principle.

6. A **cloud** is a visible mass of condensed water vapor suspended in the atmosphere.

7. A **principle** is a fundamental truth, law, doctrine, or motivating force, upon which others are based.

8. **Fusion** is the union of different things by or as if by melting.

9. A **book** is a number of sheets of paper, parchment, etc. with writing or printing on them, fastened together along one edge, usually between protective covers.

10. A **lesson** is an exercise or assignment that a student is to prepare or learn within a given time.

For each adjective in Exercises 11–18, write a definition that fits the form described in this section. (You may consult a dictionary. If a word can have several meanings, just use one of them.)

11. equilateral
12. abstract
13. canonical
14. prophetic
15. principal
16. arbitrary
17. complementary
18. complimentary

In Exercises 19–24, assuming (just for the moment) that we know what numbers are, criticize the following proposed definitions of "addition":

19. **Addition** is a way of adding numbers.

20. **Addition** is a way of combining numbers.

21. **Addition** is like $2 + 3$.

22. **Addition** is like $2 + 3 = 5$.

23. When you add $2 + 3$ and get 5, you are doing **addition** of numbers.

24. When you combine 2 and 3 to get 5, you are doing **addition** of numbers.

Review Exercises for Chapter 1

For Exercises 1–10, indicate whether the given statement is *true* or *false*.

1. Problem solving is at the heart of all mathematical endeavors.
2. The first step in trying to solve a problem is carrying out a plan.
3. Estimation plays no role in problem solving because it does not produce exact results.
4. In trying to solve a problem it is a good idea to make sure you understand all the technical terms used in it.
5. Students should be discouraged from drawing diagrams when solving problems because that interferes with their ability to think abstractly.
6. Problem solving is not very important in today's technological world because calculators and computers solve most of our problems for us.
7. Sometimes asking yourself questions about a problem can help you to solve it.
8. It has been proved that the cube cannot be duplicated by compass-and-straightedge methods.
9. A good definition must be characteristic and circular.
10. You should encourage students to use the phrase "is when" in all definitions.
11. According to Polya's *How To Solve It*, what are the four major parts of the process for solving any problem?

For Exercises 12–16, use a format similar to the "Terry dialogue" to write a series of questions and answers leading to a solution of the given exercise.

12. Exercise 5 of Section 1.2.
13. Exercise 6 of Section 1.2.
14. Exercise 7 of Section 1.2.
15. Exercise 8 of Section 1.2.
16. Exercise 9 of Section 1.2.
17. The numbers 1, 3, 6, 10, 15, ... are called **triangular numbers** because they represent the numbers of objects needed to make triangular arrays, as shown in Figure 1.10.
    (a) What are the next three triangular numbers?
    (b) What is the 100th triangular number?
    (c) What do you think would be the analogous description of "square numbers"? In particular, what are the first five square numbers and the 100th square number?
18. Using some pattern of your own, design a triangular array of numbers similar to those of Exercises 3 and 4 of Section 1.2; then make up a series of problems about your array.

```
1    3       6          10             15
o    o       o          o              o
     o o     o o        o o            o o
             o o o      o o o          o o o
                        o o o o        o o o o
                                       o o o o o
```

Figure 1.10 Triangular numbers.

In Exercises 19–23, estimate the value of each expression; then check your estimate with a calculator.

19. 19^2

20. 23.1×3.98

21. $5.07 \times 88.1 \times .49 \div 4.2$

22. $\dfrac{\sqrt{50} \times 3.26}{2.87 \times 6.9}$

23. $\dfrac{3.2^2 \times .89}{\sqrt{99.987}}$

24. What does it mean to say that a definition must be characteristic? noncircular?

In Exercises 25–28, determine whether or not the given definition is a good one. If it is not, why not?

25. A **definition** is when something is defined.

26. A **parallelogram** is a quadrilateral with both pairs of opposite sides parallel.

27. A **horse** is an animal with four legs.

28. **Equivalent** means the same as.

2.1 Statements and Negations

> Our hypothesis is that the construction of number goes hand in hand with the development of logic and that a pre-numerical period corresponds to the pre-logical level .... Logical and arithmetical operations therefore constitute a single system ... the second resulting from generalization and fusion of the first under the complementary heading of "inclusion of classes."[1]

This formidable pronouncement from Jean Piaget, one of the world's leading psychological researchers, implies that every teacher of arithmetic should understand the basic operations of the theory of reasoning, the theory called *logic*. These operations seldom appear explicitly in elementary-school text materials. Instead, the logical principles usually are taught through the language of sets, using the idea of *inclusion*, an approach that also leads to number concepts. The connection with logic often appears only in Teacher's Guides, if at all. Nevertheless, the development of a child's ability to reason may well depend on these early lessons. Therefore, you, the teacher, should understand something about the basic principles of logic.

Even at this collegiate level of sophistication there are some limitations

---

[1]Jean Piaget, *The Child's Conception Of Number*, Atlantic Highlands, NJ: Humanities Press International, Inc., 1952, p. *viii*. (London: Routledge & Kegan Paul.) Reprinted by permission.

on our study of logic. Any such study is forced to use reason to analyze reason, an awkward situation at best. Attempts to minimize that awkwardness can quickly become far too formal for our needs. In this book we choose, instead, to put up with the occasional ambiguities of the English language. Moreover, as we saw in Chapter 1, we cannot define *everything* in a self-contained logical system without running into the problem of circularity.

Statements

In order to find a simple starting point for our study, we leave to the philosophers and scientists many relevant, but difficult, questions, such as "What is truth?" and we examine, instead, just the *form* of the reasoning process. Consequently, our study of logic focuses on how the known truth or falsity of some statements can be used to guarantee the truth or falsity of others. Thus, we begin the study of formal logic by regarding the words **true** and **false** just as labels applied (somehow) to some sentences. These two words (and they are nothing more than that) are called **truth values**; it is a basic principle of our logical system that there are only two of them. (There are other systems of logic that use more than two truth values, but they are not as simple or as widely used as the system presented here.)

All reasoning is based on "statements" of one sort or another, so we consider first what makes a statement meaningful in the context of logic. A **statement** is a sentence that has a truth value; that is, a sentence that can be labeled either *true* or *false*. This truth value may come from various sources. In many early examples we shall rely on our common-sense view of reality to tell us whether a statement is true or false; in some cases the truth value is just assigned; sometimes the truth value of a compound or complex sentence can be derived from knowing the truth values of its simpler parts. In any case, a sentence *must* have a truth value in order for it to be called a statement.

Example 1 "There is printing on this page" is a true statement. "No one passes mathematics" is a false statement. □

Example 2 "Today is Tuesday" is a statement. It is either true or false, depending on when you are reading this. □

Example 3 The sentences "What's your name?" and "Close the book!" are not statements because there is no reasonable way to assign either of them a truth value. □

Standard logic is based on two fundamental assumptions about truth values:

The Law of the Excluded Middle: There are only two truth values, *true* and *false.*

The Law of Contradiction: No statement may be both true and false at the same time (in the same context).

These laws may seem too obvious to mention, but they are crucial to the proof process. Specific examples of their use will appear later in this chapter and throughout the rest of the book.

Example 4 Every sentence in the paragraph following the opening quotation of this section is a statement; each one is either true or false. (The fact that you might consider them all true stems, no doubt, from your complete faith in the accuracy of the authors. Such faith is flattering, but not always wise.) □

Example 5 The sentence "This statement is false" is *not* a statement in standard logic because it cannot have either truth value. If we consider it true, then we must accept what it says, which makes it also false, violating the Law of Contradiction. If we consider it false, then the Law of the Excluded Middle implies that the only alternative to what it says is that it is true, again violating the Law of Contradiction. □

Quantified Statements

Two types of statements deserve special consideration because of their importance in the reasoning process; they are the two basic kinds of "quantified" statements. Examples of their use sometimes appear explicitly in elementary textbooks, as Figure 2.1 shows. Moreover, an understanding of the logical principles governing them is important for the proper interpretation of many different kinds of word problems. These two kinds of quantified statements are described formally as follows:

A **universal statement** asserts that all things of a certain kind satisfy some condition.

An **existential statement** asserts the existence of at least one thing that satisfies some condition.

Example 6 "All mice are animals" and "Every building has a flat roof" are universal statements. □

Example 7 "There is a fly in my soup," "Wizards exist," and "There really are unicorns" are existential statements. □

PROBLEM SOLVING

Strategy: Classifying

Venn diagrams can help you understand the use of the words *all, some,* and *no* in some word problems.

Are all even numbers whole numbers? Are no even numbers whole numbers? Are some even numbers whole numbers?

All even numbers are whole numbers but not all whole numbers are even numbers. The Venn diagram at the right shows this.

Are all, some, or no rectangles also triangles?

No rectangles are also triangles. The Venn diagram at the right shows this.

Are all, some, or no even numbers also less than 20?

Some, but not all, even numbers are less than 20. The Venn diagram at the right shows this.

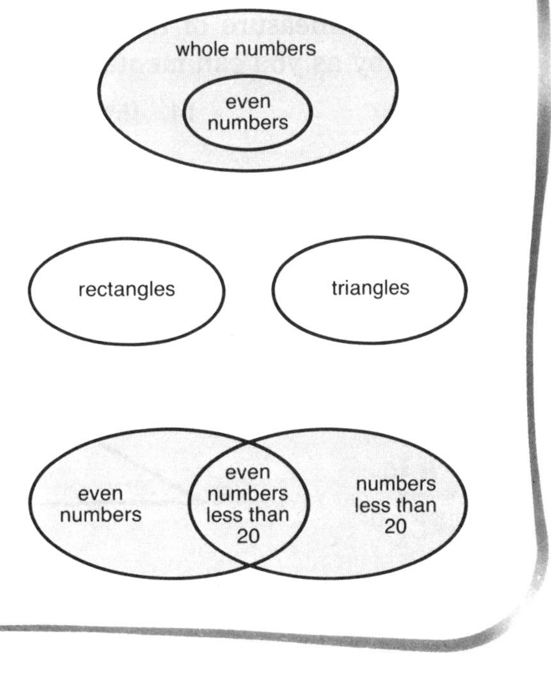

Figure 2.1 HOUGHTON MIFFLIN MATHEMATICS, Grade 6, p. 190

Note: The word *some* is used in mathematics as a shorter form of "there is at least one." It does not necessarily refer to more than one thing, nor does it rule out the possibility that *all* things might satisfy the condition being discussed. Both "Some trout are fish" and "All trout are fish" are true statements.

Example 8 "Today is Thursday" and "The moon is made of green cheese" are neither universal nor existential statements. They are called **particular statements**, because they refer to properties of specifically designated things. □

Universal, existential, and particular statements can be either true or false.

Example 9 "$x + 3 = 7$" is *not* a statement because not enough is known about x to decide whether the sentence is true or false. However:

"For all numbers x, $x + 3 = 7$"

is a false universal statement;

"There exists a number x such that $x + 3 = 7$"

is a true existential statement;

"There exists a number x such that $x + 3 = x$"

is a false existential statement;

"$4 + 3 = 7$"

is a true particular statement. □

Example 10 The sixth-grade textbook page in Figure 2.1 presents the basic quantifier words. Notice that the treatment of *some* here explicitly avoids the problem of whether its scope includes *all.* For instance, suppose a student writes "Some" as the answer to Exercise 1b. Is he/she correct? Why? It is important for the teacher to know how to handle such special cases in the event questions arise. □

Negation

Every statement has a *logical opposite.* Having said this, let us quickly caution you that "opposite" must be interpreted very carefully. For instance, "This car is going north" is *not* the *logical* opposite of "This car is going south." To reinforce that caution, we shall label this idea with a less ambiguous word defined in terms of truth values, considering the values *true* and *false* as logical opposites.

Definition The **negation** of a statement is a statement whose truth value is always opposite to that of the original statement.

Note: The word "always" in the definition of negation is equivalent to "in every circumstance." *Whenever* a statement is true its negation *must* be false, and *whenever* the statement is false its negation *must* be true.

Example 11 "The wall is completely white" and "The wall is not completely white" are negations of each other. "The wall is completely white" and "The wall is completely green" are *not* negations of each other because, although they cannot both be true at the same time, they can both be false at the same time. (Such statements are called **contraries.**) □

Notation In algebra, the task of writing numerical expressions is often simplified by using letters to represent numbers. Similarly, we shall simplify (and shorten) the writing of logical expressions by using letters, such as s, p, and q, to represent statements. If s represents a statement, then its negation will be denoted by $\sim s$.

Example 12 If s represents "Jack owns a car," then $\sim s$ represents "Jack does not own a car." Suppose s is true; that is, suppose that Jack actually owns a car. Then the statement $\sim s$ is false. On the other hand, suppose s is false. This means that Jack does not own a car, which means that $\sim s$ is true. □

The truth-value relationships among several statements may be represented conveniently by a diagram called a **truth table**. A truth table can be used to show the relationship between a statement s and its negation $\sim s$. For instance, suppose s is the statement "Jack owns a car," as in Example 12. Then $\sim s$ is "Jack does not own a car." We use a column of the table for each of the two statements (Table 2.1(a)). Because s may be either true or false, we enter both possibilities (symbolized by T and F) in the first column (Table 2.1(b)). The entries in the second column indicate that the truth value of $\sim s$ must always be the opposite of the truth value of s (Table 2.1(c)).

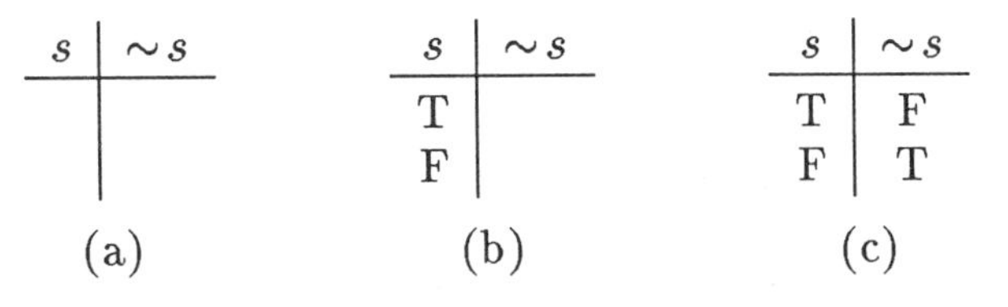

(a)

s	$\sim s$

(b)

s	$\sim s$
T	
F	

(c)

s	$\sim s$
T	F
F	T

Table 2.1 Negation

Logical Equivalence

As another example, the relationships among a statement s, its negation $\sim s$, and the negation of that, $\sim(\sim s)$, are shown in Table 2.2.

s	$\sim s$	$\sim(\sim s)$
T	F	T
F	T	F

Table 2.2 Double negation

Notice that the first and third columns have identical truth-value entries. This means that s and $\sim(\sim s)$ are the same logically, even though they might be grammatically different. If s is "Jack has a car," then $\sim(\sim s)$ can be phrased "It is not the case that Jack doesn't have a car."

Definition Two statements are **logically equivalent** (or simply **equivalent**) if they always have the same truth values.

Note: As before, "always" means "in every circumstance."

Thus, Table 2.2 shows that, for every statement s,

s and $\sim(\sim s)$ are logically equivalent.

Because they have identical truth values, equivalent statements are interchangeable at any time in any logical argument. They are just different ways to say the same thing. The choice of which form to use is merely a matter of convenience or taste.

Example 13 The negation of the universal statement "All animals have fur" is

"Not all animals have fur."

It is logically equivalent to the existential statement

"Some animals do not have fur." □

Example 14 The negation of "All mice are animals" may be phrased as "There is a mouse that is not an animal," or "Some mice are not animals," or "Not all mice are animals." The statement "No mice are animals" is *not* a negation of the first statement; it and the first statement are contraries. □

Examples 13 and 14 illustrate why we refer to *the* negation of a statement, despite previous examples of several apparently different negations of the same statement. The forms of the negation may be *grammatically* different, but they are *logically* equivalent; that is, *from the viewpoint of logic, every statement has exactly one negation.* These two examples also illustrate a basic relationship between universal and existential statements:

> The negation of a universal statement is (equivalent to) an existential statement.
>
> The negation of an existential statement is (equivalent to) a universal statement.

Example 15 "There are red fire engines" and "No fire engines are red" are negations of each other. "All birds fly" and "Some birds do not fly" are negations of each other. □

Example 16 "It is false that the wall is not green" is equivalent to "The wall is green." (Why?) □

Exercises 2.1

In Exercises 1–12, answer these questions about the given sentence:

(a) Is the sentence a statement?

(b) If it is a statement, what is its negation?

1. The sun is shining here.

2. This book is written in Greek.

3. Nobody loves me.

4. Where are you going?

5. Join the Pepsi generation.

6. You deserve a break today.

7. Meet me in St. Louis.

8. $3 + 5 = 8$

9. $3 + x = 8$

10. $3 + 5 = 7$

11. For some number x, $3 + x = 8$.

12. For some number x, $x + 2 = 2 + x$.

In Exercises 13–24:

(a) classify each statement as *universal*, *existential*, or *particular*;

(b) write (an equivalent form of) the negation of the statement.

13. Some flowers are yellow.

14. All dogs have fleas.

15. There is a unicorn on the front lawn.

16. The sun is shining.

17. Each word is important.

18. Some elephants fly.

19. Those trees are green.

20. Not all students are freshmen.

21. No man is an island.

22. Every day is Ladies' Day.

23. For every number n, $n + n = n \cdot n$.

24. For some number n, $n + n = n \cdot n$.

In Exercises 25–34, provide (an equivalent form of) the negation of each statement.

25. This is a mathematics book.

26. Today is Sunday.

27. All roses are red.

28. Some violets are blue.

29. $8 + 3 = 11$

30. 9 is less than 5.

31. No rabbits chase mice.

32. Some cars do not get good gas mileage.

33. There are at least three trees in the field.

34. There are exactly three trees in the field.

In Exercises 35–37, write two grammatically different statements that are logically equivalent to each given statement.

35. There is a rose without thorns.

36. All mathematics problems are solvable.

37. Everyone will pass the next exam.

2.2 Conjunctions and Disjunctions

When we speak with students—even with young children—we often connect two or more ideas using a combination of *and*s and *or*s. We say that it may rain *or* snow on Christmas. We say that the sun is shining *and* it is warm outside. We say that there will be an assembly today *or* tomorrow *and* another one next week. When we speak this way, we expect our students to understand these compound sentences according to clear logical rules, but we seldom examine the rules themselves. Let us examine them now.

Example 1 Suppose you take a box of marbles and say to your students, "This box contains red marbles and it contains blue marbles." Under what circumstances will this be a true statement? Its truth value can be decided by looking at the truth values of the two simple sentences involved:

p: This box contains red marbles.

q: This box contains blue marbles.

Each of these statements may be either true or false. Examining the different ways in which they can be true or false in combination with each other, we find four possibilities:

(1) p and q could both be true.
(The box contains both red marbles and blue marbles.)

(2) p could be true and q could be false.
(The box contains red marbles, but no blue ones.)

(3) p could be false and q could be true.
(The box contains blue marbles, but no red ones.)

(4) p and q could both be false.
(The box contains neither red nor blue marbles.)

If we say that "This box contains red marbles and it contains blue marbles" is a true statement, we refer only to the first of these four cases. This statement is false in the other three cases. □

This common-sense understanding of how "and" is used leads us to a general definition:

Definition The **conjunction** of two statements p and q is a statement of the form "p and q"; it is true if p and q are both true, and is false otherwise.

Now suppose you say, "This box contains red marbles or it contains blue marbles." What does it mean to claim that statement is true? This compound statement is composed of the same two simple statements as we saw before, so we can consider the same four possible truth-value combinations.

The *or* linkage between the two statements expresses the idea that one of the first three cases holds (but it doesn't specify which one). In general:

Definition The **disjunction** of two statements p and q is a statement of the form "p or q"; it is false if p and q are both false, and is true otherwise.

Note: This usage is sometimes called the *inclusive or* because it allows both statements to be true. If we had defined "p or q" to be false in that case, it would be called the *exclusive or*. This latter usage is discussed in the exercises for this section.

Notation These two logical connectives often are written symbolically—

"$p \wedge q$" represents "p and q"

"$p \vee q$" represents "p or q"

We have chosen not to use these symbols in this book because we feel that the small efficiency of replacing familiar two- and three-letter words by new single symbols does not repay the effort required to remember which is which.

The truth values for these two compound statement forms are summarized in Table 2.3. Notice that we must consider four different cases, corresponding to the four different truth-value possibilities for the *pair* of statements p, q, just as we did in Example 1. Those four pairs of possible truth values are listed in the first two columns of the table.

p	q	p and q	p or q
T	T	T	T
T	F	F	T
F	T	F	T
F	F	F	F

Table 2.3 Conjunction and Disjunction

Example 2 Consider the statements

p: Today is Friday.

q: The sun is shining.

Then

"p and q" is "Today is Friday and the sun is shining."

"p or q" is "Today is Friday or the sun is shining."

To see how the truth tables behave in this situation, observe from Table 2.3 that "p and q" is true only on sunny Fridays, but "p or q" is true every Friday, regardless of the weather, and is also true on every sunny day. □

De Morgan's Laws

At this point it is natural to ask how negation affects these compound-statement forms. In particular, how are the negations of the conjunction and disjunction of two statements related to the negations of the separate statements? Here again, we base our everyday English usage on a precise logical structure, but we seldom pay specific attention to the structure itself. For instance, if you tell someone who wants you to take a message that you do not have paper or pen, you are saying that you do not have paper *and* you do not have a pen. Often this appears as a prepositional phrase—such as "without paper or pen," meaning "without paper and without pen"—in which the preposition itself conveys the negative sense. This shift from *or* to *and* when the compound statement is negated typifies an important general principle governing the negation of conjunctions and disjunctions.

Consider the four combination forms involving negation:

$\sim(p \text{ and } q)$ $\qquad$ $\sim(p \text{ or } q)$ $\qquad$ $(\sim p) \text{ and } (\sim q)$ $\qquad$ $(\sim p) \text{ or } (\sim q)$

As usual in mathematical symbolism, we first perform the operations within the parentheses. Thus, in "$\sim(p$ and $q)$" we first consider "p and q" and then negate it; on the other hand, in "$(\sim p)$ and $(\sim q)$" we first form the negations $\sim p$, $\sim q$—then we put them together by conjunction.

Now examine Table 2.4. These truth tables give us the truth values for these four combination forms. The tables show that, as negation is distributed over the separate statements in these combinations, conjunction is changed to disjunction and vice versa.

p	q	p and q	$\sim(p$ and $q)$	p or q	$\sim(p$ or $q)$
T	T	T	F	T	F
T	F	F	T	T	F
F	T	F	T	T	F
F	F	F	T	F	T

p	q	$\sim p$	$\sim q$	$(\sim p)$ or $(\sim q)$	$(\sim p)$ and $(\sim q)$
T	T	F	F	F	F
T	F	F	T	T	F
F	T	T	F	T	F
F	F	T	T	T	T

Table 2.4 Negation with conjunction and disjunction.

By examining all the truth value possibilities, Table 2.4 actually *proves*:

> (1) "$\sim(p$ and $q)$" is logically equivalent to "$(\sim p)$ or $(\sim q)$";
> (2) "$\sim(p$ or $q)$" is logically equivalent to "$(\sim p)$ and $(\sim q)$."

These two statements are called **De Morgan's Laws**, after the 19th-century British mathematician Augustus De Morgan. They have direct counterparts involving unions, intersections, and complements of sets, as we shall see in the next chapter.

Example 3 A simple, but slightly cumbersome, example of De Morgan's Law (1) might be phrased:

> "It is false that (both) today is Friday and the sun is shining" is equivalent to "Either today is not Friday or the sun is not shining."

An example of De Morgan's Law (2) is:

> "It is false that either today is Friday or the sun is shining" is equivalent to "Today is not Friday and the sun is not shining." □

One more question about the relationships among *and*, *or*, and $\sim$ is worth pursuing at this point, if only because it is an obvious one to ask: For a single statement s, what can be said about "s and $(\sim s)$" and about "s or $(\sim s)$"? Table 2.5 provides the answer.

s	$\sim s$	s and $(\sim s)$	s or $(\sim s)$
T	F	F	T
F	T	F	T

Table 2.5 A statement with its own negation.

Notice that, regardless of the truth value of s,

> "s or $(\sim s)$" is always true,
> and
> "s and $(\sim s)$" is always false.

These are examples of two important types of logical statements:

Definition A statement that is always true, regardless of the truth values of its component parts, is called a **tautology**, and a statement that is always false is called a **contradiction.**

Using the form "s or $(\sim s)$," we can write many obvious tautologies, such as "Either today is Friday or it's not Friday." Similarly, we can construct obvious contradictions, such as "Today is Monday and it's not Monday." Some far more important uses of tautologies and contradictions appear later in this book.

HISTORICAL NOTE: BOOLE AND DE MORGAN

The symbolic treatment of logic as a mathematical system began in earnest with the work of George Boole (1815–1864) and Augustus De Morgan (1806–1871), two men who personify success in the face of adversity.

The son of an English tradesman, George Boole began life with neither money nor privilege. Nevertheless, he taught himself Greek and Latin and acquired enough of an education to become an elementary-school teacher. Boole was already twenty years old when he began studying mathematics seriously, but only eleven years later he published *Mathematical Analysis of Thought* (1847), the first of two books that have been called the beginning of pure mathematics.[2] In 1849, he became a professor of mathematics at Queens College in Dublin. Before his untimely death at the age of 49, Boole had written several other mathematics books that are now regarded as classics. His second, more famous book on logic, *The Laws of Thought* (1854), elaborates and codifies the ideas explored in the 1847 book. This symbolic approach to logic led to the development of *Boolean algebra*, the basis for modern computer-logic systems.

Augustus De Morgan was born in Madras, India, blind in one eye. Despite this disability, he graduated with honors from Trinity College in Cambridge, and was appointed a professor of mathematics at London University at the age of 22. De Morgan was a man of diverse mathematical interests who had a reputation as a brilliant, but somewhat eccentric, teacher. He wrote textbooks and popular articles on logic, algebra, mathematical history, and various other topics, and he founded the London Mathematical Society. Believing that the 19th-century separation between mathematics and logic was artificial and harmful, De Morgan worked to put many mathematical concepts on a firmer logical basis and to make logic more mathematical. He extended Boole's work in symbolic logic, contributing, among other things, the "duality" principle now known as *De Morgan's Laws*. Perhaps De Morgan was thinking of the handicap of his partial blindness when he summarized his views by saying:

> The two eyes of exact science are mathematics and logic: the mathematical sect puts out the logical eye, the logical sect puts out the mathematical eye; each believing that it can see better with one eye than with two.[3]

◇

[2]Carl B. Boyer, *A History of Mathematics*, New York: John Wiley & Sons, Inc., 1968, p. 634.

[3]Florian Cajori, *A History of Mathematics*, 2nd ed., New York: The Macmillan Company, 1919.

Exercises 2.2

In Exercises 1–10, consider the statements

p: This book is interesting.
q: I am falling asleep.

Write each of the following as a grammatically correct sentence.

1. p and q
2. p or q
3. $(\sim p)$ and q
4. p and $(\sim q)$
5. $(\sim p)$ or q
6. p or $(\sim q)$
7. $(\sim p)$ and $(\sim q)$
8. $(\sim p)$ or $(\sim q)$
9. $\sim (p$ and $q)$
10. $\sim (p$ or $q)$

In Exercises 11–14, let s represent any statement, let t represent a tautology, and let c represent a contradiction. Use truth tables to prove:

11. "s and t" is logically equivalent to s.

12. "s or c" is logically equivalent to s.

13. "s and c" is logically equivalent to c.

14. "s or t" is logically equivalent to t.

15. Exercise 11 says that any argument involving a conjunction with a tautology can be simplified by eliminating the tautology. Give the analogous interpretations of Exercises 12, 13, and 14.

In Exercises 16–22, let p, q, and r represent statements. Use truth tables to prove:

16. "$\sim \big(p$ and $(\sim q)\big)$" is logically equivalent to "$(\sim p)$ or q".

17. "$\sim \big(p$ or $(\sim q)\big)$" is logically equivalent to "$(\sim p)$ and q".

18. "p or $\sim (p$ and $q)$" is a tautology.

19. "p and $\sim (p$ or $q)$" is a contradiction.

(*Warning*: In the next three exercises, your truth table will require eight rows. Why?)

20. "$(p$ or $q)$ or r" is logically equivalent to "p or $(q$ or $r)$".

21. "p and $(q$ or $r)$" is logically equivalent to "$(p$ and $q)$ or $(p$ and $r)$".

22. "p or $(q$ and $r)$" is logically equivalent to "$(p$ or $q)$ and $(p$ or $r)$".

23. Although in mathematics *or* is usually interpreted in the *inclusive* sense unless specifically indicated otherwise, this is not always the case in everyday usage. For each of the following statements, determine from the context whether *or* is being used in the *inclusive* sense or the *exclusive* sense.

(a) The first prize for the contest is a trip to Hawaii or the cash equivalent.

(b) A student will qualify for financial aid if he/she can demonstrate financial need or high scholastic ability.

(c) Decide whether each of the following is true or false.

(d) That tree is either an oak or a maple.

(e) To receive a passing grade you either must have a class average of B or better before the final, or you must earn an A on the final exam. (Interpret both *or*'s.)

(f) He intends to join the Army or the Navy next week.

(g) To be exempt from the tax, you must be a senior citizen or have an annual income less than $10,000.

24. Use truth tables to show that the *exclusive or* is logically equivalent to each of the given expressions:

(a) $(p$ or $q)$ and $\big((\sim p)$ or $(\sim q)\big)$.

(b) $\sim \Big((p$ and $q)$ or $\big((\sim p)$ and $(\sim q)\big)\Big)$

(When pairs of parentheses lie within other such pairs, consider the innermost operations first and work outward.)

2.3 Conditional Statements

One of the most important ways to combine two statements is by the condition-consequence linkage, sometimes called the "if-then" form. As with the other compound logical statements, the formal truth-value structure coincides with normal usage, provided we think of such statements as contracts or agreements between two parties. For instance, suppose you and I have an agreement:

> If you get a perfect score on the final, then I will give you an A for the course.

Let us analyze what it means to say that this statement is true (the agreement has been kept) or false (the agreement has been broken). First of all, observe that this is a compound statement made up of two simple statements:

> p: You get a perfect score on the final.
>
> q: I give you an A for the course.

Each of these statements might be either true or false, so there are four cases to consider:

(1) p and q might both be true.
(You get a perfect score on the final and I give you an A.)

(2) p might be true and q false.
(You get a perfect score on the final, but I don't give you an A.)

(3) p might be false and q true.
(You don't get a perfect score on the final, but I still give you an A.)

(4) p and q might both be false.
(You don't get a perfect score on the final and I don't give you an A.)

Clearly, the agreement is upheld in case (1) and is broken in case (2). But what about the other two cases? Since the condition of getting a perfect score on the final is not fulfilled in those cases, I can give you an A or not, as I wish, without violating our agreement. Thus, the agreement is unbroken (true) *in all cases except (2)*. In general:

Definition A **conditional statement**, or just a **conditional**, is a statement that can be put in the form "if p, then q," where p and q are themselves statements; it is false if p is true and q is false, and it is true otherwise. The statement p is called the **hypothesis** of the conditional, and q is called the **conclusion**.

Notation The conditional "if p, then q" is symbolized by "$p \to q$", which can be read "p implies q."

Example 1 "If it is spring, then the grass is green" is a conditional statement. Its hypothesis is "It is spring"; its conclusion is "The grass is green." □

Note: The statement form "if p, then q" is logically equivalent to "q if p." The word "if" *always* labels the hypothesis.

Example 2 "I shall be happy if I get an A" is a conditional statement. Its hypothesis is "I get an A"; its conclusion is "I shall be happy." □

Example 3 "All dogs are animals" is a universal statement. It is easily rephrased as a conditional:

"If something is a dog, then it is an animal." □

Example 4 "If today is Saturday, then the moon is made of green cheese" is a conditional statement. Note that the hypothesis and conclusion need not have any real causal connection. The *form* of the statement determines whether or not it is a conditional. □

The truth-value possibilities for the conditional are summarized in Table 2.6. The four rows correspond exactly to the four cases in the example at the beginning of this section. In every conditional statement, a condition (hypothesis) is asserted to guarantee a consequence (conclusion); this assertion is false *only* if the condition is satisfied but the consequence does not follow.

p	q	$p \rightarrow q$
T	T	T
T	F	F
F	T	T
F	F	T

Table 2.6 The conditional.

If the condition is not satisfied, then the conditional statement itself is considered true "by default." In other words, *a conditional statement is true whenever its hypothesis is false.* In such cases we say the conditional is **vacuously true**. (Recall that in our logical system we *must* assign either *true* or *false* to each case.) Thus, to test whether a conditional $p \rightarrow q$ is true or false, *it suffices to assume that the hypothesis p is true.* If it is possible for q to be false at the same time, then the conditional $p \rightarrow q$ is false; otherwise, $p \rightarrow q$ is true.

The claim that a conditional statement is *false*, then, says that there are instances when the hypothesis is true and the conclusion is false. *It does not say that this is the only possible circumstance that can exist.* For example, to say that the statement "If a number ends in 3, then it is divisible by

3" is false means that *there exists* a number that satisfies the hypothesis (ends in 3) but not the conclusion (is not divisible by 3). Thus, to prove the statement false, all we have to do is find one such number.

Definition An example that proves a statement false is called a **counterexample.**

In the example of the preceding paragraph, 13 is a counterexample for the given statement. The fact that there are other numbers, such as 63, that satisfy both the hypothesis and the conclusion does not matter; if the hypothesis does not *guarantee* the conclusion, the conditional statement is false.

Example 5 [*Example 3 again*] "All dogs are animals" can be regarded as a conditional disguised as a universal statement. We consider this to be a general truth in the sense that, if Phydeau is a dog, we are confident that it is also an animal. On the other hand, if Phydeau turns out to be a cat or a boat, the fact that Phydeau may or may not also be an animal does not alter the truth of the statement, "All dogs are animals." □

Example 6 [*Example 4 again*] "If today is Saturday, then the moon is made of green cheese" is a conditional statement that is (vacuously) true six days out of every week because from Sunday through Friday both the hypothesis and the conclusion are false. Only on Saturdays is the hypothesis true, and then the falsity of the conclusion makes the entire conditional statement false. □

Converses and Related Forms

Unlike conjunctions and disjunctions, the order of the two simple statements in a conditional makes a difference. "If a figure is a square, then it is a rectangle" says something quite different from "If a figure is a rectangle, then it is a square." Conditionals involving the negations of the hypothesis and the conclusion can complicate matters further. Since these forms and the relationships among them are often useful, it is helpful to have special terms to distinguish one form from another.

As you read through the following definitions, it might help you to think of a simple example of a conditional. For instance:

p: Nancy wins. q: Tom loses.

Then the original conditional $p \rightarrow q$ is

If Nancy wins, then Tom loses.

Definition The **converse** of a conditional statement $p \rightarrow q$ is the statement

$$q \rightarrow p$$

formed by interchanging the hypothesis and the conclusion. (If Tom loses, then Nancy wins.)

The **inverse** of $p \to q$ is the statement

$$(\sim p) \to (\sim q)$$

formed by negating both the hypothesis and the conclusion. (If Nancy doesn't win, then Tom doesn't lose.)

The **contrapositive** of $p \to q$ is the statement

$$(\sim q) \to (\sim p)$$

formed by interchanging the hypothesis and the conclusion and also negating them. (If Tom doesn't lose, then Nancy doesn't win.)

We can summarize these definitions as follows:

a conditional	if p, then q	$p \to q$
its converse	if q, then p	$q \to p$
its inverse	if not p, then not q	$(\sim p) \to (\sim q)$
its contrapositive	if not q, then not p	$(\sim q) \to (\sim p)$

Example 7 *Statement*: "If it is spring, then the grass is green."

Converse: "If the grass is green, then it is spring."

Inverse: "If it is not spring, then the grass is not green."

Contrapositive: "If the grass is not green, then it is not spring." □

Universal statements can be put into conditional form. Thus, we can form the converse, inverse, and contrapositive of a universal statement. These related forms can, in turn, be translated back into a universal form, if desired.

Example 8 *Universal statement*:
"All dogs are animals."

Conditional form:
"If something is a dog, then it is an animal."

Converse:
"If something is an animal, then it is a dog."
"All animals are dogs."

Inverse:
"If something is not a dog, then it is not an animal."
"Anything that is not a dog is not an animal."

Contrapositive:
"If something is not an animal, then it is not a dog."
"Anything that is not an animal is not a dog." □

The foregoing examples suggest interesting logical connections among a conditional and its converse, inverse, and contrapositive. Since these statements are themselves conditionals, their truth values are easily determined, as shown in Table 2.7. This table shows that:

> A conditional statement and its contrapositive are logically equivalent

(because the truth values in those two columns are exactly the same). Thus, "If Nancy wins, then Tom loses" is equivalent to "If Tom doesn't lose, then Nancy doesn't win."

> The converse and the inverse are also logically equivalent statements

(for the same reason). "If Tom loses, then Nancy wins" is equivalent to "If Nancy doesn't win, then Tom doesn't lose." These interrelationships give us convenient alternative forms of conditional statements because logically equivalent statements are interchangeable. The contrapositive substitution is especially useful; examples of it will appear in the next chapter.

p	q	$p \to q$	$q \to p$	$\sim p$	$\sim q$	$(\sim p) \to (\sim q)$	$(\sim q) \to (\sim p)$
T	T	T	T	F	F	T	T
T	F	F	T	F	T	T	F
F	T	T	F	T	F	F	T
F	F	T	T	T	T	T	T

Table 2.7 Conditional, converse, inverse, and contrapositive.

Biconditionals

A true conditional may or may not have a true converse. For example, "If something is a rabbit, then it eats lettuce" is true (normally), but its converse, "If something eats lettuce, then it is a rabbit," is not. Thus, a special situation exists when a conditional $p \to q$ and its converse $q \to p$ are simultaneously true. We might describe this by saying that the truth of either of the conditions p or q guarantees the truth of the other. This is sometimes abbreviated by saying, "p is true if and only if q is true."

Definition A **biconditional statement**, or simply a **biconditional**, is a statement that can be put in the form "$(p \to q)$ and $(q \to p)$," where p and q are themselves statements. It is true precisely when p and q have the same truth values.

Notation The biconditional "p if and only if q" is denoted by "$p \leftrightarrow q$" or "p iff q."

Notice from Table 2.7 that $p \leftrightarrow q$ is true precisely when p and q have the same truth values. Thus:

> Saying that $p \leftrightarrow q$ is true is the same as asserting the logical equivalence of p and q.

Example 9 Every odd whole number is not divisible by 2, and every whole number that is not divisible by 2 is odd. Thus, "A whole number is odd" and "A whole number is not divisible by 2" are equivalent statements. That is, a whole number is odd *if and only if* it is not divisible by 2. □

A final observation from Table 2.7 is that the negation of $p \to q$ is *not* one of the three forms listed there. (How can this be observed from the table?) To construct $\sim(p \to q)$ in terms of p, q, and their negations, recall that $p \to q$ is false if and only if p is true and q is false. That is, the truth of "p and $(\sim q)$" guarantees the falsity of $p \to q$, making "p and $(\sim q)$" a likely candidate for the negation. To prove that it is, we construct the appropriate truth table (Table 2.8).

p	q	$\sim q$	$p \to q$	$\sim(p \to q)$	p and $(\sim q)$
T	T	F	T	F	F
T	F	T	F	T	T
F	T	F	T	F	F
F	F	T	T	F	F

Table 2.8 The negation of a conditional.

Example 10 A currently popular way to introduce the logical connectives to elementary-school students is by means of computer-programming statements. Proper use of the statement forms `AND`, `OR`, and `IF...THEN` in programming is based squarely on the definitions of conjunction, disjunction, and conditional. The sixth-grade textbook page shown in Figure 2.2 is an example of this. □

COMPUTER LOGIC

Programmers use logic statements when they write a plan for a program. Sentences with *IF* and *THEN* are called **conditional statements.** For a computer, the part that begins with IF is the *condition.* The part that begins with THEN is the *action.* For example:

> If the length of each side of a polygon is given, then add the lengths of the sides to determine the perimeter.

Sometimes the condition has two parts joined by *AND.* When both parts in the condition are *true,* the computer takes the action. When one part of the condition is *false,* the computer does not take the action. For example:

> If the lengths of the sides of a shape are given and they are the same, then multiply the length by the number of sides to determine the perimeter.

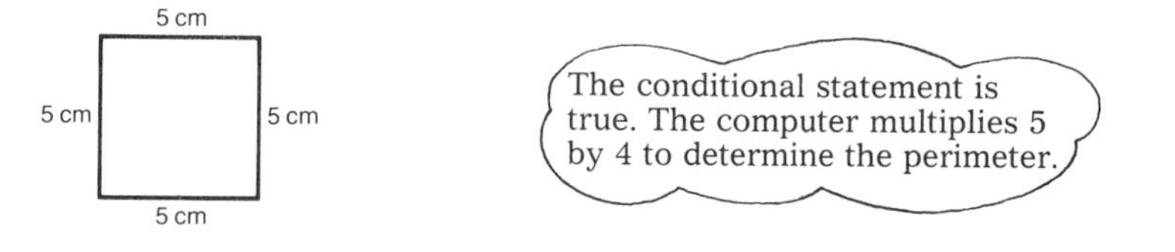

Given the conditional statement, write *true* or *false* for each example. Then write what the computer will do.

If a triangle has 3 equal sides and 3 equal angles, then write "This triangle is equilateral."

If the measure of the angle is less than 90°, then write "Acute Angle."

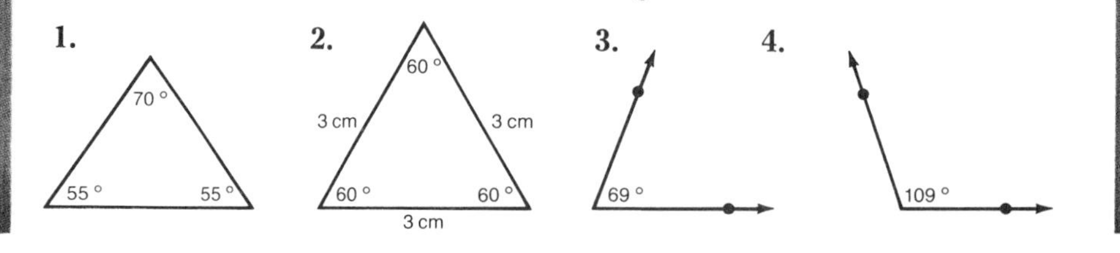

Figure 2.2 HOUGHTON MIFFLIN MATHEMATICS, Grade 6, p. 210

Exercises 2.3

In Exercises 1–10, state the hypothesis, conclusion, and converse of each statement.

1. If there is snow on the ground, then this is winter.
2. If birds fly, then fish swim.
3. Rabbits must have good eyesight if they eat carrots.
4. I'll sleep through this lecture if nobody asks a question.
5. All turkeys are birds. (*Convert to conditional form first.*)
6. All roads lead to Rome. (*Convert to conditional form first.*)
7. No airplanes fly like birds. (*Convert to conditional form first.*)
8. If there are sufficient funds and the carpenters do not strike, the building will be open by September.
9. A conditional is true if its hypothesis is false.
10. If all men are mortal, then experience is the best teacher.

11–20. State (a) the inverse and (b) the contrapositive of each statement in Exercises 1–10.

In Exercises 21–25 let p and q represent any statements. Use truth tables to prove:

21. "$p \to (\sim q)$" is logically equivalent to "$\sim(p \text{ and } q)$."
22. "$(\sim q) \to p$" is logically equivalent to "p or q."
23. "$(\sim p)$ and $\sim(p \to q)$" is a contradiction.
24. "p or $\sim(p \text{ or } q)$" is logically equivalent to "$q \to p$."
25. "p and $\sim(p \text{ and } q)$" is logically equivalent to "$\sim(p \to q)$."
26. Write the negation of "If birds have wings, then they can fly" in the form "p and $(\sim q)$."
27. Do the same as in Exercise 26 for the conditional form of "All elephants have trunks."
28. Use De Morgan's Laws and the fact that any statement s is equivalent to its double negation $\sim(\sim s)$ to write "$p \to q$" in the form of a disjunction. Check your answer by making a truth table.
29. Using Exercise 28, write the original conditionals of Exercises 26 and 29 as disjunctions.

For each conditional statement in Exercises 30–34, try to provide an instance of each of the four possible cases represented by the truth table for *if-then* statements. Based on the types of instances you find, what do you conclude about the truth or falsity of the statement?

30. If a figure is a rectangle, then it is a parallelogram. (*Note*: A parallelogram is a four-sided figure with opposite sides parallel.)
31. If a figure is a triangle, then one of its angles is a right angle.
32. If a number is divisible by 3, then the sum of its digits is divisible by 3.
33. If a number is divisible by 2 and 3, then it is divisible by 6.
34. If a number is divisible by 3 and 6, then it is divisible by 18.

2.4 Valid Arguments [optional]

We have observed that the known truth or falsity of some statements can be used to guarantee the truth or falsity of others. Whenever we cite certain statements that are regarded as being true and use them to conclude that certain other statements are true we are forming **valid arguments**; that is, we are giving a **proof**. Although proofs are often viewed as things practiced only in mathematics classes, in fact they are encountered daily—whenever one person attempts to convince another of the truth or falsity of a particular idea. The only difference between these everyday proofs and mathematical proofs is the level of logical rigor required in mathematics. In order to convince a mathematician that a statement is true, the argument used must conform to strict logical laws, known as **rules of inference**. This section describes the rules of inference that are commonly used in mathematics.

Perhaps the simplest rule of inference is derived from the logical connective *and*. If you know that the statement "p and q" is true, then you immediately know that p must be true. (You also know that q must be true.) This follows directly from the definition of *and*; the only way the compound statement "p and q" can be true is if both of its component statements are true. This rule of inference is called **simplification**.

> If "p and q" is true, then p must be true.

It is represented symbolically as

$$\frac{p \text{ and } q}{\therefore\ p}$$

where the symbol $\therefore$ (an abbreviation for "therefore") signifies the conclusion of the argument.

Example 1 If you know that the statement "Today is Monday and the sun is shining" is true, then you know that today is Monday. (You also know that the sun is shining.) □

Now let us consider the disjunction of two statements. Suppose you know that "p or q" is true. What can you conclude about the truth or falsity of p? A moment's reflection should tell you that no valid conclusion is possible because "p or q" can be true in any of three ways:

(1) when p is true and q is false,

(2) when p is false and q is true, and

(3) when both p and q are true.

If we only know that "p or q" is true, then we don't know which of these conditions prevails. Suppose, however, that besides knowing that "p or q" is true we also know that p is false (that is, $\sim p$ is true). Now we can conclude that q must be true because the falsity of p eliminates cases 1 and 3. We have already seen this rule of inference in Section 2.1; it is called the **Law of the Excluded Middle:**

> If "p or q" is true and $\sim p$ is true, then q must be true.

It is represented symbolically by

$$\begin{array}{r} p \text{ or } q \\ \sim p \\ \hline \therefore \quad q \end{array}$$

Example 2 If you know that "Today is Monday or the sun is shining" is true and you also know that today is not Monday, then you know that the sun must be shining. □

Next let us consider the conditional statement $p \rightarrow q$ ("if p, then q"). As with the disjunction, just knowing that $p \rightarrow q$ is true does not by itself allow us to conclude much about the truth or falsity of the individual statements p and q because $p \rightarrow q$ also may be true in three different ways. An examination of the truth table for $p \rightarrow q$ (Table 2.6) helps us to see this. There are three rows in which $p \rightarrow q$ is true, but in only one of them (the top row) are $p \rightarrow q$ and p both true. In this case, q is also true. Hence, if we know that $p \rightarrow q$ is true and p is also true, we can conclude that q must be true. This rule of inference is called **Modus Ponens:**

> If a conditional statement $p \rightarrow q$ is true and its hypothesis p is true, then its conclusion q must be true.

It is represented symbolically by

$$\begin{array}{r} p \rightarrow q \\ p \\ \hline \therefore \quad q \end{array}$$

Example 3 If you know that the statement "If it is Thanksgiving Day, then there are no classes" is true and you also know that it is Thanksgiving Day, you can safely conclude that there are no classes. □

Example 4 Knowing that the statements "I will wear a sweater whenever it is cold" and "It is cold" are both true statements allows us to conclude that "I will wear a sweater" is true. (Notice that "It is cold" is the hypothesis of the conditional statement.) □

Example 5 "All squares are rectangles" can be rephrased as "If something is a square, then it is a rectangle." If this conditional is true and if an object T is a square, then we know that T is a rectangle. □

Referring again to the truth table for $p \rightarrow q$, observe that there is only one row in which $p \rightarrow q$ is true and q is false (the last row). In this instance, p is also false. This fact leads to another rule of inference, called **Modus Tollens**:

> If a conditional statement $p \rightarrow q$ is true and its conclusion q is false, then its hypothesis p must be false.

It is represented symbolically by

$$\begin{array}{r} p \rightarrow q \\ \sim q \\ \hline \therefore\ \sim p \end{array}$$

Example 6 If you know that the statement "If it is Thanksgiving Day, then there are no classes" is true, but you do have classes today, then you know that today is not Thanksgiving Day. □

Example 7 Knowing that "If something is a square, then it is a rectangle" is a true statement, and also knowing that a specific object T is not a rectangle, we can conclude that T is not a square. □

Suppose you know that a conditional statement $p \rightarrow q$ is true and that its conclusion q is also true. Can you conclude anything about the truth of its hypothesis p? Examination of the truth table for $p \rightarrow q$ shows that there are two rows in which $p \rightarrow q$ and q are both true, the first and the third. In the first row, p is true; in the third row, p is false. Hence, *no valid conclusion* can be drawn about the truth value of the hypothesis. People who try to use this form of argument to conclude that p is true (as many people do!) are using an *invalid* form of reasoning, called **reasoning from the converse**.

Example 8 Although "If something is a square, then it is a rectangle" is a true statement, knowing that an object T is a rectangle does not allow us to conclude that T is a square. □

Another frequently used *invalid* form of reasoning stems from trying to reach a conclusion from the truth of a conditional statement $p \rightarrow q$ and the falsity of its hypothesis p (the truth of $\sim p$). Again, the truth table for $p \rightarrow q$ shows that it is not possible to determine the truth value of the conclusion q from this information. This invalid reasoning form is called **reasoning from the inverse.**

Example 9 If your instructor says, "If you fail the final exam, then you will fail the course," it is *not* safe to assume that passing the final exam guarantees that you will pass the course. □

It is particularly important for you as teachers to be aware of these two common forms of incorrect reasoning so that you can correct your students when they try to reason this way. The two forms may be summarized by saying:

> When a conditional statement is true, a true conclusion does *not* determine the truth value of the hypothesis, and a false hypothesis does *not* determine the truth value of the conclusion.

Symbolically,

$$\begin{array}{r} p \rightarrow q \\ q \\ \hline ? \end{array} \qquad\qquad \begin{array}{r} p \rightarrow q \\ \sim p \\ \hline ? \end{array}$$

One final observation is useful here. Recall that the biconditional statement $p \leftrightarrow q$ is logically equivalent to "$p \rightarrow q$ and $q \rightarrow p$," which is the conjunction of two conditionals. By *simplification*, if $p \leftrightarrow q$ is true, then $p \rightarrow q$ and $q \rightarrow p$ are both true. Thus, by *modus tollens*, if $p \leftrightarrow q$ is true and q is false, we can conclude that p is false.

Example 10 If we know that the statement "A whole number is divisible by 2 if and only if it is even" is true, and if we know that a particular whole number n is odd, then we also know that n is not divisible by 2. Similarly, if a given number is not divisible by 2, we know that it is not even. □

Exercises 2.4

In Exercises 1–18, consider the given statement(s) to be true, and indicate what conclusion(s) may be drawn, if any. Also state which rule of inference you used.

1. It is Sunday and the sun is shining.

2. It is cold and rainy.

3. If it is warm, then I will go swimming. It is warm.

4. Jim goes skiing if it is snowing. It is snowing.

5. If Alice owns a Raleigh, then she has a good bicycle. Alice has a good bicycle.

6. If Carl is a good student, then he has an A in mathematics. Carl does not have an A in mathematics.

7. All people with sex appeal brush their teeth with Super-Brite. George brushes his teeth with Super-Brite.

8. All good athletes are nonsmokers. Jack smokes.

9. Sue Ellen bought either chocolate or vanilla ice cream. Sue Ellen did not buy vanilla ice cream.

10. Henry is 21 or 23. If Henry is 23, then he is married. Henry is not 21.

11. Henry is 21 or 23. If Henry is 23, then he is married. Henry is not married.

12. Henry is 21 or 23. If Henry is 23, then he is married. Henry is married.

In Exercises 13–18, use the fact that a is *divisible by* b if there is some whole number c such that $b \cdot c = a$.

13. If n is divisible by 12, then it is divisible by 2 and by 3. Furthermore, n is divisible by 2 and by 3.

14. If n is divisible by 12, then it is divisible by 2 and by 3. Although n is not divisible by 2, it is divisible by 3.

15. If n is divisible by 3 and 5, then it is divisible by 15. However, n is not divisible by 15.

16. n is divisible by 18 if and only if it is divisible by 2 and 9. Moreover, n is divisible by 18.

17. n is divisible by 18 if and only if it is divisible by 2 and 9. Moreover, n is divisible by 2 and 9.

18. If n is divisible by 12, then it is divisible by 6. If n is divisible by 6, then it is divisible by 3. Furthermore, n is divisible by 12.

19. Jane has resolved that "If a candidate is for increasing the defense budget, then I will not vote for that person." At a recent press conference, when asked: "Do you favor cutting taxes and increasing the defense budget?", Candidate X replied, "No." From a logical standpoint, should Jane vote for Candidate X? Explain.

20. (a) Use a truth table to illustrate why we can conclude from the truth of statements $p \rightarrow q$ and $q \rightarrow r$ that $p \rightarrow r$ is also true. (*Hint*: You will need 8 lines in this table.) This rule of inference is called **hypothetical syllogism**; it is symbolized as

$$\begin{array}{r} p \rightarrow q \\ q \rightarrow r \\ \hline \therefore\ p \rightarrow r \end{array}$$

(b) Show how hypothetical syllogism may be used in Exercise 18.

Review Exercises for Chapter 2

In Exercises 1–10, indicate whether the given statement is *true* or *false.*

1. Any grammatically correct sentence is a statement.
2. "The car is red" and "The car is black" are negations of each other.
3. The negation of a universal statement is an existential statement.
4. Two statements are logically equivalent if they always have the same truth value.
5. If the statement "p or q" is false, then both p and q must be false.
6. If the statement "p and q" is false, then both p and q must be false.
7. The negation of "p and q" can be written "$(\sim p)$ and $(\sim q)$."
8. If p and q are both false, then the statement "$p \rightarrow q$" is false.
9. If "$p \rightarrow q$" is true, then "$q \rightarrow p$" must also be true.
10. If "$p \rightarrow q$" is true and q is false, then p must be false.

In Exercises 11–16, answer the following questions about the given sentence:

(a) Is the sentence a statement? If so, what is its truth value?

(b) If it is a statement, what is its negation, and what is the truth value of the negation?

11. Is Margaret Thatcher the Prime Minister of Great Britain?
12. Ronald Reagan was a President of the United States.
13. Spiders have polka-dot legs.
14. $2x + 1 = x$
15. For some number x, $x + 5 = 12$.
16. For every number x, $x + 5 = 12$.

In Exercises 17–21, make a truth table for the given statement. Then answer the following questions about these five statements:

(a) Which statements are tautologies?

(b) Which statements are contradictions?

(c) Which pairs statements are logically equivalent?

17. $(p \rightarrow q)$ or $(p$ and $\sim q)$
18. $(p$ and $q)$ and $(\sim p$ or $\sim q)$
19. $(q$ and $\sim p)$ or $(\sim p \rightarrow \sim q)$
20. $p \leftrightarrow (\sim p$ and $q)$
21. $(p$ or $\sim q) \rightarrow r$

In Exercises 22 and 23, form the converse, the inverse, and the contrapositive of the given (conditional) statement.

22. If the dog is furry, then it is black.
23. All igloos are made of snow.

In Exercises 24–27, form the negation of the given statement.

24. Some people are happy.
25. Philosophy is fun or history is not interesting.
26. It is not snowing and the sun is shining.
27. For all x, $x + 5 = 10$ or $2x + 1 = 7$.

28. Express the following statement as the conjunction of two conditionals:

    A number is odd if and only if it
    is not divisible by 2.

In Exercises 29–33, consider the given statements to be true and indicate what conclusion(s), if any, may be drawn from them. Also state which rule(s) of inference you used.

29. Joe will play racquetball if it is cool and not raining. It is cool and raining.

30. Joe will play racquetball if it is cool and not raining. Joe will not play racquetball.

31. All zonks are zaps. Zaps are kirps or zaps are lorps. A traz is not a kirp and it is not a lorp.

32. Sarah is 18 or 19. If Sarah is 19, then she is a college student. Sarah is not 18.

33. Sarah is 18 or 19. If Sarah is 19, then she is a college student. Sarah is not 19.

34. Suppose all of the following statements are true. What conclusion(s), if any, can be drawn about the color of Fred's house? Explain.

(a) If Joe's house is not red, then Sam's house is not black.

(b) If Joe's house is red, then Dave's house is blue.

(c) If Dave's house is blue, then Fred's house is white.

(d) Sam's house is black or Pete's house is not green.

(e) George's house is not orange and Pete's house is green.

3.1 What Is Counting?

Now that we are equipped with some basic logical tools, it is time to return to the problem posed in Chapter 1:

> How do we find a starting point for our study of elementary mathematics that is logically more basic than the concept of "number," and how then do we build a number system from that starting point?

Modern educational theorists and psychologists have been wrestling with this question in the context of how children develop their ideas of quantity and counting. Some of the most striking conclusions appear in the work of Jean Piaget and his associates. Piaget identified two major ideas whose simultaneous development is required for a child to gain a proper understanding of *number*. They are:

1. **conservation of quantity**, based on one-to-one correspondences between collections of things,[1] and

---

[1]Each thing in each collection is matched with exactly one thing in the other. A more precise definition of *one-to-one correspondence* appears in Section 5 of this chapter.

2. **seriation**, or sequential order.

In a 1953 article in *Scientific American*,[2] Piaget states:

> " Children must grasp the principle of conservation of quantity before they can develop the concept of number."

Elsewhere[3] he argues that

> "...one-to-one correspondence is obviously the tool used by the mind in comparing two sets ..."

and that this comparison process is a prerequisite for understanding conservation of quantity. But these primitive ideas require the ability to look at a collection of things simply as *things*, without regard to any of their specific characteristics, so that the only distinction left among them is the order in which they are considered.

> "Presented with a collection of items, the young child pays attention mostly to the specific physical properties of these items .... An older child abstracts all the specific properties of the objects in front of him except for their existence; thus, these objects are denied any individuality, any quality in such a way that they become equivalent to one another. The only distinction left after such an operation of abstraction is the *order* of the objects. One object is followed by another either in space, time, or, more generally, in the enumeration told by the subject. Mere counting supposes ordering, and ordering requires abstraction of individual qualities."[4]

The Counting Chant

Children encounter ordered collections early in their lives, in the form of nursery rhymes and other rote chants. Some examples are:

> " "Eeny, meeny, miney, mo, ...";
>
> " ...with a knick-knack, paddy-whack, ...";
>
> "Super-cali-fragi-listic-expi-ali-docious";
>
> "One, two, three, four, five, ..."

Yes, the last chant belongs in the list, because children often first learn it by rote as an ordered collection of meaningless sounds. There is no *a priori* connection between these noises and the concept of *number* or *quantity*; that relationship occurs later in their intellectual development.

---

[2]Jean Piaget, "How Children Develop Mathematical Concepts," *Scientific American*, Nov. 1953, p. 74.

[3]Jean Piaget, *The Child's Conception of Number*, New York: W. W. Norton & Co., 1965, p. 80. (London: Routledge & Kegan Paul; Atlantic Highlands, NJ: Humanities Press International, Inc.).

[4]From *The Essential Piaget*, edited by Howard E. Gruber and J. Jacques Voneche (p. 299).

At this first stage the **counting chant** (as we shall call it) is merely a succession of sounds without meaning, just like the other chants listed above, except for one crucial feature. Unlike the others, *this chant has no ending.* When children first learn the counting chant, they usually stop with the sound "ten." Then they learn "eleven" and "twelve," then "thirteen," "fourteen," and so on, up to "twenty." After that, a pattern of sound combinations is established that enables them to chant all the way to "a hundred" if they like (and they sometimes do). But even this is not the end; soon they learn new sounds (like "thousand" and "million") and new combination patterns like "four thousand two hundred thirty-five") that allow them to continue the chant as far as they wish.

Of course, by that time most children have begun to understand the connection between the chant and the size of a collection of objects, but that is a separate phase of development in learning to count. You, as a current student of the number systems and as a future teacher of arithmetic, must have this distinction very clear in your own mind. It is a critical component in understanding the logical development of the natural numbers and some of the difficulties children face when they are first learning to count.

Thus, in its earliest stages, "counting" is just rote chanting, and in the mind of the chanter the words are merely sounds, not related to any notion of quantity at all. That relationship, according to Piaget, comes later in a child's development and depends on conservation of quantity. This latter concept, in turn, depends on an ability to deal with collections of things (just) as collections, without focusing on specific features of the objects in those collections. For instance, before a child can move from mere chanting to actual counting, he or she must be able to see that one thing a pair of elephants and a pair of socks have in common is that they are both *pairs.* Hence, the next step in our investigation of numbers is clear: We must learn how to deal with collections of things simply as collections, or "sets," without regard to the particular features of the elements they contain.

3.2 Sets

As noted in the preceding section, an understanding of the concept of number and of the counting process depends on being able to deal with the abstract idea of *a collection of things* without considering the specific properties of the things themselves. Any collection of objects is called a **set**, and the objects themselves are called **elements** of that set. We use the term **collection** as a synonym for "set." (Notice that we do *not* have a formal definition of set here, because of the obvious circularity between the terms *set* and *collection.* It is presumed that at least one of these two words has a familiar, common-sense meaning.)

Sets and Elements

The standard notation used to write sets and elements includes the following conventions:

- Sets are usually denoted by capital English letters, such as A or B or S, and elements by small letters, such as a, b, c, etc.
- The phrase "is an element of" (or "is in") is symbolized by $\in$; thus, "$a \in S$" is read "a is an element of set S."
- The negation of the statement "$a \in S$" is denoted by "$a \notin S$" and is read "a is not an element of S." Similarly, slash marks are placed through other symbols to indicate negation.
- If a set is described by listing its elements, that list is enclosed in braces. Thus, the set of the first four letters of the alphabet is written {a, b, c, d}.

Sometimes it is inconvenient or impossible to list all the elements of a set. In that event there are two alternatives. One of them only makes sense if there is an obvious sequential pattern to the set. Then an ellipsis (...) may be used to indicate that the pattern is continued, either to a specified stopping point or indefinitely. Thus, the set of all letters of the (Roman) alphabet is {a, b, c, ..., z}, and the set of all even whole numbers is {0, 2, 4, 6, 8, ...}.

The other, more general, alternative is to specify some property or properties that characterize the elements included in the set. (By "characterize" we mean that they describe all the elements of the set and nothing else.) In this way it becomes possible to "build" the set from its description. In such cases we use a standard form of notation, called (not surprisingly) **set-builder notation:**

$$\{x \mid x \text{ has a certain property}\}$$

This is read "the set of all x such that x has a certain property." The vertical bar stands for "such that," and the variable x (or whatever expression is used in its place) symbolizes the form of a single typical element of the set.

Note: Numbers are used in the following examples as illustrations only; none of the concepts developed in this chapter depend logically on numbers in any way.

Example 1 The set of all digits can be denoted by

$$\{0, 1, 2, 3, 4, 5, 6, 7, 8, 9\}$$

or by $\{x \mid x \text{ is a digit}\}$, which is read "the set of all x such that x is a digit."

□

Example 2 The set of letters of the Roman alphabet can be denoted by {a, b, c, ..., z} or by {# | # is a letter of the Roman alphabet}. □

Example 3 The set of primary colors can be denoted by {red, yellow, blue} or by $\{x \mid x$ is a primary color$\}$. □

Example 4 $\{\frac{1}{2}, \frac{1}{4}, \frac{1}{6}, \ldots, \frac{1}{100}\}$ can also be denoted by

$$\left\{\frac{1}{n} \mid n \text{ is a positive even integer less than } 101\right\}$$ □

If a set can be described clearly enough to indicate exactly what belongs in the set and what does not, we say that the set is **well-defined**. For example, the set of all big buildings is not well-defined because "big" is subject to different interpretations in various contexts. The set {a, b, c} is well-defined, however, because it is clear exactly what is in it (the first three letters of the Roman alphabet) and what is not in it (anything else). The set {0, 1, 2, 3, ..., 100} is well-defined, provided we assume that the "..." continues the obvious pattern suggested; this set contains all whole numbers between 0 and 100, inclusive, and nothing else.

Example 5 The following are not well-defined sets. (Why not? Compare these with the sets in Example 6.)

The set of all tall people
The set of all pretty pictures
The set of all lovable puppies □

Example 6 The following sets are well-defined:

The set of all people more than six feet tall
The set of all paintings by Michelangelo
The set of all beagles less than six months old □

Example 7 Elementary arithmetic texts make extensive use of sets to teach basic number concepts in the early grades. In Figure 3.1 you see a page from a kindergarten text. At this level, pictures are used to represent sets. Observe how difficult it would be to define these sets verbally. □

The Empty Set

The set $\{x \mid x$ has a certain property$\}$ is a well-defined set provided that the statement "x has a certain property" is either true or false, unambiguously, whenever a specific thing is substituted for x. That is, the defining condition

Figure 3.1 INVITATION TO MATHEMATICS, Kindergarten, p. 32

for a well-defined set must be a *statement* (in the sense of Chapter 2) for each specific item x. Now, it is possible for the defining condition of a set to be *always* false, regardless of what is substituted for the variable. In other words, it is possible to have a set defined by a property that *nothing* fits. In this case, the set has no elements in it; nevertheless, it is still a set.

Definition The set containing no elements is called the **empty set** (or **null set**) and is denoted by $\emptyset$.

Example 8 The empty set has many possible descriptions, such as:

The set of all three-headed people
The set of all planets within 5 miles of the sun
$\{n \mid n$ is a whole number whose square is $3\}$

All of these describe the same set, $\emptyset$. □

In many elementary-school texts the empty set is denoted by $\{\ \}$ in order to emphasize visually that we have a set with nothing in it. Either $\{\ \}$ or $\emptyset$ is a correct symbol for the empty set. However, $\{\emptyset\}$ *is NOT the empty set*; it is a set containing one element, namely $\emptyset$.

Subsets

As a further example of the close connection between logic and set theory, consider the universal statement

"All turnips are vegetables,"

which can be phrased as the conditional statement

"If x is a turnip, then x is a vegetable."

When we say that this conditional is true we are claiming that anything satisfying the hypothesis must also satisfy the conclusion. In the language of sets, this means that $\{x \mid x$ is a turnip$\}$ is included within the "larger" set $\{x \mid x$ is a vegetable$\}$. Thus, a universal statement is like the set-theoretic claim that one set is included in another.

Definition A set A is a **subset** of a set B if every element of A is also an element of B. We write this as "$A \subseteq B$."

Of course, it may happen that every element of A is in B, and B contains nothing else (that is, every element of B is also in A). In this case, A and B are indistinguishable as sets because sets are just collections of things, and these two collections contain exactly the same things. This idea is used to *define* equality of sets.

Definition Two sets A and B are **equal** if A is a subset of B and B is also a subset of A. We write "$A = B$."

If A is a subset of B, but B contains other elements as well, the inclusion of A in B is said to be "proper":

Definition A set A is a **proper subset** of a set B if A is a subset of B, but A does not equal B. We write "$A \subset B$."

Notation As before, to indicate that a relation between sets does *not* hold we put a slash through the symbol for that relation. Thus, "A is not a subset of B" is written "$A \nsubseteq B$," and "A is not a proper subset of B" is written "$A \not\subset B$."

Observe the analogy between the meanings of the symbols $\subseteq$ for sets and $\leq$ for numbers. Just as "$x \leq y$" represents the disjunction "x is less than y *or* x equals y," so, too, "$A \subseteq B$" represents "A is a proper subset of B *or* A equals B."

Example 9 {a, b, c} $\subseteq$ {a, b, c, d, e}; in fact, {a, b, c} $\subset$ {a, b, c, d, e} because the two sets are not equal. □

Example 10 {a, b, c} = {c, b, a} because every element of each set is an element of the other. The order in which the elements are written does not matter; the two sets contain exactly the same elements, so the sets are equal. □

Example 11 Let $A = \{1, 2, 3, 4, 5, \ldots\}$, $B = \{x \mid x \text{ is a positive whole number}\}$, and $C = \{2, 4, 6, 8, \ldots\}$. Then the following statements are all true:

$$A \subseteq B \qquad C \subseteq B \qquad B \subseteq A \qquad A = B \qquad C \subset B \qquad A \not\subset B$$

$$B \not\subset A \qquad B \nsubseteq C \qquad A \neq C \qquad A \subseteq A \qquad A \not\subset A \qquad A = A$$ □

The Complement of a Set

Often it is just as important to know what is *not* in a set as to know what is in it. A description of a well-defined set should make that distinction clearly, but some limitation on the elements being considered is useful. For instance, if we ask for things not in the set {1, 2, 3, 4}, many different kinds of answers are possible:

$$5,\ 100,\ \frac{1}{2},\ .397, \text{ you, this book, the lions in the St. Louis Zoo}$$

and so forth. Some of these answers are more appropriate in one context, some in another. Often the appropriate kind of response is clear from the question or can be inferred from its general setting. It might be reasonable in this case to assume that the answers "you, this book, the lions in the St. Louis Zoo" are beyond the intent of the question, but it is not clear which of the others might fit. Thus, sometimes the set of all elements being considered in a particular discussion must be specified. Such a set is called the **universal set** for that discussion, and is denoted by $\mathcal{U}$. With this in mind, we can describe precisely what is not in a set.

Definition The set of all elements in the universal set that are not in a particular set A is called the **complement** of A, and is denoted by A'. In symbols,

$$A' = \{x \mid x \in \mathcal{U} \text{ and } x \notin A\}$$

Example 12 If the universal set is the set of names of days of the week and if

$$A = \{\text{Monday, Wednesday, Friday}\}$$

then

$$A' = \{\text{Sunday, Tuesday, Thursday, Saturday}\}$$

□

Example 13 Consider the set $A = \{0, 1, 2, 3, 4, 5\}$. If the universal set is all the digits, then $A' = \{6, 7, 8, 9\}$. If the universal set is all whole numbers, then

$$A' = \{6, 7, 8, 9, 10, 11, \ldots\}$$

□

Example 14 If A is the set of all elements of some universal set $\mathcal{U}$ for which a defining condition is true, then A' is the set of all elements of $\mathcal{U}$ for which this condition is false. Thus, there is a natural connection between the complement of a set and the negation of a statement. For instance, suppose $\mathcal{U}$ is the set of all trucks. If

$$F = \{x \mid x \text{ is a Ford}\}$$

then

$$F' = \{x \mid x \text{ is not a Ford}\}$$

□

Venn Diagrams

It is useful to have a way of picturing set relationships. The most common way of doing this is by using circles to draw sets as regions inside a rectangular box, as shown in Figure 3.2. The box represents the universal set. If a region is to be emphasized, it is striped, crosshatched, or shaded in some way. Such figures are called **Venn diagrams**. (They are named for John Venn, an Englishman who first used them in 1876, but are similar to figures used by the Swiss mathematician Leonhard Euler a century earlier.)

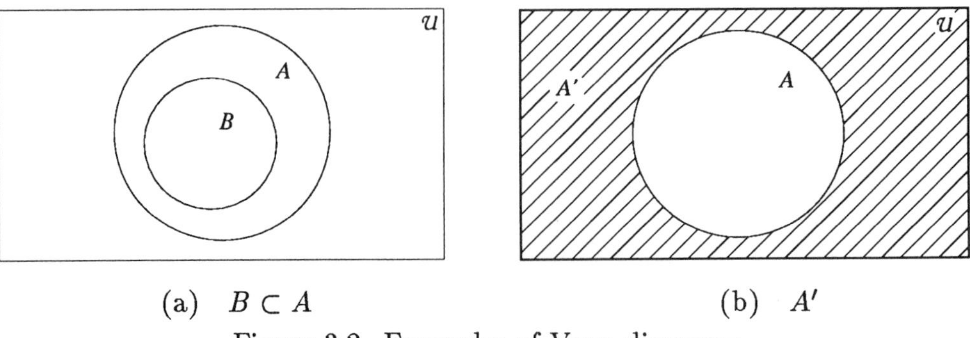

(a) $B \subset A$ (b) A'

Figure 3.2 Examples of Venn diagrams.

We end this section with several basic facts about subsets and complements. They can be proved formally, but that is unnecessary for understanding the concepts. The truth of the statements should be apparent from the informal arguments and Venn diagrams supplied.

(3.1) If A, B, and C are sets such that $A \subseteq B$ and $B \subseteq C$, then $A \subseteq C$. (See Figure 3.3.)

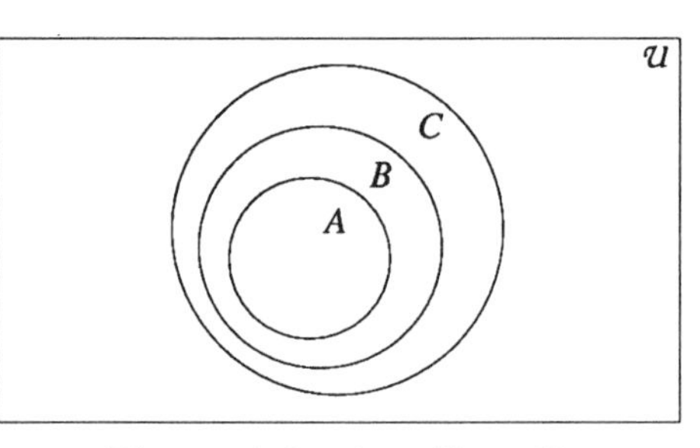

Figure 3.3 $A \subseteq B \subseteq C$

A is included in B, and B, in turn, is included in C, so A must be included in C.

(3.2) The empty set is a subset of any set.

This statement is a bit slippery. The defining condition for *subset* is a universal statement:

Every element of A is in B.

For this to be false, a counterexample would have to exist; that is, there would have to be an element of A that is not in B. Applying this observation to $\emptyset$ and any set S, we see that $\emptyset \subseteq S$ can be false *only* if there is an element of $\emptyset$ that is not in S. But, $\emptyset$ does not contain any elements at all, so this can never happen. Therefore, $\emptyset \subseteq S$ must always be true.

(3.3) Let A and B be subsets of some universal set $\mathcal{U}$. If $A \subseteq B$, then $B' \subseteq A'$. (See Figure 3.4.)

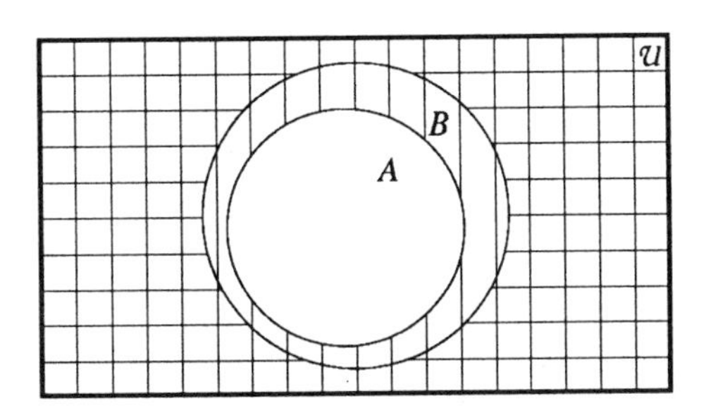

Figure 3.4 If $A \subseteq B$, then $B' \subseteq A'$.

Referring to Figure 3.4, we see that A' is the region striped vertically and B' is the region striped horizontally. Clearly, all of B' lies within A'.

(3.4) If A is any subset of some universal set $\mathcal{U}$, then $(A')' = A$. (See Figure 3.5.)

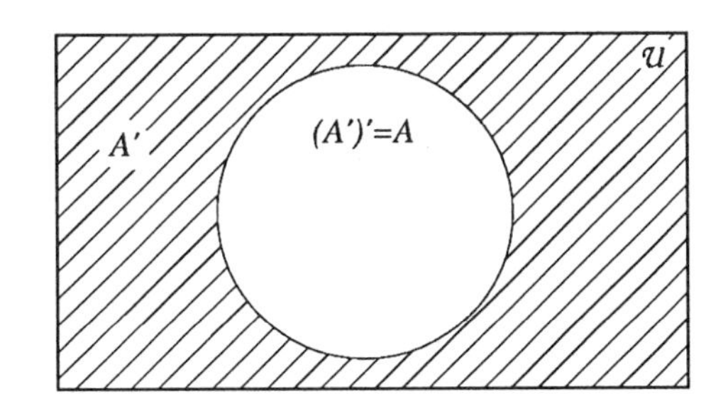

Figure 3.5 $(A')' = A$

This says that the complement of the complement of a set is the original set itself. The striped area in Figure 3.5 represents A'. The diagram clearly shows that there are only two regions in the universal set—A and A'. Thus, everything in the universal set that is not in A' must be in A; that is, $(A')' = A$.

PROBLEM-SOLVING COMMENT

The use of Venn diagrams in set theory illustrates how *drawing a diagram* can help to solve a problem. Observe how Venn diagrams make the set-theoretic facts in the rest of this chapter easier to understand. As you work through the exercises in this section and the next, try using a Venn diagram any time you run into difficulty; you'll find it helpful in most cases. ◇

Exercises 3.2

In Exercises 1–7, which of the following sets are well-defined? Why?

1. All 18-year-olds in California today.

2. All ugly freshmen.

3. $\{x \mid x \text{ is a good teacher}\}$

4. $\{x \mid x \text{ is a current U. S. Senator from Illinois}\}$

5. $\{1, 2, 3, 4, 5, \ldots, 100\}$

6. $\{1, 2, 3, 4, 5, \ldots\}$

7. $\{2, 4, 7, 3, 19, \ldots\}$

In Exercises 8–11:
(a) list the elements of the given set and
(b) find an appropriate universal set.
(There are many possible universal sets in each case; pick a "natural" one.)

8. $\{x \mid x \text{ is an odd digit}\}$

9. $\{x \mid x \text{ is a month whose name contains the letter } r\}$

10. $\{x \mid x \text{ is a state bordered by the Mississippi River}\}$

11. $\{x \mid x \text{ is a whole number and } x \leq 20\}$

In Exercises 12–16, describe each set in set-builder notation.

12. $\{1, 2, 3, 4, 5, 6, 7, 8, 9, 10, 11, 12\}$

13. {Monday, Tuesday, Wednesday, Thursday, Friday}

14. {a, e, i, o, u}

15. $\{2, 4, 6, 8, 10, \ldots\}$

16. $\{1, 3, 5, 7, \ldots\}$

17. The empty set was described in several different ways in Example 8. Give three more descriptions of it.

In Exercises 18–29, let the universal set be all the letters of the alphabet,and let $A = \{\text{a, b, c, d, e}\}$, $B = \{\text{c, d, e, f, g}\}$, and $C = \{\text{b, c, d}\}$. Label each statement *true* or *false*.

18. $A \subseteq B$ **19.** $A = B$

20. $C \subseteq A$ **21.** $C \subseteq B$

22. $C \subset A$ **23.** $C \subset B$

24. $A' \subseteq C'$ **25.** $C' \subseteq A'$

26. $C' \subset A'$ **27.** $A' = C'$

28. $B' \subseteq C'$ **29.** $C' \subseteq B'$

In Exercises 30–41, let $A = \{\text{a, b, c, d, e}\}$ and $D = \{\text{b}, \{\text{c, d}\}, \{\text{d}\}, \text{d}\}$. Label each statement *true* or *false*.

30. $\text{a} \in A$ **31.** $\text{a} \subseteq A$

32. $\{\text{a}\} \subseteq A$ **33.** $\{\text{a}\} \in A$

34. $\text{b} \in D$ **35.** $\text{b} \subseteq D$

36. $\text{d} \in D$ **37.** $\{\text{d}\} \in D$

38. $\{\text{d}\} \subseteq D$ **39.** $\{\text{c, d}\} \in D$

40. $\{\text{c, d}\} \subseteq D$ **41.** $\text{c} \in D$

In Exercises 42–44, let $A = \{1, 2, 3, 4, 5\}$. Describe A' relative to the given universal set $\mathcal{U}$.

42. $\mathcal{U} = \{0, 1, 2, 3, 4, 5\}$

43. $\mathcal{U} = \{0, 1, 2, 3, 4, 5, 6, 7, 8, 9\}$

44. $\mathcal{U} = \{x \mid x \text{ is a positive whole number}\}$

In Exercises 45–47, show that the given statement is false by finding a counterexample.

45. Every set contains at least two elements.

46. If $A \subseteq B$, then $A' \subseteq B'$.

47. If R, S, and T are sets such that $R \subseteq S$ and $T \subseteq S$, then $R \subseteq T$.

3.3 More About Sets

In the preceding section we defined the statement

"A is a subset of B"

in terms of the conditional form

"If $x \in A$, then $x \in B$"

Also, *complement* in set theory was seen to be analogous to *negation* in logic. It seems reasonable to ask if there are set-theoretic concepts analogous to conjunction and disjunction. The answer is *yes.* Many common sets of things are defined by using conjunction or disjunction: The set of all eligible voters is composed of people who are U. S. citizens *and* who are at least 18 years old; varsity athletes are students who play varsity football *or* varsity basketball *or* some other varsity sport; the possible ways of throwing 4 with a pair of dice comes from throwing two 2s *or* a 3 *and* a 1; the list could go on and on. The two main concepts presented in this section are the set-theoretic analogues of the logical connectives *and* and *or.*

Intersection

Definition The set of all elements that are both in a set A and in a set B is called the **intersection** of A and B, and is written as $A \cap B$. In symbols,

$$A \cap B = \{x \mid x \in A \textit{ and } x \in B\}$$

Example 1 If $A = \{\text{a, b, c}\}$ and $B = \{\text{a, c, e, g}\}$, then $A \cap B = \{\text{a, c}\}$. □

The intersection of two sets is usually pictured by the Venn diagram of Figure 3.6. Observe that this diagram is divided into four regions (numbered for reference). Each region corresponds to one line of the truth table for *and.* Look at Table 3.1, the truth table for "$x \in A$ and $x \in B$." The numbering in the left column of Table 3.1 matches the regions in Figure 3.6. (Do you see why?) The striped region corresponds to the line of the truth table in which the conjunction is true; the unshaded regions correspond to the lines in which the conjunction is false.

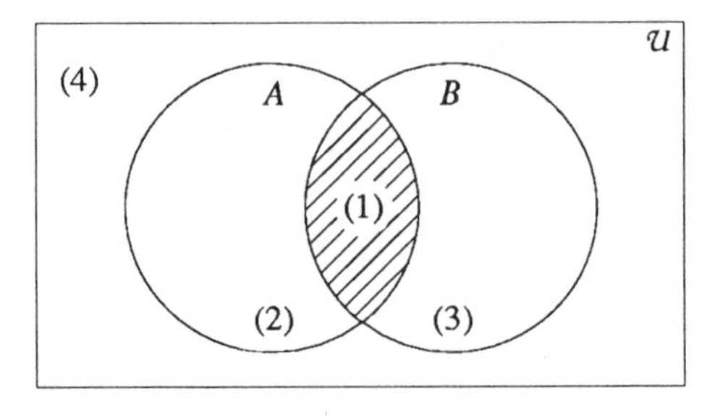

Figure 3.6 $A \cap B$

region	$x \in A$	$x \in B$	$x \in A$ and $x \in B$
(1)	T	T	T
(2)	T	F	F
(3)	F	T	F
(4)	F	F	F

Table 3.1 Intersection and conjunction.

Of course, two sets A and B may have nothing in common; that is, "$x \in A$ and $x \in B$" might be always false. For instance, we might have $A = \{1, 3\}$ and $B = \{2, 4\}$. This case is common enough to merit a name of its own.

Definition If A and B are any sets such that $A \cap B = \emptyset$, we say A and B are **disjoint**. If every pair of distinct sets in a collection of sets is disjoint, we call the collection **pairwise disjoint**.

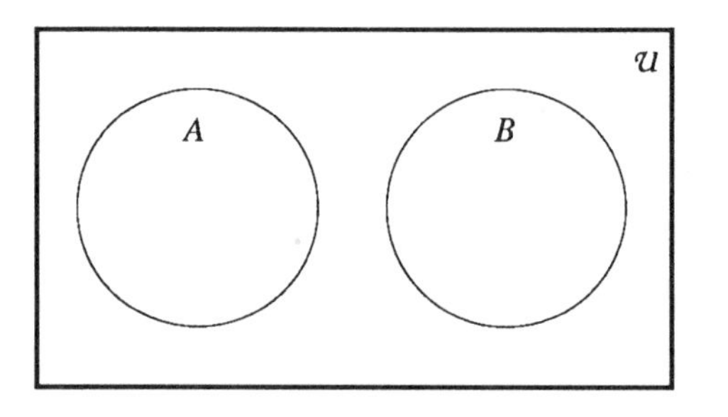

Figure 3.7 Disjoint sets A and B.

Example 2 Let $A = \{1, 2, 3\}$, $B = \{4, 5\}$, and $C = \{6, 7, 8, 9\}$. Then

$$A \cap B = \emptyset, \quad B \cap C = \emptyset, \quad \text{and } A \cap C = \emptyset$$

That is, every pair of distinct sets from the collection $\{A, B, C\}$ is disjoint. Thus, this collection of three sets is pairwise disjoint. We can ease the formality of this phrasing by simply saying: "The sets A, B, and C are pairwise disjoint." □

Example 3 Let $A = \{1, 2, 3, 4\}$, $B = \{3, 4, 5, 6\}$, and $C = \{5, 6, 7, 8\}$. Then:

$$A \cap B = \{3, 4\} \text{ and } B \cap C = \{5, 6\}, \text{ but } A \cap C = \emptyset.$$

Thus, A and C are disjoint sets, but the three sets A, B, and C are not pairwise disjoint. □

Example 4 If A is any set, then $A \cap A' = \emptyset$. This follows directly from the definition of complement. It is the set-theoretic version of the Law of Contradiction. □

Union

Definition The set of all elements that are either in a set A or in a set B is called the **union** of A and B, and is written $A \cup B$. In symbols,

$$A \cup B = \{x \mid x \in A \textit{ or } x \in B\}$$

Example 5 If $A = \{\text{a, b, c}\}$ and $B = \{\text{a, c, e, g}\}$, then $A \cup B = \{\text{a, b, c, e, g}\}$. Notice that the elements in both A and B are listed only once in the union. □

Example 6 $\{\text{a, b, c, d, e}\} \cup \emptyset = \{\text{a, b, c, d, e}\}$ □

The union of two sets is usually pictured as in Figure 3.8. The four regions of the Venn diagram correspond to the four lines of the truth table

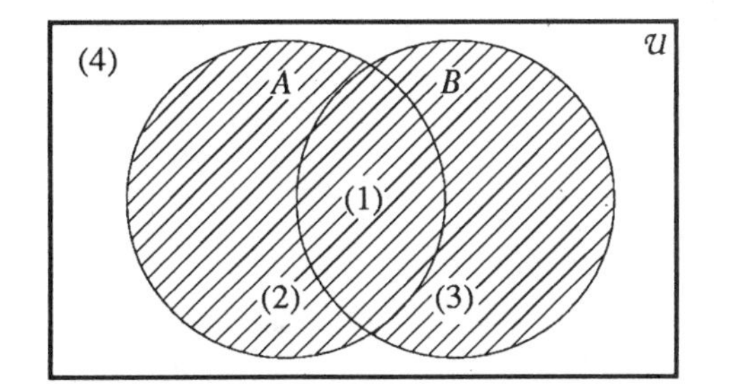

Figure 3.8 $A \cup B$

for *or*, as shown in Table 3.2. The striped part of the diagram corresponds to the lines of the truth table where the disjunction is true; the remaining part corresponds to the table line where the disjunction is false.

region	$x \in A$	$x \in B$	$x \in A$ or $x \in B$
(1)	T	T	T
(2)	T	F	T
(3)	F	T	T
(4)	F	F	F

Table 3.2 Union and disjunction.

Example 7 If A is any subset of a universal set $\mathcal{U}$, then $A \cup A' = \mathcal{U}$. This fact follows directly from the definition of complement. It is the set-theoretic version of the Law of the Excluded Middle. □

Properties of Intersection and Union

Some elementary, but useful, consequences of the definitions of intersection and union are listed here. Their justifications are left as exercises. (See Exercises 29–34.)

(3.5) If A and B are subsets of some universal set $\mathcal{U}$, then:

$$A \cap A = A \qquad A \cup A = A$$

$$A \cap \emptyset = \emptyset \qquad A \cup \emptyset = A$$

$$A \cap \mathcal{U} = A \qquad A \cup \mathcal{U} = \mathcal{U}$$

If $A \subseteq B$, then $A \cap B = A$.

If $A \subseteq B$, then $A \cup B = B$.

Since $A \cap B$ and $A \cup B$ are sets, it is natural to ask about how their complements might be expressed in terms of A' and B'. The answer to this question depends on the logical properties of *and* and *or*. Since *and* is used in defining $A \cap B$, the complement of $A \cap B$ requires the negation of an *and* statement. Similarly, *or* is used in defining $A \cup B$, so $(A \cup B)'$ must involve the negation of an *or* statement. These negations are governed by

De Morgan's Laws, as we saw in Section 2.2:

"$\sim(p$ and $q)$" is logically equivalent to "$(\sim p)$ or $(\sim q)$"

"$\sim(p$ or $q)$" is logically equivalent to "$(\sim p)$ and $(\sim q)$"

These laws are used to derive the analogous properties for sets, which are given the same name.

(3.6)

> **De Morgan's Laws**
>
> If A and B are subsets of some universal set, then
>
> $(A \cap B)' = A' \cup B'$ and $(A \cup B)' = A' \cap B'$

As an example of this connection between logic and sets, let us examine how the first of these laws is derived:

By the definition of intersection, $A \cap B = \{x \mid x \in A \textit{ and } x \in B\}$. Thus,

$(A \cap B)' = \{x \mid \sim(x \in A \textit{ and } x \in B)\}$	by the definition of complement
$= \{x \mid x \notin A \textit{ or } x \notin B\}$	by De Morgan's Laws for logic
$= \{x \mid x \in A' \textit{ or } x \in B'\}$	by the definition of complement
$= A' \cup B'$	by the definition of union

The Venn diagrams in Figure 3.9 illustrate this first law. The diagonally striped part of Figure 3.9(a) represents $(A \cap B)'$. In Figure 3.9(b), A' is the vertically striped region and B' is the horizontally striped region. $A' \cup B'$ is represented by the entire striped area, which is precisely the same region as the one in (a) that is striped. The second law may be derived and illustrated in a similar way; it is left as an exercise.

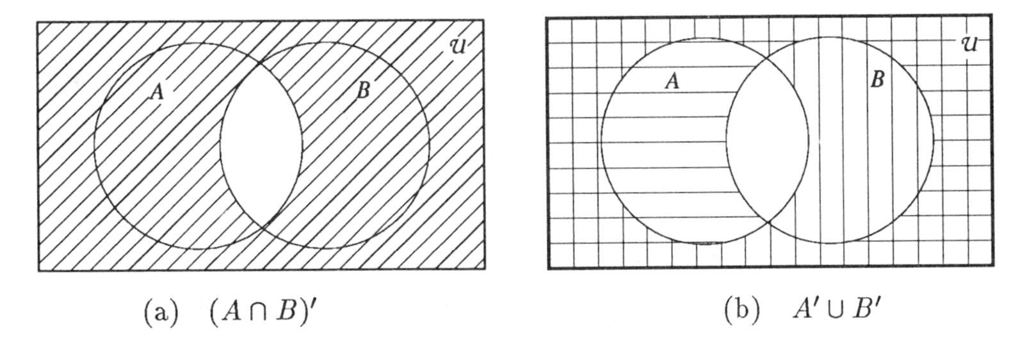

(a) $(A \cap B)'$ (b) $A' \cup B'$

Figure 3.9 The first of De Morgan's Laws.

PROBLEM-SOLVING COMMENT

De Morgan's Laws provide an example of how choosing the wrong analogy can lead us astray, whereas the correct one can be of enormous assistance. Students often find it tempting to think of set union and complement as analogous to adding and negating numbers, and in some ways that is true. However, here the number analogy would say: "Just as $-(a+b) = (-a) + (-b)$, so $(A \cup B)' = A' \cup B'$," which is *false*. We can see that it is false by constructing an example. (See Exercises 9 and 10.) Thus, although *arguing by analogy* is a powerful tool, you must always be careful to check the arguments it suggests, by *constructing examples*, by *approximating the answer*, by *looking back at the completed solution*, etc. ◇

Cartesian Product

There is another way of forming a new set from two given sets, not as obvious as union, but equally important. It is based on the idea of an **ordered pair**, a pair of elements whose order is specified. We write the ordered pair whose first element is x and whose second element is y as (x, y). Note that, because order matters, $(x, y) \neq (y, x)$ unless x and y are the same. In fact:

Definition Two ordered pairs (a, b) and (c, d) are **equal** if and only if $a = c$ and $b = d$.

Let us apply the idea of ordered pairs and some problem-solving tactics to the following problem that an elementary-school student might encounter:

> Sarah has 3 different-colored blouses—a red one, a blue one, and a green one. She also has 4 different colored skirts—a red one, a white one, a green one, and a tan one. How many different blouse/skirt outfits can she wear, and what are they?

We begin by introducing appropriate notation to simplify writing out the possible combinations. Let B represent the set of blouses, with r, b, and g standing for *red*, *blue*, and *green*, respectively, so that $B = \{r, b, g\}$. Similarly, the set S of skirts can be written as $S = \{r, w, g, t\}$. The outfit consisting of blouse x and skirt y can be represented by the ordered pair (x, y). (Notice that (r, g), which stands for the red blouse and the green skirt, is different from (g, r), which represents the green blouse and the red skirt.)

Now we might restate the problem by asking: In how many ways can an element of B be paired with an element of S? It is also helpful to make a chart (a kind of diagram) to visualize the situation better. We can form a rectangular table, listing the blouse colors down the side and the skirt colors across the top, and putting the pairs in their respective rows and columns, like this:

	r	w	g	t
r	(r,r)	(r,w)	(r,g)	(r,t)
b	(b,r)	(b,w)	(b,g)	(b,t)
g	(g,r)	(g,w)	(g,g)	(g,t)

We can see from this chart that there are 12 different outfits, and, in fact, we know exactly what they all are. We have displayed the set of all ordered pairs with first elements from B and second elements from S. This set of ordered pairs is called the *Cartesian product* of B and S. In general:

Definition The **Cartesian product**[5] of two sets A and B is the set of all ordered pairs whose first elements are from A and whose second elements are from B. It is denoted by $A \times B$. In symbols,

$$A \times B = \{(a,b) \mid a \in A \text{ and } b \in B\}$$

Example 8 If $A = \{\text{a, b, c}\}$ and $B = \{1, 2, 3, 4\}$, then

$$\begin{aligned} A \times B = \{&(\text{a}, 1), (\text{a}, 2), (\text{a}, 3), (\text{a}, 4), \\ &(\text{b}, 1), (\text{b}, 2), (\text{b}, 3), (\text{b}, 4), \\ &(\text{c}, 1), (\text{c}, 2), (\text{c}, 3), (\text{c}, 4)\} \end{aligned}$$

□

Example 9 If $A = \{1,2\}$ and $B = \{2,3,4\}$, then

$$A \times B = \{(1,2),(1,3),(1,4),(2,2),(2,3),(2,4)\}$$

and

$$B \times A = \{(2,1),(3,1),(4,1),(2,2),(3,2),(4,2)\}$$

Notice that $A \times B \neq B \times A$ because the ordered pairs in the two product sets are different: (1, 2) is not the same as (2, 1); (1, 3) is not the same as (3, 1); etc. Also,

$$A \times A = \{(1,1),(1,2),(2,1),(2,2)\}$$

$$B \times B = \{(2,2),(2,3),(2,4),(3,2),(3,3),(3,4),(4,2),(4,3),(4,4)\}$$ □

Example 10 If A is any set, then $A \times \emptyset = \emptyset$, because there are no elements to be used as second elements of the ordered pairs. Similarly, $\emptyset \times A = \emptyset$. □

---

[5]Named after René Descartes, a 17th-century French mathematician and philosopher who first developed analytic (coordinate) geometry.

We end this section with a pair of statements relating the Cartesian product to union and intersection. The first one forms the basis for what is called "the distributive law of multiplication over addition" (to be discussed later); it is easy to see from an example how it works. The second one is a natural twin of the first, with *union* replacing *intersection*. Try to construct an appropriate example for that one by mimicking Example 11.

(3.7)

> If A, B, and C are any sets, then
>
> *i.* $A \times (B \cup C) = (A \times B) \cup (A \times C)$
>
> *ii.* $A \times (B \cap C) = (A \times B) \cap (A \times C)$

HISTORICAL NOTE: Cantor's Theory of Sets

The simplicity of set theory, especially as it appears in the elementary grades, belies its profound significance within modern mathematics. In the early 1870s, the German mathematician Georg Cantor constructed a theory of arbitrary collections of things—sets—in order to deal with some paradoxical properties of infinity that had been vexing mathematicians since the time of Galileo. The very fact that Cantor dared to propose a mathematical analysis of infinity was a radical break with mathematical tradition, and the theory he proposed immediately caused intense controversy in mathematical, philosophical, and theological circles.

But Cantor's work affected mathematics itself in a decidedly positive way. His basic set theory provided a simple *unifying* approach to many different areas of mathematics. As you will see in this book, such diverse topics as the structure of the number systems, the fundamentals of geometry, and the theory of probability all are described in terms of sets. Moreover, the strange paradoxes encountered in some early extensions of his work caused mathematicians to put their logical house in order, so to speak. Their careful examination of the logical foundations of mathematics led to many new results in that area and paved the way for even more abstract unifying ideas.

The attempts of Cantor and his successors to insure that set theory was free of contradictions led to profound investigations into the foundations of mathematics. This work clarified forms of logical argument, methods of proof, and errors of syntax. In this way, modern mathematics has provided philosophy with some explicit, formal guidelines for admissible kinds of reasoning and possible logical constructions. The boundaries between religion, philosophy, and science have been brought into sharper focus as a result.[6] ◇

---

[6] A detailed, elementary explanation of Cantor's theory of infinite sets appears in Chapter 6 of William P. Berlinghoff and Kerry E. Grant, *A Mathematics Sampler,* 2nd ed., New York: Ardsley House, Publishers, Inc., 1988.

Example 11 Let us look at an example of statement *i* of (3.7), using the sets $A = \{\text{a, b}\}$, $B = \{\text{c, d}\}$, and $C = \{\text{d, e}\}$. On the one hand, $B \cup C = \{\text{c, d, e}\}$, so

$$A \times (B \cup C) = \{(\text{a, c}), (\text{a, d}), (\text{a, e}), (\text{b, c}), (\text{b, d}), (\text{b, e})\}$$

On the other hand,

$$A \times B = \{(\text{a, c}), (\text{a, d}), (\text{b, c}), (\text{b, d})\}$$

and

$$A \times C = \{(\text{a, d}), (\text{a, e}), (\text{b, d}), (\text{b, e})\}$$

so it is also true that

$$(A \times B) \cup (A \times C) = \{(\text{a, c}), (\text{a, d}), (\text{a, e}), (\text{b, c}), (\text{b, d}), (\text{b, e})\}$$ □

Exercises 3.3

In Exercises 1–12, let the universal set be

$$\{0, 1, 2, 3, 4, 5, 6, 7, 8, 9\}$$

and let $A = \{1, 2, 3\}$, $B = \{3, 4, 5, 6\}$, and $C = \{5, 6\}$. Find:

1. $A \cup B$
2. $A \cap B$
3. $A \cup C$
4. $A \cap C$
5. $B \cup C$
6. $B \cap C$
7. $A \times C$
8. $B \times C$
9. $(A \cup B)'$
10. $A' \cup B'$
11. $(A \cap B)'$
12. $A' \cap B'$

In Exercises 13–24, let the universal set consist of all the letters of the alphabet, and let $A = \{\text{a, b, c, d, e}\}$, $B = \{\text{b, d, f, h}\}$, and $C = \{\text{c, i, o}\}$. Find:

13. $A \cup C$
14. $A \cap B$
15. A'
16. $A' \cap B$
17. $(A \cup B)'$
18. $(A \cap C)'$
19. $B \cap C$
20. $(A \cup B) \cap C$
21. $A \cup (B \cap C)$
22. $A \cap (B \cap C)$
23. $(B \cap C) \times C$
24. $(A \cap B) \times C$

25. Illustrate $A \cap B'$ by shading the appropriate regions in the Venn diagram of Figure 3.10.

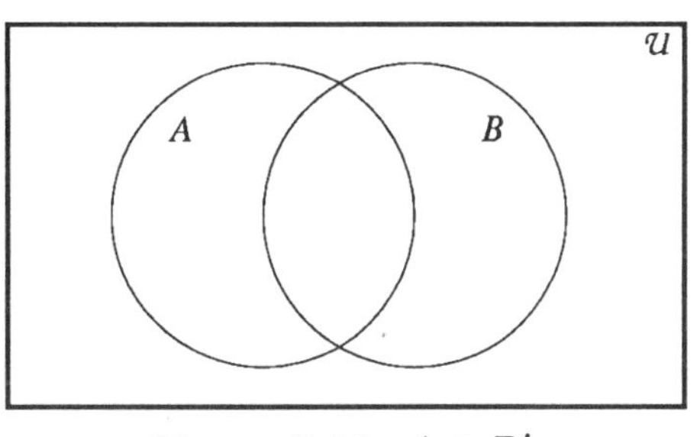

Figure 3.10 $A \cap B'$

In Exercises 26–28, illustrate the given relationships among three nonempty sets A, B, and C by using a single Venn diagram for each exercise. No region of your diagram should correspond to a set that is known to be empty.

Example: A and B are disjoint, but neither A and C nor B and C are disjoint.

Figure 3.11 is a correct illustration of this; Figure 3.12 is not because two of the regions are empty, as is indicated in the figure.

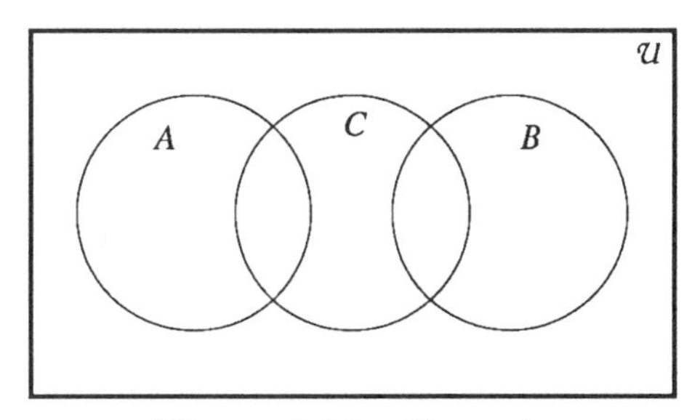

Figure 3.11 *Correct.*

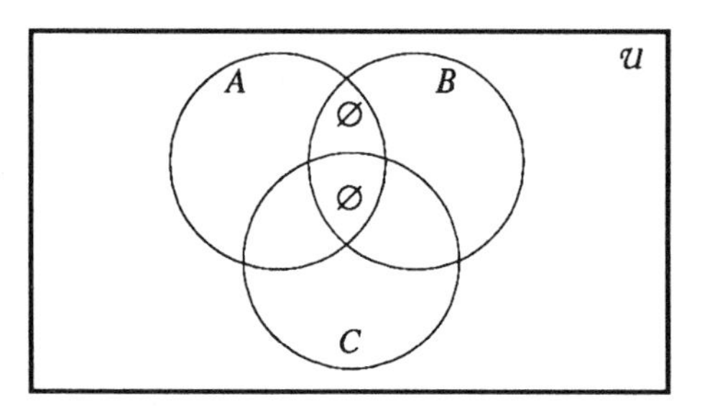

Figure 3.12 *Incorrect.*

26. $(A \cup B) \subset C$ and $A \cap B = \emptyset$

27. $A \subset B$, $A \cap C = \emptyset$, and $B \cap C \neq \emptyset$

28. A, B, and C are pairwise disjoint.

In Exercises 29–34, let A and B be subsets of some universal set $\mathcal{U}$. Show by some persuasive argument that:

29. $A \cap \emptyset = \emptyset$

30. $A \cup \emptyset = A$

31. $A \cap \mathcal{U} = A$

32. $A \cup \mathcal{U} = \mathcal{U}$

33. If $A \subseteq B$, then $A \cap B = A$

34. If $A \subseteq B$, then $A \cup B = B$

35. Illustrate statement *ii* of 3.8 using three sets of your choice.

36. If $A \cap B = \emptyset$ and $B \cap C = \emptyset$, must $A \cap C = \emptyset$? Give a counterexample to show that this is not always the case, and show by example that it is sometimes true.

37. Illustrate De Morgan's Law *ii* by using Venn diagrams.

38. Illustrate the two De Morgan's Laws using sets of your choice.

3.4 Equivalence Relations

Suppose we loaded a group of children into our bright red, late-model Ford station wagon, drove to the nearest shopping-center parking lot, and asked the children to pick out cars that were the same as ours. Chances are that, depending on their interests, one of them would pick out the red cars, another would pick out the Fords, still another the station wagons, and so on; and then there would be a big argument about whether this car or that was really "the same" as ours. Each child would have decided that sameness was characterized by one or more of the features of our car, and his or her choices would be dictated by whether or not another car had those features. Of course, no two things are truly identical in all respects, but all the cars chosen by a particular child would bear some special relationship to ours, and hence to each other. For each child, our car would be a *representative* of some particular subset of the cars in the parking lot, and all the cars in that subset would be the same; that is, they would possess some distinctive *common property*.

A generalized version of this simple notion of sameness is of fundamental importance throughout mathematics. In fact, strange as it may seem, the entire structure of the number system is based on this one concept, generalized to apply to any set. Thus, our next task is to specify which relations among things (of any sort) qualify as sameness relations. This job description may seem vague to you, and with good reason. We seek to formulate in a precise way an idea that is to be applied to a wide variety of sets about which we know nothing at this point, so there is nothing specific to be used in that formulation. Nevertheless, we can be quite precise in specifying the characteristics that we want any sameness relation to possess.

Relations

We begin by formalizing the general idea of a relation among elements of a set. The fundamental concept is very simple. Any sort of relation on a set S is some kind of rule or process by which elements of S are paired up. Remembering that the Cartesian product $S \times S$ is the set of all ordered pairs of elements of S, we can easily make a suitable definition of *relation* without the confusing vagueness of "rule or process":

Definition A **relation on a set** S is any subset of $S \times S$. (That is, it is a set of ordered pairs showing which elements of S are related.)

Example 1 Let B be a set of bricks of various solid colors. The statement "x is the same color as y" defines a relation on B; we can think of it as the set of all pairs of bricks of the same color. □

Notation If $\mathcal{R}$ is a relation on a set S, then $\mathcal{R} \subseteq \mathcal{S} \times \mathcal{S}$. However, we usually write

$$a\mathcal{R}b \qquad \text{rather than} \qquad (a,b) \in \mathcal{R}$$

to indicate that a is related to b by $\mathcal{R}$. If a is *not* related to b by $\mathcal{R}$, we write $a \not\mathcal{R}\, b$.

Example 2 Let C be the set of all cities in the United States. The statement "x is east of y" defines a relation on C. If we denote this relation by "$\mathcal{R}$," then we can write "Chicago $\mathcal{R}$ Denver" (because Chicago is east of Denver) and "New York $\mathcal{R}$ Chicago," but "Seattle $\not\mathcal{R}$ New York." □

Example 3 The statement "x is less than y" defines a relation on any set of whole numbers. For instance, on $\{1, 2, 3\}$, this relation is the set of ordered pairs

$$\{(1, 2), (1, 3), (2, 3)\}$$

because 1 is less than both 2 and 3, and 2 is less than 3. (No other ordered pair is in this relation; for instance, (2, 1) is not in the relation because 2 is not less than 1.) We usually denote this relation by $<$, so we would write

$$1 < 2, \ 1 < 3, \ 2 < 3$$ □

Example 4 Let $W = \{1, 2, 3, 4\}$ and define a relation $\mathcal{R}$ on W by "$x\mathcal{R}y$ if x and y are both even or both odd." Then $\mathcal{R}$ equals

$$\{(1, 1), (1, 3), (2, 2), (2, 4), (3, 1), (3, 3), (4, 2), (4, 4)\}$$

We may write $1\mathcal{R}3$, $2 \not\mathcal{R}\, 3$, and so forth. □

Equivalence Relations

In Examples 1 and 4, things are related by being similar or equivalent in some specific way; in Examples 2 and 3, things are related by being different in some specific way. Three simple properties of relations distinguish "sameness" (or "equivalence") from "differentness":

Definition A relation $\mathcal{R}$ on a set S is

reflexive if $x\mathcal{R}x$ for every element x in S;

symmetric if, whenever x and y are elements of S such that $x\mathcal{R}y$, then $y\mathcal{R}x$; and

transitive if, whenever x, y, and z are elements of S (not necessarily all different) such that $x\mathcal{R}y$ and $y\mathcal{R}z$, then $x\mathcal{R}z$.

A relation on S that is reflexive, symmetric, and transitive is called an **equivalence relation** on S. In this case, if $x\mathcal{R}y$, we say that x **is equivalent to** y.

In less formal terms, the preceding definitions say that a relation on a set is

reflexive if every element of that set is related to itself;

symmetric if, whenever a pair of elements is related in one order, that pair must also be related in the opposite order;

transitive if, whenever a first element is related to a second and that second element is related to a third, then the first element must be related to the third.

The relation is *an equivalence relation* if it has all three properties.

Example 5 Equality of numbers is a simple equivalence relation:

- If n is any number, then $n = n$ [reflexive]. For instance:

$$2 = 2, \quad 5 = 5, \quad 17 = 17$$

- If m and n are numbers and $m = n$, then it must also be true that $n = m$ [symmetric]. For instance:

$$\text{If } \frac{8}{2} = 4, \text{ then } 4 = \frac{8}{2}$$

- If m, n, and p are numbers such that $m = n$ and $n = p$, it follows that $m = p$ [transitive]. For instance:

$$\text{If } 1 + 4 = 5 \text{ and } 5 = 2 + 3, \text{ then } 1 + 4 = 2 + 3$$ □

Example 6 Let P be the set of all cars in a parking lot. If we call two cars related whenever they are of the same make, we have an equivalence relation on P. A second equivalence relation on P may be defined by saying that two cars are related if their hoods are of the same color(s), a third equivalence relation if they are licensed in the same state (assuming there are no foreign or diplomatic cars parked there), and so on. However, calling two cars related if both their hoods are blue is *not* an equivalence relation on P, unless all the cars in P happen to have blue hoods, because it is not reflexive. (A car with a red hood would not be related to itself by this "blue" relation.) □

Example 7 Any two geographical positions may be considered equivalent if they have the same latitude. However, if we say that Place A is related to Place B whenever A is north of B, we do not get an equivalence relation. (Why not?) □

Example 8 Congruence (same size and shape) of triangles, as studied in high-school geometry, is an equivalence relation. (Do you see why?) Is similarity (same shape) of triangles also an equivalence relation? □

Example 9 The kindergarten-text page shown in Figure 3.13 deals with an equivalence relation; what is it? Can you think of other equivalence relations that could be used with this picture? □

Partitions

Returning to the parking lot of Example 6, we can see how an equivalence relation subdivides the set on which it is defined. If we say that two cars are equivalent whenever they are of the same make, then the set of cars in the parking lot is broken down into subsets, each subset containing all cars of a particular make: all Plymouths, all Hondas, all Fords, and so forth. On the other hand, if the equivalence relation is the one determined by hood color, the subsets are: all cars with green hoods, all cars with white hoods, all cars with red hoods, and so on. In either case, the subsets we get are pairwise disjoint, and every car belongs to one of them. This example suggests an important general property of equivalence relations, which can be stated precisely once we have made some preliminary definitions. The first of these is just a natural extension of *union* to encompass more than two sets.

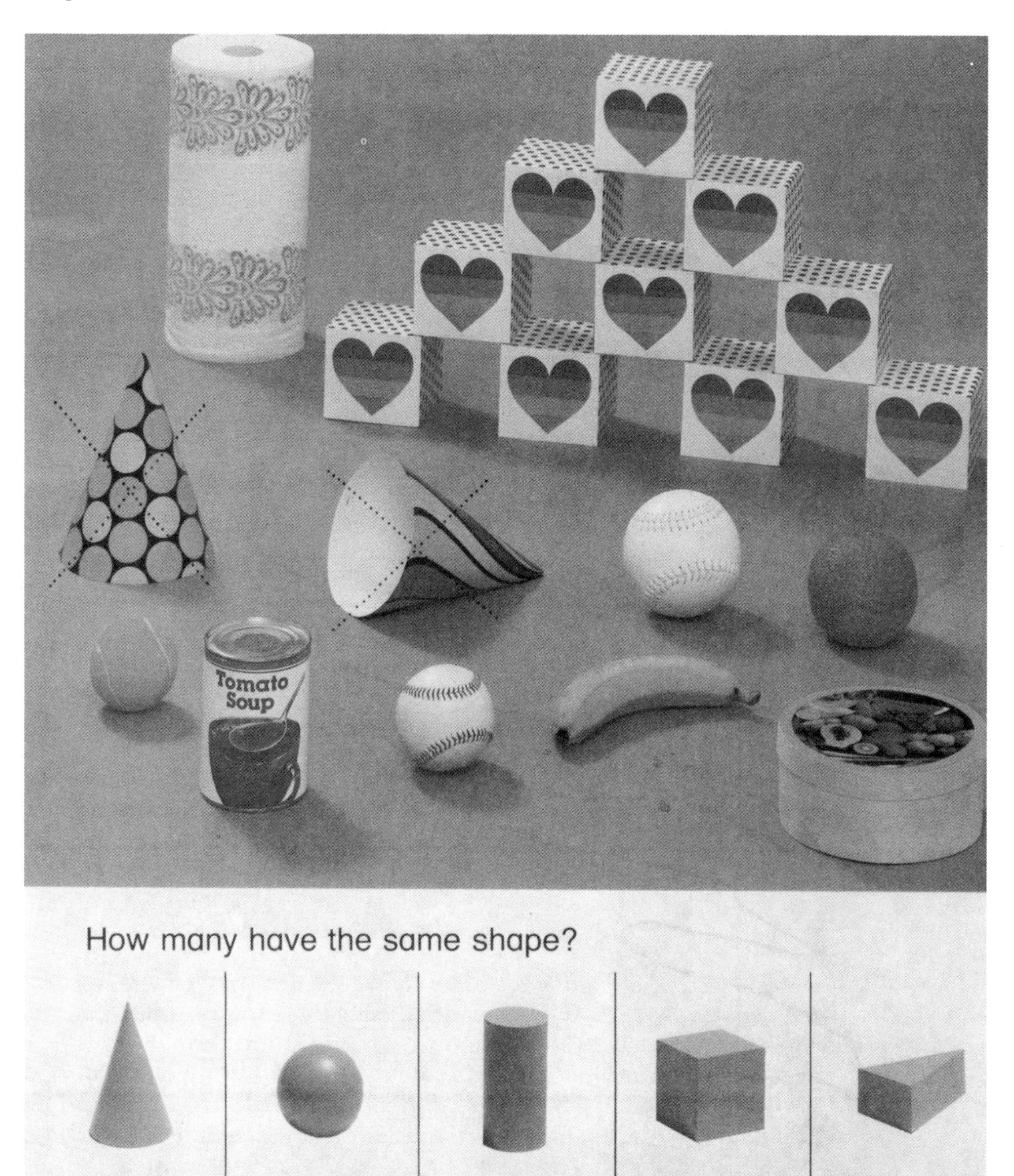

Figure 3.13 INVITATION TO MATHEMATICS, Kindergarten, p. 86

Definition If C is a collection of sets, then the **union** of all the sets in C is the set of all elements that are in at least one of the sets; that is,

$$\{x \mid x \in A \text{ for some } A \in C\}$$

Example 10 If C is the collection of the four sets {a, b, c, ..., j}, {i, j, k, ..., p}, {o, p, q, ..., v}, and {w, x, y, z}, then the union of all the sets in C is the set of all the letters in the alphabet, {a, b, c, ..., z}. □

Definition If C is a collection of nonempty subsets of a set S such that

1. C is pairwise disjoint and
2. the union of all the sets in C equals S,

then we call C a **partition** of S. Any relation on S that forms such a collection of subsets is said to **partition** S.

Example 11 The sets

$$\{3, 6, 9, \ldots, 30\}, \quad \{2, 5, 8, \ldots, 29\}, \quad \{1, 4, 7, \ldots, 28\}$$

form a partition of the set 1, 2, 3, ..., 30. By way of contrast, the sets in Example 10 do *not* form a partition of the alphabet because some of them have letters in common; that is, they are not pairwise disjoint. □

The critical fact that makes equivalence relations so useful in constructing the number systems is:

(3.8) Any equivalence relation on a nonempty set partitions that set.

The converse of Statement (3.8) is also true:

(3.9) Any partition of a set defines an equivalence relation on that set.

If we are given a partition, we can define an equivalence relation simply by declaring two sets to be related if and only if they are in the same subset of the partition. Because of this, the subsets in the partition (that is, the elements of C) are called **equivalence classes**.

Example 12 As we saw in Example 6, each equivalence class formed by the first relation on P is the set of all cars (in that lot) of a particular make. Each equivalence class of the second relation contains all cars with hoods of the same coloring.

Each equivalence class of the third relation contains all cars licensed in a particular state. □

Example 13 "Equals" on any set of symbols representing numbers is an equivalence relation. For instance, it partitions the set

$$\{2,\ 5,\ 1+1,\ 3-2,\ 1,\ 7-5,\ 8,\ 3+2\}$$

into the equivalence classes

$$\{2,\ 1+1,\ 7-5\},\ \{5,\ 3+2\},\ \{3-2,\ 1\},\ \text{and}\ \{8\}$$ □

Example 14 The kindergarten-text page shown in Figure 3.13 identifies five equivalence classes. What are they? □

Example 15 Let C be the set of all cities in the world, and define two cities to be "equivalent" if they are in the same country. (Check that this relation is reflexive, symmetric, and transitive.) This equivalence relation partitions C in a way that is particularly easy to visualize: A world map is *partitioned* (separated into nonoverlapping pieces) by the boundaries of the countries. Cities that are equivalent to each other lie within the same country, so the countries represent the equivalence classes of C. Notice that, if you pick out a city, you have automatically specified the country it is in, and hence you also know which cities are equivalent to the one you chose and which are not. □

Example 15 illustrates a fundamental property of equivalence classes that will be very useful to us. Remember that all the elements in a particular equivalence class are equivalent to each other, and two elements in different classes cannot be equivalent. Because of this,

> *any element of an equivalence class can be used to represent the entire class.*

Thus, if we are partitioning cities by country, then any U. S. city can be used to represent the class of *all* U. S. cities, and so forth. If you and I each pick a country, we can tell whether we have picked the same country just by checking to see if *any* city in my country is also in yours. For example, if Montreal is in the country I chose and also in the country you chose, then we both must have chosen Canada. The general principle is:

(3.10) | Two equivalence classes are equal if and only if they have an element in common. |

Thus, to prove that two equivalence classes are equal, it is only necessary to find one element that is in both of them.

PROBLEM-SOLVING COMMENT

In checking for equivalence relations, it is critically important to *check the definitions* of the reflexive, symmetric, and transitive properties carefully. Several key aspects of these definitions are too often overlooked by students who are just learning how to work with these ideas:

- The reflexive property is a universal statement; *every* element of the set must be related to itself.
- The symmetric and transitive properties are conditional statements. Thus, to show that they do *not* hold, you must find an example where the hypothesis is satisfied, but the conclusion is not. For instance, to say a relation is not symmetric, you must find a pair of elements that *are* related in one order *and are not* related in the other.
- The hypothesis of the transitive property is the conjunction "$x\mathcal{R}y$ and $y\mathcal{R}z$." In order for it to be true, there must be two ordered pairs of related elements, and *the second element of the first pair must be the same as the first element of the second pair.* □

A Dialogue

More generally, an examination of the definitions often will provide a ready-made plan of attack for a problem, as is shown by the following extended example, which focuses on the type of relation underlying "clock arithmetic." Let us do this with Terry, whom you met in Chapter 1. Terry is getting pretty good at this problem-solving stuff by now.

Problem: Show that the relation

$$x\mathcal{R}y \text{ if } 5 \text{ is a factor of } x - y$$

on the set $\{\ldots, -3, -2, -1, 0, 1, 2, 3, \ldots\}$, called the *integers*, is an equivalence relation, and find the equivalence classes of its partition.

Question 1: Can you restate the question by checking the key definitions?

Terry: The first part of the question asks us to show that $\mathcal{R}$ is an equivalence relation, so we must show that $\mathcal{R}$ is reflexive, symmetric, and transitive. I guess I'd better try to do that before worrying about the rest of it.

Question 2: Do you understand the definition of the relation $\mathcal{R}$?

Terry: I think so. 5 is a factor of something if 5 times some number is that thing.

Question 3: Do you mean 5 is a factor of 3 because $5 \times \frac{3}{5}$ equals 3?

Terry: No. The number you use has to be an integer. For instance, 5 is a factor of 10 because 5×2 equals 10.

Question 4: Now what?

Terry: Let's check reflexive, symmetric, and transitive. The definition of *reflexive* requires that every element be related to itself.

Let me first construct an example to see what's going on. We ought to be able to pick any number here; I'll try 17. The definition of $\mathcal{R}$ says that $17\mathcal{R}17$ provided that 5 is a factor of $17 - 17$. Now, $17 - 17 = 0$, and 5 is a factor of 0 because $5 \times 0 = 0$, so 17 is related to itself. Oh, I get it! *Any* number subtracted from itself is 0, and 5 is a factor of 0, so any number is related to itself. That means $\mathcal{R}$ is reflexive.

The definition of *symmetric* says that if a pair is related in one order, then it also must be related when we switch the elements around. Let's try a pair of numbers that are related, like... 17 and 2. $17\mathcal{R}2$ because $17 - 2$ is 15, which has 5 as a factor. Now, if I switch them around, I have to check $2 - 17$. But that's -15, which also has 5 as a factor, so $2\mathcal{R}17$. It looks like that always works. If I start with two numbers whose difference has 5 as a factor, then when I reverse them the difference just changes its sign, so it still has 5 as a factor. That means $\mathcal{R}$ is symmetric! Two down, one to go.

The definition of *transitive* requires three elements ...

Question 5: Do they all have to be different?

Terry: No, but they can be, so I'll pick an example that way because I can keep track of them better. The middle one has to be related to the other two, like this:

$$13\mathcal{R}3 \text{ because } 13 - 3 = 10 = 5 \times 2$$

and

$$3\mathcal{R}8 \text{ because } 3 - 8 = -5 = 5 \times (-1)$$

That satisfies the hypothesis. Now, the conclusion requires that 13 be related to 8; well, $13 - 8 = 5 = 5 \times 1$, so it works this time, but I wonder if I was just lucky. Let me try another example.

[Terry tried a few more cases and they all worked, but it was not obvious that this would *always* be the case. After some frustration, Terry persuaded the questioner to supply the following (slick) general argument:

$$\text{If } x\mathcal{R}y, \text{ then } x - y = 5 \cdot a \text{ for some integer } a$$
$$\text{If } y\mathcal{R}z, \text{ then } y - z = 5 \cdot b \text{ for some integer } b$$

Now,

$$x - z = (x - y) + (y - z) = 5 \cdot a + 5 \cdot b = 5 \cdot (a + b)$$

so 5 is a factor of $x - z$, implying $x\mathcal{R}z$.

By plugging some numbers into that argument, Terry was able to understand it (and was suitably impressed), but is still a little suspicious of such algebraic trickery.]

Question 6: Now that we know $\mathcal{R}$ is an equivalence relation, what's next?

Terry: I have to find equivalence classes.

Question 7: What does that mean?

Terry: I have to find sets of integers that are all related to each other. Now, every number is related to something—at least itself—so I guess I can start by choosing an integer at random and seeing what it's related to. For instance, 13 is related to 3 and 8, and 18, ... Let me put them in order: 3, 8, 13, 18, ... Oh! It's just counting by 5s, so, if I extend the pattern in both directions, this equivalence class must be

$$\{\ldots, -7, -2, 3, 8, 13, 18, 23, \ldots\}$$

In fact, I think I see the whole idea now! *An equivalence class for a number is just all the numbers I can get by counting by 5s from that number, in either direction*, so the partition must be

$$\{\ldots, -10, -5, 0, 5, 10, 15, 20, \ldots\}$$
$$\{\ldots, -9, -4, 1, 6, 11, 16, 21, \ldots\}$$
$$\{\ldots, -8, -3, 2, 7, 12, 17, 22, \ldots\}$$
$$\{\ldots, -7, -2, 3, 8, 13, 18, 23, \ldots\}$$
$$\{\ldots, -6, -1, 4, 9, 14, 19, 24, \ldots\}$$

Have I got them all? Sure! If I count vertically in this list of sets, I can see from the pattern that I get all the integers in order, so this collection of five subsets uses up the entire set of integers. That's pretty nice!

Question 8: Congratulations! That was a hard problem, and you did well. Now, before you leave it, can you see any way to generalize the solution?

Terry: Well, it seems to me that there's nothing magic about 5 here. I'll bet if we used any other positive number as the factor, like 4 or 10 or even something weird like 23, we could define an equivalence relation on the integers. If that's so, I'll bet that the number we use will tell us how many equivalence classes we get in the partition, and we might even get the equivalence classes themselves if we "skip-count" by that number.

[And Terry is right.]

Exercises 3.4

1. Construct three different relations on the set $\{1, 2, 3, 4, 5\}$.

2. Construct three different relations on the set {a, b, c, d, e, f, g}.

3. Construct three different relations on the set of all people in your class.

In Exercises 4–8, does the given statement define an equivalence relation on the set of all cities in the Western hemisphere? Why or why not?

4. x is north of y.

5. x has the same longitude as y.

6. x is less than 50 kilometers from y.

7. x and y have some residents of the same nationality.

8. x and y are on the same continent.

Exercises 9–18 define relations on the set of integers,

$$\{\ldots, -3, -2, -1, 0, 1, 2, 3, \ldots\}$$

In each case find an ordered pair of integers that are related and an ordered pair that are not. Then decide if the relation is reflexive, symmetric, or transitive. For each of these properties the relation does *not* have, provide a counterexample.

9. $x\mathcal{R}y$ if $x = y + 1$

10. $x\mathcal{R}y$ if 7 is a factor of $x - y$.

11. $x\mathcal{R}y$ if x and y have opposite signs.

12. $x\mathcal{R}y$ if 7 is a factor of both x and y.

13. $x\mathcal{R}y$ if $x \neq y$

14. $x\mathcal{R}y$ if x is greater than y.

15. $x\mathcal{R}y$ if x is not less than y.

16. $x\mathcal{R}y$ if either x or y is 7.

17. $x\mathcal{R}y$ if 7 is a factor of $x + y$.

18. $x\mathcal{R}y$ if $x + y = 0$

19. One *and only one* of the relations in Exercises 9–18 is an equivalence relation. Describe the equivalence classes of its partition. (*Hint*: Study the Problem-Solving Comment at the end of this section.)

20. Suppose we say that two triangles are related if they have an angle of the same size. (Thus, for example, all right triangles would be related to each other.) Is this an equivalence relation? Why or why not?

21. Let A be the set of all letters of the alphabet.
    (a) Form a partition of A consisting of four subsets.
    (b) The partition you made up in part (a) defines an equivalence relation on A. Give two elements of A that are equivalent by this relation and two that are not.
    (c) Form a five-subset partition of A in such a way that the resulting equivalence relation has all of the following properties:
        i. None of the letters p, q, or r are equivalent to z;
        ii. p is equivalent to q but not to r;
        iii. b is equivalent to r.
    (d) In the relation of part (c), can b be equivalent to z? To q? Why?
    (e) Using your relation of part (c), list all the elements that are equivalent to x.

22. Which of the four relations defined in Examples 1 through 4 of this section are equivalence relations? For each one that is, describe the equivalence classes of its partition.

In Exercises 23–28, determine whether the given relation $\mathcal{R}$ is reflexive, symmetric, and/or transitive. Which ones are equivalence relations on the set of all subsets of some universal set?

23. $A\mathcal{R}B$ if A is a subset of B.

24. $A\mathcal{R}B$ if A is a proper subset of B.

25. $A\mathcal{R}B$ if A is the complement of B.

26. $A\mathcal{R}B$ if A and B are disjoint.

27. $A\mathcal{R}B$ if A and B are equal.

28. $A\mathcal{R}B$ if A and B have the same number of elements. (Assume that the universal set is finite.)

29. Consider the generalized definition of *union* given in this section, and argue by analogy to construct a generalized definition for the *intersection* of an arbitrary collection of sets.

3.5 Functions

We need just one more basic idea for our mathematical tool kit — the concept of *function*. In modern mathematics this concept is as fundamental as *set*, but it has a longer, more varied history. It evolved gradually—from Nicole Oresme's idea of using graphs to represent rates of change (in the 14th century)—through the work of Leibniz in the 17th century as he adapted it to his formulation of calculus—to its precise, abstract, 20th-century definition in terms of sets, which we shall see in this section.

As is the case with many mathematical concepts that have been refined over many years, the current set-theoretic definition of *function* does not reflect the intuitive idea quite as clearly as some of its earlier formulations did. Thus, before supplying the formal definition, we shall examine some less rigorous, but perhaps more enlightening, definitions of the word.

Some Basic Terms

In everyday English, we say that a "function" of an object is something that it does. For instance, the function of a lawn mower is to cut grass. In some sense, the cutting of the grass depends upon the proper "functioning" of the lawn mower. If a teacher says to you, "In this course, the amount you learn is a function of how much work you do," you are being told that the amount you learn depends upon and varies with the work you do. In other words, the more you work, the more you learn. If your boating companion says that the water inside your leaky rowboat is a function of how fast you bail, you are being told that the faster you bail, the less water will be left in the boat. If a banker tells you that the profit from your bank account is a function of the amount you deposit, you are being told that the more you deposit, the more you will earn. This general notion of the dependence of one thing upon another—especially the dependence of one quantity upon another—is the underlying idea of the mathematical definition of *function*, but it has been sharpened and generalized.

Many mathematics textbooks define a **function** from a first set, A, to a second set, B, as any rule or process that assigns to each element of A

exactly one element of B. Thus, a function makes the choice of an element in B *depend on* the choice of an element in A. This is a reasonably good working description of the concept, but it suffers from the vagueness of not specifying exactly what qualifies as a "rule or process." However, once we see that the idea is to match each element of a first set with exactly one element of a second set, we can use *ordered pairs* to capture the idea precisely.

Definition A **function** from a set A to a set B is a subset of $A \times B$ with the property that each element of A appears first in exactly one ordered pair. The first set, A, is called the **domain** of the function.

Notation It is customary to use the letters f, g, and h to denote functions, but other symbols may be used, as well. If f is a function from A to B, we write

$$f: A \to B$$

Remember: The characteristic property of a function is that it assigns to *each* element of the domain A *exactly one* element of B.

Definition Let $f: A \to B$ be a function. If a is any domain element, the element of B that is paired with a is called the **image** of a, is written $f(a)$, and is read "f of a." The set consisting of the images of all the domain elements is called the **range** of f. In symbols,

$$\text{range of } f = \{f(a) \mid a \in A\}$$

Notice that the definition of $f: A \to B$ does *not* require that every element of B must be used, nor does it prohibit using elements of B as images more than once. This means that the range of f may be a proper subset of B, and it may be a much smaller set than A.

Example 1 The assignment of a course grade to each student in a class is a function from the set of all students in that class to the set of all possible grades. For instance, suppose $G = \{\text{A, B, C, D, F, W}\}$ is the set of possible grades. If Nancy is in a class K and if we let $g: K \to G$ denote the grading function, then we can symbolize the fact that Nancy got a "B" grade by

$$g(\text{Nancy}) = \text{B}$$

If Ralph also got a "B" grade, we can write $g(\text{Ralph}) = \text{B}$. If nobody failed, then "F" is not an element of the range. If we say that the range of g is the set $\{\text{A, B, C, D}\}$, we are saying that

i. every student got one of these four grades, and
ii. each of these grades was assigned to at least one student. □

Example 2 Figure 3.14 shows three functions f, g, and h from {a, b, c, d} to {1, 2, 3, 4}. The domain of all of these functions is {a, b, c, d}. The arrows indicate that $f(\text{a}) = 2$, $g(\text{a}) = 3$, $h(\text{a}) = 1$, $f(\text{b}) = 1$, $g(\text{b}) = 3$, etc. The range of f is {1, 2, 4}; the range of g is {3}; the range of h is {1, 2, 3, 4}. □

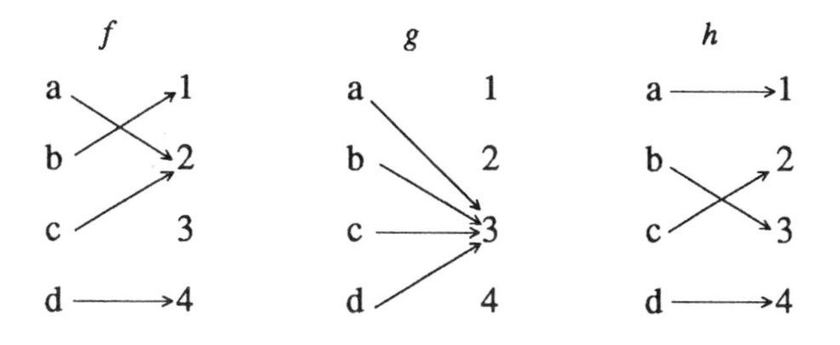

Figure 3.14 Three functions.

Example 3 Let I denote the set of integers. If we define $f\colon I \to I$ by $f(x) = 2x$, then f is a function whose range is the set of all even integers. (Why?) Some of its images are

$$f(1) = 2, \quad f(6) = 12, \quad f(-10) = -20$$

If we define $g\colon I \to I$ by $g(x) = x + 3$, then g is a function whose range is all of I. (Why?) Some of its images are

$$g(1) = 4, \quad g(6) = 9, \quad g(-10) = (-7)$$

□

Example 4 The sixth-grade text page shown in Figure 3.15 exhibits a good practical application of a function. The picture shows a thermometer with two temperature scales, Celsius and Fahrenheit, arranged to imply a relation between them. Each temperature reading on the Celsius scale corresponds to the Fahrenheit temperature directly to the right of it. Two such pairings are (0°C, 32°F) and (100°C, 212°F). It is helpful to know a formula that determines the pairs. Let us write that formula as a function F which has the set of Celsius temperatures as its domain and the set of Fahrenheit temperatures as its range:

$$F(x) = \frac{9}{5}x + 32$$

To find what Fahrenheit temperature is paired with the Celsius temperature 20°, we just substitute for x:

$$F(20°) = \frac{9}{5} \cdot 20 + 32 = 36 + 32 = 68°$$

□

Temperature

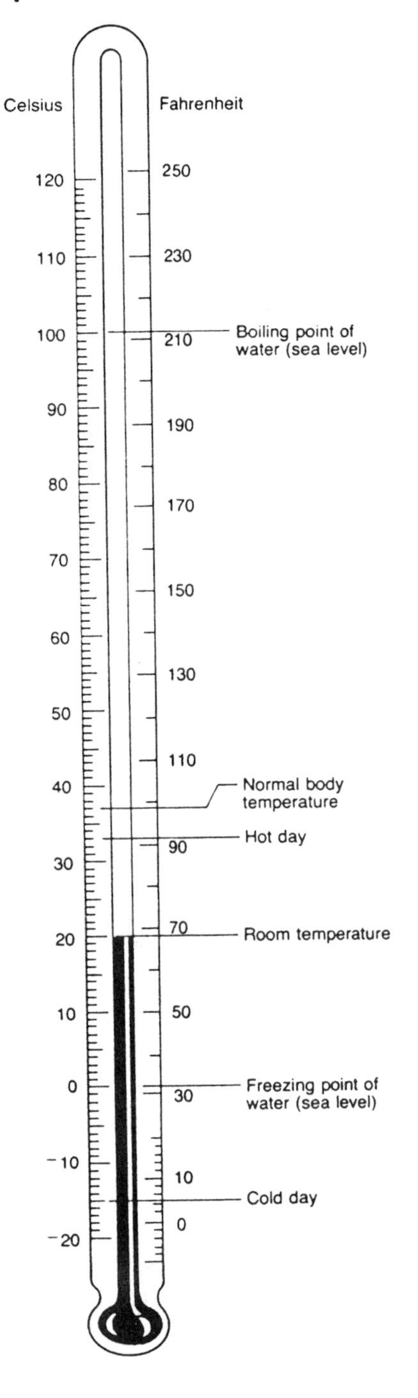

Mr. Carletti, the plant engineer at Loch Raven High School, uses a computer to regulate the temperature of the building.

A. This thermometer shows a room temperature of 20 degrees *Celsius* (20°C).

B. Is the more sensible temperature for cold milk 40°F or 80°F?

40°F is more sensible, since 80°F is above room temperature.

Try

a. Give the Celsius temperature shown on the thermometer.

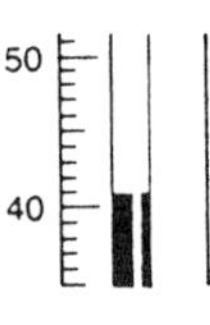

b. *Estimation* Choose the more sensible temperature for hot soup.

20°C 80°C

Practice Give the Celsius temperature shown on each thermometer.

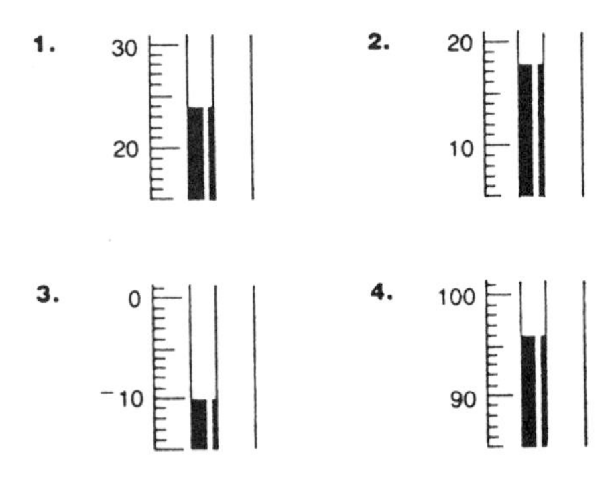

Figure 3.15 INVITATION TO MATHEMATICS, Grade 6, p. 178

One-to-one Correspondence

The temperature function of Example 4 is particularly nice because every number on each scale is paired with exactly one number on the other. Thus, it is "reversible," in that either set can be used as the domain of a function whose pairings are defined by the matching between the temperature scales. Figure 3.16 shows that not every function is reversible. If the arrows of f (shown in that diagram) are reversed, then $\{1, 2, 3, 4\}$ becomes the domain of the new relation, and for that to be a function, every element of $\{1, 2, 3, 4\}$ must have exactly one image. However, it is clear from the diagram that 3 has no image, and 2 has two images.

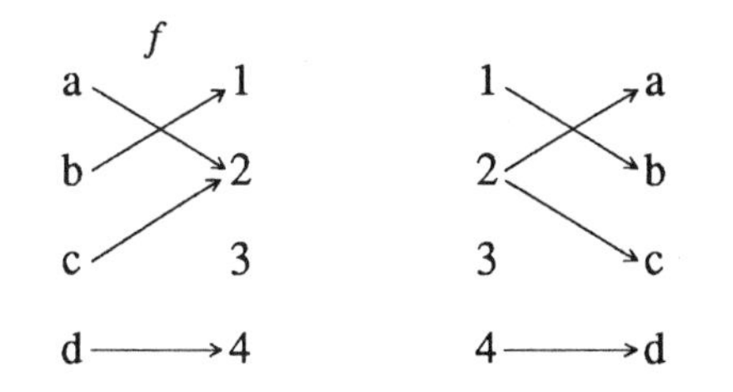

Figure 3.16 A function that is not reversible.

Functions that are reversible are important enough to have a special name.

Definition A **1-1 correspondence** (or **one-to-one correspondence**) between two sets A and B is a function from A to B with the property that each element of B appears second in exactly one ordered pair.

This definition says that a 1-1 correspondence between two sets A and B is any process that matches each element of A with exactly one element of B and each element of B with exactly one element of A. We might say that the existence of a 1-1 correspondence between two sets means that they are the same size (as collections of things). This is the topic of a later discussion.

Example 5 Both of the functions f and g in Figure 3.17 are 1-1 correspondences between $\{a, b, c, d\}$ and $\{\$, \%, \#, \&\}$. □

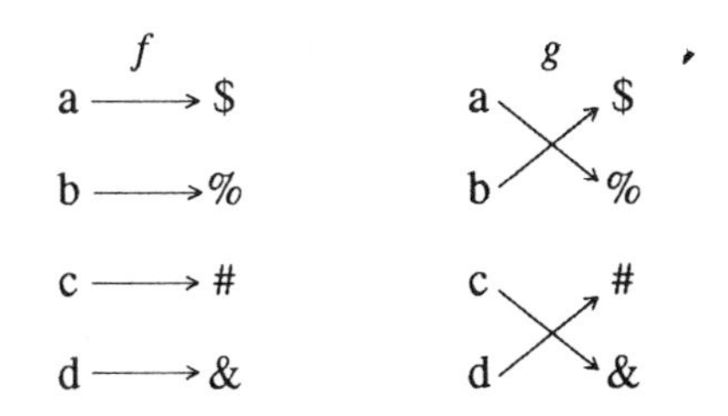

Figure 3.17 Two 1-1 correspondences.

Example 6 Only one of the functions in Figure 3.14 is a one-to-one correspondence; it is h. □

Example 7 The function that matches each integer with its double (that is, $f(x) = 2x$) is a 1-1 correspondence between the set of all integers and the set of all even integers. □

Counting

Now we have in hand the two basic concepts identified by Piaget as essential to an understanding of counting. We have examples of abstract ordered sets (*seriation*) and of 1-1 correspondences between sets (*conservation of quantity*). In Section 3.1 we saw that the seriation concept alone led to counting as a rote chanting process with no necessary relationship to quantity. In this sense,

$$\{\text{"one," "two," "three," "four," "five," } \ldots\}$$

is just an ordered set of sounds, which can be extended endlessly in a particular order, without referring to any notion of quantity or size. But that is not "real" counting as we know it. Real counting has something to do with the number of elements in a set, and to make that connection properly we need the idea of 1-1 correspondence.

Think, for example, how children would count the days of the week. They would probably say something like:

> "Sunday, that's one. Monday is two, Tuesday is three, Wednesday is four, Thursday is five, Friday is six, and Saturday is seven. That shows that there are seven days in a week."

In the language of this chapter, we might say that those children established a 1-1 correspondence between the set of days of the week and part of the counting chant, beginning at "one" and assigning chant words in order until all the days had been paired with words. The last chant word used was declared to be the size of the set.

This example typifies the general process of counting. Let us describe the process in slightly more formal terms to tie it to the mathematical machinery we have built:

> Given a set S of elements we wish to count, we set up a 1-1 correspondence between S and a subset of the counting-chant set (an ordered set of noises or symbols) beginning with the first chant element and using them all in order until each element of S has been assigned a chant image; but there are still chant

elements left over. If this can be done, we call S a **finite set** and use the chant element assigned last to symbolize the "size" (or "number of elements") of the set. (The empty set is also considered finite.) If this cannot be done—that is, if the correspondence with the counting chant up to a specific element of that chant never exhausts all of S—we call S an **infinite set**.

With this description of counting we are ready to tackle the question posed in Chapter 1: What is a number? The key to a mathematically satisfactory answer lies in the observation that two finite sets can be put in 1-1 correspondence if and only if a counting function assigns them the same size. This allows us to use counting to define an equivalence relation ("same size") on finite sets, and the equivalence classes of that relation (the sizes) are the numbers we are looking for. The next chapter provides a detailed development of this idea.

Exercises 3.5

In Exercises 1–6, list five ordered pairs that are determined by the given statement. Does the statement describe a function? Why or why not? (In each case the elements mentioned first are considered to be in the first set.)

1. Each word in this sentence is paired with its initial letter.
2. Each word in this sentence is paired with its final letter.
3. The initial letter of each word in this sentence is paired with the final letter of that word.
4. Each word in this sentence is paired with the number of letters it contains.
5. Each number of $\{1, 2, 3, 4, 5, 6, 7, 8, 9, 10\}$ is paired with every word in this sentence having that number of letters.
6. Each letter of the alphabet is paired with a word in this sentence if the letter is contained in that word.

For Exercises 7–15, state whether or not the given diagram describes a function from the first listed set to the second. If it is a function, is it a 1-1 correspondence? Why or why not?

7. a → 3, b → 1, c → 4, d → 3

8. a → 2, b → 1, b → 3, c → 2, d → 4

9. a → 2, b → 3, c → 4, d → 4

10. a → 2, b → 2, c → 2, d → 2

11. b → 1, b → 2, b → 3, b → 4

12. a → 1, b → 2, c → 3, d → 4

13. 1 → 1, 2 → 3, 3 → 3, 4 → 3

14. 1 → 2, 2 → 1, 3 → 4, 4 → 3

15. 1 → 1, 2 → 3, 3 → 4, 4 → 2, 5 → 4

Exercises 16–18 refer to Example 4, in which F denotes the temperature function that converts Celsius to Fahrenheit.

16. Determine $F(50)$, $F(15)$, $F(\text{-}10)$, and $F(75)$.

17. Find x in each case:

(a) $F(x) = 40$ (b) $F(x) = 23$
(c) $F(x) = 176$

18. Let C be the function that matches each number of the Fahrenheit scale with the corresponding number of the Celsius scale. Find each of the following:

(a) $C(32)$ (b) $C(212)$ (c) $C(68)$
(d) $C(140)$ (e) $C(23)$ (f) $C(176)$

19. Let I denote the set of all integers. Define $f: I \to I$ by assigning to each integer the number one more than its double. (In symbols, $f(n) = 2n + 1$, for each $n \in I$.) Find:

(a) $f(1)$ (b) $f(5)$ (c) $f(8)$
(d) $f(100)$ (e) $f(0)$ (f) $f(-7)$

20. Do the same as in Exercise 19 for the function $g: I \to I$ defined by

$$g(n) = n^2 - 2$$

For Exercises 21–24, define three different 1-1 correspondences between the two given sets:

21. {#, \$, 0, c, K} and {1, 2, 3, 4, 5}

22. {a, b, c, d, e, f} and itself

23. $\{1, 2, 3, 4, \ldots\}$ and $\{-1, -2, -3, -4, \ldots\}$

24. $\{1, 2, 3, 4, \ldots\}$ and $\{2, 3, 4, 5, \ldots\}$

25. Let us define a counting function from the set of all words in this sentence to the counting chant. Let S be the set of words in the previous sentence, and call the function h. Find:

(a) h(from)
(b) h(sentence)
(c) h(chant)
(d) the "number of elements" in S

Review Exercises for Chapter 3

For Exercises 1–10, indicate whether the given statement is *true* or *false*.

1. The set of all good singers is a well-defined set.

2. $\emptyset$, { }, and $\{\emptyset\}$ all are symbols for the empty set.

3. Two sets are equal if each set is a subset of the other.

4. The union of two sets is the set of all elements in both sets.

5. Intersection of sets is to conjunction of statements as union of sets is to disjunction of statements.

6. An equivalence relation is any relation that is reflexive, symmetric, and transitive.

7. An equivalence relation is any relation that is reflexive, symmetric, or transitive.

8. A relation $\mathcal{R}$ on a set S is symmetric if, for all $x, y \in S$, $x\mathcal{R}y$ and $y\mathcal{R}x$.

9. A function from a set X to a set Y is a subset of $X \times Y$ with the property that each element of Y appears first in exactly one ordered pair.

10. Every 1-1 correspondence is a function.

In Exercises 11–15, list or otherwise specify the elements of each set:

11. $\{x \mid x \text{ is a whole number}, x > 7, \text{ and } x \leq 11\}$

12. $\{x \mid x$ is the name of a month that is spelled using the letter "z"$\}$

13. $\{x \mid x$ is the name of a day of the week that is spelled using the letter "r"$\}$

14. $\{x \mid x$ is a vowel of the alphabet and comes after the letter "t"$\}$

15. $\{x \mid x$ is a vowel of the alphabet or comes after the letter "t"$\}$

In Exercises 16–31, let

$A = \{3, 4, 5, 6\}, \quad B = \{3, 4, 5\}, \quad C = \{4, 5\},$

$D = \Big\{\text{e}, \{\text{d}, \text{e}, \text{f}\}, \{\text{d}\}\Big\}, \quad E = \Big\{\text{d}, \{\text{e}\}, \emptyset\Big\}$

Indicate whether the given statement is *true* or *false*.

16. $A \subset B$
17. $B \subset A$
18. $B \subseteq A$
19. $C \subset A$
20. $2 \notin A$
21. $5 \in C$
22. $\{3\} \in A$
23. $\{5\} \subset C$
24. $\{d\} \in D$
25. $\{d\} \subset D$
26. $\{e\} \in D$
27. $\{d, e, f\} \subset D$
28. $\emptyset \in E$
29. $\emptyset \subset E$
30. $\{\emptyset\} \subset E$
31. $\Big\{\{e\}\Big\} \subset E$

In Exercises 32–41, let the universal set be

$$\{10, 20, 30, \ldots, 100\}$$

and let $A = \{30, 50, 60\}$, $B = \{30, 50, 60, 70\}$, $D = \{40, 60\}$, and $E = \{50, 60, 70, 80\}$. List the elements of each set.

32. $A \cap B$
33. $A \cup B$
34. $A \cup \emptyset$
35. B'
36. $D \times A$
37. $B \cap D'$
38. $(B \cap D)'$
39. $B \cap (A \cup D)$
40. $(A \cap E) \times D$
41. $(A \cup D') \cap (B \cap E)$

42. Use a universal set and two nondisjoint sets of your choice to illustrate that $(A \cup B)' = A' \cap B'$.

43. Draw a Venn diagram that shows two sets A and B with $A \cap B \neq \emptyset$, and shade the region that corresponds to $(A \cap B)'$.

In Exercises 44–47, indicate whether the given relation is reflexive, symmetric, or transitive. Are any of them equivalence relations?

44. $\mathcal{R}$ is defined on the set $\{1, 6, 9\}$ and $\mathcal{R} = \{(6,9), (1,9), (9,6), (9,1)\}$.

45. $\mathcal{R}$ is defined on the set of all 1989 Boston Marathon runners, and $x\mathcal{R}y$ means x "was as fast as" y.

46. $\mathcal{R}$ is defined on the set of all 1989 Boston Marathon runners, and $x\mathcal{R}y$ means x "came in ahead of" y.

47. $\mathcal{R}$ is defined on the set {red, rod, rot} and $x\mathcal{R}y$ means x "has at least two letters in common with" y.

48. $\mathcal{R}$ is defined on the set {red, blue, run, bad, back, pink, pick, born, pub, zoo}, and $x\mathcal{R}y$ means x "has the same first letter as" y. Partition S into its equivalence classes.

In Exercises 49–53, let $A = \{\text{a, b, c}\}$ and $B = \{\text{e, f, g}\}$. Draw diagrams similar to those in Figure 3.15 to represent a relation from A to B that satisifes the given condition. If no diagram is possible, write "impossible."

49. A relation that is not a function.

50. A function that is not a 1-1 correspondence.

51. A relation that is a 1-1 correspondence.

52. A relation that is a 1-1 correspondence, but is different from your answer to the preceding exercise.

53. A relation that is a 1-1 correspondence, but is not a function.

4.1 What Is A Whole Number?

Preliminary Note: Up to now we have assumed an informal familiarity with some types of numbers, just for the sake of having a few simple examples. None of our previous work *required* the use of numbers, however, so we may at this point consider ourselves to be at the very beginning of the development of the number systems. Thus,

> *we shall no longer assume any prior knowledge of numbers, but instead shall build all the numerical concepts we need step by step.*

Equivalent Sets

Figure 4.1 displays a typical activity that is used with kindergarten students. They are asked to match the elements of one set with those of another, thereby setting up a 1-1 correspondence. The purpose of this kind of "pre-number concept" activity is to help them develop a sense of what it means to say that two sets "have the same number of elements" or "are the same size," without actually being able to count them yet. This activity represents the conceptual heart of our development of the whole-number system. We begin by formalizing the concept of "same size":

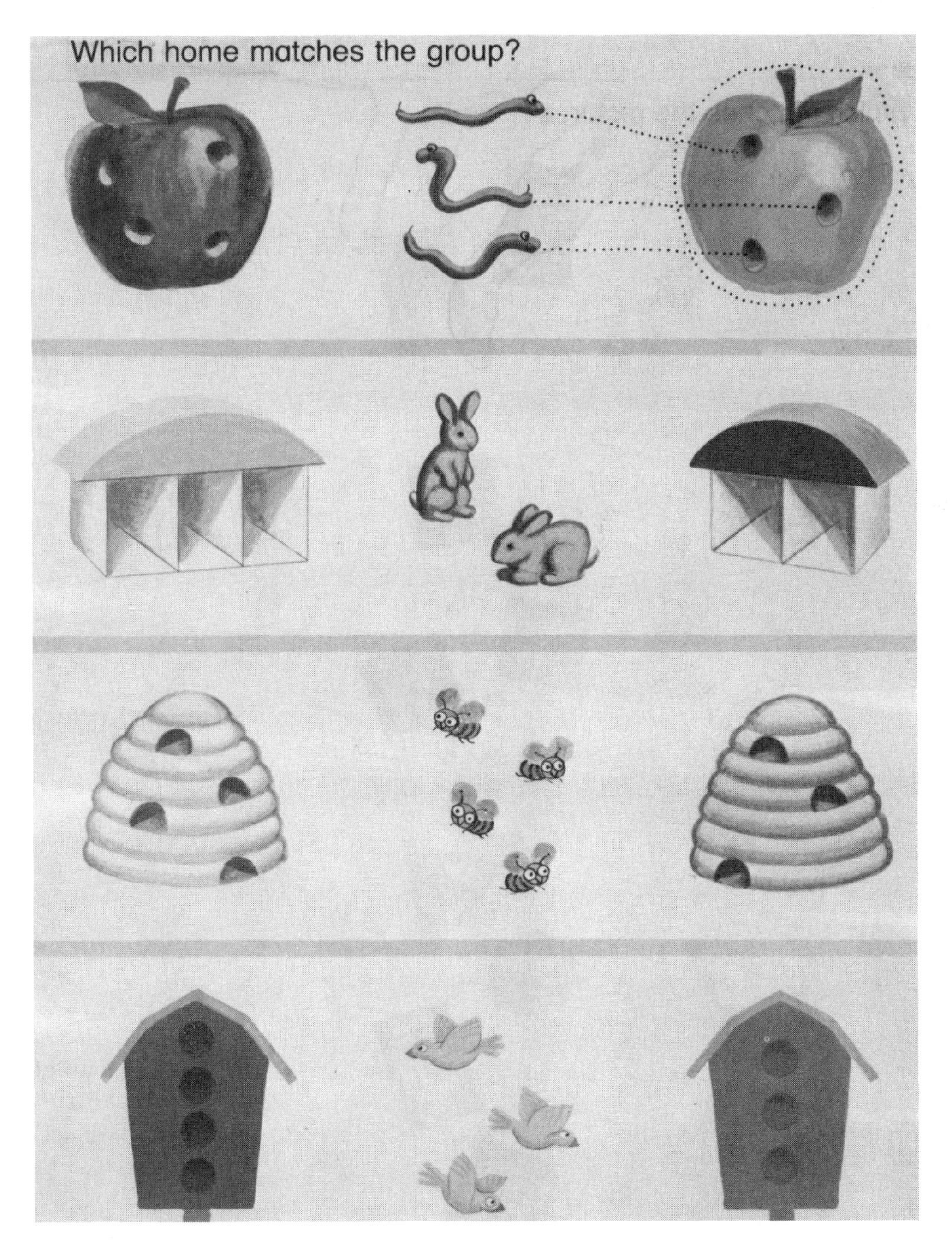

Figure 4.1 INVITATION TO MATHEMATICS, Kindergarten, p. 16

Definition Two sets are **equivalent** if they can be put in 1-1 correspondence.

Example 1 {a, b, c, d} and {\$, %, #, &} are equivalent sets because

{ a, b, c, d }
↕ ↕ ↕ ↕
{ \$, %, #, & }

is a 1-1 correspondence between them. There are many other 1-1 correspondences between these sets. For instance:

(In fact, 24 different 1-1 correspondences can be set up between them; can you find all 24?) Neither set is equivalent to {X, Y, Z}, because it is impossible to set up a 1-1 correspondence between {a, b, c, d} and {X, Y, Z} or between {\$, %, #, &} and {X, Y, Z}. □

Example 2 *Equal sets are always equivalent.* If two sets contain exactly the same elements (regardless of order), then matching each element in one copy of that set with itself in the other is an obvious 1-1 correspondence between the sets. For instance:

{ a, b, c, d, e }

{ a, c, e, b, d } □

Notation If A and B are equivalent sets, we write $A \sim B$.

Notice that the definition of equivalent sets is an *existential* statement; two sets are equivalent if *there exists* a 1-1 correspondence between them. There may be many such correspondences; in order to assert the equivalence of the two sets all we have to do is find one 1-1 correspondence.

PROBLEM-SOLVING COMMENT

Figure 4.1 and Examples 1 and 2 show how diagrams can be used to convey ideas effectively. The simple matching diagrams are as appropriate for elementary-school children as they are for us. What aspects of such diagrams must you focus on to decide if the correspondence is one-to-one? What would a correspondence that is not one-to-one look like? ◇

"Wait a minute!" we hear you say. "Didn't we already use *equivalent* to mean something else back in Chapter 3? How can you give the same word two different meanings?"

In Section 3.4 *equivalent* was used to denote things that are related by an equivalence relation. The present use of the word actually preserves that meaning. The *1-1 correspondence* relation is an equivalence relation on any collection of sets, so this use of the word is a (very important) special case of that general meaning. To see why this is true, we check the three defining properties of an equivalence relation.

> REFLEXIVE: Any set can be put in 1-1 correspondence with itself by matching each element of the set with itself.
>
> SYMMETRIC: If two sets are in 1-1 correspondence with each other, then each element of either set is matched with exactly one element of the other. Thus, if A is in 1-1 correspondence with B, then B is also in 1-1 correspondence with A.
>
> TRANSITIVE: If there is a 1-1 correspondence between two sets S and T and also a 1-1 correspondence between that second set T and a third set V, then we have a situation like that of Figure 4.2. Each element of S is matched with exactly one element of V just by following the pairings through the intermediate set T. For instance, Figure 4.2 shows that a is matched with \$, b with o, and c with #.

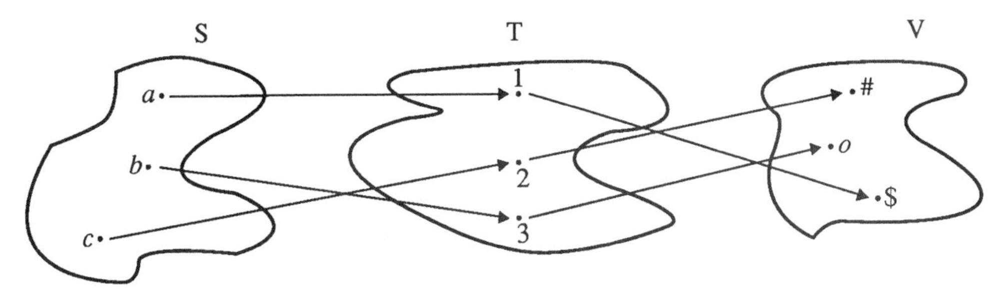

Figure 4.2 Transitivity of 1-1 correspondence

Thus, we have shown:

(4.1)

1-1 correspondence is an equivalence relation on any collection of sets.

One-to-one correspondence on the collection of all finite sets partitions that collection into (disjoint) equivalence classes. Each class contains sets that can be put in 1-1 correspondence with each other. For example,

$\{*, o, \%\}$ is in the class containing all those sets (and only those sets) that can be put in 1-1 correspondence with $\{*, o, \%\}$. We describe this class in symbols as

$$\{S \mid S \sim \{*, o, \%\}\}$$

In this class are the sets

{a, b, c}   {X, Y, Z}   {$, &, #}   {"one," "two," "three"}

and so on. All these sets share a common size property, and no sets outside this class have that specific property. Moreover, any set in the class can be used to represent the whole class (and hence the size). Property (3.11) guarantees that

$$\{S \mid S \sim \{\text{a, b, c}\}\}$$

for instance, is the same class of sets.

Cardinality

We often say that equivalent sets have "the same number of" elements, thereby suggesting that a number is a property shared by sets that can be put in 1-1 correspondence. But this "common property" notion is a little too vague to provide a basis for counting and arithmetic. For instance, just saying that "three" is a property shared by {a, b, c} and {#, ?, !}, and "four" is a property of {w, x, y, z} and {$, %, *o*, &} does not provide much insight into how to multiply three times four.

However, if we can specify the collection of *all* sets possessing a particular property, then we have "captured" the property, in a sense. That is, the property *is defined by* the collection of all the things that share it. That is, we capture the idea of the "number" of elements in a set by using the equivalence class of all sets sharing that property, and we know which sets they are by means of 1-1 correspondences. Thus, the *number* of elements in {a, b, c} *is*

$$\{S \mid S \sim \{\text{a, b, c}\}\}$$

Definition If A is any nonempty finite set, the **number** of elements in A (or the **cardinality** of A) is the collection of all sets that are equivalent to A. Any number formed in this way is called a **natural number** (or a **counting number**), and the set A is called a **representative set** for that number.

Note: A truly rigorous treatment of this idea would require some careful axiomatic distinctions among sets, collections, classes of sets, etc., to avoid the paradoxes that plagued set theory in the early 1900s. The formal details of such a construction would only obscure the natural connection between the theory and the underlying common-sense idea, so we shall not pursue them. It suffices to note that a consistent development of this approach to number can be carried out.

Notation

The number of elements in a set A is denoted by $n(A)$. In symbols,

$$n(A) = \{S \mid S \sim A\}$$

(Literally, this is read, "n of A is the collection of all sets S such that S is equivalent to A.")

The set of all natural numbers is denoted by $\mathbf{N}$. The definition of these numbers bears careful examination; it is the focal point of most of what we have done up to now. It says that *each* natural number is a collection of sets; in fact,

(4.2) The natural numbers are the equivalence classes determined by 1-1 correspondence on the collection of all nonempty finite sets.

Now that we have all these different classes, how shall we name them? We know that any set in an equivalence class can serve as a representative for that class, so we might simply pick at random a set from each class to represent it. But we can do better than that. Thanks to our memorized counting chant, there is a ready-made, nicely ordered collection of representative sets, one in each equivalence class:

{"one"}
{"one," "two"}
{"one," "two," "three"}
{"one," "two," "three," "four"}
⋮

We use these initial (ordered) subsets of the counting chant, called **counting sets**, as the standard representative sets for the natural numbers. Each of these counting sets is determined from the chant just by knowing where it ends, so we can name the counting sets and the classes they represent by using their last chant words. Thus, the natural numbers are:

$$\underline{\text{one}} = \{S \mid S \sim \{\text{“one”}\}\}$$
$$\underline{\text{two}} = \{S \mid S \sim \{\text{“one,” “two”}\}\}$$
$$\underline{\text{three}} = \{S \mid S \sim \{\text{“one,” “two,” “three”}\}\}$$
$$\underline{\text{four}} = \{S \mid S \sim \{\text{“one,” “two,” “three,” “four”}\}\}$$

and so forth.

Notation In this book we emphasize the distinction between the chant words themselves and their use in representing numbers by putting the words in quotation marks when they are used simply as chant noises; when those words are used to represent the natural numbers they are underlined.

It is customary to represent the counting words (in all their different uses) by the familiar symbols 1, 2, 3, 4, etc., with combinations of the basic symbols used to represent numbers beyond nine. This custom, while convenient, can be ambiguous or confusing at times. For instance, the symbol "5" can have several distinct meanings:

- It may mean the chant word "five," representing nothing but a sound between "four" and "six" in the chant;
- it may mean the natural number five, standing for a class of equivalent sets;
- it may be just the "numeral" symbol 5 itself, a figure drawn or printed on a piece of paper.

Often these meanings overlap in a particular context, making it difficult to keep them distinct. Usually children (and adults) encounter 1, 2, 3, 4, ... as a confused chant-number-numeral mixture. Young children do not distinguish explicitly among these three meanings, even when they have a proper intuitive understanding of numerical ideas. It is essential for you as a teacher to have the distinctions among chant word, number, and numeral clear in your own mind so that you can focus on the root of any difficulty a particular child might be having in learning about numbers.

In summary, when we say that the set **N** of natural numbers is

$$\{1, 2, 3, 4, 5, \ldots\}$$

we mean that **N** *is an infinite set of equivalence classes, each class represented by a counting set in it, and that representation is abbreviated by a symbol for the last word in the chant set!*

Example 3 Figure 4.3 displays a kindergarten exercise designed to help children associate the number symbols 1, 2, and 3 with sets containing one, two, or three elements. Of course, if sets of birds were the only sets used to illustrate

Figure 4.3 INVITATION TO MATHEMATICS, Kindergarten, p. 22

these numbers, children might quickly infer that 1, 2, and 3 always referred to birds. To guard against such mistaken assumptions, children must be provided with a wide variety of example sets as they form their early number concepts. As they learn about <u>three</u>, for example, they should be shown sets of three crayons, three books, three cars, etc., *and also* sets of three elements that have no obvious common feature. In this way, they will gradually focus on the single property *all* such sets share, "threeness." □

One finite set has been left out of this scheme so far—the empty set. The reason for this omission is that the natural numbers are based on counting elements in sets, and $\emptyset$ contains no elements to count. Nevertheless, $\emptyset$ is a finite set; so it seems appropriate to define the class $\{S \mid S \sim \emptyset\}$ as a number.

Definition **Zero** is the class of all sets that are equivalent to the empty set.

Notice that, even though the whole number $\emptyset$ is defined by the same process as the other whole numbers, the resulting equivalence class is quite different. Infinitely many sets can be put in 1-1 correspondence with the representative set {"one," "two," "three"}, for instance, so $\underline{\text{three}}$ is a very large collection of sets. However, the only set equivalent to $\emptyset$ is $\emptyset$ itself, so the equivalence class $\underline{\text{zero}}$ only contains one set; that is,

$$\underline{\text{zero}} = \{\emptyset\}$$

Definition The set of **whole numbers** is the set consisting of zero and all the natural numbers.

Notation We denote $\underline{\text{zero}}$ by 0 and the set of whole numbers by **W**. Thus,

$$\mathbf{W} = \{0, 1, 2, 3, 4, \dots\} = \{0\} \cup \mathbf{N}$$

Exercises 4.1

1. Set up a 1-1 correspondence between {a, b, c, d, e} and {$, %, #, o, ?}.
2. Set up a 1-1 correspondence between {$, %, #, o, ?} and {v, w, x, y, z}.
3. Illustrate the transitive property of $\sim$ by using your answers to Exercises 1 and 2 to define a 1-1 correspondence between {a, b, c, d, e} and {v, w, x, y, z}.
4. List the equivalence classes of the partition induced by $\sim$ on the following collection of sets:
   { {c, a, t}, {m, o, u, s, e}, {s, y, m, b, o, l}, {s, q, u, a, r, e}, {p, r, i, n, t}, {d, o, g}, {b, i, r, d}, {y, o, u}, {f, i, s, h}, {f, o, r, e, s, t}, {m, e} {a, r, t, i, c, h, o, k, e}, {u, s} }
5. List four different sets that represent the whole number 6.

For Exercises 6–10, find two different representative sets for the given whole number:

6. 1 **7.** 2 **8.** 7
9. 10 **10.** 25

HISTORICAL NOTE: TRANSFINITE NUMBERS

We have seen that the whole numbers can be regarded as size classifications for finite sets. In the 1870s, the German mathematician Georg Cantor extended this idea of *number* to infinite sets, as well. He removed the restriction of finiteness on representative sets, defining the **cardinal number** of a set A (any set at all) to be

$$\{S \mid S \sim A\}$$

Of course, when A is finite, its cardinal number is just the whole number that expresses how many elements A contains. Prior to Cantor's time, mathematicians thought that extending this idea to infinite sets would be useless because, after all, the fact that an infinite set has no "last" element ought to imply that all infinite sets could be put in 1-1 correspondence with each other. However, in 1879 Cantor proved a surprising theorem, which says, in essence:

> The set of all subsets of a given set cannot be put in 1-1 correspondence with the given set.

Here the phrase "given set" expresses the startling fact that Cantor's result applies to *any* set, even infinite ones. Because the "size" of a set is defined in terms of 1-1 correspondence, this theorem asserts that there are different sizes of infinity! Cantor called these different sizes of infinite sets **transfinite numbers**.

In later sections of this chapter we shall see how the development of whole-number arithmetic depends entirely on combinations and comparisons of representative sets. The fact that the transfinite numbers are constructed just like the whole numbers (by using representative sets) allows this whole-number arithmetic to be extended directly to a "transfinite arithmetic," in which sizes of infinite sets can be added, multiplied, and ordered. The definitions of addition and multiplication are exactly the same for transfinite numbers as they are for whole numbers, but the results often are peculiarly different. ◇

For Exercises 11–14, list the counting set that is equivalent to the given set:

11. {s, e, t} **12.** {\$, 2, #, 4, z}

13. {5, 0, 11, 23} **14.** {A, ∅, B, q, 9}

For Exercises 15–32, let the universal set be all letters of the (Roman) alphabet, and let

$$\begin{aligned} A &= \{\text{a, c, t, o, r, s}\} \\ B &= \{\text{b, o, t, h, e, r}\} \\ C &= \{\text{c, a, t}\} \end{aligned}$$

15. $n(A) =$ ___ **16.** $n(C) =$ ___

17. $n(A \cup B) =$ ___ **18.** $n(A \cap B) =$ ___

19. $n(A \cup C) =$ ___ **20.** $n(C \cap B) =$ ___

21. $n(A' \cap C) =$ ___ **22.** $n(B' \cap C) =$ ___

23. $n(C \times C) =$ ___ **24.** $n(A \times B) =$ ___

25. $n\big((B \cap C) \times C\big) =$ ___

26. $n\big(B \cap (C \times C)\big) =$ ___

27. Does $A = B$? **28.** Is $A \sim B$?

29. Is $A \sim C$? **30.** Is $A = C$?

31. Does $n(A) = n(B)$?

32. Does $n(A) = n(C)$?

For Exercises 33–56, answer *true* or *false*:

33. $\{\text{a, b, c, d}\} \in 4$ **34.** $\{\text{a, b, c, d}\} = 4$

35. $\{\text{a, b, c, d}\} \sim 4$ **36.** $\{\text{a, b, c, d}\} \subseteq 4$

37. $4 \in \{a, b, c, d\}$

38. $n\Big(\{a, b, c, d\}\Big) \in 4$

39. $n\Big(\{a, b, c, d\}\Big) = 4$

40. $n\Big(\{a, b, c, d\}\Big) \sim 4$

41. $\{a, b, c, d\} = \{$"one," "two," "three," "four"$\}$

42. $\{a, b, c, d\} \sim \{$"one," "two," "three," "four"$\}$

43. $\{a, b, c, d\} \in \{$"one," "two," "three," "four"$\}$

44. $\{a, b, c, d\} \subseteq \{$"one," "two," "three," "four"$\}$

45. $0 = \emptyset$

46. $0 \in \emptyset$

47. $\emptyset \in 0$

48. $\emptyset \subseteq 0$

49. $0 \subseteq \emptyset$

50. $0 = \{\emptyset\}$

51. $\emptyset \in \{\emptyset\}$

52. $\emptyset \subseteq \{\emptyset\}$

53. $\{\emptyset\} \in 0$

54. $\{\emptyset\} \sim \{0\}$

55. $\{\emptyset\} \in 1$

56. $\{\emptyset\} = \{0\}$

57. (a) Give an example of two sets A and B such that
$$A \times B \neq B \times A$$
then set up a 1-1 correspondence between these two Cartesian product sets.

(b) Justify the statement:
For any two sets A and B,
$$A \times B \sim B \times A$$

(c) What does this tell you about the cardinalities of $A \times B$ and $B \times A$?

4.2 Addition and Subtraction of Whole Numbers

The example in Figure 4.4 is taken from a widely used first-grade mathematics book:

This example contains the seeds of many fundamental ideas about whole numbers. Some of them we have already seen: 3 and 4 are sizes of sets, and they can be represented by different sets of things—children, apples, toys, or any sets of mixed objects; the sizes relate to the counting sets by 1-1

Figure 4.4 INVITATION TO MATHEMATICS, Grade 1, p. 74

correspondences; the counting chant can be used to order the objects in the sets. However, some of the other concepts hinted at here are not yet part of our construction. In the expression "$4 + 3 = 7$," the symbols 4, 3, and 7 are by now old friends; they represent well-defined equivalence classes of sets, called whole numbers. The symbols + and =, on the other hand, invite further attention because their meanings are not yet formally established.

Equality of Whole Numbers

We have seen = before, but its formal use has been restricted to equality between sets. That usage also extends to whole numbers, because whole numbers are just collections of sets. In particular, to say that two whole numbers a and b are **equal** means that every set in the equivalence class a is also in the class b, and vice versa.

Now, if $a = n(A)$ and $b = n(B)$, then a set S is in the equivalence class a if $S \sim A$, and S is in b if $S \sim B$. This means that a set S is in both a and b if (and only if) $A \sim B$ (because $\sim$ is symmetric and transitive). Thus, we have established an important fact:

(4.3) Two whole numbers are *equal* if and only if they have *equivalent* representative sets. In symbols,

$$n(A) = n(B) \text{ if and only if } A \sim B$$

Example 1 $n(\{\text{a, b, c}\}) = n(\{\text{p, q, r}\})$ because $\{\text{a, b, c}\}$ is equivalent to $\{\text{p, q, r}\}$. □

Example 2 If $x = n(A)$ and $A \sim \{\text{a, b, c, d}\}$, then $x = 4$. Informally, we are saying that A must contain 4 elements because it can be put in 1-1 correspondence with a 4-element set. □

Operations

Now that we know what "=" means when used with whole numbers, let us focus on "+." If you have heard addition called an "arithmetic operation," but are not quite sure exactly what that means, maybe this will help. Do you recall using flash-cards to learn addition and multiplication facts? That was an activity designed to help you quickly associate pairs of numbers with individual numbers, depending on the symbol used with the pairs. For example, if "+" was between 3 and 2, you answered "5," but if "×" was between 3 and 2, you answered "6." For a particular symbol—"+" or "×"—each pair of whole numbers was matched with exactly one whole number—the "sum" or the "product." That is, addition and multiplication are processes that assign specific numbers to each *pair* of whole numbers.

In the language of Section 3.5, then, addition and multiplication are *functions* from the set $\mathbf{W} \times \mathbf{W}$ of all (ordered) pairs of whole numbers to the set $\mathbf{W}$ of (single) whole numbers. This idea of matching ordered pairs with individual elements is the key to the notion of an *operation* on any set.

Definition An **operation** on a set S is a function from $S \times S$ to S.[1]

(**Remember**: By the definition of *function*, this definition says that an operation on a set S matches *each* ordered pair of elements of S with a single element of S.)

Notation Operations are usually denoted by symbols such as $*$, $+$, $\circ$, $\times$, etc. We shall use $*$ as the generic symbol for an operation. If $*$ is an operation on a set S, then the element of S that is matched with the ordered pair (x, y) is written $x * y$.

Example 3 Matching each ordered pair with its first element is an operation on any set; the diagram

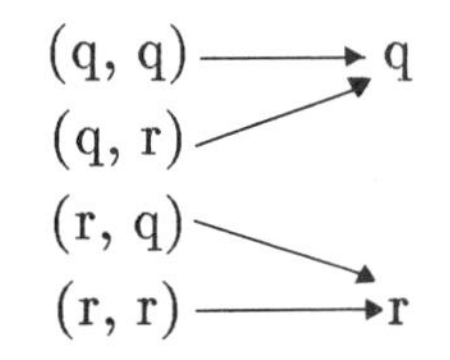

describes this operation on the set {q, r}. Denoting this operation by $*$, we would write

$$q * q = q \qquad q * r = q \qquad r * q = r \qquad r * r = r$$ □

Example 4 Let $\mathcal{S}$ be the collection of all subsets of some universal set $\mathcal{U}$. The union of any two subsets of $\mathcal{U}$ is also a subset of $\mathcal{U}$, so "$\cup$" is an operation on $\mathcal{S}$. Intersection is also an operation on $\mathcal{S}$, but Cartesian product is not. (Why not?) □

Addition of Whole Numbers

The formal definition of whole-number addition is foreshadowed by the way first graders are introduced to adding numbers. Figure 4.5 illustrate a typical

[1]Such operations are usually called *binary operations* because they combine two elements at a time. This is the only kind of operation we shall consider formally, so the word *binary* has been omitted.

approach. To see how $3 + 2 = 5$, the children are encouraged to find a set of 3 objects, another set of 2 objects, combine the two sets, and then count how many objects there are all together. In (our) more formal terms, they

Figure 4.5 INVITATION TO MATHEMATICS, Grade 1, p. 27

find representative sets for 3 and 2, form the union of the two sets, then determine the number of elements in the union.

This is precisely how we shall define the operation of addition on **W**, except that we must explicitly require the representative sets to be disjoint. We make this requirement to avoid situations such as

$$\{a, b, c\} \cup \{c, d\} = \{a, b, c, d\}$$

where the union of a 3-element set and a 2-element set contains 4, rather than 5, elements. That disjointness occurs implicitly in the early grades when distinct physical objects—apples, blocks, crayons—are used as examples in combining sets, but later problem-solving situations often require recognizing this restriction explicitly. Consider, for example, the following problem:

> A third-grade class has two committees—a bulletin-board committee of 5 members and a recess committee of 8 members. How many students are on at least one committee?

This problem cannot be solved until it is known how many students are on both committees, and only if the committees are disjoint will the solution be $5 + 8$.

Definition Let a and b be whole numbers, with representative sets A of a and B of b chosen so that A and B are disjoint. The **sum** of a and b is the whole number represented by $A \cup B$; we write this number as $a + b$. In symbols:

$$a + b = n(A \cup B)$$

where $a = n(A)$, $b = n(B)$, and $A \cap B = \emptyset$. The numbers a and b are called **addends** or **summands** of the sum; the process of forming sums is called **addition.**

Example 5 To form $3 + 2$, choose disjoint representative sets, such as $\{a, b, c\} \in 3$ and $\{d, e\} \in 2$. Then

$$\begin{aligned} 3 + 2 &= n\Big(\{a, b\} \cup \{c, d, e\}\Big) \\ &= n\Big(\{a, b, c, d, e\}\Big) \\ &= 5 \end{aligned}$$

The answer, 5, is found by counting the elements of $\{a, b, c, d, e\}$, the union set. Notice how this example mirrors the lesson given to first graders in Figure 4.5.

In the early grades, the disjointness of the representative sets can be handled implicitly by choosing 3 apples and 2 pears, then counting all the pieces of fruit in the two sets together, or by putting 3 red crayons and 2 green crayons in the same hand, then counting all the crayons in that hand, and so forth. □

Commutativity and Associativity

If you ask young children to explore whole number addition by using blocks, beads, or other "counters" (to form representative sets) on a large variety of examples, before long some of them are likely to complain,

> "I don't need to do $7+6$ because I just did $6+7$ and I *know* the answers will be the same."

In fact, they are observing a fundamental property of addition—that the order in which the numbers are taken does not affect the answer. After all, it doesn't matter whether the set of 6 blocks is to the right or to the left of the set of 7 blocks; when you push them together and count, the result is still 13 blocks. You have known this for so long, you may take it for granted, but it is one of the most important properties of addition. Even your young students will find it useful; it cuts the number of basic addition facts they have to memorize almost in half! In the early grades, this property is sometimes called the "order property"; in the higher grades it is formally called the *commutative law.*

Definition An operation $*$ on a set S is **commutative** if

$$a * b = b * a$$

for all elements a, b in S.

Example 6 Union and intersection of sets are commutative operations: If an element is in either a set A or a set B, then it is also in either B or A; thus, $A \cup B = B \cup A$. Similarly, an element in two sets A and B is in both of them regardless of the order in which the sets are taken, so $A \cap B = B \cap A$. □

Example 6 points the way to commutativity of addition of whole numbers. Applying the commutativity of union to the definition of the sum of whole numbers a and b, we get

$$a + b = n(A) + n(B) = n(A \cup B) = n(B \cup A) = n(B) + n(A) = b + a$$

Thus, we can add two whole numbers in either order and obtain the same sum. In other words:

(4.4)

> Addition of whole numbers is commutative. That is, for any a, $b \in \mathbf{W}$,
> $$a + b = b + a$$

Operations have been defined in terms of *pairs* of elements. When combining three or more elements, we must use two at a time. If the operation is not commutative, the order in which the elements are taken is important. But even if the order is kept fixed, there are still two ways to proceed. If a, b, and c are three elements, taken in that order, we could first combine a and b and then combine the result with c—written as $(a * b) * c$—or we could combine b and c and then combine a with that result—written as $a * (b * c)$. Of course, an operation is easier to use if we don't have to worry about this "grouping" problem; these "nicer" operations have a special name:

Definition An operation $*$ on a set S is **associative** if

$$(a * b) * c = a * (b * c)$$

for all elements a, b, c in S.

Example 7 Union and intersection of sets are associative operations; for any sets A, B, and C,

$$\begin{aligned}(A \cup B) \cup C &= A \cup (B \cup C)\\ (A \cap B) \cap C &= A \cap (B \cap C)\end{aligned}$$

The definitions of $\cup$ and $\cap$ are completely independent of how the sets are grouped, as you can see by using any three (small) sets to check these statements. □

Example 8 We can define a nonassociative operation, say #, on {q, r} as follows:

$$\mathrm{q} \# \mathrm{q} = \mathrm{r} \qquad \mathrm{q} \# \mathrm{r} = \mathrm{r} \qquad \mathrm{r} \# \mathrm{q} = \mathrm{q} \qquad \mathrm{r} \# \mathrm{r} = \mathrm{q}$$

To show that # is not associative we only need to find *one* triple of elements that yields different answers when grouped differently:

$$(\mathrm{q} \# \mathrm{r}) \# \mathrm{q} = \mathrm{r} \# \mathrm{q} = \mathrm{q}$$

but

$$\mathrm{q} \# (\mathrm{r} \# \mathrm{q}) = \mathrm{q} \# \mathrm{q} = \mathrm{r}$$ □

Associativity of set union is the key to associativity of whole-number addition. By the definition of *sum*, for any three whole numbers a, b, and c,

$$\begin{aligned}(a+b)+c &= \big(n(A)+n(B)\big)+n(C)\\ &= n(A\cup B)+n(C)\\ &= n\big(A\cup B)\cup C\big)\\ &= n\big(A\cup(B\cup C)\big)\\ &= n(A)+n(B\cup C)\\ &= n(A)+\big(n(B)+n(C)\big)\\ &= a+(b+c)\end{aligned}$$

In other words:

(4.5)

> Addition of whole numbers is associative. That is, for any $a, b, c \in \mathbf{W}$,
>
> $$(a+b)+c = a+(b+c)$$

This means that expressions such as $3+4+5$ can be written without ambiguity. Associativity and commutativity together imply that we can add a string or column of numbers in any order or according to any grouping without affecting the answer.

Example 9 The fact that addition is independent of grouping and order is often used to simplify adding long strings of numbers. For instance,

$$2+3+4+5+6+7+8$$

can be added quickly if we group by 10s:

$$\begin{aligned} & (2+8) + (3+7) + (4+6) + 5\\ =\ & 10 + 10 + 10 + 5\\ =\ & 35\end{aligned}$$

Associativity and commutativity of addition are the properties that permit this kind of rearrangment. □

Subtraction of Whole Numbers

An addition statement such as $4+3=7$ can be turned into a question in two essentially different ways. We might ask

$$4+3=? \qquad \text{or} \qquad 4+?=7$$

Pedagogical research has found that these apparently similar questions seem to involve two different levels of comprehension in a child. An accurate understanding of $4 + 3 = ?$ tends to occur at an earlier stage of development than an understanding of the "missing addend" question $4 + ? = 7$.[2] In the structural development of the whole-number system, these twin questions suggest processes with quite different mathematical properties. The first form,

$$a + b = ?$$

is simple addition. We have already seen how to answer questions like this for any whole numbers a and b by using representative sets. The second form,

$$a + ? = b$$

suggests the process known as *subtraction*.

Subtraction need not always be viewed as a missing-addend process. Figure 4.6 shows another typical way to introduce subtraction to first graders. Like addition, it uses representative sets. The second example on that page illustrates the question $5 - 2 = ?$ by removing two ladybugs from a set of five ladybugs. To describe this process formally using representative sets, we would need **set difference**, defined by

$$A - B = \{x \mid x \in A \textit{ and } x \notin B\}$$

This seems to be the more natural way to define subtraction for youngsters, particularly because so many "real-world" subtraction problems can be described in this way.

However, almost at the same time that students experience this type of subtraction activity, they are also taught the "inverse" relationship between addition and subtraction. This is often suggested by phrases such as, "You can always check your subtraction problem by using addition." As we noted previously, this means that for every subtraction question of the form $a - b = ?$ there is a corresponding addition question $b + ? = a$ with the same answer. For instance, the "ladybug" illustration could be altered to describe the question in "missing addend" form, by saying: "There are 2 ladybugs on a leaf. How many more ladybugs must crawl onto the leaf so that there are 5 ladybugs altogether?" These two types of problems depict different situations, and it is important for children to learn that a solution to one is also a solution to the other.

Thus, when treating subtraction formally, we may choose whether to define the process *directly* in terms of representative sets or *inversely* in terms

---

[2]For a more detailed discussion of this topic, see Jean Piaget and Alina Szeminska, *The Child's Conception of Number*, New York: W. W. Norton & Co., 1965, pp. 86–87, and Richard W. Copeland, *How Children Learn Mathematics*, 2nd ed., New York: Macmillan Publishing Co., 1974, pp. 90–92.

of addition. We choose the latter approach here because the logical structure of the number systems is tidier and easier to generalize if subtraction is considered as the "opposite" of addition.

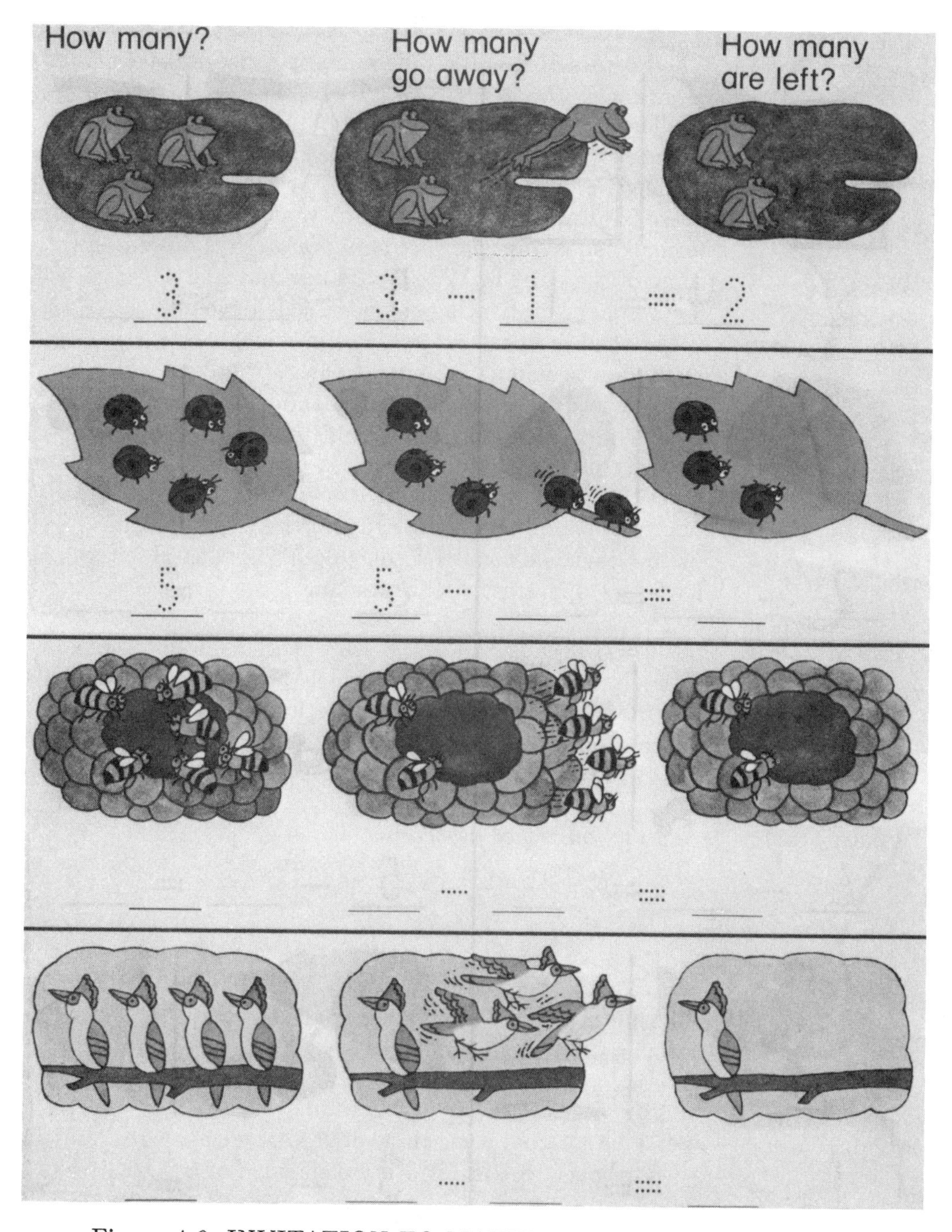

Figure 4.6 INVITATION TO MATHEMATICS, Grade 1, p. 123

Definition Let a and b be whole numbers. If there exists a whole number x such that $b + x = a$, then x is called the **difference** of a **minus** b; we write this number as $a - b$. In symbols,

$$a - b = x \quad \text{if} \quad b + x = a$$

The process of forming differences is called **subtraction**.

There is one very important thing to notice here: The definition of *difference* is a conditional statement form and its hypothesis cannot always be made true in **W**. If 2 and 5 are the whole numbers a and b, for example, there is no whole number x such that $5 + x = 2$. (See Example 10.) Thus, formally speaking, *the subtraction process is not an operation on* **W** because some of the pairs in $\mathbf{W} \times \mathbf{W}$ are not matched with any whole numbers. In fact, a main motivating theme of this development of the number systems is the need to construct systems in which subtraction and division (defined in the next section) are operations.

Since we are headed toward making these processes into operations, from time to time we shall "cheat" a little in our terminology, as in Examples 10 and 11. These examples discuss commutativity and associativity of subtraction, even though those two properties are formally defined just for operations. In such cases, rather than risk getting bogged down in fussy distinctions, we count on you to see the natural extension of the properties to these "operations-in-the-making."

Example 10 $5 - 2 = 3$ because $3 + 2 = 5$. However, $2 - 5$ does not exist (in **W**) because there is no whole number x such that $5 + x = 2$. (If there were, then the union of a representative set for 5 and some other set would have to contain only two elements, which is impossible.) This example also shows that *subtraction is not commutative.* □

Example 11 *Subtraction is not associative*:

$$(3 - 2) - 1 = 1 - 1 = 0 \qquad \text{but} \qquad 3 - (2 - 1) = 3 - 1 = 2$$ □

Zero

The number zero plays a special role with respect to addition of whole numbers. Although it is a role we take for granted, it is of fundamental importance in the theory of arithmetic.

Definition Let $*$ be an operation on a set S. An element z in S is called an **identity** with respect to $*$ if, for every element a of S,

$$a * z = a \qquad \text{and} \qquad z * a = a$$

It is easy to see that zero plays the role of identity for addition of whole numbers. For instance, if we take 4 as a typical example of a whole number, then, using representative sets {a, b, c, d} and $\emptyset$, we have

$$4 + 0 = n\Big(\{a, b, c, d\} \cup \emptyset\Big) = n\Big(\{a, b, c, d\}\Big) = 4$$

By commutativity of addition, $0 + 4 = 4$, as well. This example illustrates the general argument that justifies Statement (4.6).

(4.6) Zero is the identity for addition of whole numbers; it is called the **additive identity** for **W**.

We say that zero is *the* identity because it is the *only one* for addition of whole numbers. To see why this is true, suppose that x were another whole number with the identity property for addition. Then $x + 0 = 0$ because 0 is a whole number. But, by Statement (4.6), it is also true that $x + 0 = x$, so x *must* equal 0.

Exercises 4.2

For Exercises 1–8, consider the subsets

$$\begin{aligned} A &= \{a, b, c, d, e\} \\ B &= (b, d\} \\ C &= \{c, d, e, f\} \end{aligned}$$

of the universal set of all letters of the (Roman) alphabet. Justify each statement.

1. $n(A) = n(B \cup C)$

2. $n(B \times B) = n(C)$

3. $n(A' \cap C) = n(C \cap B)$

4. $n(B \times C) = n(C \times B)$

5. $n(A' \cap B) = n(\emptyset)$

6. $n(A) = n(A \cup B)$

7. $n(B \cup \emptyset) = n(B)$

8. $n(A \cup C) = n\Big((A \cap C) \times B\Big)$

For Exercises 9–17, compute directly from the definition of sum (by using representative sets):

9. $1 + 1$ **10.** $2 + 3$

11. $7 + 4$ **12.** $2 + 2$

13. $5 + 0$ **14.** $0 + 6$

15. $(2 + 3) + 4$ **16.** $2 + (3 + 4)$

17. $0 + 0$

For Exercises 18–22, show how commutativity and associativity of addition can be used to make the given sum easier to compute. (*Hint*: Look at Example 11.)

18. $2 + 4 + 6 + 8 + 10$

19. $11 + 13 + 15 + 17 + 19$

20. $2 + 2 + 5 + 3 + 5 + 3 + 6 + 4$

21. $28 + 5 + 39 + 12 + 11 + 15$

22. $73 + 45 + 11 + 27 + 86 + 9 + 14 + 15$

For Exercises 23–31, use the definition of difference to compute the given difference or give a reason why it cannot be done.

23. $3 - 1$ **24.** $7 - 2$ **25.** $3 - 6$

26. $4 - 4$ **27.** $6 - 3$ **28.** $1 - 3$

29. $3 - 0$ **30.** $0 - 3$ **31.** $0 - 0$

For Exercises 32–36, illustrate the given statement using sets of your choice.

32. Union of sets is commutative.

33. Union of sets is associative.

34. Intersection of sets is commutative.

35. Intersection of sets is associative.

36. Difference of sets is not commutative.

In Exercises 37–42, you are given a process $*$, defined in terms of addition and subtraction, that works on ordered pairs of nonzero whole numbers. If the process $*$ is commutative or associative, give an example to illustrate that fact; if not, provide a counterexample.

37. $a * b = a + b - 1$ **38.** $a * b = a + 1 - b$

39. $a * b = a + b + 1$ **40.** $a * b = a - b + a$

41. $a * b = a + b - a$ **42.** $a * b = a + b + a$

43. Justify the following statement:

For any finite sets A and B (not necessarily disjoint),

$$n(A) + n(B) = n(A \cup B) + n(A \cap B)$$

(*Hint*: Draw a Venn diagram.)

44. (a) Show by example that Cartesian product is not a commutative process; that is, give an example of sets A and B such that $A \times B \neq B \times A$.

(b) For your example in part (a), is it true that $n(A \times B) = n(B \times A)$? Is this true for *any* two sets? Why or why not?

4.3 Multiplication and Division of Whole Numbers

Multiplication of Whole Numbers

As Figure 4.7 illustrates, repeated addition of whole numbers is often used to introduce the concept of multiplication. We might describe the idea behind the picture of the twelve bicycle bells in Practice Exercise 2 as follows:

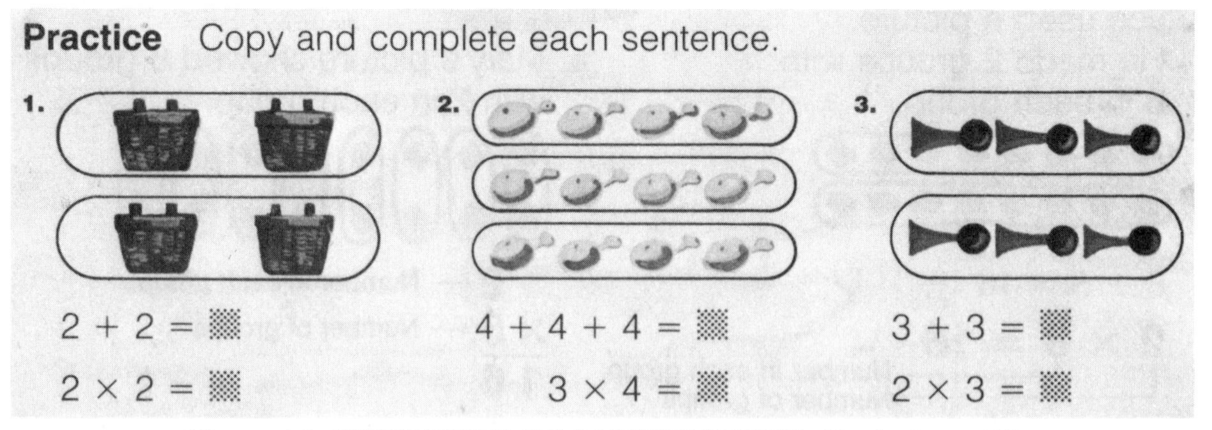

Figure 4.7 INVITATION TO MATHEMATICS, Grade 3, p. 153

If we label the four bells in any one row

$$\{1\text{st}, 2\text{nd}, 3\text{rd}, 4\text{th}\}$$

then there is a copy of that set for each of the three rows. Moreover, any of the twelve bells in that array can be identified by specifying its row and then its position in that row. For instance, the bell in the upper right-hand corner is "row 1, 4th bell," the next-to-last bell in the middle row is "row 2, 3rd bell," etc. That is, the bells may be "labeled" by the array

(1, 1st)	(1, 2nd)	(1, 3rd)	(1, 4th)
(2, 1st)	(2, 2nd)	(2, 3rd)	(2, 4th)
(3, 1st)	(3, 2nd)	(3, 3rd)	(3, 4th)

This is just the Cartesian product of the set {1, 2, 3} of rows with the set {1st, 2nd, 3rd, 4th} of bell positions.

One way to count the bells is to observe that there are 3 rows with 4 bells in each row, and then simply to add $4 + 4 + 4$; that is, to treat the product $3 \cdot 4$ as the sum of repeated addends. This is an example of an approach to multiplication often used in elementary school:

(4.7) If a and b are whole numbers, then

$$a \cdot b = b + b + \ldots + b$$

where there are a addends in this sum.

Our slightly more formal way of defining multiplication uses this approach, observing that the set-theoretic basis for repeated addition is the Cartesian product.

Definition Let a and b be whole numbers, with representative sets A of a and B of b. The **product** of a and b is the whole number represented by $A \times B$. We write this number as $a \cdot b$, or simply as ab. In symbols:

$$\text{If } a = n(A) \text{ and } b = n(B), \text{ then } a \cdot b = n(A \times B)$$

The numbers a and b are called **factors** of the product ab; the process of forming products is called **multiplication.**

Example 1 To form $2 \cdot 3$, choose representative sets, such as {x, y} for 2 and {p, q, r} for 3. Then

$$\begin{aligned} 2 \cdot 3 &= n\Big(\{x, y\} \times \{p, q, r\}\Big) \\ &= n\Big(\{(x, p), (x, q), (x, r), (y, p), (y, q), (y,r)\}\Big) \\ &= 6 \end{aligned}$$

The answer is found by counting the pairs—that is, by matching the set of pairs with the counting chant to determine its number. □

Example 2 The representative sets chosen for multiplication do not have to be disjoint. For instance, $2 \cdot 3$ can also be represented by

$$\{x, y\} \times \{x, y, z\} = \Big\{(x, x), (x, y), (x, z), (y, x), (y, y), (y, z)\Big\}$$

and this set also contains six ordered pairs. □

Example 3 In Section 3.3 we examined this problem:

> Sarah has 3 different blouses and 4 different skirts; how many different blouse/skirt outfits can she wear?

Its solution provides an excellent example of multiplication in terms of Cartesian product. We could also reason that, since every blouse could be worn with each of the 4 skirts, she has $4 + 4 + 4 = 12$ outfits, thus illustrating the repeated-addition interpretation. □

Commutativity and Associativity

Pursuing Example 3 a little further, we might just as easily say that every skirt could be worn with each of the 3 blouses, so there are $3+3+3+3 = 12$ outfits. To elementary-school students it is not immediately apparent that 3 addends of 4 is the same as 4 addends of 3, but after exploring various examples of this kind, they are likely to ask if multiplication, like addition, is commutative. That question can easily lead to asking about the other "nice" property we have seen, associativity.

(4.8)

Multiplication of whole numbers is commutative. That is, for any $a, b \in \mathbf{W}$, $$a \cdot b = b \cdot a$$

To see why this is true, let sets A and B represent the numbers a and b, and apply the definition of product:

$$a \cdot b = n(A) \cdot n(B) = n(A \times B)$$
$$b \cdot a = n(B) \cdot n(A) = n(B \times A)$$

Now, if A and B are distinct nonempty sets, then $A \times B \neq B \times A$. However, these two sets of pairs are always *equivalent* because the matching of each pair (x, y) in $A \times B$ with its reverse pair (y, x) in $B \times A$ is a 1-1 correspondence between $A \times B$ and $B \times A$. That is, $A \times B \sim B \times A$, so $n(A \times B)$ equals $n(B \times A)$.

Example 4 If $A = \{\text{x, y}\}$ and $B = \{\text{p, q, r}\}$, then

$$\begin{array}{rcllllllll} A \times B & = & \{ & (\text{x, p}), & (\text{x, q}), & (\text{x, r}), & (\text{y, p}), & (\text{y, q}), & (\text{y, r}) & \} \\ & & & \updownarrow & \updownarrow & \updownarrow & \updownarrow & \updownarrow & \updownarrow & \\ B \times A & = & \{ & (\text{p, x}), & (\text{q, x}), & (\text{r, x}), & (\text{p, y}), & (\text{q, y}), & (\text{r, y}) & \} \end{array}$$

is a 1-1 correspondence, showing that $2 \cdot 3 = 3 \cdot 2$. □

(4.9)

> Multiplication of whole numbers is associative. That is, for any $a, b, c \in \mathbf{W}$,
> $$(a \times b) \times c = a \times (b \times c)$$

To justify Statement (4.9) we could let sets A, B, and C represent the numbers a, b, and c, respectively, and show that the sets $(A \times B) \times C$ and $A \times (B \times C)$ are always equivalent. Instead of writing out the details of this argument, we rely on Example 5 to illustrate the general idea.

Example 5 Let us use the sets $A = \{\text{p, q}\}$, $B = \{\text{r, s}\}$, and $C = \{\text{t, u}\}$ to compare $(A \times B) \times C$ with $A \times (B \times C)$. We list the elements of the product sets in columns:

$(A \times B) \times C$		$A \times (B \times C)$
((p, r), t)	$\longleftrightarrow$	(p, (r, t))
((p, r), u)	$\longleftrightarrow$	(p, (r, u))
((p, s), t)	$\longleftrightarrow$	(p, (s, t))
((p, s), u)	$\longleftrightarrow$	(p, (s, u))
((q, r), t)	$\longleftrightarrow$	(q, (r, t))
((q, r), u)	$\longleftrightarrow$	(q, (r, u))
((q, s), t)	$\longleftrightarrow$	(q, (s, t))
((q, s), u)	$\longleftrightarrow$	(q, (s, u))

Thus, the sets $(A \times B) \times C$ and $A \times (B \times C)$ are equivalent because they can be put in 1-1 correspondence. But they are not equal; for example, the ordered pair $\big((p, r), t\big)$ is not the same as the pair $\big(p, (r, t)\big)$. (To see that these two ordered pairs are unequal, ask yourself: What is the first element of each pair?) □

Because multiplication is a binary operation, a three-factor product such as $3{\cdot}4{\cdot}5$ must be interpreted either as $(3{\cdot}4){\cdot}5$ or as $3{\cdot}(4{\cdot}5)$. But associativity of multiplication says that these two products are equal, so $3 \cdot 4 \cdot 5$ can be written without ambiguity. That is, because all groupings of the factors yield the same final product, the associative property allows us to write a product of three (or more) factors without parentheses.

Example 6 $(2 \cdot 5) \cdot 4 = 10 \cdot 4 = 40$ and $2 \cdot (5 \cdot 4) = 2 \cdot 20 = 40$, so we can simply write $2 \cdot 5 \cdot 4 = 40$. □

Example 7 Commutativity and associativity of whole-number multiplication allow us to multiply long strings of numbers without worrying about order or grouping. Thus, the product $3 \cdot 5 \cdot 4 \cdot 7 \cdot 5$ can be rearranged as

$$\big((5 \cdot 5) \cdot 4\big) \cdot (3 \cdot 7) = (25 \cdot 4) \cdot 21 = 100 \cdot 21 = 2100$$ □

Division of Whole Numbers

If you compare the definition of *product* with the definition of *sum* given in Section 4.2, you should notice an analogy between the use of Cartesian product here and union there. This observation helps to predict how the development of multiplication should proceed. The fact that both operations are commutative and associative strengthens the parallel. This analogy allows us to use Section 4.2 as a "road map," showing the way to multiplication questions that are obvious analogues of previous addition questions.

As an example of this analogy, observe that the analogue of the sum $4 + 3$ is the product $4 \cdot 3$, so the addition problem $4 + 3 = ?$ is analogous to the multiplication problem $4 \cdot 3 = ?$ Now, recall that in Section 4.2 we looked at inverse ("missing addend") questions for addition problems, such as $4 + ? = 7$, and related them to subtraction problems, such as $7 - 4 = ?$ This suggests analogous inverse (or "missing factor") questions for multiplication, such as $4 \cdot ? = 12$, which are related to division questions, such as $12 \div 4 = ?$ This, in turn, points to a definition.

Definition Let a and b be whole numbers, with $b \neq 0$. If there exists a whole number x such that $b \cdot x = a$, then x is called the **quotient** of a **divided by** b; we write this number as $a \div b$ or a/b. In symbols,

$$a \div b = x \quad \text{if} \quad b \cdot x = a$$

The process of forming quotients is called **division**.

PROBLEM-SOLVING COMMENT

See how easy that was? Once you notice an *analogy* between a new idea and one whose development is already known, the old one practically dictates how to develop the new one. Compare the development of division in this section with that of subtraction given in Section 4.2. ◇

Notice that the definition of *quotient* is a conditional form and its hypothesis cannot always be made true in **W**. For example, if a and b are 2 and 5, respectively, there is no whole number x such that $5 \cdot x = 2$. [See Exercise 4(a).] Thus, *division is not an operation on* **W** because some pairs in $\mathbf{W} \times \mathbf{W}$ are not matched with any whole numbers. The desire to make division an operation leads to the construction of the rational-number system (in Chapter 9). It is also worth noting that division is not defined using representative sets. It is a process derived from the definition of *product*, and can be considered the "inverse" of multiplication.

Example 8 $6 \div 2 = 3$ because $3 \cdot 2 = 6$. However, $2 \div 6$ does not exist (in **W**) because there is no whole number x such7 that $6 \cdot x = 2$. [See Exercise 4(b).] Thus, *division is not commutative.* □

Example 9 *Division is not associative*, even when all the required quotients exist:

$$(12 \div 6) \div 2 = 2 \div 2 = 1 \quad \text{but} \quad 12 \div (6 \div 2) = 12 \div 3 = 4$$ □

Example 10 Figure 4.8 shows how the relationship between division and multiplication is introduced informally to third graders. □

One and Zero

Continuing our analogy with the development of addition, the next question is: Does **W** contain an identity for multiplication? As you know, it does.

(4.10) One is the identity for multiplication of whole numbers; it is called the **multiplicative identity** for **W**.

To see why this is true, let us look at a typical example. Pick some whole number, say 4, and use {a, b, c, d} as its representative set. Then, using {#} to represent 1, we have

Figure 4.8 INVITATION TO MATHEMATICS, Grade 3, p. 264

$$\begin{aligned} 4 \cdot 1 &= n\Big(\{\text{a, b, c, d}\} \times \{\#\}\Big) \\ &= n\Big(\big\{(\text{a}, \#), (\text{b}, \#), (\text{c}, \#), (\text{d}, \#)\big\}\Big) \end{aligned}$$

The Cartesian-product set contains exactly one pair for each element in {a, b, c, d}, so the set of letters is equivalent to the set of ordered pairs. That is,

$$4 \cdot 1 = n\Big(\{\text{a, b, c, d}\}\Big) = 4$$

Of course, you would not use this level of formality with third graders. You can capture the same idea by saying something like: If you have 1 box of 4 bells, then you have 4 bells.

At this point it is natural to ask how the additive identity, 0, behaves with respect to multiplication. You probably know the answer to this from elementary school, but you may not know the reason for it. If a is any whole number and if we choose a representative set A for a, then

$$0 \cdot a = n(\emptyset) \cdot n(A) = n(\emptyset \times A)$$

But $\emptyset \times A$ is just $\emptyset$, so $0 \cdot a$ must be 0. Therefore:

(4.11)

> The product of 0 and any whole number is 0.
> That is, for any $a \in \mathbf{W}$, $a \cdot 0 = 0$.

Property 4.11 is the key to understanding the distinction between dividing a number *by zero* and dividing *zero by* a number, a source of confusion for many students of arithmetic. Just applying the definition of quotient carefully in each case:

> If a is any nonzero whole number, then 0 can be divided by a; the result is 0:
>
> $$0 \div a = x \quad \text{if} \quad a \cdot x = 0$$

Now, Statement (4.11) tells us that $a \cdot x = 0$ if $x = 0$; moreover, x cannot be anything else. (See Exercise 34.) Thus, $0 \div a = 0$.

> However, a CANNOT be divided by 0:
>
> $$a \div 0 = x \quad \text{if} \quad 0 \cdot x = a$$

But Statement (4.11) says that $0 \cdot x = 0$, regardless of what x is. That is, no number x will make $0 \cdot x = a$ true, so $a \div 0$ makes no sense.

Distributivity

Finally, we must examine an important link between multiplication and addition, called the *distributive property (of multiplication over addition).* It is the basis for many arithmetic and algebraic techniques, including multiplication of several-digit numbers and factoring of polynomials. We first provide a general definition, stated in terms of two arbitrary operations, because distributivity applies to many areas beyond arithmetic. Then we restate the distributive property as it applies to multiplication and addition of whole numbers.

Definition If $*$ and $\#$ are two operations on a set S, then $*$ is said to be **distributive over** $\#$ if

$$a * (b \# c) = (a * b) \# (a * c)$$

and

$$(b \# c) * a = (b * a) \# (c * a)$$

for any three elements $a, b, c \in S$. The first equation is called the **left distributive law**; the second is the **right distributive law.**

(4.12) Multiplication of whole numbers is distributive over addition. That is, for any $a, b, c \in \mathbf{W}$,

$$a \cdot (b + c) = (a \cdot b) + (a \cdot c)$$

and

$$(b + c) \cdot a = (b \cdot a) + (c \cdot a)$$

Property (4.12) simply says that you can multiply a number by the sum of two others either by adding those two other numbers and then multiplying (as on the left side of the equations) or by multiplying the summands by the first number separately and then adding (as on the right side).

Example 11 Figure 4.9 shows the commutative, associative, and distributive laws as they appear in a sixth-grade textbook. Notice, in particular, how the distributive property allows us to break a two-digit factor into *tens* and *ones* before multiplying:

$$8 \cdot 14 = 8 \cdot (10 + 4) = 8 \cdot 10 + 8 \cdot 4 = 80 + 32 = 112$$

This sort of thing will be an important part of our discussion of arithmetic processes in Chapter 6. □

Example 12 In using the formula for finding the perimeter of a rectangle, the distributive property permits us to add the length and width first, then double the sum:

$$P = 2l + 2w = 2(l + w)$$

□

Properties of Multiplication

When you multiply numbers, you can change the order of the factors and get the same product. This is the *commutative property of multiplication.*

```
    36         123
 × 123       ×  36
   108         738
   720        3690
  3600       4,428
 4,428
```

When you multiply numbers, you can change the grouping of the factors and get the same product. This is the *associative property of multiplication.*

(18 × 5) × 2	**18 × (5 × 2)**
90 × 2	**18 × 10**
180	**180**

When you add first then multiply, as in 8 × (10 + 4), you can write (8 × 10) + (8 × 4) instead and get the same answer. This is the *distributive property.*

8 × (10 + 4)	**(8 × 10) + (8 × 4)**
8 × 14	**80 + 32**
112	**112**

These properties can help you do the computation mentally.

(500 × 96) × 2

(96 × 500) × 2

96 × (500 × 2)

96 × 1,000 = 96,000

4 × 98 (Think of 98 as 100 − 2.)

(4 × 100) − (4 × 2)

400 − 8 = 392

Mental Math

Figure 4.9 INVITATION TO MATHEMATICS, Grade 6, p. 59

The distributive laws can be proved directly from the definitions of sum and product. The general idea of the argument should be clear from the following typical example.

To justify the statement $2 \cdot (1 + 3) = (2 \cdot 1) + (2 \cdot 3)$, choose representative sets for the three numbers, then apply the definitions in the order dictated by the parentheses:

For $2 = n\Big(\{a, b\}\Big)$, $1 = n\Big(\{c\}\Big)$, and $3 = n\Big(\{d, e, f\}\Big)$,

$$\begin{aligned} 2 \cdot (1+3) &= n\Big(\{a,b\}\Big) \cdot n\Big(\{c\} \cup \{d,e,f\}\Big) \\ &= n\Big(\{a,b\} \times \{c,d,e,f\}\Big) \\ &= n\Big(\big\{(a,c),(a,d),(a,e),(a,f),(b,c),(b,d),(b,e),(b,f)\big\}\Big) \\ &= 8 \end{aligned}$$

$$\begin{aligned} (2 \cdot 1) + (2 \cdot 3) &= n\Big(\{a,b\} \times \{c\}\Big) + n\Big(\{a,b\} \times \{d,e,f\}\Big) \\ &= n\Big(\big\{(a,c),(b,c)\big\} \cup \big\{(a,d),(a,e),(a,f),(b,d),(b,e),(b,f)\big\}\Big) \\ &= n\Big(\big\{(a,c),(b,c),(a,d),(a,e),(a,f),(b,d),(b,e),(b,f)\big\}\Big) \\ &= 8 \end{aligned}$$

Thus,

$$2 \cdot (1+3) = 8 = (2 \cdot 1) + (2 \cdot 3)$$

PROBLEM-SOLVING COMMENT

Basic operation tables provide a fertile area for encouraging even very young students to *look for patterns* and *generalize* what they find. The visual symmetries of the tables make some general properties—notably commutativity and the identity behavior of 0 and 1—quite easy to spot. Students might well be encouraged to formulate some of these general properties in their own words before you present them in a formal lesson. Their attempts to state the properties accurately can be checked by *constructing examples*. Their general statements can be sharpened by *introducing appropriate notation*, a worthwhile class-participation exercise, especially for higher grade levels. ◇

Exercises 4.3

1. Verify that multiplication of whole numbers satisfies the definition of an operation.

For Exercises 2–10, compute directly from the definition of a product (by using representative sets), then by using repeated addition, if possible:

2. $3 \cdot 5$ **3.** $4 \cdot 2$ **4.** $4 \cdot 4$

5. $2 \cdot 6$ **6.** $7 \cdot 1$ **7.** $1 \cdot 1$

8. $0 \cdot 3$ **9.** $(2 \cdot 3) \cdot 4$ **10.** $2 \cdot (3 \cdot 4)$

For Exercises 11–14, show how commutativity and associativity of multiplication can be used to make the given product easier to compute. (*Hint*: Look at Example 7.)

11. $2 \cdot 3 \cdot 5 \cdot 11$

12. $5 \cdot 17 \cdot 2 \cdot 5 \cdot 2$

13. $25 \cdot 13 \cdot 2 \cdot 8$

14. $2 \cdot 2 \cdot 2 \cdot 2 \cdot 5 \cdot 5 \cdot 5 \cdot 5$

15. Prove that $3 \cdot 1 = 3$ and $1 \cdot 4 = 4$ by using arguments like the one used to justify Statement (4.10).

16. Use reasoning analogous to that of Example 12 of Section 4.2 to show that there is no whole number x such that $5 \cdot x = 2$.

17. Show *directly from the definitions* of a quotient and a product that $2 \div 6$ cannot be a whole number.

For Exercises 18–26, use the definition of a quotient to compute the given quotient or give a reason why it cannot be done.

18. $6 \div 3$	19. $10 \div 2$	20. $3 \div 6$
21. $2 \div 10$	22. $5 \div 5$	23. $4 \div 1$
24. $1 \div 4$	25. $0 \div 7$	26. $7 \div 0$

27. Give an example to show that addition of whole numbers is not distributive over multiplication.

For Exercises 28–33, illustrate the given statement using (small) sets of your choice.

28. Union of sets is distributive over intersection.

29. Intersection of sets is distributive over union.

30. Cartesian product of sets is distributive over intersection.

31. Cartesian product of sets is distributive over union.

32. Union of sets is not distributive over Cartesian product.

33. Intersection of sets is not distributive over Cartesian product.

34. Prove that the product of two nonzero whole numbers cannot be 0. (*Hint*: Use the definition of a product.)

35. Explain why $0 \div 0$ is not a meaningful expression in arithmetic.

36. Use reasoning analogous to the argument following Statement (4.6) of Section 4.2 to show that 1 is the only identity for multiplication.

37. (a) Using any three sets A, B, and C,
    i. set up a 1-1 correspondence between $A \times B$ and $B \times A$;
    ii. set up a 1-1 correspondence between $(A \times B) \times C$ and $A \times (B \times C)$.
    (b) What properties of whole number multiplication are illustrated by part (a)?

In Exercises 38–42, for each of the proposed properties:

(a) express it as a symbolic statement, and
(b) determine if it is true for all "reasonable" values of the variables.
(c) If the statement is always true, provide three examples; if not, provide at least one counterexample.

38. Left distributivity of multiplication over subtraction

39. Right distributivity of multiplication over subtraction

40. Left distributivity of addition over multiplication

41. Left distributivity of division over addition

42. Right distributivity of division over addition

4.4 Multiples and Powers

At the beginning of the previous section, multiplication of whole numbers was introduced as repeated addition. In particular, we saw that

$$n \cdot b = \underbrace{b + b + \ldots + b}_{n \text{ summands}}$$

That is, the sum of n copies of a number b can be written concisely as a product. Some standard terminology accompanies this notation.

Definition Let b be any whole number. For any natural number $n \geq 2$, the sum of n copies of b is called a **multiple** of b (specifically, the **nth multiple** of b), and it is written $n \cdot b$. In this situation n is called a **multiplier** of b.

This notation and language is analogous to a familiar "shorthand" for repeated multiplication:

Definition Let b be any whole number. For any natural number $n \geq 2$, the product of n copies of b is called a **power** of b (specifically, the **nth power** of b), and it is written b^n. In this situation n is called an **exponent** of b. Thus,

$$b^n = \underbrace{b \cdot b \cdot \ldots \cdot b}_{n \text{ factors}}$$

We also define $b^1 = b$.

Properties of Exponents

Understanding the origin of a concept often helps to clarify the properties of that concept. For instance, if we consider whole-number multiplication as repeated addition, then distributivity of multiplication over addition becomes almost obvious:

$$3 \cdot 5 = 5 + 5 + 5 \quad \text{and} \quad 4 \cdot 5 = 5 + 5 + 5 + 5$$

so

$$(3 \cdot 5) + (4 \cdot 5) = (5 + 5 + 5) + (5 + 5 + 5 + 5) = (3 + 4) \cdot 5$$

In general, if n copies of some number b are added to m copies of that same number b, then the sum must involve $n + m$ copies of b:

$$nb + mb = \underbrace{(b + b + \ldots + b)}_{n \text{ summands}} + \underbrace{(b + b + \ldots + b)}_{m \text{ summands}} = (n + m)b$$

A similar distributive property holds for repeated multiplication:

$$5^3 = 5 \cdot 5 \cdot 5 \quad \text{and} \quad 5^4 = 5 \cdot 5 \cdot 5 \cdot 5$$

so

$$5^3 \cdot 5^4 = (5 \cdot 5 \cdot 5) \cdot (5 \cdot 5 \cdot 5 \cdot 5) = 5^{3+4}$$

In general,

$$b^n \cdot b^m = \underbrace{(b \cdot b \cdot \ldots \cdot b)}_{n \text{ factors}} \cdot \underbrace{(b \cdot b \cdot \ldots \cdot b)}_{m \text{ factors}} = b^{n+m}$$

Let us restate this important principle without the intermediate step:

(4.13)

> For any whole number b and any natural numbers n and m,
> $$b^n \cdot b^m = b^{n+m}$$

Example 1

$$12 \cdot 7 + 8 \cdot 7 = (12 + 8) \cdot 7 = 20 \cdot 7$$

Analogously,

$$7^{12} \cdot 7^8 = 7^{12+8} = 7^{20}$$

□

Example 2 $5^3 + 5^4$ is *not* equal to 5^{3+4}. In fact,

$$5^3 + 5^4 = (5 \cdot 5 \cdot 5) + (5 \cdot 5 \cdot 5 \cdot 5) = 125 + 625 = 750$$

but

$$5^{3+4} = (5 \cdot 5 \cdot 5) \cdot (5 \cdot 5 \cdot 5 \cdot 5) = 125 \cdot 625 = 78,125$$

□

There are several other easy and useful consequences of the exponent definition. Although we have developed enough machinery to formulate general proofs of these properties, they are probably better understood by introducing them with specific examples. (This is an approach similar to what you will use when you teach exponents to elementary-school students.) The proofs are straightforward generalizations of these examples and are omitted.

Consider first the product $7^5 \cdot 8^5$. By the definition of exponent, we can write

$$7^5 \cdot 8^5 = (7 \cdot 7 \cdot 7 \cdot 7 \cdot 7) \cdot (8 \cdot 8 \cdot 8 \cdot 8 \cdot 8)$$

Associativity and commutativity of multiplication allow us to rearrange the product on the right like this:

$$(7 \cdot 8) \cdot (7 \cdot 8) \cdot (7 \cdot 8) \cdot (7 \cdot 8) \cdot (7 \cdot 8)$$

that is, $(7 \cdot 8)^5$. In general:

(4.14)

> For any whole numbers a and b and natural number n,
> $$a^n \cdot b^n = (a \cdot b)^n$$

Consider next a power of a power of a number. For example, 8 equals 2^3, so we can write 8^5 as $(2^3)^5$:

$$(2^3)^5 = (2 \cdot 2 \cdot 2) \cdot (2 \cdot 2 \cdot 2) \cdot (2 \cdot 2 \cdot 2) \cdot (2 \cdot 2 \cdot 2) \cdot (2 \cdot 2 \cdot 2)$$

By associativity of multiplication, we can ignore the parentheses in the product on the right and just use 15 (the total number of 2s) as the exponent, so $(2^3)^5 = 2^{3 \cdot 5}$. In general:

(4.15) For any whole number b and natural numbers n and m,

$$(b^n)^m = b^{nm}$$

Finally, consider the quotient $7^5 \div 7^3$. The definition of quotient implies that this is 7^2, because $7^3 \cdot 7^2 = 7^5$. That is, in order to change $7 \cdot 7 \cdot 7$ to $7 \cdot 7 \cdot 7 \cdot 7 \cdot 7$ by multiplying, we need two more 7s as factors. In general, if we have a problem of the form $b^n \div b^m$, where m is less than n, then the answer is the product of as many b's as are needed to change b^m to b^n by multiplying.

(4.16) For any natural numbers b, n, and m, where $n - m$ is a natural number,[3]

$$b^n \div b^m = b^{n-m}$$

Example 3 By Statement (4.14),

$$2^7 \cdot 5^7 = (2 \cdot 5)^7 = 10^7 = 10{,}000{,}000$$ □

Example 4 By Statement (4.15),

$$32^4 = (2^5)^4 = 2^{20}$$ □

Example 5 By Statement (4.16),

$$3^{11} \div 3^5 = 3^{11-5} = 3^6$$ □

Zero as an Exponent

You might have noticed that in all these rules we have explicitly avoided using 0 as an exponent. That's not surprising, because the exponent is the number that "counts" the repeated factors, and 0 doesn't "count" anything. Nevertheless, there is an easy extension of the exponent concept that allows us to define in a useful way what it means to raise a number to the zeroth power.

For example, consider 7^0. If we are going to *define* this new mathematical animal, we get to *choose* what we want it to mean. What shall we base our choice on? Looking back at the discussion of exponents in this section,

[3]This law actually holds for *all* pairs of whole numbers m and n, but we have not yet developed negative numbers or the meaning of negative exponents. Discussion of the more general case is deferred until a later chapter.

it appears that Statement (4.13) is a key fact in making exponents work efficiently for us. It makes sense, then, to want this law to work even for the exponent 0. We would like, for instance,

$$7^3 \cdot 7^0 = 7^{3+0}$$

But $3 + 0$ is just 3, so we are actually facing the question

$$7^3 \cdot ? = 7^3$$

and the answer to this is the multiplicative identity, 1. In other words, in order for (4.13) to hold we *must* define 7^0 to equal 1. In general, then:

Definition For any nonzero whole number b: $b^0 = 1$.

(Notice that 0^0 is not defined. For an explanation of why this is so, see Exercise 57.)

Example 6 Now that we know how to interpret the exponent 0, Statement (4.16) can be extended to the case where $m = n$. Thus,

$$7^3 \div 7^3 = 7^{3-3} = 7^0 = 1 \qquad \square$$

PROBLEM-SOLVING COMMENT

The laws of exponents can be useful in *approximating* powers of numbers. In the following example, the symbol $\approx$ stands for "is approximately equal to."

$$2^8 = 256 \approx 250$$

so

$$2^{10} = 2^8 \cdot 2^2 \approx 250 \cdot 4 = 1000$$

Then

$$2^{20} = (2^{10})^2 \approx 1000^2 = 1,000,000$$

Similarly,

$$3^4 = (3^2)^2 = 9^2 = 81 \approx 80$$

so

$$3^{12} = (3^4)^3 \approx 80^3 = 8^3 \cdot 10^3 = (2^3)^3 \cdot 1000 = 2^9 \cdot 1000$$
$$= 2 \cdot (2^8) \cdot 1000 \approx 2 \cdot 250 \cdot 1000 = 500,000$$

(The exact value of 3^{12} is 531,441.)

However, approximations involving exponents must be handled with care—a small initial difference raised to larger and larger powers diverges from the true value *very* rapidly! For instance, if we assume $9 \approx 10$, then

$9^3 \approx 10^3 = 1000$, but actually $9^3 = 729$
$9^5 \approx 10^5 = 100{,}000$, but actually $9^5 = 59{,}049$
$9^7 \approx 10^7 = 10{,}000{,}000$, but actually $9^7 = 4{,}782{,}969$ ◇

Exercises 4.4

For Exercises 1–18, use the properties of exponents presented in this section to evaluate the given expression.

1. $3 \cdot 12 + 7 \cdot 12$ **2.** $4 \cdot 26 + 5 \cdot 26$

3. $23^5 \cdot 23^7$ **4.** $9^8 \cdot 9^2$ **5.** $7^{10} \cdot 3^{10}$

6. $2^9 \cdot 5^9$ **7.** $(5^8)^2$ **8.** $(25^3)^5$

9. $3^7 \div 3^2$ **10.** $5^{10} \div 5$ **11.** 73^0

12. 1^0 **13.** $2^5 + 2^5$ **14.** $4^3 \cdot 2^4$

15. $10^5 \div 5^4$ **16.** $(5^2)^3 \div 5^3$

17. $(7^5 + 14^{10})^0$ **18.** $2^4 \cdot 5^4 \div 10^2$

In Exercises 19–44, determine whether the given statement is *true* or *false*.

19. $3^4 + 3^2 = 3^6$ **20.** $3^4 \cdot 3^2 = 3^6$

21. $3^4 \cdot 3^2 = 3^8$ **22.** $5^2 \cdot 5^7 = 5^{14}$

23. $5^2 \cdot 5^7 = 5^9$ **24.** $5^2 \cdot 5^7 = 10^9$

25. $5^2 \cdot 5^7 = 25^9$ **26.** $7^3 \cdot 7^3 = 14^3$

27. $7^3 \cdot 7^3 = 7^9$ **28.** $7^3 \cdot 7^3 = 7^6$

29. $7^3 \cdot 7^3 = 14^6$ **30.** $7^3 \cdot 7^3 = 49^6$

31. $7^3 \cdot 7^3 = 49^3$ **32.** $(2^3)^5 = 2^8$

33. $(2^3)^5 = 2^{15}$ **34.** $(2^3)^5 = 8^5$

35. $9^6 \div 9^2 = 9^3$ **36.** $9^6 \div 9^2 = 9^4$

37. $9^6 \div 9^2 = 1^4$ **38.** $9^6 \div 9^2 = 1^3$

39. $9^6 \div 3^2 = 9^4$ **40.** $9^6 \div 3^2 = 9^5$

41. $9^6 \div 9^6 = 9^0$ **42.** $9^6 \div 9^6 = 0$

43. $9^6 \div 9^6 = 1^0$ **44.** $27^0 = 3^0$

In Exercises 45–56, simplify each expression, if possible, assuming all letters represent nonzero whole numbers.

45. $3a + 3b$ **46.** $3^a + 3^b$ **47.** $3^a \cdot 3^b$

48. $a^3 \cdot b^3$ **49.** $(3^n)^2$ **50.** $3^x \cdot 2^x$

51. $3^x \cdot 2^y$ **52.** $3^{2n} \div 3^2$ **53.** $3^{n+2} \div 3^2$

54. $3^a \div 3^a$ **55.** $a^3 \div a^3$ **56.** $a^3 \div b^3$

57. Near the end of this section there is an explanation of why $7^0 = 1$. In a similar way, explain why $5^0 = 1$; then try such an explanation for 0^0. Do you arrive at a similar conclusion? Why or why not?

For Exercises 58–60, mimic the Problem-Solving Comment at the end of this section to estimate each of the following numbers; then compare your estimate with the exact value.

58. 2^{20} **59.** 3^{20} **60.** 12^{10}

4.5 Ordering The Whole Numbers

We began our study of numbers by considering the two principles identified by Piaget as central to understanding the concept of number—conservation of quantity, and seriation. One of these led from 1-1 correspondences to classes of equivalent sets, called the whole numbers. The other led to a counting chant, providing a convenient collection of representative sets and names for the numbers. Each of these two major ideas carries with it a natural sense of order that leads to an order relation on the set **W** of whole numbers. The two approaches begin differently, but they yield the same final result.

As we saw at the beginning of Chapter 3, a counting chant is just an unlimited succession of sounds—

"one," "two," "three," "four," "five," ...

—and the main feature of this chant is the order in which these sounds occur: "one" comes before "two," "two" before "three," and so forth. This ordering of the chant sounds may be borrowed in the obvious way to order the set of whole numbers. That is, we can order the set **W** by agreeing that one whole number, a, comes **before** another, b, if and only if the chant name for a precedes the chant name for b in the counting process. (We must also agree that zero comes before one.)

This ordering is natural and convenient; children and adults alike think of numerical order in this way most of the time. Many everyday uses of the order concept for whole numbers require no more than this intuitive chant-based notion. However, some important properties of order used in arithmetic, algebra, and other areas of mathematics are not immediately obvious from such an informal understanding. They are more easily understood if we approach the idea of whole-number order in a slightly different way, beginning with 1-1 correspondences.

Order and 1-1 Correspondences

If two sets *cannot* be put in 1-1 correspondence, it is natural to think of the set that "runs out" first in any attempted 1-1 matching as the smaller one. (It is always the same set that runs out, regardless of the 1-1 correspondence used.) Rephrasing this idea provides a useful formal definition:

Definition A whole number a is **less than** a whole number b if a representative set for a is equivalent to a proper subset of a representative set for b. In this case, we also say that b is **greater than** a, and we write $a < b$ or $b > a$.

These separate ideas of *before* and *less than* actually yield the same ordering of **W**. To see why, recall the basic link between the chant and 1-1 correspondences. If one chant word comes before another, then the chant set ending with that first word is in some sense smaller than the chant set ending with the second one. For example, "three" comes before "five" in the chant, so the chant set

{"one," "two," "three"}

is a proper subset of the chant set

{"one," "two," "three," "four," "five"}

Thus, a whole number a is less than a whole number b if and only if the chant name for a occurs before the chant name for b in the counting process.

Example 1 To show that 2 is less than 3, consider {o, k} and {x, y, z} as representatives for 2 and 3, respectively. Clearly, {o, k} is equivalent to {x, y}, which is a proper subset of {x, y, z}. We can conclude that $2 < 3$. □

Example 2 By looking at the example "$3 < 5$" discussed previously, it should be easy to see how *before* and *less than* give the same ordering of **W**. If we choose other representative sets, say {x, y, z} for 3 and {a, b, c, d, e} for 5, then the equivalence of one set to a proper subset of the other comes right out of the counting process:

{	x,	y,	z		}	
	↕	↕	↕			
{	"one,"	"two,"	"three,"	"four,"	"five"	}
	↕	↕	↕	↕	↕	
{	a,	b,	c,	d,	e	}

□

Example 3 Figures 4.10 and 4.11 show how the two approaches to order are introduced at the kindergarten level. Figure 4.10 uses comparison of sets to illustrate the meaning of "less." Figure 4.11 uses the serial ordering 0, 1, 2, 3, ... without referring to sizes of sets because the child is asked to determine which box has less, even though some of the boxes are covered. □

Order and Addition

The definition of *less than* can be reformulated using addition. Property (4.17) is logically equivalent to the definition of *less than*, and often it is more convenient to use. We shall regard the two statements as interchangeable.

(4.17) A whole number a is less than a whole number b if and only if there is a nonzero whole number d such that $a + d = b$.

Example 2 illustrates how Statement (4.17) follows from the way *less than* is defined. It shows that $3 < 5$ is justified by

$$\{x, y, z\} \sim \{a, b, c\} \subset \{a, b, c, d, e\}$$

Thus, the number that must be added to 3 to get 5 is represented by the set {d, e}.

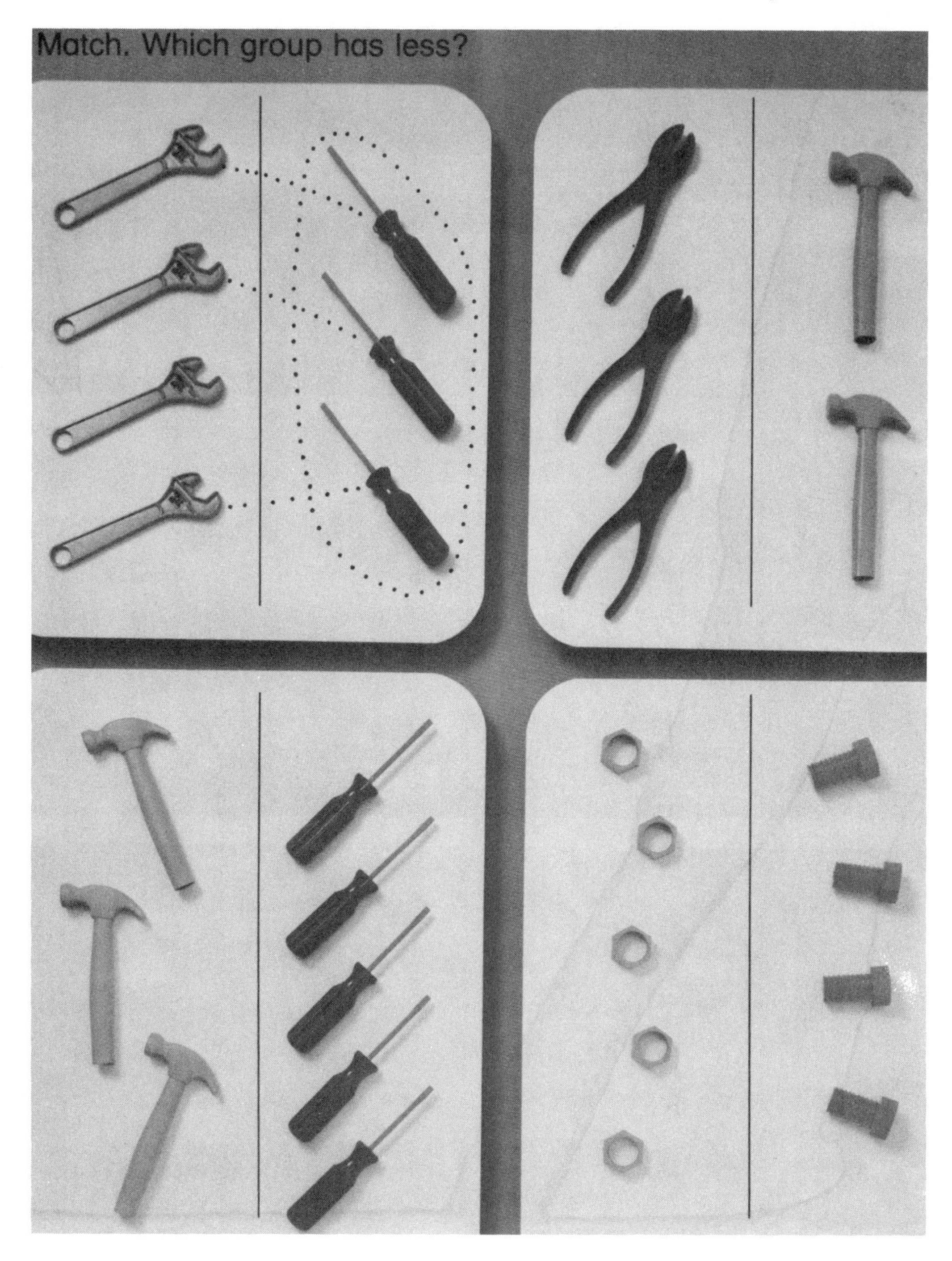

Figure 4.10 INVITATION TO MATHEMATICS, Kindergarten, p. 13

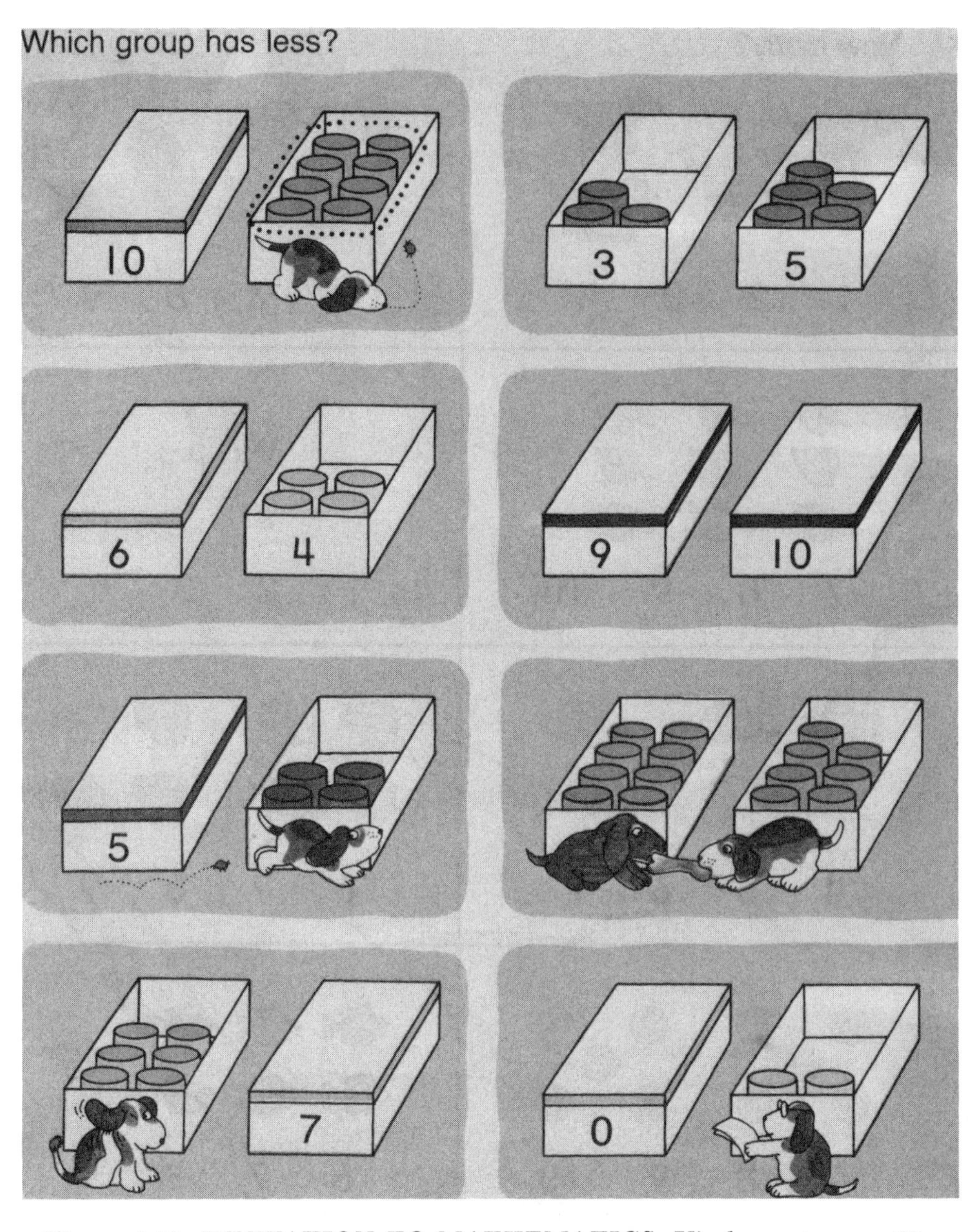

Figure 4.11 INVITATION TO MATHEMATICS, Kindergarten, p. 63

In general, $n(A) < n(B)$ if and only if A is equivalent to some proper subset S of B. The elements of B that are *not* in this proper subset form a nonempty subset of B. (See Figure 4.12.)

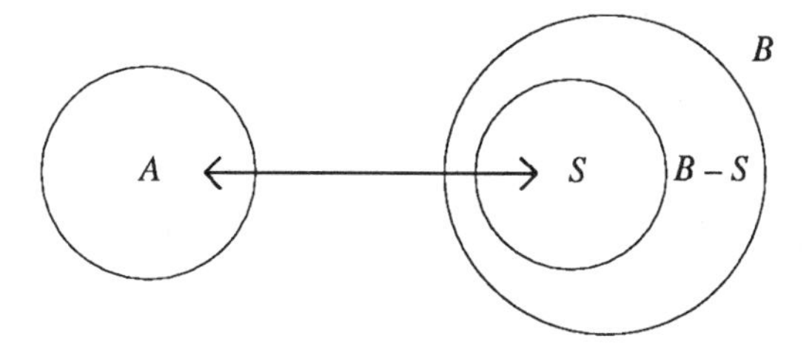

Figure 4.12 $A \sim S \subset B$

Properties of Order

The first and most obvious property of the relation $<$ on **W** is an old friend—transitivity. If a first number is less than a second and if that second number is less than a third, then the first number is less than the third:

(4.18) **Transitive Property for** $<$: If a, b, and c are whole numbers such that $a < b$ and $b < c$, then $a < c$.

The transitive property follows easily from Statement (4.17). For example:

$$2 < 5 \text{ because } 3 \text{ is a whole number and } 2 + 3 = 5$$
$$5 < 9 \text{ because } 4 \text{ is a whole number and } 5 + 4 = 9$$

To show $2 < 9$, we must find a whole number x such that $2 + x$ equals 9. The obvious choice for x is $3 + 4$ because the first summand, 3, gets us part of the way (from 2 to 5) and the second summand, 4, gets us the rest of the way (from 5 to 9). A general proof of the Transitive Property mimics this example exactly:

$$a < b \text{ implies } a + d = b \text{ for some nonzero whole number } d$$
$$b < c \text{ implies } b + e = c \text{ for some nonzero whole number } e$$

Then $d + e$ is a nonzero whole number and

$$a + (d + e) = (a + d) + e = b + e = c$$

Thus, $a < c$, by Property (4.17).

Transitivity gives meaning to a useful basic word:

Definition If three whole numbers a, b, and c have the property that $a < b$ and $b < c$, then we say b is **between** a and c, and write $a < b < c$.

Your conditioned mathematical reflexes by now should respond to the word *transitive* by calling to mind *reflexive* and *symmetric*. A moment's reflection reveals that $<$ is neither reflexive nor symmetric, and hence it is not an equivalence relation. Nevertheless, something useful is suggested by these properties.

(4.19) **The Trichotomy Law:** If a and b are whole numbers, then exactly one of the following is true:

$$a = b \qquad a < b \qquad a > b$$

The Trichotomy Law is justified by going back to the definition of *less than.* If a and b are any two whole numbers with representative sets A and B, respectively, then either A is equivalent to B or it is not. In the former case, the numbers are equal. In the latter case, either A can be matched with a proper subset of B or vice versa, implying either $n(A) < n(B)$ or $n(A) > n(B)$; that is, either $a < b$ or $a > b$.

The Trichotomy Law tells us that the whole numbers can be ordered in a simple way, beginning with the smallest whole number, 0. We can picture the whole numbers arranged as points on a line, from smaller to larger as we progress from left to right, with a fixed space between each number and its successor, as in Figure 4.13. Not surprisingly, such an arrangement is called a **number line.** In the very early grades, the number line sometimes is represented as a train with a number on each car, starting with 1, the smallest *counting* number, on the locomotive.

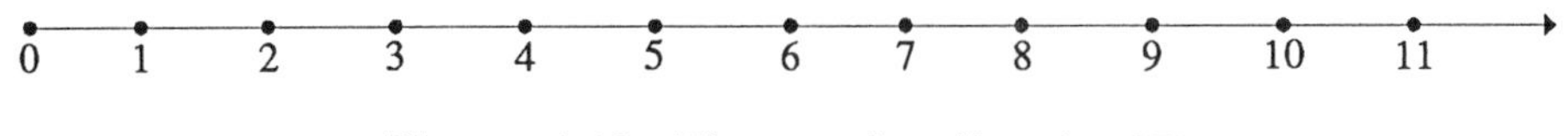

Figure 4.13 The number line for **W**

Less Than Or Equal To

It is often useful to consider a numerical relation that rules out only one of the three cases of the Trichotomy Law. For instance, we might want to allow one number to be less than *or* equal to another, thereby saying only that the first number cannot be greater than the second. (Recall from Chapter 2 that an *or* statement is true whenever either part is true.) These two-case statements are abbreviated in a standard way:

Notation The symbol $\leq$ means *less than or equal to.* Thus, "$a \leq b$" means *either* $a < b$ *or* $a = b$. Similarly, $\geq$ means *greater than or equal to.* Thus, "$a \geq b$" means *either* $a > b$ *or* $a = b$.

PROBLEM-SOLVING COMMENT

This is a good example of how *appropriate notation* has been introduced to represent some important ideas efficiently. To solve problems involving inequalities, one must check the definitions that correspond to these symbols. In particular, notice that the definitions of $\leq$ and $\geq$ are *or* statements, and the notation for *between* is an *and* statement:

$1 < x < 5$ means "1 is less than x *and* x is less than 5."

$2 \leq x \leq 7$ means "$2 < x$ or $2 = x$, *and* $x < 7$ or $x = 7$."

Notice also that the transitivity of $<$ and $\leq$ is expressed by this notation. Thus, an expression such as $3 < x < 1$ makes no sense because 3 is not less than 1, and a number cannot be both greater than 3 and less than 1. ◇

Order and Arithmetic

Finally, let us look at how the relation $<$ behaves with respect to addition, multiplication, and subtraction. These next three properties should be familiar; they all follow from Statement (4.17). In all of them, consider a, b, and c to be arbitrary whole numbers.

(4.20) $\quad a < b \quad \text{implies} \quad a + c < b + c.$

Justification: $a < b$ implies $a + d = b$ for some nonzero whole number d. Then $(a + c) + d = (a + d) + c = b + c$. Therefore, $a + c < b + c$.

(4.21) $\quad a < b \quad \text{implies} \quad ac < bc$ for any nonzero whole number c.

Justification: $a < b$ implies $a + d = b$ for some nonzero whole number d. Then $ac + dc = (a + d)c = bc$. Therefore, $ac < bc$.

(4.22) $\quad a < b \quad \text{if and only if} \quad b - a$ is a nonzero whole number.

Justification: $a < b$ implies $a + d = b$ for some nonzero whole number d. The definition of difference tells us immediately that this number d is the difference $b - a$.

Properties of the Whole Numbers

This completes our development of the whole-number system. As a review, the structural properties of the system are summarized below. The statement numbers used in this listing are the same as the ones used when these properties first appeared in the chapter.

The set **W** of whole numbers is the set of all equivalence classes of finite sets formed by the equivalence relation *1-1 correspondence*. This set is symbolized by

$$\mathbf{W} = \{0, 1, 2, 3, 4, \ldots\}$$

Two operations, addition and multiplication, are defined on **W**. They have the following properties:

Both operations are commutative: For all $a, b \in \mathbf{W}$,

(4.4) $a + b = b + a$ (4.8) $a \cdot b = b \cdot a$

Both operations are associative: For all $a, b, c \in \mathbf{W}$,

(4.5) $(a + b) + c = a + (b + c)$ (4.9) $(a \cdot b) \cdot c = a \cdot (b \cdot c)$

Both operations have identity elements—0 for addition, 1 for multiplication. For all $a \in \mathbf{W}$,

(4.6) $0 + a = a + 0 = a$ (4.10) $1 \cdot a = a \cdot 1 = a$

Multiplication is (left and right) distributive over addition: For all a, b, $c \in \mathbf{W}$,

(4.12) $a \cdot (b + c) = (a \cdot b) + (a \cdot c)$ and $(b + c) \cdot a = (b \cdot a) + (c \cdot a)$

The order relation "less than," denoted by $<$, is defined on **W**. It has the following properties:

(4.18) Transitivity: For all $a, b, c \in \mathbf{W}$,
if $a < b$ and $b < c$, then $a < c$

(4.19) Trichotomy: For any $a, b \in \mathbf{W}$, exactly one is true:

$$a = b \qquad a < b \qquad a > b$$

(4.20) Addition preserves order: For all $a, b, c \in \mathbf{W}$,

$$a < b \quad \text{implies} \quad a + c < b + c$$

(4.21) Nonzero multiplication preserves order: For all $a, b, c \in \mathbf{W}$, if $c \neq 0$, then

$$a < b \quad \text{implies} \quad a \cdot c < b \cdot c$$

The processes of subtraction and division can be defined for whole numbers, but it is not always possible to carry them out. Consequently, subtraction and division are not operations on $\mathbf{W}$.

Exercises 4.5

1. Show by example that the relation $<$ is neither reflexive nor symmetric.

2. Is the relation $\leq$ reflexive? Is it symmetric? Justify your answers.

For Exercises 3–12, use the definitions given in this section to justify each statement.

3. $4 < 7$
4. $3 > 2$
5. $0 < 5$
6. $3 \leq 5$
7. $4 \leq 9$
8. $2 \geq 1$
9. $7 \geq 5$
10. $6 \leq 6$
11. $8 \geq 8$
12. 4 is between 3 and 6.

13. Prove that 0 is less than every natural number. [*Hint*: Use Property (4.17).]

14. Justify the statement
    For all $a, b, c \in \mathbf{W}$, $a < b$ implies $a \cdot c \leq b \cdot c$.

    How does this statement differ from (4.21)?

For Exercises 15–22, decide whether or not the given statement is true for all $a, b, c \in \mathbf{W}$. If it is, give an example; if it is not, provide a counterexample.

15. If $a \leq b$ and $b \leq c$, then $a \leq c$.
16. If $a \leq b$ and $b \leq c$, then $a < c$.
17. If $a < b$ and $b < c$, then $a \leq c$.
18. If $a < b$ and $b \leq c$, then $a \leq c$.
19. If $a \leq b$ and $b < c$, then $a < c$.
20. If $a \leq b$ and $b \leq c$, then $a = c$.
21. If $a \leq b$ and $b = c$, then $a \leq c$.
22. If $a \leq b$ and $b = c$, then $a = c$.

Review Exercises for Chapter 4

For Exercises 1–10, indicate whether the given statement is *true* or *false*.

1. If two sets are equal, then they are equivalent.

2. If two sets are equivalent, then they are equal.

3. The natural numbers are the equivalence classes determined by the *1-1 correspondence* relation on the collection of all nonempty finite sets.

4. All infinite sets are of the same "size."

5. The sum of two whole numbers is represented by the union of any two representative sets of those numbers.

6. An operation $*$ on a set S is associative if $a * b = b * a$ for all $a, b \in S$.

7. Zero is the identity for addition of whole numbers.

8. The product of two whole numbers is represented by the Cartesian product of any two representative sets of those numbers.

9. Zero can be divided by any nonzero whole number, but no whole number can be divided by zero.

10. A whole number a is less than a whole number b if there is a whole number c such that $a + c = b$.

In Exercises 11–20, give examples of sets that satisfy the specified condition(s). If it is not possible to satisfy the conditions, write "impossible."

11. Two sets that are equivalent, but not equal.

12. Two sets that are equivalent and equal.

13. Two sets that are equal, but not equivalent.

14. Two sets A and B such that $A \subset B$ and $A \sim B$.

15. A set that is a representative set for 6, but is not a counting set.

16. A counting set for 4.

17. A representative set for 0.

18. A counting set for 3 that is not a representative set for 3.

19. Two sets A and B such that $n(A) = 2$, $n(B) = 5$, and $n(A \cup B) = 5$.

20. Two sets A and B such that $n(A) = 2$, $n(B) = 5$, and $n(A \cup B) = 7$.

21. Use the definition of whole-number addition to show that $3 + 4 = 7$.

22. Use the definition of whole-number multiplication to show that $3 \times 4 = 12$.

23. Use the definition of whole-number subtraction to show that $7 - 5 = 2$.

24. Use the definition of whole-number division to show that $12 \div 4 = 3$.

25. Explain why $0 \div 3$ is a legitimate number and why $3 \div 0$ is undefined.

In Exercises 26–35:

(a) If the statement is true, illustrate by an example;

(b) If the statement is false, provide a counterexample.

26. The whole numbers are closed under addition.

27. The whole numbers are closed under subtraction.

28. The whole numbers are commutative under division.

29. The whole numbers are associative under multiplication.

30. The whole numbers are associative under subtraction.

31. There is an identity element for multiplication of whole numbers.

32. There is an identity element for subtraction of whole numbers.

33. For those whole numbers for which subtraction is defined, multiplication is distributive over subtraction.

34. Whole-number addition is distributive over multiplication.

35. For those whole numbers for which division is defined, division is right distributive over addition.

In Exercises 36–40, simplify, if possible, and leave the answers in exponential form.

36. $5^3 \cdot 5^4$ **37.** $3^{10} \cdot 2^{10}$ **38.** $(7^3)^5$

39. $4^8 \div 4^3$ **40.** $3^3 \cdot 5^2$

41. Explain why $4^0 = 1$.

42. Use the definition of *less than* (using representative sets) to show that $4 < 9$.

5.1 Number versus Numeral

In Chapter 4 the set of whole numbers was carefully defined. Thus, the question "What is the number three?" originally posed in Chapter 1 finally has been answered; we now know it is the equivalence class

$$\left\{S \mid S \sim \{\text{“one,” “two,” “three”}\}\right\}$$

Recall that the word three is used in more than one way. When it denotes a chant word, we put it in quotation marks—"three"—and when it represents the natural number, we underline it—three.

Of course, in other languages different words are used to represent this same number: "tres" in Spanish, "drei" in German, "trois" in French, and so on. Indeed, hundreds of words are used to represent the number we refer to as three. Besides using words to represent this number, we often use other symbols. A small child might use three fingers; a teacher might use a set of three checkers, while writing on the board the symbol 3; a young student of arithmetic might use $1 + 2$, or $8 - 5$, or $15/5$; an older student might use $(-1)(-3)$ or $\sqrt{9}$, or even $\log_{10} 1000$. Many, many different symbols can be used to represent the same number; each of these symbols is called a *numeral.*

Definition A **numeral** is a symbol (or combination of symbols) used to represent a number.

In general, a symbol is a numeral for the number three if it can be used to bring to someone's mind the concept associated with three. There is only one number three, but there are many numerals for three. This is analogous to our method of identifying people. For example, there is only one person who was the first president of the United States, but there are many symbols to cause us to think of him. We might use the name-labels George Washington, George, General Washington, G. W., the Father of our Country, or we might use more general symbols, such as a portrait of him or even a cherry tree. None of these is the person himself, but each one causes us to think of this individual. Thus, they are different symbols that represent the same person.

Example 1 "Four," "cuatro," "vier," and "quatre" all are numerals representing the number four. (Words are symbols.) □

Example 2 4, IV, $2+2$, $9-5$, $\frac{1}{5} \times 20$, $\frac{8}{2}$, $(-2)^2$, $\log 10{,}000$, and $\sqrt{16}$ all are numerals representing the number four. □

Example 3 The kindergarten-text page in Figure 5.1 shows how children are taught to form the numerals we commonly use. At this level, teachers usually do not expect the children to distinguish between number and numeral, so children usually are asked to "write the numbers." However, it is important for you, the teacher, to understand the distinction. A lesson to teach the number four is quite different from a lesson to teach the numeral for four, each with its own difficulties and student expectations. □

Example 4 All the numerals cited in Examples 1 and 2 are visual symbols. There are also numerals that are not visual. For instance, the sounds used to say the words in Example 1 are auditory numerals; they also represent the number four. Even the magnetic traces used to store numbers in a calculator or computer are numerals. □

Exercises 5.1

1. Find at least ten different numerals for the number thirty-six.

2. Which of the figures given here expresses the larger number? Which figure is the larger numeral? Explain.

   8   **6**

3. Comment on the following proposed definition of equality of numbers:

   Two numbers x and y are *equal* if x and y are names for the same number.

4. (a) Suggest some activities that would be suitable for teaching the number four.

   (b) Suggest some activities that would be suitable for teaching 4, the usual numeral for four.

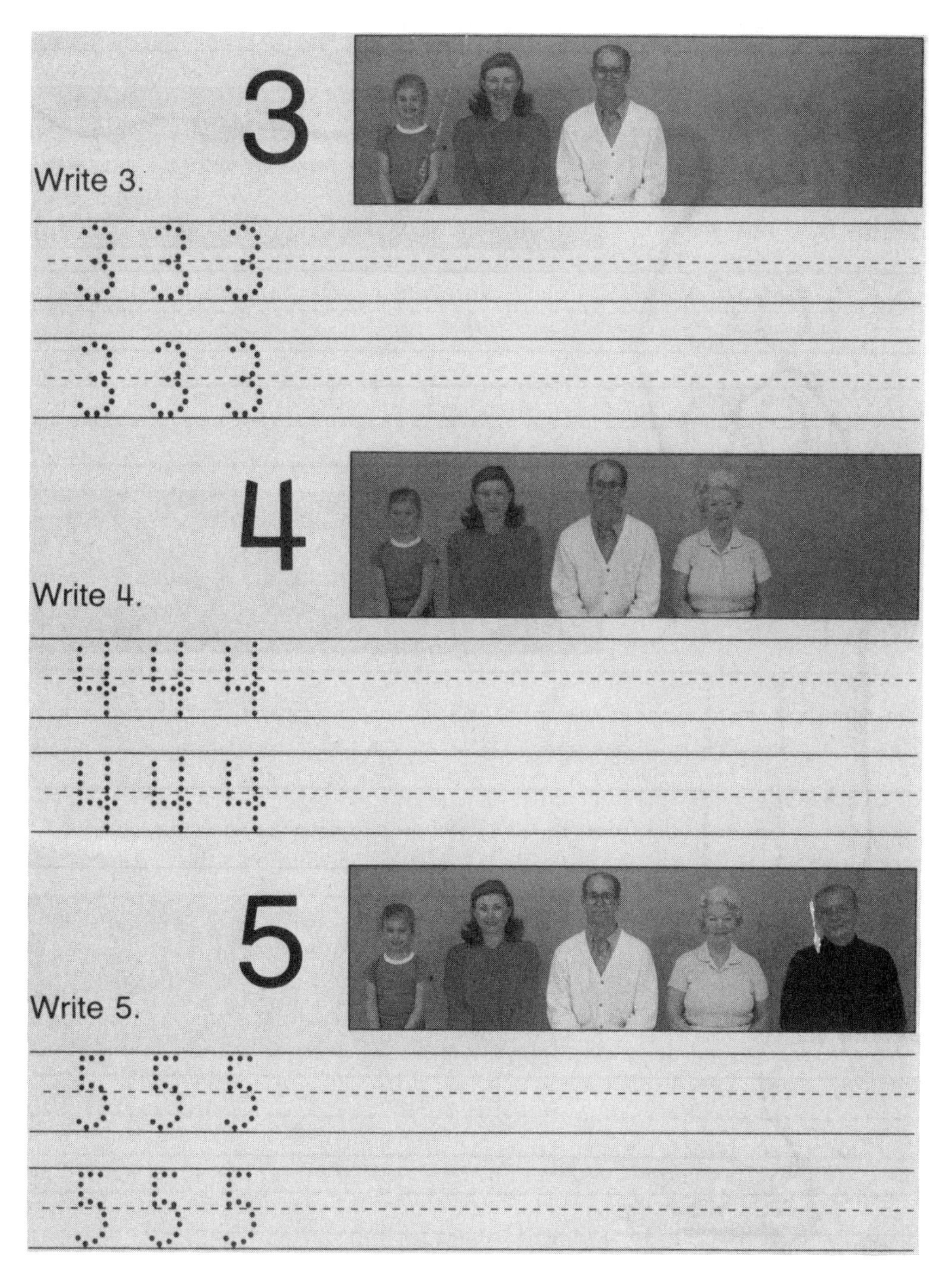

Figure 5.1: INVITATION TO MATHEMATICS, Kindergarten, p. 38

5.2 Numeration Systems, Past and Present

Throughout history, man has been faced with the problem of representing the whole numbers by symbols. Constructing a totally new symbol for each number was impractical, so it was necessary to invent a few basic symbols and devise a method whereby combinations of these symbols could be used to represent all the whole numbers. Such methods are called **numeration systems**. Before discussing our familiar system, we consider a few numeration systems of the past, because their features can be used to shed some light on how our system works.

PROBLEM-SOLVING COMMENT

What follows in this section is a brief account of humanity's quest to solve one of the classical problems of civilization, the development of an efficient symbol system for doing calculations. Although various cultures developed different ways of writing numbers, the fundamental problem-solving question was always that of *introducing appropriate notation.* The difficulty lay in determining what was "appropriate" in each case. This depended largely upon the needs and the sophistication level of each culture. ◇

The Tally System

One of the most primitive numeration systems was a **tally system**, where there was only one basic symbol—the tally or stroke, | . This was the numeral for the number one. The other numbers were represented by the appropriate number of tallies—

|| ||| |||| |||||

—and so on. This system has obvious drawbacks. Later civilizations improved on it by inventing more basic symbols and using different methods of symbol combinations to represent the whole numbers. The methods they used can be classified according to some fundamental characteristics, the first of which is *additivity.*

Definition A numeration system is **additive** if the number represented by a combination of basic symbols is determined by adding in some way the numbers represented by the basic symbols.

Example 1 The tally system is additive because the value of the basic symbol (one) is added (several times) to obtain the value of the combination. Thus,

$$|||| \quad \text{represents} \quad 1+1+1+1=4$$ □

The Egyptian System

The Egyptians arrived at a more practical system of numeration by inventing other symbols and using the additive principle. Believed to date back as far as 3400 B.C., these symbols were hieroglyphic in nature; some of them are shown in Table 5.1. ("Hieroglyphic" refers to "picture writing," the method of using pictures to express ideas.)

The basic symbols are used in combinations to represent all other numbers by adding the values of the symbols in a particular combination to determine the value of that combination. One of the earliest records using this system is a royal Egyptian mace, dating from 3100 B.C. A record of a successful military campaign, the mace contains many numerals representing numbers in the hundreds of thousands and even in the millions. It is currently on exhibit in a museum at Oxford, England.

Symbol	Interpretation	Number represented
𓏺	stroke	1
𓎆	heel bone	10
𓍢	coiled rope	100
𓆼	lotus flower	1000
𓂭	pointed finger	10,000
𓆐	tadpole	100,000
𓁨	astonished man	1,000,000

Table 5.1 Egyptian hieroglyphic numerals.

Example 2 𓍢 𓎆 𓏺𓏺𓏺 represents $100 + 10 + 1 + 1 + 1 = 113$. □

Example 3 𓆐 𓆐 𓆼 𓆼 𓆼 𓎆 𓎆 𓎆 𓎆 𓏺 represents

$100,000 + 100,000 + 1000 + 1000 + 1000 + 10 + 10 + 10 + 10 + 1 = 203,041$ □

Example 4 The order in which the symbols appear does not change the value of the combination. Thus, both 𓎆𓍢 𓏺𓏺𓏺 and 𓏺 𓎆 𓏺 𓍢 𓏺 represent the number 113. □

The Babylonian System

From the Tigris and Euphrates river valleys came the Babylonian numeration system, which used two wedge-shaped (*cuneiform*) symbols, 𒁹 and 𒌋. It is believed that this uring the period roughly from 2000 B.C. to 200

B.C. Evidence of this system was found on clay tablets, some of which still exist in various museum collections, including the Plimpton Collection at Columbia University and the Yale Collection. Besides being additive, the Babylonian system uses the *position* of the symbols to determine the value of a symbol combination, employing a method that also involves multiplication. Their system is a *sexagesimal* (base sixty) system. For the numbers 1 to 59, combinations of the two basic symbols are used additively, with each 𒁹 representing one and each 𒌋 representing ten.

Example 5 𒌋𒌋𒁹𒁹𒁹 represents $10+10+1+1+1=23$. □

Example 6 𒌋𒁹𒁹𒁹𒁹𒁹 represents $10+1+1+1+1+1=15$. □

The numbers from 60 to 3599 are represented by two combinations of symbols, the second combination placed to the left of the first one and separated from it by a space. The value of the number represented is determined by using the additive method to obtain the values of each combination, then multiplying the value of the left combination by 60 and adding that product to the value of the right combination.

Example 7 𒌋𒁹𒁹 𒌋𒌋𒌋𒁹 represents

$$(10+1+1)\cdot 60+(10+10+10+1)$$

which equals $12\cdot 60+31=720+31=751$. □

Example 8 𒁹𒁹 𒌋𒌋𒁹𒁹𒁹𒁹 represents

$$(1+1)\cdot 60+(10+10+1+1+1+1)$$

which equals $2\cdot 60+24=120+24=144$. □

Similarly, numbers 3600 ($=60^2$) or greater are represented by using more combinations of the two basic wedge shapes, placed further to the left and each separated from the others by spaces. Each single combination value is multiplied by an appropriate power of 60—the combination on the far right is multiplied by 60^0 ($=1$), the second from the right by 60^1, the third from the right by 60^2, and so on.

Example 9 𒁹𒁹 𒌋𒁹 𒌋𒌋𒁹𒁹𒁹 represents

$$\begin{aligned}(1+1)\cdot 60^2+(10+1)\cdot 60+(10+10+1+1+1) &= 2\cdot 3600+11\cdot 60+23\\ &= 7200+660+23\\ &= 7883\end{aligned}$$

□

Definition A numeration system is **positional** if the number represented by a combination of the basic symbols depends on the position of the symbols in the combination. Such a system is also called a **place-value system.**

Definition A numeration system is **multiplicative** if the number represented by a combination of basic symbols is obtained by multiplying in some way the values of those symbols.

HISTORICAL NOTE: THE BABYLONIAN BASE SIXTY

An interesting question is why the Babylonians chose the number sixty as the base for their numeration system. This question puzzled historians for many years. One persuasive explanation was given by Otto Neugebauer in 1934.[1] He claimed that the importance of the number sixty was derived from the Babylonian monetary system, in which it was desirable to have two units, a small denomination (*shekels*) and a larger one (*minas*). (This is analogous to our pennies and dollars.) The question arose: How many shekels should equal one mina? Whatever number was chosen needed to be divisible by both 2 and 3 because fractions such as $\frac{1}{2}$, $\frac{1}{3}$, and $\frac{2}{3}$ were commonly used in everyday Babylonian transactions. Thus, the smallest possible number to use would be the common multiple 6. However, to set one mina equal to only 6 shekels would not make the difference between the two units large enough for practical purposes. Therefore, it was decided to multiply six by ten, setting one mina equal to sixty shekels. With 60 as the basis for their monetary system, it was natural for the Babylonians to develop their numeration system in the same way.

Other mathematical historians propose a related, but somewhat more theoretical, explanation. They have observed that the Babylonian notation for fractions was confined to those with denominators that were powers of the base. (This is like restricting our numeration system to finite decimal expressions.) The variety of fractions that can be written easily in this way depends on the number of different divisors the base has: the more divisors there are, the easier it is to write fractions. Now, 60 is a relatively small number with many "useful" divisors—

$$2, 3, 4, 5, 6, 10, 12, 15, 20, 30$$

This variety of divisors made it easy to write the fractions

$$\frac{1}{2}, \frac{1}{3}, \frac{1}{4}, \frac{1}{5}, \frac{1}{6}, \frac{1}{10}, \frac{1}{12}, \frac{1}{15}, \frac{1}{20}, \frac{1}{30}$$

in the Babylonian notational system, so 60 was a convenient choice for the base.

Whatever their reasons, the Babylonians' choice of a numeration base has survived for four thousand years to become part of our own arithmetic. We still measure time as they did, dividing each hour into 60 minutes and each minute into 60 seconds, and our measurement of angles in degrees comes directly from the Babylonians, who subdivided the full circle into $6 \cdot 60$, or 360, parts. ◇

---

[1]Karl Menninger, *Number Words and Number Symbols, A Cultural History of Symbols,* Cambridge, MA: MIT Press, 1969, pp. 162–165.

Example 10 The Babylonian numeration system is positional because the power of 60 by which the value of a combination is multiplied depends on the position of that combination—whether it is the first combination on the right, second from the right, etc. This system is also multiplicative because combination values are multiplied by powers of 60.

The Egyptian system is neither positional nor multiplicative. □

There is one major difficulty with the Babylonian system: It is not always obvious how much space is intended between symbol combinations. For instance, it is not clear how 𒁹 𒌋 should be interpreted; it could be

$$1 \cdot 60^2 + 10, \text{ or } 1 \cdot 60^3 + 10 \cdot 60^2, \text{ or } 1 \cdot 60 + 10, \text{ or } 1 \cdot 60^2 + 10$$

etc. The Mayan civilization of South America developed a numeration system similar to that of the Babylonians, but free from this spacing difficulty.

The Mayan System

Like the Babylonians, the Mayans had two basic symbols—a dot "·" for the number one, and a bar "——" for the number five. Thus, one through ten were written like this:

· ·· ··· ···· — $\overset{\cdot}{—}$ $\overset{\cdot\cdot}{—}$ $\overset{\cdot\cdot\cdot}{—}$ $\overset{\cdot\cdot\cdot\cdot}{—}$ ═

The Mayans also used groupings of the basic symbols to represent the larger numbers. These combinations, arranged vertically, rather than horizontally, are evaluated additively. The lowest grouping represents single units; the value of the second grouping is multiplied by 20, the value of the third by $18 \cdot 20$, the value of the fourth by $18 \cdot 20^2$, the value of the fifth by $18 \cdot 20^3$, and so on. Thus, the Mayan system is essentially *vigesimal* (base twenty), except for the peculiar use of 18. The spacing difficulty of the Babylonian system was circumvented by the invention of a symbol for zero, "⊖". This numeral is used to indicate when a grouping position is to be skipped, thus removing the ambiguity inherent in the Babylonian system.

Example 11 $\overset{\cdot\cdot\cdot\cdot}{\equiv}$ represents $5 + 5 + 5 + 1 + 1 + 1 + 1 = 19$. □

Example 12 The following Mayan numeral represents 946:

··	=	$(1+1) \cdot (18 \cdot 20)$	=	720
$\overset{\cdot}{═}$	=	$(5+5+1) \cdot 20$	=	220
$\overset{\cdot}{—}$	=	$5+1$	=	+ 6
				946

□

Example 13

$$
\begin{array}{rlrr}
= & (5+1+1)\cdot(18\cdot 20^2) & = & 50{,}400 \\
= & (5+1)\cdot(18\cdot 20) & = & 2{,}160 \\
= & 0\cdot 20 & = & 0 \\
= & 5+5+1+1 & = & +\quad 12 \\
\hline
 & & & 52{,}572
\end{array}
$$

□

It should be clear from this discussion that the Mayan system is additive, positional, and multiplicative. You might wonder why the Mayans disrupted the pattern of the base-twenty system by using $18 \cdot 20$, rather than $20 \cdot 20$, to determine the value of the third group of symbols. It is believed that this was because the Mayan calendar was based on a 360-day year and, of course, $18 \cdot 20 = 360$.

The Roman System

Most people are familiar with the Roman numeration system, in use by 300 B.C. Its basic symbols and their corresponding values (written in our usual notation) are listed in Table 5.2.

Symbol	Value
I	1
V	5
X	10
L	50
C	100
D	500
M	1000

Table 5.2 Roman numerals.

Example 14 The Roman numeration system is additive because the values of the basic symbols are added to determine the value of the numeral. Thus,

$$\mathrm{CCLXXXI} = 100 + 100 + 50 + 10 + 10 + 10 + 1 = 281$$

□

Furthermore, when a bar is placed over a set of symbols, it indicates that the value of these symbols is to be multiplied by 1000. Thus, the Roman numeration system is also multiplicative.

Example 15

$$\overline{\mathrm{V}} = 5 \cdot 1000 = 5000 \quad \text{and} \quad \overline{\overline{\mathrm{L}}} = 50 \cdot 1000 \cdot 1000 = 50{,}000{,}000$$

Thus,

$$\overline{\mathrm{VII}}\mathrm{CLXV} = 7000 + 100 + 50 + 10 + 5 = 7165$$

□

The Roman system is also **subtractive** (although this was a later development), in that the numbers represented by some combinations of basic symbols are found by subtracting the value of one symbol from another. If a basic symbol in a numeral has a smaller value than the one immediately to its right, then the smaller value is subtracted from the larger one to get the value of the pair. For instance,

$$\text{IV} = 5 - 1 = 4$$

There are some restrictions on this subtractive feature, designed to avoid ambiguity. Only symbols representing powers of ten may be subtracted, and they can only be paired with the next two larger symbols:

I can be paired with V and X, but not with L, C, D, or M
X can be paired with L and C, but not with D or M
C can be paired only with D and M

By this method, no more than three adjacent copies of the same basic symbol are needed in any Roman numeral.

Example 16 XLIX = 49 because XL = 50 − 10 = 40 and IX = 10 − 1 = 9 □

Example 17 MCMXCIV = 1000 + 900 + 90 + 4 = 1994 □

The Hindu-Arabic System

The best-known numeration system in the Western world, and the one that rapidly is becoming the most widely used system in the entire world, is the Hindu-Arabic numeration system. It was invented by the Hindus in India about 250 B.C. and was introduced into Western Europe by the Arabs about A.D. 800. This system is positional, additive, and multiplicative. It has ten basic symbols—0, 1, 2, 3, 4, 5, 6, 7, 8, and 9—called **digits**, which represent the numbers zero, one, two, ..., nine, respectively.

The reason why the number ten was originally "chosen" as the base for this system appears to be more biological than logical. Research into the history of counting indicates that this numeration system, like many others, emerged from finger counting, so it was natural that the base number should correspond to the number of fingers we human beings have. Indeed, the very word we use for the basic symbols reflects this fact: *digitus* is the Latin word for *finger*.

To determine the value of a combination of the digits, the value of each digit is multiplied by a power of ten, depending on the position of that digit in the combination:

The value of the digit farthest to the right is multiplied by 10^0;
the value of the second digit from the right is multiplied by 10^1;
the value of the third digit from the right is multiplied by 10^2;

and so forth. In general, the value of the nth digit from the right is multiplied by 10^{n-1}. (Thus, the fourth digit from the right is multiplied by $10^{4-1} = 10^3 = 1000$, etc.) The number represented by the combination is obtained by adding these products together. Because there are ten basic symbols and each digit value is multipled by some power of ten, this system is referred to as a **base-ten** numeration system.

If a_n, $a_{n-1}, \ldots$, a_1, a_0 are digits, then the numeral $a_n a_{n-1} \ldots a_1 a_0$ is called the **standard form** of the number represented, and the numeral

$$a_n \cdot 10^n + a_{n-1} \cdot 10^{n-1} + \ldots + a_1 \cdot 10^1 + a_0 \cdot 10^0$$

is called the **expanded form** of that number. a_0 is the **units digit**, a_1 the **tens digit**, a_2 the **hundreds digit**, and so on.

Example 18 The numeral 2389 is in standard form. Its expanded form is

$$2 \cdot 10^3 + 3 \cdot 10^2 + 8 \cdot 10^1 + 9 \cdot 10^0$$

The units digit is 9, the tens digit is 8, the hundreds digit is 3, and the thousands digit is 2. □

Example 19

$$3 \cdot 10^4 + 0 \cdot 10^3 + 9 \cdot 10^2 + 8 \cdot 10^1 + 7 \cdot 10^0$$

is a numeral in expanded form. Its corresponding standard form is 30987 or 30,987. (As you probably know, sometimes commas are used to separate numerals more than three or four digits long into groups of three digits for easier reading. These commas have no effect on the value of the numeral.)

Because $10^0 = 1$ (the multiplicative identity), and $10^1 = 10$, $10^2 = 100$, $10^3 = 1000$, etc., the expanded form may also be written as

$$3 \cdot 10^4 + 0 \cdot 10^3 + 9 \cdot 10^2 + 8 \cdot 10 + 7$$

or as

$$3 \cdot 10{,}000 + 9 \cdot 100 + 8 \cdot 10 + 7$$

or even as

$$30{,}000 + 900 + 80 + 7$$

Your choice of one form rather than another will depend on how you wish to use it, particularly in relation to the grade and ability level of your students. □

Example 20 Figure 5.2 shows a page of a fifth-grade textbook that discusses place value and the use of expanded numerals. This particular text series does not distinguish between number and numeral at this level, but some other series do. Do you think it is important to have fifth graders recognize the difference between number and numeral? Why or why not? Notice also the type of expanded form used at this level. □

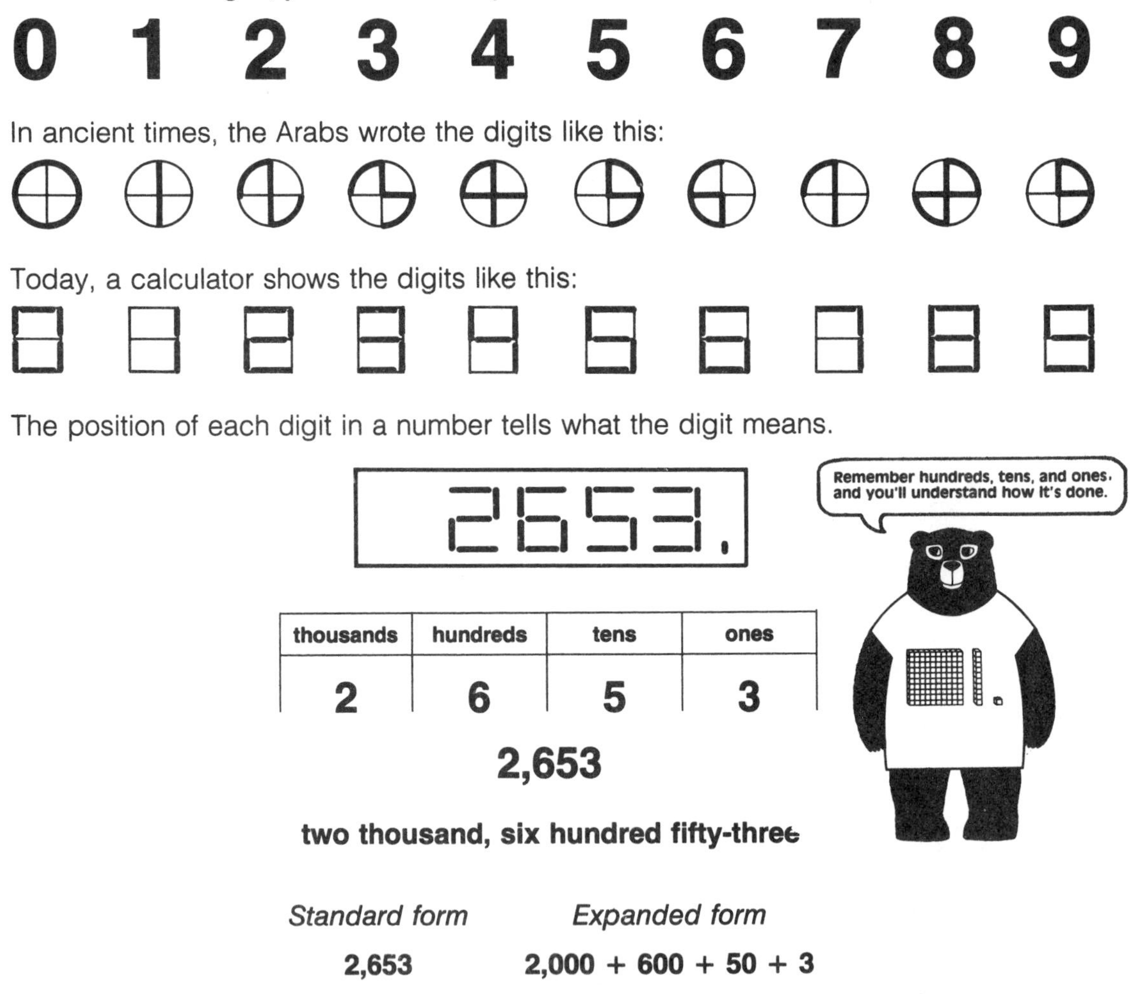

Digits and Place Value

Digits are the building blocks of numbers.
With these ten digits, you can write any number.

0 1 2 3 4 5 6 7 8 9

In ancient times, the Arabs wrote the digits like this:

Today, a calculator shows the digits like this:

The position of each digit in a number tells what the digit means.

thousands	hundreds	tens	ones
2	6	5	3

2,653

two thousand, six hundred fifty-three

Standard form	*Expanded form*
2,653	**2,000 + 600 + 50 + 3**

Figure 5.2 INVITATION TO MATHEMATICS, Grade 5, p. 2

Summary

Table 5.3 summarizes the contents of this section by comparing the various numeration systems with respect to the characteristics we have discussed.

System	Additive	Sub-tractive	Multi-plicative	Positional	Zero Symbol
Tally	yes	no	no	no	no
Egyptian	yes	no	no	no	no
Babylonian	yes	no	yes	yes	no
Mayan	yes	no	yes	yes	yes
Roman	yes	yes	yes	?	no
Hindu-Arabic	yes	no	yes	yes	yes

Table 5.3 The numeration systems.

Exercises 5.2

For Exercises 1–8, express the given numeral in standard Hindu-Arabic form.

1. ∩∩∩||

2. 𓆼𓆼𓍢𓁨 ∩𓍢𓍢||

3. 𒌋𒌋𒌋𒌋𒁹𒁹𒁹

4. 𒌋𒌋𒁹𒁹 𒌋 𒌋𒌋𒁹𒁹

5.
· ·
———
· · ·
———
———

6.
· ·
———
(shell)
· · · ·
———
·

7. MCMXLVI

8. DCLIX

For Exercises 9–12, express the given Hindu-Arabic numeral in Egyptian form.

9. 1,023,324 **10.** 548

11. 173 **12.** 2,312,431

For Exercises 13–16, express the given Hindu-Arabic numeral in Babylonian form.

13. 613 **14.** 8472

15. 3878 **16.** 157

For Exercises 17–20, express the given Hindu-Arabic numeral in Mayan form.

17. 260 **18.** 1293

19. 649 **20.** 27,151

For Exercises 21–24, express the given Hindu-Arabic numeral in Roman form.

21. 124 **22.** 182

23. 2,376,451 **24.** 278,000,000

For Exercises 25–30, write the expanded form of the given numeral, using exponential form for all powers of 10.

25. 238 **26.** 7856

27. 2,000,351 **28.** 35,906

29. 472,398 **30.** 222,000

For Exercises 31–35, write the standard form of the given numeral.

31. $5 \times 10^4 + 6 \times 10^3 + 8 \times 10^2 + 4 \times 10^1 + 2 \times 10^0$

32. $7 \times 10^6 + 8 \times 10^1$

33. $4 \times 10^4 + 0 \times 10^3 + 3 \times 10^2 + 7 \times 10^1 + 5 \times 10^0$

34. $2 \times 10^7 + 3 \times 10^3 + 9 \times 10 + 8$

35. $4 \times 10^5 + 2 \times 10^4 + 7 \times 10^3 + 3 \times 10^2$

36. Note that in Table 5.3 there is a "?" in the place that says whether the Roman numeration system is positional. Give an argument to support a "yes" answer, then give an argument to support a "no" answer.

5.3 Numeration Systems in Other Bases

As we have seen, the Hindu-Arabic numeration system is a base-ten system. Aside from the anthropological reason (the fact that humans have ten fingers), there appears to be no real advantage to using ten as the number around which the construction of the Hindu-Arabic system is based. You might even reason that, if we are to tie the choice to our biological characteristics, then five, twenty, or two would be just as reasonable numbers to use as a base. In fact, any whole number greater than one can be used as the base for a numeration system like ours.

PROBLEM-SOLVING COMMENT

Once you understand the Hindu-Arabic numeration system, it is relatively easy to develop numeration systems in other bases by *analogy*. In fact, one reason mathematics educators advocate the study of numeration systems in other bases is to capitalize on this analogy, which extends from the numerals themselves to the arithmetic processes that employ them. They (and we) believe that studying the arithmetic of these unfamiliar, but analogous, systems will enhance a student's understanding of the usual Hindu-Arabic system and its arithmetic. ◇

Let us design a numeration system beginning with just the basic symbols 1, 2, 3, 4, and 5, representing the numbers one, two, three, four, and five, respectively. How can we represent six, seven, eight, nine, etc., in this new system? At this point we have several options. We could, for instance, design a system that is strictly additive, using any combination of basic symbols whose values add up to the number being represented. For example, six could be expressed as 5 + 1, or 4 + 2, or 3 + 3, so we might use combinations such as 51 or 42 or 33. But this really is not a practical way of representing larger numbers, such as two thousand one hundred thirty-four. If we make our system positional, we can improve it considerably.

To represent six, seven, eight, nine, ..., we could start by observing that seven is six plus one, eight is six plus two, nine is six plus three, and so on. Thus, if we allow the value of a second symbol (to the left of the first) in the combination to be multiplied by six,

$$\underline{\text{seven}} \text{ would be } 11\ (= 1 \cdot \underline{\text{six}} + 1)$$
$$\underline{\text{eight}} \text{ would be } 12\ (= 1 \cdot \underline{\text{six}} + 2)$$
$$\underline{\text{nine}} \text{ would be } 13\ (= 1 \cdot \underline{\text{six}} + 3)$$

etc. Likewise,

$$45 \text{ would represent } 4 \cdot \underline{\text{six}} + 5 = \underline{\text{twenty-nine}},$$
$$51 \text{ would represent } 5 \cdot \underline{\text{six}} + 1 = \underline{\text{thirty-one}}$$

Now, how do we represent six? Let's see—that's just one six, so why not just use 1 by itself? But then we would not know whether 1 represents one or six. This problem is similar to the one we saw in the Babylonian system. How did the Mayans avoid this difficulty? They introduced a symbol for zero. We shall do this, too. If we choose "0" as the basic symbol for zero, then six can be expressed as

$$10 \quad (= 1 \cdot \underline{\text{six}} + 0)$$

Similarly, twelve can be expressed as 20 $(= 2 \cdot \underline{\text{six}} + 0)$, and so on.

We call this numeration system a **base-six** system because it uses six basic symbols—0, 1, 2, 3, 4, 5—and expresses quantities in terms of multiples of six. To avoid ambiguity, we shall use the subscript "six" to indicate when a numeral is written in the base-six system. If no subscript word is used and you are not otherwise told what system the symbols are in, you may assume the numeral is in base ten, our usual system. Thus,

$$10_{six} = 6 \qquad 11_{six} = 7 \qquad 12_{six} = 8$$

and so on. We read these new numerals as "one-zero base six," "one-one base six," "one-two base six," etc.

Let us compare the two systems by listing base-ten numerals in one line and the corresponding base-six numerals below them:

base ten:	1,	2,	3,	4,	5,	6,	7,	8,	9,	10,	11,	12,
base six:	1,	2,	3,	4,	5,	10,	11,	12,	13,	14,	15,	20,
base ten:	13,	14,	15,	16,	17,	18,	19,	20,	21,	22,	23,	24,
base six:	21,	22,	23,	24,	25,	30,	31,	32,	33,	34,	35,	40,
base ten:	25,	26,	27,	28,	29,	30,	31,	32,	33,	34,	35	
base six:	41,	42,	43,	44,	45,	50,	51,	52,	53,	54,	55	

The largest number expressible by two symbols in base six is the number represented by 55_{six}, which is $5 \cdot \underline{\text{six}} + 5$; that is, thirty-five. How can we express thirty-six? Well, $36 = 6^2$, so we just follow the pattern suggested by our earlier use of the zero symbol; that is, we let the third symbol from the right be 1 and the other two be 0:

$$100_{six} = 1 \cdot \underline{\text{six}}^2 + 0 \cdot \underline{\text{six}} + 0 = 36_{ten}$$

Actually, the agreement that numerals without subscript are in base ten allows us to write this unambiguously as

$$100_{six} = 1 \cdot 6^2 + 0 \cdot 6 + 0 = 36$$

Continuing the pattern,

$$101_{six} = 1 \cdot 6^2 + 0 \cdot 6 + 1 = 37$$
$$102_{six} = 1 \cdot 6^2 + 0 \cdot 6 + 2 = 38$$
$$\cdots$$
$$110_{six} = 1 \cdot 6^2 + 1 \cdot 6 + 0 = 42$$
$$\cdots$$
$$123_{six} = 1 \cdot 6^2 + 2 \cdot 6 + 3 = 51$$

Extending our comparative list a bit further, we get:

base ten:	36,	37,	38,	39,	40,	41,	42,	43,	44,	45,	46,	47,
base six:	100,	101,	102,	103,	104,	105,	110,	111,	112,	113,	114,	115,

base ten:	48,	49,	50,	...,	71,	72,	73,	...,	214,	215
base six:	120,	121,	122,	...,	155,	200,	201,	...,	554,	555

The largest number expressible by a three-digit numeral in base six is written 555_{six}. The next number is written 1000_{six}; it is

$$1 \cdot 6^3 + 0 \cdot 6^2 + 0 \cdot 6 + 0 = 216$$

By now, you probably can see how the pattern continues, permitting us to represent numbers as large as we please using only the six basic digits 0, 1, 2, 3, 4, 5. In general, when these digits are used in combination, the value of the nth digit from the right is multiplied by 6^{n-1}, and the number represented by the combination is just the sum of those products. Numerals in base six are evaluated by multiplying each single-symbol value by some power of six (depending on the symbol's position in the numeral) and then adding the products together. Thus, the base-six system, like the Hindu-Arabic (base-ten) system, is additive, multiplicative, and positional.

Example 1 $415_{six} = 4 \cdot 6^2 + 1 \cdot 6^1 + 5 \cdot 6^0 = 144 + 6 + 5 = 155$ □

Example 2 As in the usual Hindu-Arabic system, we use the terms "standard form" and "expanded form" to describe numerals in any base. However, the expanded form of a numeral is often written using the base-ten symbol for the base number itself because that form is easier to understand. Thus, $3{,}541{,}230_{six}$ is a base-six numeral in standard form; its expanded form is

$$3 \cdot 6^6 + 5 \cdot 6^5 + 4 \cdot 6^4 + 1 \cdot 6^3 + 2 \cdot 6^2 + 3 \cdot 6^1 + 0 \cdot 6^0$$

Computing this, we get the base-ten numeral 184,338. □

The general procedure for expressing numbers in a system with base less than ten is to use as many of the digits 0, 1,..., 9 as are needed to express the numbers less than the base. For the base-five system we use the digits 0, 1, 2, 3, and 4, and we indicate numerals written in base five by using the subscript "five." Thus, 1204_{five} is a numeral in base five. To determine the corresponding numeral in base ten, we use expanded notation:

$$1204_{five} = 1 \cdot 5^3 + 2 \cdot 5^2 + 0 \cdot 5 + 4 = 125 + 50 + 0 + 4 = 179$$

For bases greater than ten, additional symbols must be created to represent the numbers from ten to one less than the base number. (See Example 5.)

Example 3

$$341_{five} = 3 \cdot 5^2 + 4 \cdot 5 + 1 = 75 + 20 + 1 = 96$$

$$341_{seven} = 3 \cdot 7^2 + 4 \cdot 7 + 1 = 147 + 28 + 1 = 176$$

$$341 = 3 \cdot 10^2 + 4 \cdot 10 + 1 = 300 + 40 + 1 = 341$$

□

Example 4 216_{five} is not a numeral in base five because it contains a digit greater than 5. Nor, for that matter, is 215_{five}, which contains the digit 5. To be a numeral in base n, all digits must have values less than n. □

Example 5 To express a number in base twelve we can use the digits 0, 1, 2,..., 9, but we need two more symbols, for ten and eleven. It is common to use T and E, respectively. Thus,

$$\begin{aligned} \text{T}83_{twelve} &= 10 \cdot 12^2 + 8 \cdot 12 + 3 \\ &= 1440 + 96 + 3 \\ &= 1539 \end{aligned}$$

$$\begin{aligned} 73\text{E}7\text{T}_{twelve} &= 7 \cdot 12^4 + 3 \cdot 12^3 + 11 \cdot 12^2 + 7 \cdot 12 + 10 \\ &= 145{,}152 + 5184 + 1584 + 84 + 10 \\ &= 152{,}014 \end{aligned}$$

□

We have seen how to convert a numeral in any base other than ten to a base-ten numeral. The next question is: How do we reverse the process? That is, how can we convert the base-ten numeral for some number to a numeral representing that same number in some other base? Since we are most familiar with the base-six system from our work in this section, let us consider first a conversion problem from base-ten into base-six.

Suppose we wish to write the base-ten numeral 4835 as a base-six numeral representing the same number, four thousand eight hundred thirty-five. We must use some combination of the digits 0, 1, 2, 3, 4, 5, and each digit in the combination will represent a multiple of a power of six. We first need to know the largest power of six that is contained in 4835, so we begin by constructing a list of successive powers of six:

$$6^0 = 1, \quad 6^1 = 6, \quad 6^2 = 36, \quad 6^3 = 216, \quad 6^4 = 1296, \quad 6^5 = 7776$$

Now, 6^5 is greater than 4835, so we know that there are no fifth powers of six contained in our number. But there is at least one fourth power of six in 4835. In fact, by dividing we can see that there are 3 fourth powers of six in 4835, with a remainder of 947:

$$\begin{array}{r} 3 \\ 1296\,\overline{)4835} \\ \underline{3888} \\ 947 \end{array}$$

How many third powers of six are contained in 947? Dividing by 216 (which is 6^3), we get 4 third powers of six and a remainder of 83:

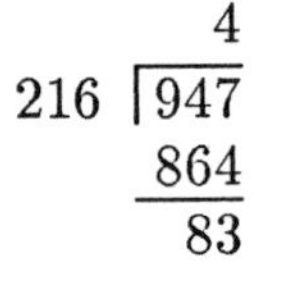

To find how many second powers of six are contained in the remainder 83, we divide again, this time by 36:

$$\begin{array}{r} 2 \\ 36\,\overline{)83} \\ \underline{72} \\ 11 \end{array}$$

This remainder, 11, can be expressed as one six with a (final) remainder of 5. This series of divisions shows that we can write 4835 as

$$3 \cdot 6^4 + 4 \cdot 6^3 + 2 \cdot 6^2 + 1 \cdot 6 + 5$$

This is the expanded form for a base-six numeral, corresponding to the standard form $34{,}215_{six}$.

Example 6 Convert 1107 to a base-six numeral.

(1) $6^3 = 216$ is the highest power of six less than 1107.

$$\begin{array}{r} 5 \\ 216\,\overline{)1107} \\ \underline{1080} \\ 27 \end{array}$$

(2) Now divide 27 by $6^2 = 36$:

$$\begin{array}{r|l} & 0 \\ \cline{2-2} 36 & 27 \\ & \underline{0} \\ & 27 \end{array}$$

(3) Now divide 27 by $6^1 = 6$:

$$\begin{array}{r|l} & 4 \\ \cline{2-2} 6 & 27 \\ & \underline{24} \\ & 3 \end{array}$$

(4) The expanded form of 1107 for base six, then, is

$$5 \cdot 6^3 + 0 \cdot 6^2 + 4 \cdot 6 + 3$$

corresponding to the standard form 5043_{six}. □

Example 7 Convert 851 to a base-twelve numeral.

(1) $12^2 = 144$ is the highest power of <u>twelve</u> less than 851, so we divide 144 into 851:

$$\begin{array}{r|l} & 5 \\ \cline{2-2} 144 & 851 \\ & \underline{720} \\ & 131 \end{array}$$

(2) Divide 131 by 12:

$$\begin{array}{r|l} & 10 \\ \cline{2-2} 12 & 131 \\ & \underline{120} \\ & 11 \end{array}$$

(3) Thus, the expanded form of 851 for base twelve is

$$5 \cdot 12^2 + 10 \cdot 12 + 11$$

(Notice that this expression is still in base ten.) The standard form of this numeral in base twelve is obtained by writing 5, 10, and 11 as single digits:

$$5\text{TE}_{twelve}$$ □

Numeration Bases and Computers

With the advent of computer technology, the ability to express numbers in bases two, eight, and sixteen has become increasingly important. Initially, base two—called the **binary system**—was used to express numbers and words in computer code. Base two was a convenient choice because, at the most primitive level, electronic computers respond to two electrical states, *on* and *off*. (This is actually somewhat oversimplified, but not in an essential way for our purposes. In fact, the two electrical states these days are more often *high voltage* and *low voltage*, but the key fact for us is that there are just *two* states, so we shall view them as *on* and *off*.) Now, if we use *on* to represent 1 and *off* to represent 0, then any numeral in base two can be represented by a finite sequence of these electrical states. For example, the number thirteen, which is 1101_{two}, is represented by *on-on-off-on*.

Since every whole number can be written in base two, this means that computers are able to "read" all the whole numbers. Therefore, computers are able to "read" *anything that can be expressed by using whole numbers*, and therein lies the key to word processing, spreadsheet calculations, and all the more sophisticated things modern computers do. If each symbol on a keyboard is assigned a different whole number, then a computer will be able to distinguish, remember, and manipulate strings of all those symbols—words, sentences, formulas, and the like—just by translating their whole-number strings into strings of *on*s and *off*s by means of the 1s and 0s (called **bits**) in base two.

The simplicity of the binary system carries with it one major drawback to its use. Because only two digits are allowed, even relatively small numbers require long strings of 0s and 1s in base two, making them cumbersome to read and write. For example, the binary representation of the decimal numeral 2493 is the twelve-bit string 100110111101_{two}. This difficulty is eased by using an "intermediate" numeration system, one with a large enough base to make numeral expressions more compact. The two common intermediate systems are based on eight (the **octal** system) or sixteen (the **hexadecimal** or **hex** system). The greater variety of primary symbols in each of these systems makes numeral expressions shorter than in base two, and the fact that they are both based on *powers of two* makes it easy to convert to and from the base-two system. In particular, notice that, in the hexadecimal system, some of the digits are letters. This stems from the fact that a base-sixteen system must have sixteen single-digit symbols. It is standard practice to use the ten familiar digits of our decimal system and the letters A, B, C, D, E, F to stand for the numbers ten through fifteen, respectively.

Since a computer relies on binary representation, the number of bits used or available in a situation is usually counted by powers of two. The standard minimum unit of memory size is a **byte**, which is 8 bits. When the memory size of a computer is listed, it is usually expressed as "so-many K" bytes of capacity. One **K** is 2^{10}, or 1024, bytes. The fact that 1024 is approximately 1000 explains the choice of the symbol K; it is an abbreviation for the metric prefix *kilo-*, which stands for *one thousand.* Thus, a 640K computer has a memory storage capacity of approximately 640,000 bytes; more precisely, it has

$$640 \times 1024 \text{ bytes} = 655{,}360 \text{ bytes} = 5{,}242{,}880 \text{ bits}$$

of binary memory.

The use of the byte (rather than the bit) as the unit for counting memory storage is directly related to base sixteen. As we saw at the beginning of this section, a computer can store and manipulate any information that is expressed in terms of binary numerals. Now, anything that can be typed on a keyboard can be expressed this way, provided that each distinct key stroke has its own binary code. A standard code for this was established in 1968 by the American National Standards Institute. It is called the **ASCII** code, an acronym for "American Standard Code for Information Interchange." The code assigns to each separate keyboard symbol (and to 32 special control functions, such as carriage return) a number from 0 to 127. Expressed in hexadecimal form, these numbers require only two digits each, so when they are converted to binary form they require at most eight digits. Thus, every item in ASCII code can be stored as a single byte of memory.

Part of the ASCII code is shown in Table 5.4. Example 8 uses the information in this table to illustrate how the ASCII code permits words and other symbols to be stored as binary numerals (and thereby as electrical states).

Example 8 If you type

```
KEEP IT SIMPLE!
```

on your computer keyboard, each of the fifteen keystrokes is translated into ASCII (hexadecimal) code:

4B 45 45 50 20 49 54 20 53 49 4D 50 4C 45 21

Each of these code numerals is converted to an eight-bit binary numeral and is stored electronically as a single byte of memory. Thus,

4B is converted to 0100 1011

45 is converted to 0100 0101

50 is converted to 0101 0000

and so forth. These binary digits are all strung together; however, the computer treats each group of eight bits as a separate symbol, so the binary representation of your message (with commas inserted to separate bytes) is

01001011,01000101,01000101,01010000,00100000,01001001,
01010100,00100000,01010011,01001001,01001101,01010000,
01001100,01000101,00100001 □

Key	Decimal	Hex	Octal	Binary
<space>	32	20	40	00100000
!	33	21	41	00100001
...	...	...	...	...
,	44	2C	54	00101100
-	45	2D	55	00101101
.	46	2E	56	00101110
...	...	...	...	...
A	65	41	101	01000001
B	66	42	102	01000010
C	67	43	103	01000011
D	68	44	104	01000100
E	69	45	105	01000101
F	70	46	106	01000110
G	71	47	107	01000111
H	72	48	110	01001000
I	73	49	111	01001001
J	74	4A	112	01001010
K	75	4B	113	01001011
L	76	4C	114	01001100
M	77	4D	115	01001101
N	78	4E	116	01001110
O	79	4F	117	01001111
P	80	50	120	01010000
Q	81	51	121	01010001
R	82	52	122	01010010
S	83	53	123	01010011
T	84	54	124	01010100
U	85	55	125	01010101
V	86	56	126	01010110
W	87	57	127	01010111
...	...	...	...	...

Table 5.4 Part of the ASCII Code.

Exercises 5.3

1. Beginning with the numeral 342_{six}, list the next twenty-five numerals in the base-six system.

2. Beginning with the numeral 33_{five}, list the next fifteen numerals in the base-five system.

3. Beginning with the numeral 101_{two}, list the next twelve numerals in the base-two system.

4. Beginning with the numeral 17_{twelve}, list the next ten numerals in the base-twelve system.

For Exercises 5–14, write the expanded form of the given numeral for the base specified, then write a base-ten numeral (in standard form) that represents the same number.

5. 321_{five} **6.** 4201_{five}

7. 321_{eight} **8.** 4201_{eight}

9. $TE83_{twelve}$ **10.** $100E_{twelve}$

11. $100{,}111_{two}$ **12.** $100{,}001{,}111_{two}$

13. 3502_{six} **14.** $52{,}210_{six}$

For Exercises 15–18, convert the given base-ten numeral to (a) base two, (b) base five, (c) base six, and (d) base twelve. Write both the expanded form and the standard form in each case.

15. 38 **16.** 125 **17.** 856 **18.** 3485

19. How many single digits are needed in a base-sixteen numeration system? Why? What is the largest number that can be written with two digits in this base?

For Exercises 20–25, convert the given decimal numeral to binary, octal, and hexadecimal numerals.

20. 25 **21.** 99 **22.** 100

23. 128 **24.** 250 **25.** 1234

For Exercises 26–29, determine, both approximately and precisely, how many bytes and bits are described by the given phrase.

26. 16K bytes **27.** 64K bytes

28. 128K bytes **29.** 1024K bytes

For Exercises 30–33, use Table 5.4 to write the given message in ASCII code and then in binary code.

30. HELP!

31. LEWIS CARROLL

32. THE MAD HATTER

33. THROUGH THE LOOKING GLASS

For Exercises 34–37, use Table 5.4 to translate the given ASCII message into English.

34. 41 4C 49 43 45

35. 54 57 45 45 44 4C 45 44 55 4D

36. 54 48 45 20 4D 4F 43 4B 20 54 55 52 54 4C 45

37. 4F 46 46 20 57 49 54 48 20 48 49 53 20 48 45 41 44 21

For Exercises 38–42, translate the given binary message into ASCII code and then into English.

38. 01010001,01010101,01000101,01000101, 01001110

39. 01010111,01001111,01001110,01000100, 01000101,01010010,01001100,01000001, 01001110,01000100

40. 01010100,01001000,01000101,00100000, 01000011,01001000,01000101,01010011, 01001000,01001001,01010010,01000101, 00100000,01000011,01000001,01010100

41. 01010100,01010111,01000001,01010011, 00100000,01000010,01010010,01001001, 01001100,01001100,01001001,01000111, 00101100,00100000,01000001,01001110, 01000100,00101110,00101110,00101110

42. 01001111,01000110,00100000,01000011, 01000001,01000010,01000010,01000001, 01000111,01000101,01010011,00101101, 00101101,01000001,01001110,01000100, 00100000,01001011,01001001,01001110, 01000111,01010011,00101101,00101101

Review Exercises for Chapter 5

For Exercises 1–10, indicate whether the given statement is *true* or *false*.

1. We most likely have the same number system as the Chinese.
2. We most likely have the same numeration system as the Chinese.
3. The Egyptian hieroglyphic numeration system is positional.
4. The Babylonian and Mayan numeration systems are positional.
5. The best-known numeration system in the Western world is the Hindu-Arabic system.
6. The Mayan and Hindu-Arabic systems have a symbol for zero, whereas the Egyptian, Babylonian, and Roman systems do not.
7. In the Roman numeration system you always subtract the value of a symbol from the value of the symbol to its immediate right.
8. The largest number expressible as a three-digit numeral in base eight is 888_{eight}.
9. There is no practical application of numeration systems in other bases.
10. In a base-sixteen numeration system you need single-digit symbols to represent the numbers <u>zero</u> through <u>sixteen</u>.

For Exercises 11–18, represent the given numeral in standard Hindu-Arabic form.

11. 𓍢𓍢 𓎆 𓆐 𓆼 || 12. 𓁨 𓂭 𓍢 𓎆𓎆𓎆 |

13. ◁◁◁ ◁▽▽ ◁▽▽▽▽

14. ◁◁◁▽▽

15. ·
═
··

16. ··
═
⊖
·
═

17. MMMCDLIV 18. $\overline{\text{DXLIX}}$

For Exercises 19 and 20, express the given Hindu-Arabic numeral in Egyptian form.

19. 357 20. 3818

For Exercises 21 and 22, express the given Hindu-Arabic numeral in Babylonian form.

21. 98 22. 103

For Exercises 23 and 24, express the given Hindu-Arabic numeral in Mayan form.

23. 20 24. 782

For Exercises 25 and 26, express the given Hindu-Arabic numeral in Roman form.

25. 295 26. 21,682

For Exercises 27–31, convert the given numeral to base ten.

27. 3112_{five} 28. $22{,}525_{six}$

29. 2013_{eight} 30. $2T94_{twelve}$

31. $1{,}101{,}000{,}010_{two}$

For Exercises 32–36, convert the given base-ten numeral to the specified base.

32. 160 to base two 33. 2903 to base five

34. 1454 to base six 35. 4565 to base eight

36. 12,310 to base twelve

For Exercises 37 and 38, convert the given decimal (base-ten) numeral to binary, octal, and hexadecimal forms.

37. 34 38. 126

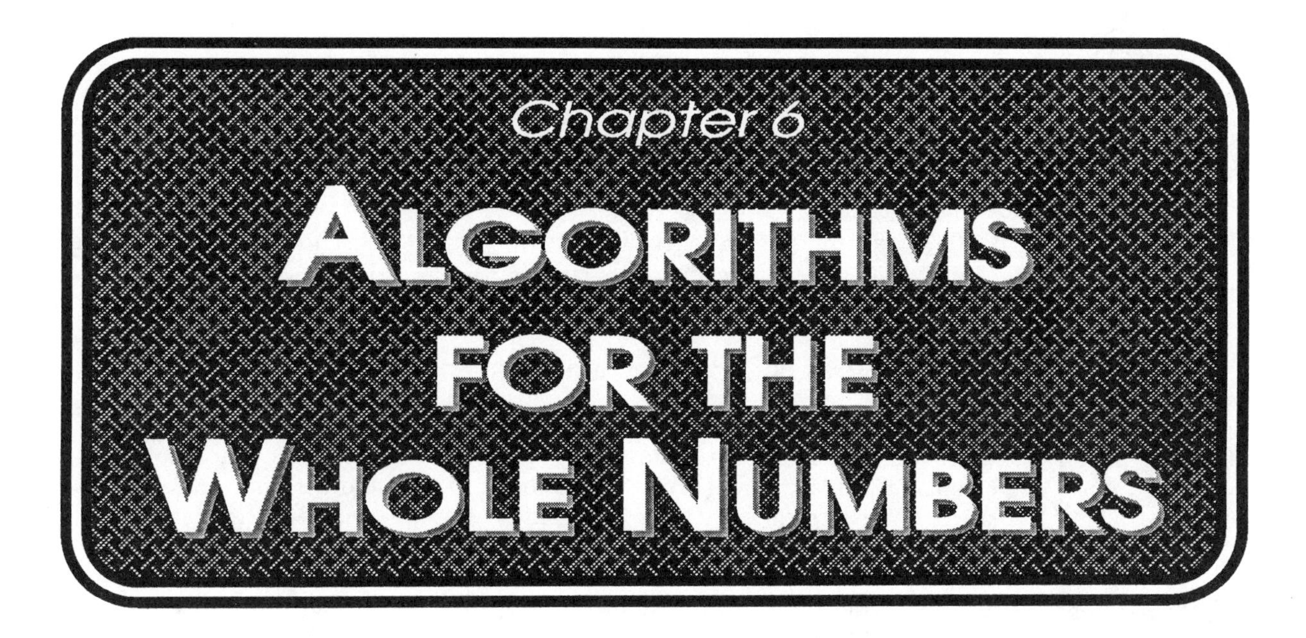

6.1 What Is an Algorithm?

An **algorithm** is a set of rules for solving a problem in a finite number of steps. There are many algorithms that are used in the various branches of mathematics. The advantage of developing an algorithm is that it provides a systematic, efficient method for solving a particular problem. Often, however, that very efficiency masks the underlying reasoning. Consequently, students may learn algorithms without understanding why they work and, as a result, regard mathematics as a memorized set of complicated steps with little relation to reality and with no logical explanation. One aspiring teacher, when asked how the division algorithm could be presented in a reasonable and understandable way, replied, "I didn't know it was supposed to be understandable." Indeed, it—and all the algorithms of elementary mathematics—are supposed to be understandable. Your job will be to make them understandable for your students. In order to do this, you have to understand them yourself; our job in this chapter is to help you do just that.

No algorithms are more fundamental to mathematics than the ones that deal with the four operations of arithmetic—addition, subtraction, multiplication, and division. As we discuss these algorithms, it is important to keep in mind two points. First, it is *not* our purpose to teach you *how* to use them. We assume you already know how to use these algorithms with acceptable accuracy and speed. Our purpose here is to analyze the algorithms, explaining *why* they do what they are supposed to do. Second, these algorithms are not unique. That is, there is more than one algorithm for each of the operations being discussed. In most instances, we analyze the algorithm most frequently taught in elementary school. In the exercises you will see some alternative algorithms for the arithmetic operations.

HISTORICAL NOTE: ALGORITHMS AND ALGEBRA

The word *algorithm* has a long, rich history. It originated from the name of a 9th-century Moslem scholar, Mohammed ibn Musa al-Khowarizmi, who wrote two important mathematics books, one on arithmetic and the other on algebra. Both books were translated into Latin by 12th-century European scholars. In translation, the title of the first book became *Algorithmi de numero Indorum* ("Al-Khowarizmi on Indian Numbers"). This book was one of the main sources from which Europe learned about the Hindu-Arabic numeration system. It became quite well-known, sometimes as *liber Algorismi* ("the book of Al-Khowarizmi"). Thus, the name Al-Khowarizmi in its various Latinized forms—notably *algorithm* and *algorism*—gradually became synonymous with arithmetic manipulations of any sort.

In modern times *algorithm* has become the more common form, and its meaning is somewhat restricted. It now refers to finite, often repetitive, processes for finding numerical answers or constructing mathematical objects. Interest in algorithms has blossomed with the coming of the computer age because such processes are ideal for computer-assisted solutions to problems. Thus, the ancient Arab master of arithmetic has been joined by modern electronic partners in the search for constructible answers to mathematical questions.

Al-Khowarizmi's other book has also influenced the language of modern mathematics. That book, devoted entirely to the study of linear and quadratic equations, was entitled *al-jabr w'al muqabalah* ("Restoration and Opposition"). It, too, was very popular in translation; soon both the book and the mathematical techniques it described became known by the Latin version of the key word in its title—*algebra*. ◇

6.2 The Addition Algorithm

Why do we have algorithms for addition? The reason is efficiency. If we did not have an algorithm for adding two numbers, we would have to

1. memorize all possible sums of all possible pairs of whole numbers, or
2. manipulate representative sets of the two numbers using the definition of addition from Chapter 4.

While either of these methods works well for $2 + 3$, neither alternative is practical for such problems as $35 + 79$ or $2846 + 9753$. Hence, we develop an algorithm, a process that enables us to find the sum of two large numbers by adding pairs of the single-digit numbers 0, 1, 2, ..., 9. Thus, a *prerequisite* for using the addition algorithm is knowing the sums of all pairs of the digits 0 through 9. Table 6.1 shows these 100 basic addition facts; it is called the **basic addition table.**

+	0	1	2	3	4	5	6	7	8	9
0	0	1	2	3	4	5	6	7	8	9
1	1	2	3	4	5	6	7	8	9	10
2	2	3	4	5	6	7	8	9	10	11
3	3	4	5	6	7	8	9	10	11	12
4	4	5	6	7	8	9	10	11	12	13
5	5	6	7	8	9	10	11	12	13	14
6	6	7	8	9	10	11	12	13	14	15
7	7	8	9	10	11	12	13	14	15	16
8	8	9	10	11	12	13	14	15	16	17
9	9	10	11	12	13	14	15	16	17	18

Table 6.1 Commutativity in the basic addition table.

Imagine the consternation of a second or third grader who is told to memorize these 100 facts! Maybe you remember your own feelings when you were confronted with this problem. However, there really are far fewer than 100 sums to memorize, provided we remember certain properties of addition.

Notice that the addition table has a symmetry about the main diagonal of the table (the line of sums extending from the top left to the bottom right). The numbers that are the same perpendicular distance from a point on the diagonal are the same. This observation is simply a visual confirmation of the commutativity of addition. The sum for $2+3$ is above the diagonal, whereas the sum for its commutative "twin," $3+2$, is below. Likewise, the sum for $3+7$ is above the diagonal, whereas the sum for $7+3$ is below, and so on for every pair of distinct whole numbers. (See Table 6.1.) Thus, if we learn the top (triangular) half of the table, we automatically will have learned the bottom half, as well. Consequently, the top half and the main diagonal contain all the information of the entire table. This means that the 100 facts to be memorized have been reduced to 55.

We can reduce the rote-memory task even further by observing that the top line of sums is identical to the line of addends just above it. This is a visual confirmation of the Additive-Identity Property of zero:

$$0 + a = a \text{ for every whole number } a$$

Knowing this property means that the top line of the table does not have to be memorized either, leaving just 45 sums to be learned. Even if there were no other justification for teaching the properties of addition early in grade school, the fact that knowing two properties allows a student to reduce the number of basic sums to be memorized from 100 to 45 would be reason enough.

PROBLEM-SOLVING COMMENT

The foregoing discussion is a good illustration of how the problem of remembering a large quantity of related facts can be handled by *looking for a pattern*. By connecting the observed patterns in the addition table with the known properties of addition (commutativity and the existence of an identity element), students can memorize the necessary basic facts more efficiently. Moreover, they can *generalize this solution* to find similar memory-aid patterns for the *analogous* problem, learning the basic multiplication facts. ◇

Addition Without Carrying

Now let us consider the Addition Algorithm. There are actually two algorithms here; one is typically referred to as "addition without carrying" and the other, "addition with carrying." (The former is a special case of the latter). First, we examine the algorithm that does not require carrying, as in adding 25 and 34, for example. The algorithm for this sum looks like this:

$$\begin{array}{r} 2\ 5 \\ +\ 3\ 4 \\ \hline 5\ 9 \end{array}$$

Breaking the process into separate steps, we have:

Step 1. "Line up" the digits, beginning at the right.

Step 2. Add the units digits; this sum is the units digit of the final sum.

Step 3. Add the tens digits; this sum is the tens digit of the final sum.

An **analysis** of such an algorithm presents the computation so that each of these steps is justified or supported by a reason, which may be any of the addition properties we have encountered in previous chapters. Of course, we also use the basic addition facts shown in Table 6.1. The analysis in this case is as follows:

$\boxed{25 + 34}$

1. $= (2 \cdot 10 + 5) + (3 \cdot 10 + 4)$ Expanded form

2.	$= (2 \cdot 10 + 3 \cdot 10) + (5 + 4)$	Commutativity and associativity of +
3.	$= (2 \cdot 10 + 3 \cdot 10) + 9$	Basic addition table
4.	$= (2 + 3) \cdot 10 + 9$	Distributivity of · over +
5.	$= 5 \cdot 10 + 9$	Basic addition table
6.	$= 59$	Standard form of numeral

Steps 1 and 2 of the analysis correspond to Step 1 of the algorithm, lining up the digits. Analysis Step 3 justifies Step 2 of the algorithm, adding the units digits. Steps 4 and 5 add the tens digits. Step 6 explains why the sums of the units digits and tens digits become, respectively, the units and tens digits of the sum.

Addition With Carrying

Now let us consider the addition algorithm that employs "carrying." Here is the form of a typical example, followed by the steps of the algorithm:

$$\begin{array}{rcc} & 1 & \\ & 4 & 8 \\ + & 3 & 7 \\ \hline & 8 & 5 \end{array}$$

Step 1. Line up the digits.

Step 2. Add the units digits. (The sum is 15.)

Step 3. Put the 5 of the units-digit sum in the units place of the sum.

Step 4. Carry the 1 of the units-digit sum to the top of the tens column.

Step 5. Add the three digits in the tens column.

Step 6. Put this sum in the tens place of the sum.

Analysis:

$\boxed{48 + 37}$

1.	$= (4 \cdot 10 + 8) + (3 \cdot 10 + 7)$	Expanded form
2.	$= (4 \cdot 10 + 3 \cdot 10) + (8 + 7)$	Commutativity and associativity of +
3.	$= (4 \cdot 10 + 3 \cdot 10) + 15$	Basic addition table
4.	$= (4 \cdot 10 + 3 \cdot 10) + (1 \cdot 10 + 5)$	Expanded form
5.	$= [(1 \cdot 10 + 4 \cdot 10) + 3 \cdot 10] + 5$	Commutativity and associativity of +
6.	$= [(1 + 4) + 3] \cdot 10 + 5$	Distributivity of · over + (twice)

7.	$= [5 + 3] \cdot 10 + 5$	Basic addition table
8.	$= 8 \cdot 10 + 5$	Basic addition table
9.	$= 85$	Standard form of numeral

Steps 1 and 2 of the analysis again justify the first step of the algorithm, and Step 3 of the analysis justifies the second step of the algorithm. Analysis Steps 4 and 5 justify the carrying process in Algorithm Step 4. Analysis Steps 6, 7, and 8 justify Algorithm Step 5; the final step of the analysis justifies Algorithm Steps 3 and 6. Addition of any two whole numbers by this algorithm can be analyzed in a similar way. The number of steps may increase to accommodate more digits and more carrying, but the theory is exactly the same, as you can see from the following example.

Example 1 $2359 + 826$ is usually computed as shown here:

$$\begin{array}{cccc} {}^1 & & {}^1 & \\ 2 & 3 & 5 & 9 \\ + & 8 & 2 & 6 \\ \hline 3 & 1 & 8 & 5 \end{array}$$

As you might expect, its analysis is almost twice as long as the previous one. Instead of writing it out formally, we shall use a format often found in elementary school texts. See if you can justify each step as you work through it.

$$\begin{array}{rcl} 2359 & = & 2 \cdot 1000 + 3 \cdot 100 + 5 \cdot 10 + 9 \\ +\ 826 & = & 8 \cdot 100 + 2 \cdot 10 + 6 \\ \hline & & 2 \cdot 1000 + 11 \cdot 100 + 7 \cdot 10 + 15 \\ & = & 2 \cdot 1000 + 10 \cdot 100 + 1 \cdot 100 + 7 \cdot 10 + 10 + 5 \\ & = & 2 \cdot 1000 + 1 \cdot 1000 + 1 \cdot 100 + 7 \cdot 10 + 1 \cdot 10 + 5 \\ & = & 3 \cdot 1000 + 1 \cdot 100 + 8 \cdot 10 + 5 \\ & = & 3185 \end{array}$$

□

PROBLEM-SOLVING COMMENT

As you start to analyze algorithms on your own, you might find it helpful to start with simpler problems and to reason backwards from the desired conclusion. After you have analyzed a few simple examples, look for a pattern in your work and generalize the solution. A little practice with these techniques should convince you that analyzing algorithms is not as difficult as it may seem at first. ◇

Addition in Other Bases

This same addition algorithm can be used for adding numbers written in other bases. The process is identical, except that we use the basic addition facts for the particular base being considered. For instance, if we wish to add in base five, we need an addition table for 0, 1, 2, 3, and 4, as shown in Table 6.2. Using these basic facts, we can add base-five numerals "in columns," just as in base ten.

+	0	1	2	3	4
0	0	1	2	3	4
1	1	2	3	4	10
2	2	3	4	10	11
3	3	4	10	11	12
4	4	10	11	12	13

Table 6.2 The basic addition table in base five.

Example 2 The algorithm for adding $34_{five} + 23_{five}$ is

$$\begin{array}{rrr} & {\scriptstyle 1} & \\ & 3 & 4 \\ + & 2 & 3 \\ \hline 1 & 1 & 2 \end{array}$$

The analysis for this algorithm parallels the analysis for $48 + 37$ in base ten, shown previously. (Writing it out in detail might be an instructive exercise for you.) □

Example 3 $2433_{five} + 2142_{five}$ is computed as follows:

$$\begin{array}{rrrrr} & {\scriptstyle 1} & {\scriptstyle 1} & {\scriptstyle 1} & \\ & 2 & 4 & 3 & 3 \\ + & 2 & 1 & 4 & 2 \\ \hline 1 & 0 & 1 & 3 & 0 \end{array}$$

□

PROBLEM-SOLVING COMMENT

Algorithms are useful in helping us calculate quickly and efficiently. However, no matter how careful you are, there is always a chance of making errors. You can guard against this by *approximating the answer* before doing the calculation, then comparing the computed answer with the estimated one. A glaring discrepancy between the two signals a possible error.

One easy way to approximate calculations is to "round off" your numbers to multiples of powers of ten. If your rounding-off skills are a bit rusty, you might polish them up by looking ahead to Section 10.7, where that topic is discussed in detail. ◇

Exercises 6.2

For Exercises 1–4, fill in the missing reasons and steps in the given analysis.

1. $46 + 32$

$= (4 \cdot 10 + 6) + (3 \cdot 10 + 2)$ ____________

$= (4 \cdot 10 + 3 \cdot 10) + (6 + 2)$ ____________

$= (4 \cdot 10 + 3 \cdot 10) + 8$ ____________

$= (4 + 3) \cdot 10 + 8$ ____________

$= 7 \cdot 10 + 8$ ____________

$= 78$ ____________

2. $253 + 614$

$= (2 \cdot 100 + 5 \cdot 10 + 3) +$ ________ Expanded form

$= (2 \cdot 100 + 6 \cdot 100) +$ ______ $+ (3 + 4)$ ____________

$= (2 \cdot 100 + 6 \cdot 100) +$ ______ $+ 7$ ____________

$= (2 + 6) \cdot 100 + (5 + 1) \cdot 10 + 7$ ____________

$= 8 \cdot 100 + 6 \cdot 10 + 7$ ____________

$=$ ______ Standard form of numeral

3. $29 + 35$

$= (2 \cdot 10 + 9) + (3 \cdot 10 + 5)$ ____________

$= (2 \cdot 10 + 3 \cdot 10) + (9 + 5)$ ____________

$= (2 \cdot 10 + 3 \cdot 10) + 14$ ____________

$= (2 \cdot 10 + 3 \cdot 10) + (1 \cdot 10 + 4)$ ____________

$= [(1 \cdot 10 + 2 \cdot 10) + 3 \cdot 10] + 4$ ____________

$=$ ____________ Distributivity of $\cdot$ over $+$

$=$ ____________ Basic addition table

$= 64$ ____________

4. $73 + 51$

$=$ ____________ Expanded form

$=$ ____________ Associativity and commutativity of $+$

$= (7 \cdot 10 + 5 \cdot 10) +$ ____ Basic addition table

$= (7 + 5) \cdot 10 + 4$ ____________

$= 12 \cdot 10 + 4$ ____________

$= (1 \cdot 10 + 2) \cdot 10 + 4$ ____________

$= [(1 \cdot 10) \cdot 10 + 2 \cdot 10] + 4$ ____________

$= 1 \cdot 100 + 2 \cdot 10 + 4$ ____________

$=$ ______ Standard form of numeral

For Exercises 5–8, begin by estimating the given sum. Then analyze the use of the addition algorithm in each case, indicating which steps of the analysis justify "lining up" the digits, which justify adding the units digits, and which justify the carrying process (when it applies).

5. $17 + 32$ **6.** $240 + 539$

7. $68 + 25$ **8.** $665 + 378$

For Exercises 9–12, begin by estimating the sum. Then write out the computation in the elementary-analysis form of Example 1.

9. $324 + 513$ **10.** $813 + 627$

11. $1234 + 987$ **12.** $8376 + 4462$

13. Some students are taught an algorithm for the carrying process that involves carrying to the bottom of the next column, rather than to the top. Thus, in Exercise 7 we would add as follows:

$$\begin{array}{r} 6\,8 \\ +\,\underset{1}{2}\,5 \\ \hline 9\,3 \end{array}$$

Analyze this algorithm, and compare it to the analysis of Exercise 7.

For Exercises 14–18, find the given sum by using the base-five addition table (Table 6.2) and the Addition Algorithm.

14. $23_{five} + 14_{five}$ **15.** $213_{five} + 3402_{five}$

16. $332_{five} + 421_{five}$ **17.** $4432_{five} + 3344_{five}$

18. $321_{five} + 23402_{five}$

19. Construct an addition table for base six. (Save it for use later in this chapter.)

For Exercises 20–24, use the base-six addition table from Exercise 19 to compute the given sum.

20. $352_{six} + 542_{six}$ **21.** $12345_{six} + 53402_{six}$

22. $15_{six} + 243_{six}$ **23.** $3230_{six} + 4452_{six}$

24. $443025_{six} + 213352_{six}$

For Exercises 25 and 26, add in *each* of the following bases: (a) six, (b) seven, (c) eight, (d) nine, (e) ten, (f) eleven, (g) twelve.

25.
$$\begin{array}{r} 35 \\ 12 \\ +\,34 \\ \hline \end{array}$$

26.
$$\begin{array}{r} 514 \\ 341 \\ +\,253 \\ \hline \end{array}$$

6.3 The Subtraction Algorithm

Like the one for addition, the subtraction algorithm can be viewed as two algorithms—one that does not require borrowing and one that does. Again, the first is a special case of the second. As it was for addition, the first step for subtraction is to line up the digits, but it is more difficult to justify this step for subtraction. For addition, after expressing the numbers in expanded form, we rearranged the addends, using commutativity and associativity. The problem here is that the subtraction process is neither commutative nor associative. The way around this difficulty depends on some additional properties of the subtraction process.

First consider the sum $4 + (7 - 4)$. By commutativity of addition, this can be rewritten as $(7 - 4) + 4$, which clearly should equal 7, because we are first subtracting 4 and then adding it back. This argument can be made more rigorous by using the definition of subtraction:

$$a - b = x \text{ if and only if } b + x = a$$

Apply this definition to the example, letting a represent 7, b represent 4, and x represent $(7 - 4)$. We get

$$7 - 4 = (7 - 4) \text{ if and only if } 4 + (7 - 4) = 7$$

Now, it is obvious that $7 - 4$ equals $(7 - 4)$, so the expected subtraction answer, 7, must be correct. In general:

(6.1) If a, b, and $a - b$ are whole numbers, then $b + (a - b) = a$.

Example 1

$$\begin{array}{ccccc} 7 & + & (9-7) & = & 9 \\ 7 & + & 2 & = & 9 \\ & 9 & & = & 9 \end{array}$$

□

Now let us consider the difference of two sums—for instance,

$$(8 + 6) - (3 + 2)$$

We want to show that this can be arranged to equal the sum of two differences; in particular, we want to show that

$$(*) \qquad (8 + 6) - (3 + 2) = (8 - 3) + (6 - 2)$$

We use Property (6.1), with $a = 8 + 6$ and $b = 3 + 2$. Now, Property (6.1) tells us that $a - b$ is $(8 - 3) + (6 - 2)$, provided that this sum added to b equals a. Let us form this second sum and see what we get. By the commutativity and associativity of addition,

$$(3 + 2) + [(8 - 3) + (6 - 2)] = [3 + (8 - 3)] + [2 + (6 - 2)]$$

which equals $8 + 6$, because, by Property (6.1),

$$3 + (8 - 3) = 8 \quad \text{and} \quad 2 + (6 - 2) = 6$$

Therefore, $(*)$ is true. The form of Statement $(*)$ is the general property we seek.

(6.2) The **Sum-Difference Property** of Whole Numbers:
If a, b, c, d, $a - c$, and $b - d$ are whole numbers, then

$$(a + b) - (c + d) = (a - c) + (b - d)$$

Note: The name *Sum-Difference Property* is not a standard term in the mathematical literature, but it serves well as a memory-jogging abbreviation of the essential idea of (6.2).

Proof: (This is just like the previous numerical example.) By the definition of subtraction, $a - b = x$ if and only if $b + x = a$. Thus,

$$(a+b)-(c+d)=x \quad \text{if and only if} \quad (c+d)+x=a+b$$

To verify that x is $(a-c)+(b-d)$, we add this expression to $c+d$ and observe that we get the required result:

$$\begin{aligned}
&(c+d)+[(a-c)+(b-d)] \\
&\quad = [c+(a-c)] + [d+(b-d)] && \text{by associativity and commutativity of } + \\
&\quad = a + b && \text{by Property (6.1)}
\end{aligned}$$

Notice that, although subtraction is neither commutative nor associative, the Sum-Difference Property allows us to reorder and regroup a subtraction problem in a special way; *it tells us how to express the difference of two sums as the sum of two differences.* This property is vital for our analysis of the subtraction algorithm.

Example 2 Here are two ways of seeing how the Sum-Difference Property works:

$$\begin{array}{ccccccccc}
(30+9) & - & (10+2) & = & (30-10) & + & (9-2) \\
39 & - & 12 & = & 20 & + & 7 \\
 & 27 & & = & & 27 &
\end{array}$$

That is,

$$(30+9)-(10+2)=39-12=27$$

and

$$(30-10)+(9-2)=20+7=27$$

□

Example 3 The Sum-Difference Property does not apply directly to the difference

$$(9+5)-(2+6)$$

because, when we rewrite it as $(9-2)+(5-6)$, the second summand, $5-6$, is not a whole number. □

Example 4 By applying the Sum-Difference Property twice to the expression

$$(6+7+12)-(2+4+8)$$

we can obtain

$$\begin{array}{ccccccccc}
(6+7+12) & - & (2+4+8) & = & (6-2) & + & (7-4) & + & (12-8) \\
25 & - & 14 & = & 4 & + & 3 & + & 4 \\
 & 11 & & = & & & 11 & &
\end{array}$$

□

One more property is needed for our analysis of the subtraction algorithm. Consider $2\cdot 7 - 2\cdot 4$, the difference of two products. By repeated addition, this can be written as $(7+7)-(4+4)$. By the Sum-Difference

Property, this difference of sums can be expressed as $(7-4)+(7-4)$, which is the same as $2\cdot(7-4)$. Thus,

$$2\cdot 7 - 2\cdot 4 = 2\cdot(7-4)$$

Similarly,

$$\begin{aligned} 3\cdot 9 - 3\cdot 5 &= (9+9+9)-(5+5+5) \\ &= (9-5)+(9-5)+(9-5) \\ &= 3\cdot(9-5) \end{aligned}$$

In fact, this kind of rearrangement always works. That is, *multiplication of whole numbers is distributive over subtraction*:

(6.3) If a, b, c, and $b-c$ are whole numbers, then
$a\cdot(b-c) = a\cdot b - a\cdot c$ and $(b-c)\cdot a = b\cdot a - c\cdot a$

Example 5 Here is an instance of the first equation of Property (6.3), the left distributive property; the right distributive property can be illustrated similarly.

$$\begin{aligned} 7\cdot(4-3) &= 7\cdot 4 - 7\cdot 3 \\ 7\cdot 1 &= 28-21 \\ 7 &= 7 \end{aligned}$$

□

Finally, before we can use and analyze the subtraction algorithm, we must assume that the basic subtraction facts are known. The definition of subtraction relates each subtraction fact to some addition fact, so there really are no new basic facts to be memorized. For example, we know the subtraction fact $5-3=2$ once we know that $3+2=5$. (Although it is not *mathematically* necessary to learn new facts for subtraction, teachers often find it *pedagogically* sound to teach the subtraction facts separately because some students do not easily grasp the relationship between subtraction and addition.)

Subtraction Without Borrowing

Let us first consider a problem that does not require borrowing. For instance, the algorithm for $48-36$ works as follows:

$$\begin{array}{rr} & 4\ 8 \\ - & 3\ 6 \\ \hline & 1\ 2 \end{array}$$

Step 1. Line up the digits.

Step 2. Subtract the units digits. This answer is the units digit of the difference.

Step 3. Subtract the tens digits. This answer is the tens digit of the difference.

Step 4. Thus, the difference is 12.

Now let us analyze this process:

$\boxed{48 - 36}$

1.	$= (4 \cdot 10 + 8) - (3 \cdot 10 + 6)$	Expanded form
2.	$= (4 \cdot 10 - 3 \cdot 10) + (8 - 6)$	Sum-Difference Property
3.	$= (4 \cdot 10 - 3 \cdot 10) + 2$	Basic subtraction fact
4.	$= (4 - 3) \cdot 10 + 2$	Distributivity of $\cdot$ over $-$
5.	$= 1 \cdot 10 + 2$	Basic subtraction fact
6.	$= 12$	Standard form

Steps 1 and 2 of the analysis explain the first step of the algorithm—lining up the digits. The third analysis step justifies the second algorithm step—subtraction of the units digits. The fourth and fifth analysis steps justify the third algorithm step—subtraction of the tens digits. The sixth analysis step shows that the differences of the units and tens digits are the units and tens digits, respectively, of the difference.

Subtraction With Borrowing

Now let us consider $73 - 28$, a subtraction problem that requires borrowing. *Note*: We do *not* claim that the wording of the steps in the following example is pedagogically sound. It is presented in this way because such wording is commonly used in the classroom. In Exercise 7 you will be asked to reword the statements so that they more closely correspond to the analysis.

$$\begin{array}{rcc} & \overset{6}{\not 7} & {}^{1}3 \\ - & 2 & 8 \\ \hline & 4 & 5 \end{array}$$

Step 1. Line up the digits.

Step 2. You cannot subtract 8 from 3, so you borrow 1 from 7, making it a 6, and change the 3 to 13.

Step 3. Subtract 8 from 13. This is the units digit of the difference.

Step 4. Subtract 2 from 6; this is the tens digit of the difference.

Step 5. The answer is 45.

Here is the analysis:

$73 - 28$		
1.	$= (7 \cdot 10 + 3) - (2 \cdot 10 + 8)$	Expanded form
2.	$= [(6 + 1) \cdot 10 + 3] - (2 \cdot 10 + 8)$	Basic addition fact
3.	$= [(6 \cdot 10 + 1 \cdot 10) + 3] - (2 \cdot 10 + 8)$	Distributivity of $\cdot$ over $+$
4.	$= [6 \cdot 10 + (1 \cdot 10 + 3)] - (2 \cdot 10 + 8)$	Associativity of $+$
5.	$= (6 \cdot 10 + 13) - (2 \cdot 10 + 8)$	Standard form
6.	$= (6 \cdot 10 - 2 \cdot 10) + (13 - 8)$	Sum-Difference Property
7.	$= (6 \cdot 10 - 2 \cdot 10) + 5$	Basic subtraction fact
8.	$= (6 - 2) \cdot 10 + 5$	Distributivity of $\cdot$ over $-$
9.	$= 4 \cdot 10 + 5$	Basic subtraction fact
10.	$= 45$	Standard form

Notice that the lining up of digits is not completely justified until the sixth step of the analysis; Steps 2–5 explain the "borrowing" (or "regrouping") process.

Example 6 Figure 6.1 shows two examples from a first-grade text. Notice how the borrowing process is illustrated. How does this compare with a formal algorithm analysis? Can you write out an analysis for each problem? □

Example 7 Another kind of subtraction problem that frequently gives students trouble is one in which they have to "borrow across a zero." Here is an example, again justified in elementary-textbook style.

$$\begin{array}{cccc} & 6 & 9 & \\ & \not{7} & {}^{1}\not{0} & {}^{1}4 \\ - & 2 & 5 & 6 \\ \hline & 4 & 4 & 8 \end{array}$$

$$\begin{array}{rcl} 704 & = & 7 \cdot 100 + 0 \cdot 10 + 4 \;=\; 6 \cdot 100 + 1 \cdot 100 + 4 \\ -\,256 & = & 2 \cdot 100 + 5 \cdot 10 + 6 \;=\; 2 \cdot 100 + 5 \cdot 10 + 6 \\ & = & 6 \cdot 100 + 10 \cdot 10 + 4 \;=\; 6 \cdot 100 + 9 \cdot 10 + 1 \cdot 10 + 4 \\ & = & 2 \cdot 100 + 5 \cdot 10 + 6 \;=\; 2 \cdot 100 + 5 \cdot 10 + 6 \\ & = & 6 \cdot 100 + 9 \cdot 10 + 14 \\ & = & 2 \cdot 100 + 5 \cdot 10 + 6 \\ & = & 4 \cdot 100 + 4 \cdot 10 + 8 \end{array}$$

□

Figure 6.1 INVITATION TO MATHEMATICS, Grade 1, p. 302

Subtraction in Other Bases

Subtraction in other bases is done using the same algorithm, along with the appropriate addition table (to derive the basic subtraction facts). For instance, using the addition table for base five given in Table 6.2 on page 173,

$$4213_{five} - 1424_{five}$$

is done as follows:

$$\begin{array}{ccccc} & {}^{3} & {}^{1} & {}^{1}0 & \\ & \not{4} & \not{2} & \not{1} & {}^{1}3 \\ - & 1 & 4 & 2 & 4 \\ \hline & 2 & 2 & 3 & 4 \end{array}$$

The addition table in Table 6.2 is used as follows: To subtract 4_{five} from 13_{five}, locate 4 in the extreme left column. Read across that row until you locate 13 (in the lower-right corner), then read up that column to the top row. The number there is 4, so

$$13_{five} - 4_{five} = 4_{five}$$

Similarly, to subtract 2_{five} from 10_{five}, locate 2 in the left column, read across that row until you find 10, then read up that column until you find 3 at the top. Thus,

$$10_{five} - 2_{five} = 3_{five}$$

Example 8 The algorithm applied to $4002_{five} - 1343_{five}$ looks like this:

$$\begin{array}{rcccc} & \overset{3}{\not 4} & \overset{4}{{}^{1}\not 0} & \overset{4}{{}^{1}\not 0} & {}^{1}2 \\ - & 1 & 3 & 4 & 3 \\ \hline & 2 & 1 & 0 & 4 \end{array}$$

□

Exercises 6.3

For Exercises 1–3, fill in the missing reasons and steps in the given analysis.

1. $96 - 42$

$= (9 \cdot 10 + 6) - (4 \cdot 10 + 2)$ ______________

$= (9 \cdot 10 - 4 \cdot 10) + (6 - 2)$ ______________

$= (9 \cdot 10 - 4 \cdot 10) + 4$ ______________

$= (9 - 4) \cdot 10$ ______________

$= 5 \cdot 10 + 4$ ______________

$= 54$ ______________

2. $78 - 25$

= ______________ Expanded form

= ______________ Sum-Difference Property

= ______________ Definition of subtraction

= ______________ Distributivity of · over −

= ______________ Definition of subtraction

= ______________ Standard form

3. $84 - 29$

= ______________ Expanded form

$= [(7 + 1) \cdot 10 + 4] - (2 \cdot 10 + 9)$ ______________

$= [(7 \cdot 10 + 1 \cdot 10) + 4] - (2 \cdot 10 + 9)$ ______________

$= [7 \cdot 10 + (1 \cdot 10 + 4)] - (2 \cdot 10 + 9)$ ______________

$= (7 \cdot 10 + 14) - (2 \cdot 10 + 9)$ ______________

= ______________ Sum-Difference Property

= ______________ Basic subtraction fact

$= (7 - 2) \cdot 10 + 5$ ______________

$= 5 \cdot 10 + 5$ ______________

= ________ Standard form

For Exercises 4–7, begin by estimating the difference. Then analyze the use of the subtraction algorithm in each case, indicating which steps of the analysis justify the "lining up" of digits and the borrowing process, where appropriate.

4. $86 - 25$ **5.** $795 - 431$

6. $94 - 56$ **7.** $234 - 159$

For Exercises 8–11, begin by estimating the difference. Then write out the computation in the elementary analysis form of Examples 6 and 7.

8. $76 - 43$ **9.** $73 - 55$

10. $534 - 278$ **11.** $802 - 357$

For Exercises 12–16, use the base-five addition table (Table 6.2) and the Subtraction Algorithm to compute the difference.

12. $43_{five} - 31_{five}$ **13.** $42_{five} - 23_{five}$

14. $4231_{five} - 3033_{five}$

15. $14332_{five} - 4444_{five}$

16. $30212_{five} - 21334_{five}$

17. Analyze the use of the subtraction algorithm in Exercise 13.

For Exercises 18 and 19, use the addition table for base six (constructed in Exercise 19 of Section 6.2) and the Subtraction Algorithm to compute the difference.

18. $5231_{six} - 54_{six}$

19. $45302_{six} - 33215_{six}$

20. Reword the algorithm steps for $73-28$ as described in this section so that they more appropriately reflect the ideas of the analysis. (It might be useful to consult an elementary arithmetic text.)

For Exercises 21 and 22, subtract in *each* of the following bases: (a) six, (b) seven, (c) eight, (d) nine, (e) ten, (f) eleven, (g) twelve.

21.
$$\begin{array}{r} 4321 \\ -\ 1234 \\ \hline \end{array}$$

22.
$$\begin{array}{r} 4042 \\ -\ 1315 \\ \hline \end{array}$$

6.4 The Multiplication Algorithm

In order to analyze the multiplication algorithm, we need two preliminary ideas—(1) a table of basic multiplication facts, and (2) a property dealing with the product of any whole number and a power of ten. Table 6.3 shows the multiplication table for the numbers 0 through 9. As in the case of addition, there are 100 formally different facts in the table. You will be asked in an exercise to determine how many of these actually have to be memorized.

Powers of Ten

Now let us turn to multiplying by powers of ten. For instance, the product $487 \cdot 10^3$ can be rewritten as

$$(4 \cdot 10^2 + 8 \cdot 10 + 7) \cdot 10^3$$

·	0	1	2	3	4	5	6	7	8	9
0	0	0	0	0	0	0	0	0	0	0
1	0	1	2	3	4	5	6	7	8	9
2	0	2	4	6	8	10	12	14	16	18
3	0	3	6	9	12	15	18	21	24	27
4	0	4	8	12	16	20	24	28	32	36
5	0	5	10	15	20	25	30	35	40	45
6	0	6	12	18	24	30	36	42	48	54
7	0	7	14	21	28	35	42	49	56	63
8	0	8	16	24	36	40	48	56	64	72
9	0	9	18	27	36	45	54	63	72	81

Table 6.3 The basic multiplication table.

which, by the distributivity of · over + together with a law of exponents, equals

$$4 \cdot 10^5 + 8 \cdot 10^4 + 7 \cdot 10^3$$

Because of the additive-identity property and the behavior of zero in multiplication, this last expression equals

$$4 \cdot 10^5 + 8 \cdot 10^4 + 7 \cdot 10^3 + 0 \cdot 10^2 + 0 \cdot 10 + 0$$

the expanded form of the numeral 487,000. Thus, multiplying 487 by 10^3 results in the annexation of three zeros to the right end of the numeral 487. (Why do we say "annexation of" rather than "addition of"?) This example illustrates the following general principle, which will be used in analyzing the multiplication algorithm. We call it the **Powers-of-Ten Product Property:**

> The product of any whole number a and the nth power of ten can be formed by annexing n zeros to the right end of the standard-form numeral that represents a.

Example 1

$$\begin{aligned} 37 \cdot 10^2 &= 3700 \\ 2915 \cdot 10^5 &= 291,500,000 \\ 720 \cdot 10^2 &= 72,000 \end{aligned}$$

□

To clarify the analysis, we present the multiplication algorithm in two parts. We first consider multiplying an n-digit whole number by a 1-digit whole number. The process used in this case will be called the **n-by-1 Multiplication Algorithm.** Then we discuss multiplying any n-digit whole number by any m-digit whole number; that process is called the **n-by-m Multiplication Algorithm.**

The n-by-1 Algorithm

As a first example, consider the multiplication problem $63 \cdot 7$. In this case, the 2-by-1 algorithm works as follows:[1]

$$\begin{array}{r} 6\ 3 \\ \times\ 7 \\ \hline 4\ 4\ 1 \end{array}$$

Step 1. Place the 1-digit numeral under the 2-digit numeral.

Step 2. 3 times 7 is 21; write the 1 and carry the 2.

Step 3. 6 times 7 is 42, plus 2 is 44.

Step 4. Thus, the product is 441.

Analysis:

$\boxed{63 \cdot 7}$

1.	$= (6 \cdot 10 + 3) \cdot 7$	Expanded form
2.	$= (6 \cdot 10) \cdot 7 + 3 \cdot 7$	Distributivity of $\cdot$ over $+$
3.	$= (6 \cdot 7) \cdot 10 + 3 \cdot 7$	Associativity and commutativity of $\cdot$
4.	$= 42 \cdot 10 + 21$	Basic multiplication table
5.	$= 420 + 21$	Powers-of-ten Product Property
6.	$= 441$	Addition Algorithm

Observe that in the last step this algorithm uses the Addition Algorithm; that explains why the multiplication process involves "carrying."

The analysis for any product that uses the n-by-1 Multiplication Algorithm is similar to the one just given. In fact, increasing the size of n does not even increase the number of steps needed in the analysis.

Example 2 The excerpt from a third-grade text shown in Figure 6.2 illustrates one way to introduce the n-by-1 Multiplication Algorithm. Can you do a formal analysis of the problem shown there? □

---

[1]We have been using the dot (·) notation for multiplication when writing the expression horizontally. Consistent with almost all grade-school textbooks, we shall use the cross (×) for multiplication in algorithm form.

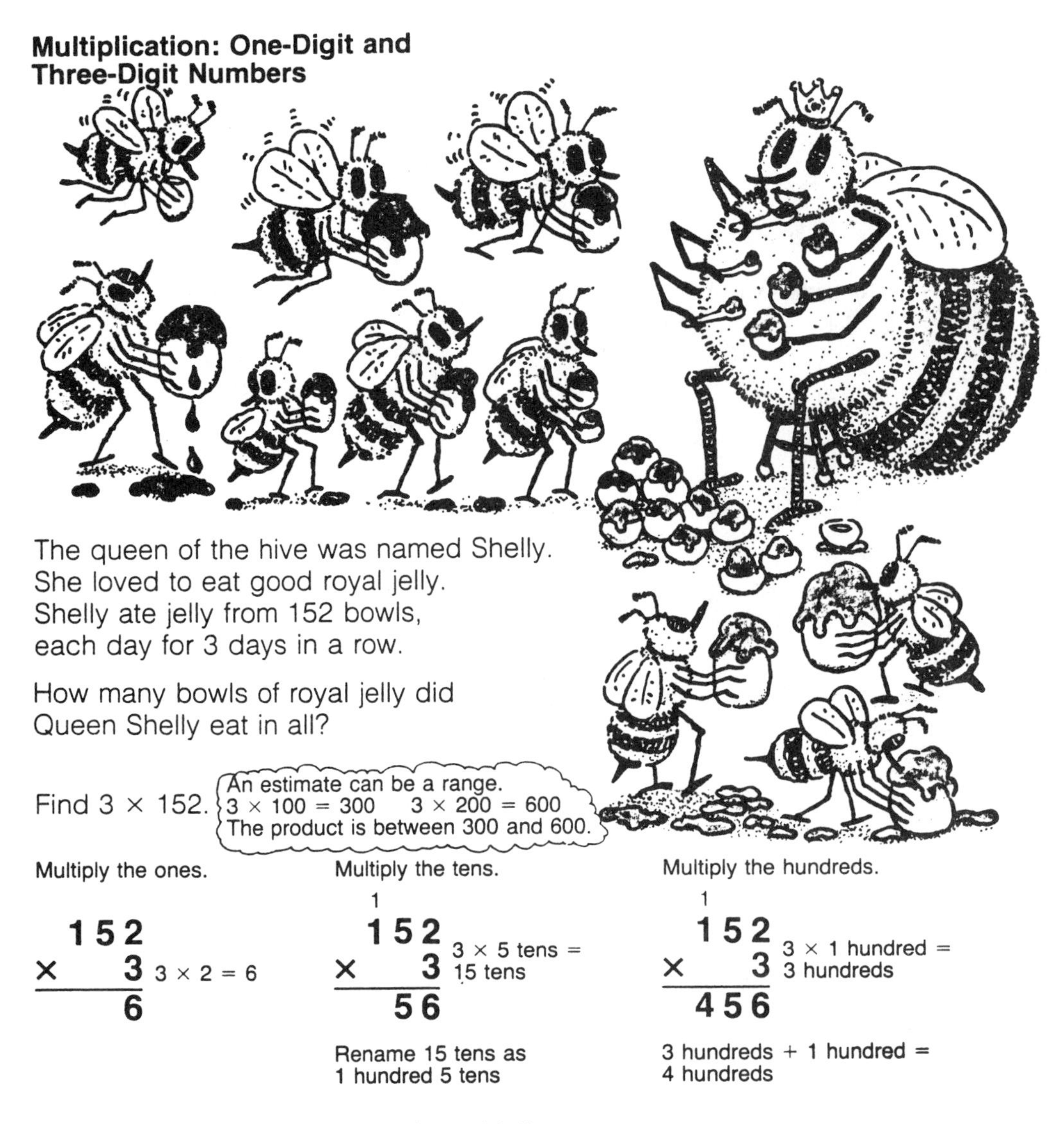

Multiplication: One-Digit and Three-Digit Numbers

The queen of the hive was named Shelly.
She loved to eat good royal jelly.
Shelly ate jelly from 152 bowls,
each day for 3 days in a row.

How many bowls of royal jelly did Queen Shelly eat in all?

Find 3 × 152.

An estimate can be a range.
3 × 100 = 300   3 × 200 = 600
The product is between 300 and 600.

Multiply the ones.

$$\begin{array}{r} 152 \\ \times \quad 3 \\ \hline 6 \end{array}$$

3 × 2 = 6

Multiply the tens.

$$\begin{array}{r} {}^{1} \\ 152 \\ \times \quad 3 \\ \hline 56 \end{array}$$

3 × 5 tens = 15 tens

Rename 15 tens as 1 hundred 5 tens

Multiply the hundreds.

$$\begin{array}{r} {}^{1} \\ 152 \\ \times \quad 3 \\ \hline 456 \end{array}$$

3 × 1 hundred = 3 hundreds

3 hundreds + 1 hundred = 4 hundreds

Queen Shelly ate 456 bowls of royal jelly.

Figure 6.2 INVITATION TO MATHEMATICS, Grade 3, p. 198

The n-by-m Algorithm

Now let us consider an example of the n-by-m algorithm for multiplication. It may appear more complicated than the n-by-1 algorithm, but its analysis really is not, because the process uses algorithms that have already been analyzed.

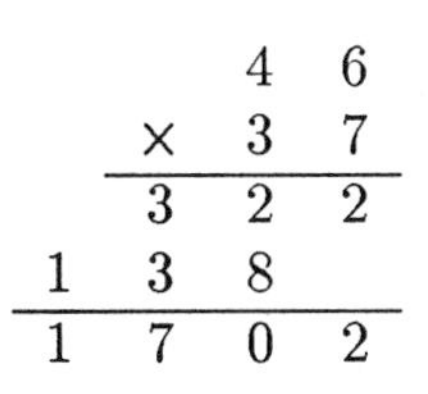

Step 1. Multiply 46 by 7. Write the result 322 below the line.

Step 2. Multiply 46 by 3. Write the result 138 below 322, shifting it one space to the left.

Step 3. Add the two lines together.

As you read the analysis, pay particular attention to the reason for shifting the second product one space to the left of the first one.

Analysis:

$\boxed{46 \cdot 37}$

1.	$= 46 \cdot (3 \cdot 10 + 7)$	Expanded form
2.	$= 46 \cdot (3 \cdot 10) + 46 \cdot 7$	Distributivity of $\cdot$ over $+$
3.	$= (46 \cdot 3) \cdot 10 + 46 \cdot 7$	Associativity of $\cdot$
4.	$= 138 \cdot 10 + 322$	n-by-1 Multiplication Algorithm
5.	$= 1380 + 322$	Powers-of-Ten Product Property
6.	$= 1702$	Addition Algorithm

It should be apparent from this analysis that the n-by-m Multiplication Algorithm is really just a combination of the n-by-1 Multiplication Algorithm and the Addition Algorithm. The reason for shifting the product $46 \cdot 3 = 138$ over one space to the left is that we are not really multiplying 46 by 3, but by $3 \cdot 10$. Thus, we must add 1380, not just 138, to 322. Because students often forget to shift the products appropriately, many teachers have the students actually put in the zeros, so that the preceding problem would look like this:

$$\begin{array}{rrrr} & & 4 & 6 \\ & \times & 3 & 7 \\ \hline & 3 & 2 & 2 \\ 1 & 3 & 8 & 0 \\ \hline 1 & 7 & 0 & 2 \end{array}$$

Multiplication in Other Bases

To multiply numbers in other base systems, we can use the same algorithm. We just need the addition and multiplication tables for the particular base being used. The next two examples use the addition and multiplication tables for base five, repeated here for convenience.

+	0	1	2	3	4
0	0	1	2	3	4
1	1	2	3	4	10
2	2	3	4	10	11
3	3	4	10	11	12
4	4	10	11	12	13

×	0	1	2	3	4
0	0	0	0	0	0
1	0	1	2	3	4
2	0	2	4	11	13
3	0	3	11	14	22
4	0	4	13	22	31

Table 6.4 Base-Five Operation Tables.

Example 3 In base five:

$$\begin{array}{r} 134 \\ \times \quad 4 \\ \hline 1201 \end{array}$$

Analysis:

$\boxed{134_{five} \cdot 4_{five}}$

1.	$= (1 \cdot 5^2 + 3 \cdot 5 + 4) \cdot 4$	Expanded form
2.	$= (1 \cdot 5^2) \cdot 4 + (3 \cdot 5) \cdot 4 + 4 \cdot 4$	Distributivity of · over +
3.	$= (1 \cdot 4) \cdot 5^2 + (3 \cdot 4) \cdot 5 + 4 \cdot 4$	Associativity and commutativity of ·
4.	$= 4 \cdot 5^2 + 22_{five} \cdot 5 + 31_{five}$	Basic multiplication table
5.	$= 400_{five} + 220_{five} + 31_{five}$	Multiplication by powers of five (in base five)
6.	$= 1201_{five}$	Addition Algorithm □

Example 4 In base five:

$$\begin{array}{r} 132 \\ \times \quad 24 \\ \hline 1133 \\ 314 \\ \hline 4323 \end{array}$$

□

Exercises 6.4

1. Keeping in mind the discussion in Section 6.2 of basic addition facts, determine how many of the 100 basic multiplication facts have to be memorized if you make use of the properties of multiplication.

For Exercises 2–5, fill in the missing reasons and steps in the given analysis.

2. $46 \cdot 8$

$= (4 \cdot 10 + 6) \cdot 8$ ________________

$= (4 \cdot 10) \cdot 8 + 6 \cdot 8$ ____________

$= (4 \cdot 8) \cdot 10 + 6 \cdot 8$ ____________

$= 32 \cdot 10 + 48$ ____________

$= 320 + 48$ ____________

$= 368$ ____________

3. $658 \cdot 7$

$= (6 \cdot 10^2 + 5 \cdot 10 + 8) \cdot 7$

____

$=$ ____________ Distributivity of $\cdot$ over $+$

$= (6 \cdot 7) \cdot 10^2 + (5 \cdot 7) \cdot 10 + 8 \cdot 7$

____

$=$ ____________ Basic multiplication facts

$= 4200 + 350 + 56$ ____________

$= 4606$ ____________

4. $27 \cdot 76$

$= 27 \cdot (7 \cdot 10 + 6)$ ____________

$= 27 \cdot (7 \cdot 10) + 27 \cdot 6$ ____________

$= (27 \cdot 7) \cdot 10 + 27 \cdot 6$ ____________

$= 189 \cdot 10 + 162$ ____________

$= 1890 + 162$ ____________

$= 2052$ ____________

5. $658 \cdot 37$

$=$ ____________ Expanded form

$= 658 \cdot (3 \cdot 10) + 658 \cdot 7$

____

$= (658 \cdot 3) \cdot 10 + 658 \cdot 7$

____

$= 1974 \cdot 10 + 4606$

____

$=$ ____________ Powers-of-Ten Product Property

$= 24{,}346$ ____________

For Exercises 6–13, first estimate the given product, then analyze the algorithm for computing that product. Show where the steps of the analysis justify the "shifting" of (partial) products, when appropriate.

6. $43 \cdot 6$ **7.** $257 \cdot 8$

8. $23 \cdot 41$ **9.** $98 \cdot 35$

10. $522 \cdot 45$ **11.** $368 \cdot 872$

12. $5872 \cdot 239$ **13.** $35{,}789 \cdot 23{,}675$

For Exercises 14–17, use the base-five addition and multiplication tables given in this section to compute each of the following products by the Multiplication Algorithm.

14. $21_{five} \cdot 3_{five}$ **15.** $42_{five} \cdot 3_{five}$

16. $43_{five} \cdot 32_{five}$ **17.** $143_{five} \cdot 23_{five}$

18. Analyze the algorithms for Exercises 15 and 17.

19. Analyze the algorithm given in Example 4.

For Exercises 20–23, construct a multiplication table for base six and use it, along with the base-six addition table you constructed earlier, to work out the given product.

20. $21_{six} \cdot 4_{six}$ **21.** $35_{six} \cdot 4_{six}$

22. $43_{six} \cdot 25_{six}$ **23.** $214_{six} \cdot 34_{six}$

For Exercises 24 and 25, multiply in *each* of the following bases: (a) six, (b) seven, (c) eight, (d) nine, (e) ten, (f) eleven, (g) twelve.

24. $\begin{array}{r} 234 \\ \times\ 3 \\ \hline \end{array}$ **25.** $\begin{array}{r} 135 \\ \times\ 24 \\ \hline \end{array}$

26. Write the basic addition and multiplication tables for base two. Use these tables to compute the following products in base two.

(a) $\begin{array}{r} 1101 \\ \times\ 11 \\ \hline \end{array}$ (b) $\begin{array}{r} 11011 \\ \times\ 1011 \\ \hline \end{array}$

6.5 The Division Algorithm

Of the four arithmetic algorithms, the one for division seems to be the most difficult for teachers to present in a reasonable, understandable way. Its complexity tends to obscure how the process works and how division actually is being accomplished. Consequently, before we analyze the Division Algorithm we shall spend some time deriving it, so that the ideas you learn here may provide a basis for your understandable presentation of it when you teach.

As we noted before, there is more than one algorithm for each of the arithmetic processes. In this section we present four algorithms for division, the fourth one being the "usual" algorithm taught in elementary school. As we progress from one to another, you should notice that each is more efficient and practical than its predecessor. However, each algorithm also becomes less understandable than the preceding one; that is, it becomes less and less apparent that the process yields the results of a division problem.

Division and Subtraction

Let us begin by observing a useful relationship between division and subtraction. In our earlier work we saw how reasoning by analogy led us to develop the definition of division in terms of multiplication in the same way that subtraction was defined in terms of addition. If we pursue this analogy further, we can argue that, since multiplication can also be defined in terms of repeated addition, we ought to be able to view division as repeated subtraction. For example, by the definition of division we know that

$$8 \div 4 = 2 \quad \text{if and only if} \quad 4 \cdot 2 = 8$$

Moreover, we know that

$$8 = 4 \cdot 2 = 4 + 4$$

But

$$\text{if } 8 = 4 + 4, \text{ then } 8 - 4 = 4$$

by the definition of subtraction. Applying that definition a second time, we see that

$$(8 - 4) - 4 = 0$$

Therefore, we can say that $8 \div 4 = 2$ because subtracting *two* 4s from 8 results in zero. Thus, to find the quotient $8 \div 4$, rather than asking "4 times what equals 8?" we could ask "*How many* 4s must be subtracted from 8 to get 0?"

All four division algorithms are based on this view of division as repeated subtraction. The first and simplest one is merely a restatement of the method just described.

Division Algorithm I: Repeated Subtraction of the Divisor

To find the quotient $15 \div 3$, we subtract 3s from 15 until we reach 0. Then we count the number of 3s subtracted. In this case, five 3s have been subtracted from 15. Therefore,

$$15 \div 3 = 5$$

$$\begin{array}{rl} 15 & \\ -\ 3 & \leftarrow \text{1st } 3 \\ \hline 12 & \\ -\ 3 & \leftarrow \text{2nd } 3 \\ \hline 9 & \\ -\ 3 & \leftarrow \text{3rd } 3 \\ \hline 6 & \\ -\ 3 & \leftarrow \text{4th } 3 \\ \hline 3 & \\ -\ 3 & \leftarrow \text{5th } 3 \\ \hline 0 & \end{array}$$

Now, we have already seen that, given two whole numbers a and b, there is not always a whole number x such that $a \div b = x$. However, a slight modification of Algorithm I enables us to deal with such problems. If we try to use the algorithm on $17 \div 6$, for example, we find that we can subtract two 6s; but then we are left with 5, from which we cannot subtract 6 and still get a whole number. In this case we say that $17 \div 6$ equals 2 *with a remainder of* 5, and we write this as

$$\begin{array}{r} 17 \\ -\ 6 \\ \hline 11 \\ -\ 6 \\ \hline 5 \end{array}$$

$$17 \div 6 = 2\,R\,5$$

Definition If a, b, q, and r are whole numbers, with $b \neq 0$, then a **divided by** b equals q **with a remainder of** r if

$$a = q \cdot b + r \quad \text{and} \quad r < b$$

We say that a is the **dividend**, b is the **divisor**, q is the **quotient**, and r is the **remainder**, and we write

$$a \div b = q\,R\,r$$

Example 1 Seven 28s can be subtracted from 212 before the remainder, 16, is less than 28. Thus,

$$212 = 7 \cdot 28 + 16$$

This yields the corresponding division statement

$$212 \div 28 = 7\,R\,16$$

where 212 is the dividend, 28 is the divisor, 7 is the quotient, and 16 is the remainder. □

Example 2 $30,853 = 82 \cdot 376 + 21$ can be rewritten as

$$30,853 \div 376 = 82\,R\,21$$

By the commutativity of multiplication, we have

$$30,853 = 376 \cdot 82 + 21$$

so we can also say

$$30,853 \div 82 = 376 \; R \; 21$$

Notice that such an interchange of quotient and divisor cannot be done in Example 1, because the remainder 16 is not less than 7. □

Example 3 $990 = 25 \cdot 38 + 40$ does not define a quotient-remainder statement because 40 is not less than either 25 or 38. □

Example 4 $32 = 4 \cdot 8$ can be written as $32 = 4 \cdot 8 + 0$, so

$$32 \div 8 = 4 \; R \; 0$$

Of course, by the earlier definition of division of whole numbers, we can simply write

$$32 \div 8 = 4$$

□

Example 5 Algorithm I can be used to find the quotient and remainder for $129 \div 35$ as shown at the right. Here, three 35s are subtracted from 129; therefore,

$$129 \div 35 = 3 \; R \; 24$$

$$\begin{array}{r} 129 \\ -\ 35 \\ \hline 94 \\ -\ 35 \\ \hline 59 \\ -\ 35 \\ \hline 24 \end{array}$$

□

If we were to use Algorithm I for $21{,}624 \div 37$, we would eventually find the quotient and remainder, but that approach would not be very practical. The process would begin and end as shown here. If we actually were to complete the process, we would have subtracted five hundred eighty-four 37s, one at a time, ending (after much time and effort) with 16 left over:

$$21{,}624 \div 37 = 584 \; R \; 16$$

$$\begin{array}{r} 21624 \\ -\quad 37 \\ \hline 21587 \\ -\quad 37 \\ \hline 21550 \\ -\quad 37 \\ \hline 21513 \\ \vdots \\ 53 \\ -\ 37 \\ \hline 16 \end{array}$$

Thus, although Division Algorithm I is easy to understand, it is not very practical for division problems involving large quotients. The next algorithm eases this difficulty.

Division Algorithm II: Subtracting Multiples of the Divisor

This algorithm also uses repeated subtraction, but, instead of having us subtract the divisor over and over again, it allows us to subtract the divisor quantity "in bunches." We use the preceding problem, $21{,}624 \div 37$, as a typical example of how this algorithm works. We want to subtract large quantities of the divisor and, for added efficiency, we shall make the quantities multiples of powers of ten. First, notice that we can subtract one hundred 37s, but we cannot subtract one thousand 37s. To find the largest multiple of 100 37s that can be subtracted, we make a list of hundreds of 37s:

$$\begin{aligned} 100 \cdot 37 &= 3700 \\ 200 \cdot 37 &= 7400 \\ &\vdots \\ 500 \cdot 37 &= 18500 \\ 600 \cdot 37 &= 22200 \end{aligned}$$

Now, $18{,}500 < 21{,}624 < 22{,}200$, so we can subtract 500 37s, but we cannot subtract 600 37s. Thus,

$$\begin{array}{r} 21624 \\ -\ 18500 \\ \hline 3124 \end{array}$$

Because no more hundreds of 37s can be subtracted, we look for the largest multiple of ten 37s that can be subtracted. As before, we make a partial list of these multiples:

$$\begin{aligned} &\vdots \\ 80 \cdot 37 &= 2960 \\ 90 \cdot 37 &= 3330 \end{aligned}$$

Since $2960 < 3124 < 3330$, we can subtract 80 37s:

$$\begin{array}{r} 3124 \\ -\ 2960 \\ \hline 164 \end{array}$$

No more tens of 37 can be subtracted, so we now look at units of 37.

$$\begin{aligned} &\vdots \\ 4 \cdot 37 &= 148 \\ 5 \cdot 37 &= 185 \end{aligned}$$

Because $148 < 164 < 185$, we can subtract 4, but not 5, 37s:

$$\begin{array}{r} 164 \\ -\ 148 \\ \hline 16 \end{array}$$

We are left with 16, and we have subtracted a total of

$$500 + 80 + 4 = 584 \text{ 37s}$$

This example of Algorithm II can be displayed more compactly, as follows:

$$\begin{array}{rrl} 21624 & & \\ -\underline{18500} & 500 & \\ 3124 & & \\ -\underline{2960} & 80 & \text{number of 37s subtracted} \\ 164 & & \\ -\underline{148} & \underline{4} & \\ 16 & 584 & \end{array}$$

Thus,

$$21{,}624 \div 37 = 584\ R\ 16$$

Example 6 Use Algorithm II to divide 70,872 by 52:

$$\begin{array}{rrl} 70872 & & \\ -\underline{52000} & 1000 & \\ 18872 & & \\ -\underline{15600} & 300 & \\ 3272 & & \text{number of 52s subtracted} \\ -\underline{3120} & 60 & \\ 152 & & \\ -\underline{104} & \underline{2} & \\ 48 & 1362 & \end{array}$$

Therefore,

$$70{,}872 \div 52 = 1362\ R\ 48$$

□

Algorithm II differs from Algorithm I in that large numbers of divisors are subtracted at each stage, using the greatest possible multiples of a power of ten in each case. Thus, Algorithm II is really quite a shortcut! Notice that many of the trial multiples of 37 that helped us to select 500, 80, and 4 are left out of the final form because much of that work normally would be done by mental calculation or estimation. We can make the process even more efficient just by rearranging the numbers used. Algorithm III does this.

Division Algorithm III: Rearranged Subtraction of Multiples

In this algorithm the long-division symbol

$$\overline{\big)\quad}$$

is introduced, and the dividend is placed within it:

$$37\overline{\big)21624}$$

Then we proceed as in Algorithm II, but we write *above the dividend* how many multiples of the divisor were subtracted. We put the multiples of the divisor over the appropriate digits of the dividend—units over units, tens over tens, hundreds over hundreds, and so forth. For example,

```
        4  \
       80   } = 584
      500  /
 37 |21624
  - 18500
     3124
   - 2960
      164
    - 148
       16
```

Algorithm III closely resembles the familiar long-division algorithm. In fact, some teachers present this one as the final algorithm taught.

Example 7 Here is a rearrangement of Example 6 in the form of Algorithm III:

```
        2  \
       60   |
      300   } = 1362
     1000  /
 52 |70872
  - 52000
    11872
  - 15600
     3272
   - 3120
      152
    - 104
       48
```

Thus,

$$70{,}872 \div 52 = 1362\ R\ 48$$

□

Example 8 The use of 0s in the quotient sometimes causes students unnecessary trouble. In such cases the form of Division Algorithm III may help to ease this difficulty.

```
       5
     200
 21 |4311
     4200
      111
      105
        6
```

□

Division Algorithm IV: The "Usual" Long-Division Algorithm

The fourth and final algorithm is very similar to Algorithm III, but it uses an additional observation. The digits in the quotient representing units, tens, hundreds, etc. correspond to the multiples of units, tens, hundreds, etc. used in each subtraction step. The only other adjustment is to replace unnecessary zeros by blank spaces.

Consider, once again, the example we have been using to illustrate these algorithms, presented now in familiar form:

```
      584   R 16
 37 |21624
     185
      312
      296
       164
       148
        16
```

Observe that, when you multiply 37 by 5 here, you are really multiplying by 500. When you subtract 185 from 216 and then "bring down a 2," you really are subtracting 18,500 from 21,624, and the 2 "comes down" because you are subtracting 0 from 2. These are the kinds of things that elementary-school students often puzzle over and that cause them to view the Division Algorithm as a series of mysterious steps apparently having little justification and little relationship to the concept of division.

Because it appears to be so far removed from the division concept, Algorithm IV is the least understood of the division algorithms. Nevertheless, it is the most efficient of them; it calculates the quotient in the quickest way. We have presented its evolution from the three previous algorithms so that you can see how it, too, is nothing more than a very efficient way of doing repeated subtraction. Among other things, this means that the Division Algorithm requires no separate analysis; its justification follows from the algorithms for subtraction and multiplication.

PROBLEM-SOLVING COMMENT

Currently many elementary-textbook series precede the introduction of the long-division algorithm by a discussion of *estimating* quotients. See, for example, a fifth-grade treatment of this topic in Figure 6.3. Do you see how it informally introduces the student to Division Algorithm II? ◇

ESTIMATING QUOTIENTS

There are 728 seats in 4 sections of a stadium. About how many seats are there per section?

If you round to the greatest place value or if you use front-end estimation, the estimated quotient may be difficult to find: $728 \div 4 \longrightarrow 700 \div 4$.

In cases like this, it's easier to estimate the quotient by using **compatible numbers,** or numbers that divide easily.

$4\overline{)728}$ Try: $\begin{array}{r}200\\4\overline{)800}\end{array}$

The exact answer must be less than 200 because 728 is less than 800.

The exact answer is 182 seats and 182 is close to but less than 200. $\begin{array}{r}182\\4\overline{)728}\end{array}$

You can estimate $648 \div 74$ by thinking $630 \div 70 = 9$ or $640 \div 80 = 8$.

Figure 6.3: HOUGHTON MIFFLIN MATHEMATICS, Grade 5, p. 142

Division in Other Bases

Division in other bases can be done by using any one of the four Division Algorithms, in conjunction with the appropriate addition and multiplication tables. Example 9 illustrates how Division Algorithm IV is used in base five.

Example 9 In this example the subscript "five" has been omitted in the algorithm format, but it is to be understood that all numerals are in base five.

$$\begin{array}{r} 1240 \\ 3\,\overline{)4322} \\ \underline{3} \\ 13 \\ \underline{11} \\ 22 \\ \underline{22} \\ 02 \\ \underline{0} \\ 2 \end{array} \qquad\qquad \begin{array}{r} 3233 \\ 24\,\overline{)144321} \\ \underline{132} \\ 123 \\ \underline{103} \\ 202 \\ \underline{132} \\ 201 \\ \underline{132} \\ 14 \end{array}$$

$4322 \div 3 = 1240\ R\ 2$ $\qquad$ $144{,}321 \div 24 = 3233\ R\ 14$ □

Exercises 6.5

For Exercise 1–8, use the definition of division with remainder to write the given division statement as an equivalent multiplication statement. Specify the dividend, the divisor, the quotient, and the remainder.

1. $52 \div 8 = 6\ R\ 4$
2. $213 \div 17 = 12\ R\ 9$
3. $5698 \div 251 = 22\ R\ 176$
4. $14{,}222 \div 547 = 26$
5. $8658 \div x = 149\ R\ 16$
6. $233 \div y = 18\ R\ x$
7. $2379 \div x = z\ R\ y$
8. $w \div z = y\ R\ x$

For Exercises 9–22, use the definition of division with remainder to write the given multiplication statement as an equivalent division statement, if possible. If two division statements are possible, give both; if there is no possible division statement, answer "none." Specify the dividend, the divisor, the quotient, and the remainder.

9. $17 = 2 \cdot 7 + 3$
10. $626 = 17 \cdot 35 + 31$
11. $3729 = 68 \cdot 54 + 57$
12. $14{,}486 = 157 \cdot 92 + 42$
13. $804 = 31 \cdot 25 + 29$
14. $642 = 7 \cdot 84 + 54$
15. $1036 = 37 \cdot 28$
16. $2712 = 4 \cdot 528 + 600$
17. $26{,}071 = 128 \cdot 203 + 87$
18. $1434 - 54 \cdot 26 = 30$
19. $587 - 65 = 6 \cdot 87$
20. $654 \cdot 5 + 8 = 3278$
21. $878 = 2342 - 24 \cdot 61$
22. $15 = 1443 - 51 \cdot 28$

For Exercises 23–26, estimate the given quotient before computing it. Then compute it using Division Algorithms II, III, and IV.

23. $108{,}075 \div 42$
24. $10{,}032 \div 14$

25. 34,854 ÷ 17 **26.** 75,649 ÷ 346

27. Without carrying out Algorithm I completely, apply it to Exercises 23–26 by showing the first three subtractions and the final step in each case.

28. Solve the problem 188 ÷ 34 by using each of the four Division Algorithms.

29. Because of the zeros in the quotient, students frequently make errors in using the usual Division Algorithm to solve problems like Exercise 25. What suggestions can you provide to minimize these errors?

For Exercises 30–33, use Division Algorithm IV and the appropriate addition and multiplication tables to find the quotient.

30. $33{,}204_{five} \div 3_{five}$

31. $4{,}321{,}304_{five} \div 42_{five}$

32. $45{,}213_{six} \div 4_{six}$

33. $541{,}032_{six} \div 35_{six}$

For Exercises 34 and 35, use Division Algorithm IV to divide in *each* of the following bases: (a) six, (b) seven, (c) eight, (d) nine, (e) ten, (f) eleven, (g) twelve.

34. $5\overline{)2043}$ **35.** $32\overline{)1450}$

For Exercises 36 and 37, use Division Algorithm IV to divide in base two.

36. $11\overline{)10101}$ **37.** $101\overline{)1101101}$

Review Exercises for Chapter 6

For Exercises 1–10, indicate whether the given statement is *true* or *false*.

1. A set of rules for solving a problem in a finite number of steps is called a logarithm.
2. The knowledge of properties of whole numbers can reduce the number of rote addition and multiplication facts to be memorized.
3. One should never do an algorithm for a whole-number operation without also doing the analysis.
4. The purpose of the analysis of an algorithm is to provide a rational explanation to justify the steps of the algorithm.
5. There is a different addition algorithm for each numeration base.
6. Before using an algorithm for computation it is useful to estimate the answer.
7. The Division Algorithm can be described as a very efficient method for subtracting large quantities of the divisor from the dividend.
8. The Multiplication Algorithm makes use of the Addition Algorithm.
9. In the answer to the division problem $a \div b$, the remainder r should always be smaller than a.
10. When multiplying 342 by 10,000, students would be expected to use the 3-by-5 Multiplication Algorithm.

In Exercises 11–13,
(a) estimate the answer;
(b) compute with the appropriate algorithm; and
(c) analyze the algorithm.

11. 37 + 26 **12.** 52 × 7 **13.** 76 × 43

14. Fill in the missing reasons.

$72 - 27$

$= (7 \cdot 10 + 2) - (2 \cdot 10 + 7)$ ________________

$= [(6 + 1) \cdot 10 + 2] - (2 \cdot 10 + 7)$ ______________________

$= [(6 \cdot 10 + 1 \cdot 10) + 2] - (2 \cdot 10 + 7)$ ______________________

$= [6 \cdot 10 + (1 \cdot 10 + 2)] - (2 \cdot 10 + 7)$ ______________________

$= (6 \cdot 10 + 12) - (2 \cdot 10 + 7)$ ______________________

$= (6 \cdot 10 - 2 \cdot 10) + (12 - 7)$ ______________________

$= (6 \cdot 10 - 2 \cdot 10) + 5$ ______________________

$= (6 - 2) \cdot 10 + 5$ ______________________

$= 4 \cdot 10 + 5$ ______________________

$= 45$ ______________________

15. Illustrate the four stages of the Division Algorithm using $55{,}173 \div 67$.

In Exercises 16–19, use the base-five addition and multiplication tables and the appropriate algorithm to compute the answer.

16. $231_{five} + 344_{five}$ **17.** $421_{five} - 133_{five}$
18. $431_{five} \times 41_{five}$ **19.** $4211_{five} \div 32_{five}$

In Exercises 20–23, use the base-six addition and multiplication tables and the appropriate algorithm to compute the answer.

20. $342_{six} + 455_{six}$ **21.** $532_{six} - 244_{six}$
22. $542_{six} \times 52_{six}$ **23.** $5322_{six} \div 43_{six}$

7.1 Factors

Chapter 4 presented the theoretical underpinnings of the whole numbers and their operations; Chapter 5 showed how our Hindu-Arabic numeration system allows us to write large numbers with ease; and Chapter 6 examined "mechanical" ways of computing sums, differences, products, and quotients. In this chapter we return to the theory of whole numbers to examine some fundamental properties of products, quotients, divisors, and multiples. There are two reasons for our interest in these ideas, one old and one new. The new one is that precise definitions of these concepts are prerequisites for constructing the system of rational numbers (in Chapter 9). The old reason is that for thousands of years questions about these ideas have held a special fascination for mathematicians, puzzle enthusiasts, and just plain curious people. Perhaps the simplest explanation for why numbers have entertained so many people for so many centuries is that interesting questions about numbers are very easy to ask, even when their answers are hard to find.

This common curiosity about numbers is a valuable, but often neglected, tool for you as a teacher of young children. With a little ingenuity, you can use some of the simple ideas presented here to get students of any age to explore the world of mathematical ideas on their own, to share some of the excitement of discovering new ideas without first being required to master a maze of specialized techniques. Many of these "paths to the unknown" begin from a spot close to home—the concept of *a factor*.

Definition Let a and b be natural numbers. We say a is a **factor** of b if there exists a natural number c such that $a \cdot c = b$. In this case we also say that a is a **divisor** of b, a **divides** b, b is a **multiple** of a, or b is **divisible** by a.

Example 1 3 is a factor of 15 because there is a natural number, 5, such that $3 \cdot 5 = 15$. We can also say:

3 is a divisor of 15
3 divides 15
15 is a multiple of 3
15 is divisible by 3 □

Example 2 16 is a multiple of 2 because 8 is a natural number such that $2 \cdot 8 = 16$. We can also say that 2 is a factor of 16, 2 is a divisor of 16, 16 is divisible by 2, or 2 divides 16. □

Example 3 6 is *not* a factor of 3 because there is no natural number c such that $6 \cdot c = 3$. (Note that 6 is a *multiple* of 3.) □

Example 4 The set of factors of 20 is $\{1, 2, 4, 5, 10, 20\}$. The set of multiples of 20 is $\{20, 40, 60, 80, \ldots\}$. □

"I can't think of any interesting questions about factors," says Terry (whom you met in Chapter 1). "That seems like a pretty simple, but dull, idea." (A challenge too tempting to ignore!)

Question 1: What is the sum of all the factors of 6?

Terry: 1 + 2 + 3 + 6; that's 12. Dull.

Question 2: Can you think of another number whose factors add up to 12?

Terry: How about 11? Its only two factors are 11 and 1. Still dull.

Question 3: Find a number whose factors add up to 10.

Terry: Sure.... I can't think of any; that's strange. I guess there aren't any.

Question 4: Are you sure? Convince me that there aren't any.

Terry: Well, any number whose factors add up to 10 must be smaller than 10, so just add up the factors of each of those numbers and you'll see that none of them work.

Question 5: O.K. Is there any number smaller than 10 that is not the sum of the factors of a number?

Terry (scribbling on an old envelope): $3 = 1 + 2$, the sum of the factors of 2; $4 = 1 + 3$; $6 = 1 + 5$; $7 = 1 + 2 + 4$; $8 = 1 + 7$. The sum of the factors of 1 is 1, so it works out, too. I can't get 2, 5 or 9 to work, though.

Question 6: It's interesting, at least to me, that 5 and 10 don't "work" in this way. Do you suppose that the other multiples of 5 don't work, either?

Terry: ...(scribbling again)...15 works; it's $1+2+4+8$, which is the sum of the factors of 8.... But I can't get 20 to work. I wonder if 30 works?

Terry scribbles on, in pursuit of the possible pattern implied by the last question, which, we note with some satisfaction, was posed *by Terry*. Perhaps factors are not so dull, after all. The broader, harder, more interesting questions emerging here are: What numbers are *not* the sums of all factors of some (other) number? Are there only a few of them, or are there infinitely many? If there are a lot of them, do they fit a pattern? There is a lot for Terry to explore, with intriguing possibilities at all levels of sophistication.

The pedagogical lesson in this for you, as a future teacher, is that interest in a simple idea comes from a provocative sequence of questions. Knowing how to ask such questions when you teach mathematics depends on knowing the mathematical ideas so thoroughly that you can turn and twist them in lots of different, sometimes unusual ways. This is a major reason for studying the theory of arithmetic, rather than being content with just knowing how to do it. Children are naturally curious, but much of that curiosity may be wasted unless you—the teacher—can direct it down promising paths for effective learning.

Example 5 Here is another chain of questions that starts with adding the factors of a number and leads to some profound ideas. We pose these questions here for you to see how easily the recognition of a simple pattern can be expanded into an entire line of inquiry.

What is the sum of the factors of 2? (3)

What is the sum of the factors of 2^2? (7)

What is the sum of the factors of 2^3? (15)

What is the sum of the factors of 2^4? (31)

What is the sum of the factors of 2^5? (63)

Compare your answers with the powers of 2:

$$\begin{array}{lcccccc} \text{Powers of 2:} & 2, & 4, & 8, & 16, & 32, & 64, \quad \ldots \\ & & \nearrow & \nearrow & \nearrow & \nearrow & \nearrow \\ \text{Sums of factors:} & 3, & 7, & 15, & 31, & 63, & \ldots \end{array}$$

What pattern do you see? Can you write a general formula for the sums of the factors of powers of 2?

Historical Note: Perfect Numbers

Interest in the sums of factors of numbers dates back at least to the days of the ancient Greeks. In particular, they called a number **perfect** if it was found to be the sum of all its divisors except for itself. The smallest perfect number is $6 = 1 + 2 + 3$. The next one is 28:

$$28 = 1 + 2 + 4 + 7 + 14$$

A discussion of the theory of perfect numbers appears in Euclid's *Elements* in 300 B.C., along with the first four perfect numbers, the only ones known at that time:

$$6,\ 28,\ 496,\ 8128$$

The *Elements* contains a proof that every number of the form

$$2^{n-1} \cdot (2^n - 1)$$

must be perfect whenever $2^n - 1$ is prime (has no divisors other than itself and 1), and that no other *even* numbers can be perfect. Thus,

$$28 = 4 \cdot 7 = 2^{3-1} \cdot (2^3 - 1)$$

is perfect because 7 is prime, and there are no other perfect numbers between 28 and 496, which is $2^4 \cdot (2^5 - 1)$.

Despite such a good description of the form of even perfect numbers, the Greeks were unable to find any beyond 8128, probably because their numeration system was too awkward to permit convenient calculations with large numbers. The next perfect number was not found until A.D. 1536, when Hudalrichus Regius showed that $2^{13} - 1$ is prime, thereby insuring that

$$33,350,336 = 2^{12} \cdot (2^{13} - 1)$$

is perfect. Considering that he was probably using Roman numerals and an abacus, this was quite an achievement!

By the late 1500s the Hindu-Arabic numeration system had become common in Europe, so division with large numbers was a feasible (but tedious) process. The sixth perfect number was found in 1603, but the next two were not found until 1772. The largest perfect number found by hand calculation is the tenth one, $2^{88} \cdot (2^{89} - 1)$, verified in 1914. The next one was found in 1952 with the aid of a computer. Between 1952 and 1989 another eighteen perfect numbers have been found, the largest of which is

$$2^{216,090} \cdot (2^{216,091} - 1)$$

If printed in ordinary newspaper type, the 130,100 digits that represent this number would fill a line as long as 3 football fields!

Computers have helped to extend the list of known perfect numbers, but they have not solved the mystery. There still is no formula for finding even perfect numbers, and no one knows whether or not any odd perfect numbers exist. Such answers still lie hidden in the theory of numbers, waiting to be uncovered by just the right combination of human logic and intuition. ◇

Can you find a similar pattern for the sums of the factors of powers of 3? of 4? of 5?

Can you find a general pattern for the sums of the factors of any fixed number? of some fixed numbers? □

Exercises 7.1

For Exercises 1–17, fill in the blanks with *all* the appropriate phrases from among

is a factor of, is a multiple of,
is divisible by, is a divisor of,
divides

that make the sentence true. If none of the phrases are appropriate, write *none*.

1. 32 _______ 8 **2.** 15 _______ 70

3. 7 _______ 21 **4.** 10 _______ 10

5. 125 _______ 25 **6.** 17 _______ 1

7. 1 _______ 35 **8.** 13 _______ 39

9. 42 _______ 8 **10.** 29 _______ 29

11. 1 _______ every natural number.

12. Every natural number _______ 1.

13. Every natural number _______ itself.

14. If a is a factor of b, then a _______ b and b _______ a.

15. If $a > b$, then it is not true that a _______ b, but it could be true that a _______ b.

16. Every even number _______ 2, and 2 _______ every even number.

17. Every odd number _______ some even number, but no even number _______ any odd number.

18. List three different factors of 12.

19. List three different multiples of 12.

20. How many different factors does 12 have?

21. How many different multiples does 12 have?

22. Find a natural number that has exactly one factor.

23. Give 3 examples of natural numbers that have exactly two different factors.

24. Give 3 examples of natural numbers that have exactly three different factors.

For Exercises 25–30, find the set of factors of the given number.

25. 15 **26.** 17 **27.** 32

28. 36 **29.** 70 **30.** 100

For Exercises 31–33, find the set of multiples of the given number.

31. 5 **32.** 25 **33.** 18

7.2 Rules of Divisibility

You can always see whether a is a factor of b by using the Division Algorithm (Section 6.5) to divide b by a. However, for numbers written in the base-ten system, there are some shortcuts, called **Rules of Divisibility**, that can be used to help you answer such questions more quickly. These rules follow directly from the expanded form of numbers written in base ten. A proof of the first rule is presented (as optional material) to show you exactly how such rules depend on the base-ten numeration form. Proofs of the others are similar in spirit; they have been omitted because their details would take us unnecessarily far afield. Remember:

> *These rules may not work for numbers written in other bases because they depend on the digital representation of the numbers.*

Divisibility Rule for 2: A number is divisible by 2 if and only if its last digit is 0, 2, 4, 6, or 8. (The "last digit" of a number is the digit farthest to the right.)

Definition Numbers that are divisible by 2 are called **even numbers**; numbers that are not divisible by 2 are called **odd numbers**.

Example 1 18, 20, 32, 144, and 356 are all divisible by 2, so they are even numbers. 41, 23, and 75 are not divisible by 2; they are odd numbers. □

Proof of the Rule for 2 [optional]:

The expanded form of any number written in base ten is

$$a_n \cdot 10^n + a_{n-1} \cdot 10^{n-1} + \ldots + a_1 \cdot 10 + a_0$$

where a_0, a_1, ..., a_n are digits from the set $\{0, 1, \ldots 9\}$. (See Chapter 5.) Now, all the summands except for a_0 (the last digit) have 10 as a factor, and 2 is a factor of 10; so 2 must be a factor of all those summands. By the distributive law, then, 2 can be "factored out" of all the terms except the last one, implying that the sum

$$s = a_n \cdot 10^n + a_{n-1} \cdot 10^{n-1} + \ldots + a_1 \cdot 10$$

is even. But the original number is just $s + a_0$, which is even if and only if a_0 is even—that is, if and only if a_0 is 0, 2, 4, 6, or 8.

Example 2 As an example of the reasoning in the proof of the Rule for 2, consider 378:

$$\begin{aligned} 378 &= 370 + 8 \\ &= 37 \cdot 10 + 8 \\ &= (37 \cdot 2 \cdot 5) + 8 \\ &= 2 \cdot (37 \cdot 5) + 2 \cdot 4 \end{aligned}$$

Clearly, $2 \cdot (37 \cdot 5)$ is even because 2 is a factor, and $8 = 2 \cdot 4$ is even, so the sum, which can be written $2 \cdot (37 \cdot 5 + 4)$, has a factor of 2 and hence is even. On the other hand, if the last digit had been odd, 2 could not have been a factor of the entire sum. Thus, $379 = 370 + 9$ is odd. □

Divisibility Rule for 3: A number is divisible by 3 if and only if the sum of its digits is divisible by 3.

Example 3 345 is divisible by 3 because $3 + 4 + 5 = 12$ and 12 is divisible by 3. On the other hand, 1873 is not divisible by 3 because $1 + 8 + 7 + 3 = 19$ and 19 is not divisible by 3. □

Example 4 For large numbers you may sometimes want to apply a rule more than once. To determine whether 9,786,984 is divisible by 3, we find the sum of the digits:

$$9 + 7 + 8 + 6 + 9 + 8 + 4 = 51$$

We can see that 51 is divisible by 3 by adding its digits: $5 + 1 = 6$; clearly, 6 is divisible by 3. Thus, 51 is also divisible by 3, which implies that the original number, 9,786,984, is divisible by 3. □

Example 5 To see why the Rule for 3 works, study the following typical example:

$$\begin{aligned} 5271 &= 5 \cdot 1000 + 2 \cdot 100 + 7 \cdot 10 + 1 \\ &= 5 \cdot (999 + 1) + 2 \cdot (99 + 1) + 7 \cdot (9 + 1) + 1 \\ &= 5 \cdot 999 + 5 + 2 \cdot 99 + 2 + 7 \cdot 9 + 7 + 1 \\ &= (5 \cdot 999 + 2 \cdot 99 + 7 \cdot 9) + (5 + 2 + 7 + 1) \\ &= 9 \cdot (5 \cdot 111 + 2 \cdot 11 + 7) + (5 + 2 + 7 + 1) \end{aligned}$$

In the last line, we see the original number written in the form of two major summands:

$$9 \cdot (\;—\;) + (\;—\;)$$

The first of these summands is divisible by 9 and so also by 3. Thus, the entire sum will be divisible by 3 if and only if the second major summand is. But that second summand is just the sum of the original digits—$(5 + 2 + 7 + 1)$. (This example should also tell you the Divisibility Rule for 9.) □

Here are some other divisibility rules that you might find convenient from time to time. You might also find it instructive to try to justify them using the standard form of numerals written in base ten.

Divisibility Rule for 4: A number is divisible by 4 if and only if the number formed by its last two digits is divisible by 4.

Example 6 Both 732 and 4128 are divisible by 4 because 32 and 28 are divisible by 4. Neither 426 nor 5413 is divisible by 4 because neither 26 nor 13 is. □

Divisibility Rule for 5: A number is divisible by 5 if and only if its last digit is either 5 or 0.

Example 7 35, 120, and 425 are all divisible by 5, whereas 23, 82, and 108 are not. □

Divisibility Rule for 6: A number is divisible by 6 if and only if it is divisible by both 2 and 3.

Example 8 78 is divisible by 6 because it is divisible by 2 (its last digit is 8) and by 3 ($7 + 8 = 15$, which is divisible by 3). 130 is not divisible by 6 because, although it is divisible by 2, it is not divisible by 3 ($1 + 3 + 0 = 4$). □

Divisibility Rule for 8: A number is divisible by 8 if and only if the number formed by its last three digits is divisible by 8.

Example 9 7064 and 4,283,256 are divisible by 8 because 064 and 256 are; 3110 is not divisible by 8 because 110 is not. □

Divisibility Rule for 9: A number is divisible by 9 if and only if the sum of its digits is divisible by 9.

(*Note*: To see why this rule works, examine Example 5.)

Example 10 45,216 is divisible by 9 because $4 + 5 + 2 + 1 + 6 = 18$, and 18 is divisible by 9. On the other hand, 8341 is not divisible by 9 because $8 + 3 + 4 + 1 = 16$, and 16 is not divisible by 9. □

Divisibility Rule for 10^n: A number is divisible by 10^n if and only if its last n digits are zeros.

Example 11 70 is divisible by $10 = 10^1$ because its last digit is a zero. 18,000 is divisible by $1000 = 10^3$ because its last three digits are zeros. 18,000 is also divisible by 10 and 100, but not by 10,000. □

Divisibility Rule for 11: A number is divisible by 11 if and only if the difference between the sum of its 1st, 3rd, 5th, ... digits and the sum of its 2nd, 4th, 6th, ... digits is divisible by 11.

Example 12 187 is divisible by 11 because $1+7 = 8$ and $8 - 8 = 0$, which is divisible by 11. Also, 8272 is divisible by 11 because $(8 + 7) - (2 + 2) = 11$. However, 3285 is not divisible by 11 because $3 + 8 = 11$ and $2 + 5 = 7$, and the difference, $11 - 7 = 4$, is not divisible by 11. □

Example 13 Let us put all these rules to work on one large number:

286,051,547,100

— is divisible by 2 because 0 is even;

— is divisible by 3 because the sum of its digits, 39, is divisible by 3;

— is divisible by 4 because 00 is divisible by 4;

— is divisible by 5 because it ends in 0;

— is divisible by 6 because it is divisible by both 2 and 3;

— is not divisible by 8 because 100 is not divisible by 8;

— is not divisible by 9 because the sum of its digits, 39, is not divisible by 9;

— is divisible by 10 and 100 (but not by 1000) because its last two digits (but not its last three) are 0s;

— is divisible by 11 because 11 is a divisor of

$$(2 + 6 + 5 + 5 + 7 + 0) - (8 + 0 + 1 + 4 + 1 + 0)$$

which is $25 - 14 = 11$. □

These divisibility rules make it easy to find all the factors of any natural number less than 144. We use a simple observation to help us:

> If one factor of a number n is found, then we can immediately find a second factor by dividing n by the first one.

In other words, *factors of a number come in pairs.*

Consider the problem of finding all factors of 36. Because 1 is a factor of every number, we have our first pair of factors (1 and 36). The divisibility rule for 2 tells us that 2 is a factor of 36, so dividing by 2 gives us the next pair (2 and 18). Using the rules for 3 and 4 and dividing gives us our third and fourth pairs (3 and 12, 4 and 9). The rule for 5 fails, but the one for 6 works, giving us the pair 6 and 6. Listing the pairs in the order in which they were found, we see that as the left factor increases, the right one decreases:

1 and 36
2 and 18
3 and 12
4 and 9
6 and 6

Thus, the factors in the pairs approach each other in size. If we were to look for another pair of factors, the next factor would have to be larger than 6. The factor paired with it would have to be smaller than 6, so we would already have found it (because we had tested for *all* possible factors less than or equal to 6). In particular, the next factor larger than 6 is 9, giving us the pair 9 and 4, which we already found. Thus, we have all the factors of 36—namely, 1, 2, 3, 4, 6, 9, 12, 18, and 36.

This example can be generalized to a principle that tells us when to stop looking for factors of a number.

> To find the set of all factors of any number n, continue to test n for divisibility by successive numbers as long as the squares of those numbers are less than or equal to n.

In other words, you need only test numbers less than or equal to the square root of n.

Example 14 Find all the factors of 80: Since

$$8^2 = 64 < 80 < 81 = 9^2$$

we must test 80 for divisibility by all numbers up to and including 8, but not by 9 or by any number larger than that. We proceed to find the factors in pairs, using a divisibility rule to find the first number, then dividing to find the second one. The pairs are

1 and 80 2 and 40 4 and 20 5 and 16 8 and 10

Therefore, all the factors of 80 are

1, 2, 4, 5, 8, 10, 16, 20, 40, and 80 □

Exercises 7.2

For Exercises 1–6, test the given number for divisibility by 2, 4, and 8.

1. 2374
2. 1288
3. 5,291,136
4. 5723
5. 350,352
6. 2,468,842

For Exercises 7–12, test the given number for divisibility by 3 and 9.

7. 546
8. 6120
9. 1176
10. 1183
11. 5544
12. 304,050

For Exercises 13–18, test the given number for divisibility by 5, 10, 100, and 1000.

13. 960 **14.** 1245

15. 83,000 **16.** 96,005

17. 12,300 **18.** 50,001

For Exercises 19–24, test the given number for divisibility by 6.

19. 140 **20.** 15,120

21. 858 **22.** 4641

23. 7350 **24.** 232,323

For Exercises 25–30, test the given number for divisibility by 11.

25. 462 **26.** 101,409

27. 27,846 **28.** 46,189

29. 829,081 **30.** 9,182,712

For Exercises 31–36, test the given number for divisibility by the numbers 2 through 11.

31. 3080 **32.** 1386

33. 1320 **34.** 630

35. 27,720 **36.** 131,071

37. Using the divisibility tests for 2, 4, and 8 to guide you, make a conjecture about a reasonable divisibility test for 16. Test your conjecture.

38. Using the divisibility tests for 3 and 9 to guide you, make a conjecture about a reasonable divisibility test for 27. Test your conjecture.

For Exercises 39–43, label the given statement *true* or *false*.

39. Any number that is divisible by 3 and 4 must be divisible by 12.

40. Any number that is divisible by 3 and 6 must be divisible by 18.

41. Any number that is divisible by 2 and 9 must be divisible by 18.

42. Any number that is divisible by 3 and 8 must be divisible by 24.

43. Any number that is divisible by 4 and 6 must be divisible by 24.

44. Make a chart with 10 columns, labeled 1, 2, 3, ..., 10. Place each of the numbers from 1 to 50 into a column according to the number of distinct factors it has. For example, 36 goes into column 9 because it has 9 distinct factors.

PROBLEM-SOLVING COMMENT

Exercises 37 and 38 illustrate the interaction among the tactics of *reasoning by analogy, looking for patterns,* and *constructing examples* in forming useful conjectures. If we rely only on analogy and patterns, we might be led to similar conjectures in these two exercises. Only when those conjectures are tested by constructing examples does it become clear that the one for Exercise 37 is reinforced but the one for Exercise 38 is refuted. (Remember: Although many "successful" examples may reinforce your confidence in a universal conjecture, they never *prove* it; however, a single counterexample is sufficient to *disprove* it.)

Here is an additional problem to give you a chance to sharpen these tactical tools: Using Exercises 39–43 to guide you, consider the question:

> Under what condition is a number that is divisible by two other numbers also divisible by their product?

Construct examples, look for patterns, form a conjecture, and then construct more examples to test your conjecture. ◇

FOR DISCUSSION

We observed at the beginning of this section that the rules of divisibility stated here work only for numbers written in base ten. Which of these rules hold for numbers written in base four? in base five? in base six? Can you construct other convenient rules of divisibility for these bases? What about for other bases?

7.3 Prime Factors

Exercise 44 of Section 7.2 asked you to classify the numbers 1 through 50 according to the number of distinct factors each one has. If you did that exercise, your answer probably looked something like Table 7.1.

Notice that only one of the numbers from 1 to 50 is in the first column. The only number with exactly one factor is 1 because any number n can be expressed as $1 \cdot n$; thus, if n is not 1, then n must have at least two distinct factors. The numbers with exactly two factors are especially important because all other numbers (those with three or more factors) can be "composed" from them by multiplication. Thus, we name two different types of numbers as follows:

Definition A natural number is **prime** (or **a prime number**) if it has exactly two distinct factors; it is **composite** (or **a composite number**) if it is greater than 1 and is not prime.

To find out whether a number n is prime or composite, it is only necessary to see if n has more than two factors—that is, if there is some factor other than 1 and n.

Number of factors:	1	2	3	4	5	6	7	8	9	10
	1	2	4	6	16	12		24	36	48
		3	9	8		18		30		
		5	25	10		20		40		
		7	49	14		28		42		
		11		15		32				
		13		21		44				
		17		22		45				
		19		26		50				
		23		27						
		29		33						
		31		34						
		37		35						
		41		38						
		43		39						
		47		46						

Table 7.1 Numbers of distinct factors for 1–50.

Example 1 Column 2 of Table 7.1 lists all the prime numbers less than 50. All the numbers in columns 3 through 10 are composite numbers. □

Example 2 51 is a composite number; the Divisibility Rule for 3 tells us that 3 is a factor of 51. □

Example 3 61 is a prime number; its only factors are 1 and 61. Remember that, since $7^2 < 61 < 8^2$, you only have to test 61 for factors from 1 through 7. □

Examples 2 and 3 typify one way of determining whether a given natural number is prime or composite. That process works quite well when applied to one number at a time. However, if you are interested in finding *all* the primes less than or equal to a given number, there is a more efficient process, called the **Sieve of Eratosthenes**. It is named for its inventor, Eratosthenes of Alexandria (c. 276–194 B.C.), a Greek astronomer, geographer, philosopher, and mathematician. The process simply eliminates all multiples of each prime in turn, leaving only the primes.

For instance, suppose we want to find all the primes less than 100. We begin by listing all the natural numbers from 2 to 100. (See Table 7.2.) Now, 2 is a prime number, but all multiples of 2 are composites; so we cross all other even numbers off the list. The next number on the list, 3, must be prime, but we can cross off all multiples of 3 after it. (Some of these multiples were already crossed off when we eliminated the multiples of 2.) The next number still on the list is 5, which *must* be prime. (Why?) We eliminate the remaining multiples of 5 from the list (starting with 25, because 10, 15, and 20 have already been crossed off), then move on to the

	2	3	~~4~~	5	~~6~~	7	~~8~~	~~9~~	~~10~~
11	~~12~~	13	~~14~~	~~15~~	~~16~~	17	~~18~~	19	~~20~~
~~21~~	~~22~~	23	~~24~~	~~25~~	~~26~~	~~27~~	~~28~~	29	~~30~~
31	~~32~~	~~33~~	~~34~~	~~35~~	~~36~~	37	~~38~~	~~39~~	~~40~~
41	~~42~~	43	~~44~~	~~45~~	~~46~~	47	~~48~~	~~49~~	~~50~~
~~51~~	~~52~~	53	~~54~~	~~55~~	~~56~~	~~57~~	~~58~~	59	~~60~~
61	~~62~~	~~63~~	~~64~~	~~65~~	~~66~~	67	~~68~~	~~69~~	~~70~~
71	~~72~~	73	~~74~~	~~75~~	~~76~~	~~77~~	~~78~~	79	~~80~~
~~81~~	~~82~~	83	~~84~~	~~85~~	~~86~~	~~87~~	~~88~~	89	~~90~~
~~91~~	~~92~~	~~93~~	~~94~~	~~95~~	~~96~~	97	~~98~~	~~99~~	~~100~~

Table 7.2 The Sieve of Eratosthenes.

next remaining number, the prime 7. Since $2 \cdot 7$, $3 \cdot 7$, $4 \cdot 7$, $5 \cdot 7$, and $6 \cdot 7$ have been crossed off already, the first multiple of 7 to be crossed off here is $7 \cdot 7$, or 49.

Now, the next prime left is 11. In attempting to cross off all multiples of 11, we see that $2 \cdot 11, 3 \cdot 11, 4 \cdot 11, \ldots, 9 \cdot 11$ have already been crossed off, and all other multiples of 11 exceed 100. Therefore, all numbers that remain on the list must be prime. Thus, we have found all prime numbers less than 100, namely

$$2, 3, 5, 7, 11, 13, 17, 19, 23, 29, 31, 37, 41,$$
$$43, 47, 53, 59, 61, 67, 71, 73, 79, 83, 89, 97$$

In general, if n is the prime number being considered, then the first multiple of n to be crossed off is n^2, and all numbers less than n^2 remaining on the list are prime. (The reason this works was discussed in Section 7.2, just before Example 14.) Perhaps now you can see why this process is called a "sieve"; the prime numbers are just sifted out as the multiples fall away.

If we examine the list of primes less than 100, some interesting patterns appear. First, it is apparent that 2 is the only even prime. Next, notice that 3, 5, and 7 form a set of three consecutive odd primes. This is called a **prime triplet**. It can be shown that no other prime triplet exists. (Exercise 17 will help you see why this is true.) The list also contains several pairs of consecutive odd primes, such as 11 and 13, 17 and 19, and 41 and 43. Such pairs are called **twin primes**. Other twin primes are 101 and 103 and also 137 and 139. It is natural to ask whether the set of twin primes is finite or infinite; the answer to this question is not yet known. A more basic question is whether the set of primes itself is finite or infinite. Euclid proved that this set is infinite (around 300 B.C.); Exercises 29–31 address this question. Other interesting questions about prime numbers also appear in the exercises.

Factorizations

Our main interest in primes centers about the role they play in factoring composite numbers. In fact, this role is so important that the statement that describes it is called The Fundamental Theorem of Arithmetic. In order to state the theorem succinctly, we must first define some terms.

Definition A **factorization** of a whole number n is a product of two or more whole numbers that equals n. Two factorizations are said to be **the same** if the only difference between them is the order in which the factors appear or the number of 1s appearing as factors.

Example 4 $1 \cdot 60$, $2 \cdot 30$, $2 \cdot 3 \cdot 10$, and $2 \cdot 3 \cdot 5 \cdot 2$ all are different factorizations of 60, but $5 \cdot 4 \cdot 3$, $3 \cdot 4 \cdot 5$, and $1 \cdot 4 \cdot 3 \cdot 5$ are considered to be the same factorization. □

Example 5 If a factor appears more than once in a factorization, we can make good use of exponents. For instance, the factorization $2 \cdot 2 \cdot 2 \cdot 2 \cdot 3 \cdot 3 \cdot 5 \cdot 7 \cdot 7 \cdot 7$ can be expressed as

$$2^4 \cdot 3^2 \cdot 5 \cdot 7^3$$ □

By the very nature of their definitions, it is obvious that one fundamental difference between prime and composite numbers is the number of different factorizations they have. Since a prime has exactly two factors (1 and itself), it follows that every prime p has exactly one factorization, namely $1 \cdot p$. On the other hand, since any composite number has three or more distinct factors, it must have at least two different factorizations. For instance, $1 \cdot 6$ and $2 \cdot 3$ are two different factorizations of 6. However, *a composite number has only one factorization made up solely of primes*, which is called, naturally enough, a **prime factorization**. A rigorous proof of this fact is not at all easy to explain, but if you try a few examples you should be able to convince yourself that the statement itself is plausible. (It was proved by Carl Friedrich Gauss in 1801.) Much of what we will be discussing in the rest of this chapter and in some of the later ones will be based on this idea.

The Fundamental Theorem of Arithmetic:

Every composite number has exactly one prime factorization, provided that the order of the factors is ignored.

Example 6 The composite number 48 has many different factorizations, of which $2 \cdot 24$, $12 \cdot 2 \cdot 2$, and $2^2 \cdot 3 \cdot 4$ are only a few. The prime factorization of 48 is $2^4 \cdot 3$ (because all the factors are primes). □

Example 7 $2^2 \cdot 3 \cdot 55$ is not a prime factorization of 660 because 55 is not a prime. If we replace 55 by its prime factorization, $5 \cdot 11$, then we get the prime factorization of 660, namely $2^2 \cdot 3 \cdot 5 \cdot 11$. □

There is an algorithm for finding the prime factorization of a number n:

> Find the smallest prime that divides n. Divide n by this prime and check to see if the quotient is prime. If so, you are done. If not, repeat the process on the quotient until you get a quotient

that is prime. The prime numbers you divided by together with the final (prime) quotient make up the prime factorization of n.

(*Remember*: The largest prime you have to divide by is the largest one less than or equal to the square root of the number n; if no such prime works, then the number n is prime.)

Example 8 Find the prime factorization of 72:

$$\begin{array}{r|l} 2 & 72 \\ \hline 2 & 36 \\ \hline 2 & 18 \\ \hline 3 & 9 \\ \hline & 3 \end{array}$$

Thus, the prime factorization of 72 is $2 \cdot 2 \cdot 2 \cdot 3 \cdot 3$, or $2^3 \cdot 3^2$. □

Example 9 Find the prime factorization of 231:

$$\begin{array}{r|l} 3 & 231 \\ \hline 7 & 77 \\ \hline & 11 \end{array}$$

Thus, the prime factorization of 231 is $3 \cdot 7 \cdot 11$. □

Example 10 Find the prime factorization of 535:

$$\begin{array}{r|l} 5 & 535 \\ \hline & 107 \end{array}$$

Now, $7^2 = 49 < 107 < 121 = 11^2$, so we need only try to divide 107 by the primes 2, 3, 5, and 7. None of these is a factor of 107. Thus, we can conclude that 107 is a prime, so the prime factorization of 535 is $5 \cdot 107$. □

Example 11 Figure 7.1 shows a sixth-grade text page that presents prime factorizations. What property of numbers is referred to in the fifth paragraph? What part of that property is omitted? Why do you suppose the authors chose to do that? □

PROBLEM-SOLVING COMMENT

In Figure 7.1, notice how *diagrams* are used to clarify the process of finding prime factors. The "factor tree" shows the successive stages of the process clearly, emphasizing how different intermediate factorizations all lead to the same prime factorization. ◇

PRIME AND COMPOSITE NUMBERS

A number which is a multiple of 2 is an **even number.** Some even numbers are 0, 2, 4, 6, 8, 10, . . .

A number which is not a multiple of 2 is an **odd number.** Some odd numbers are 1, 3, 5, 7, 9, 11, . . .

A number which has only two factors, 1 and itself, is a **prime number.** The numbers 2, 3, 5, 7, 11, 13, 17, . . . are prime numbers.

A number which has more than two factors is called a **composite number.** The numbers 4, 6, 8, 9, 10, 12, 14, 15, . . . are composite numbers. Notice that 0 and 1 are neither prime nor composite.

Any composite number can be written as the product of prime numbers. This written form is called the **prime factorization** of a number.

To determine the prime factorization of a number such as 60, you can make a factor tree. First, write the number as the product of any two factors.

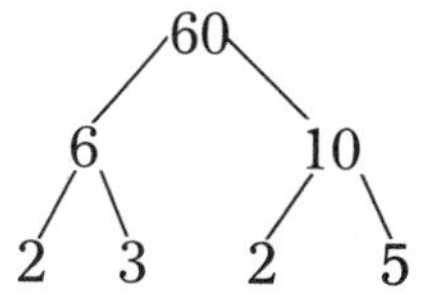

Then, if those factors are not prime numbers, write each of them as the product of any two factors.

Keep going until you have only prime numbers. All of these prime numbers are factors. The prime factorization of 60 is $2 \times 2 \times 3 \times 5$.

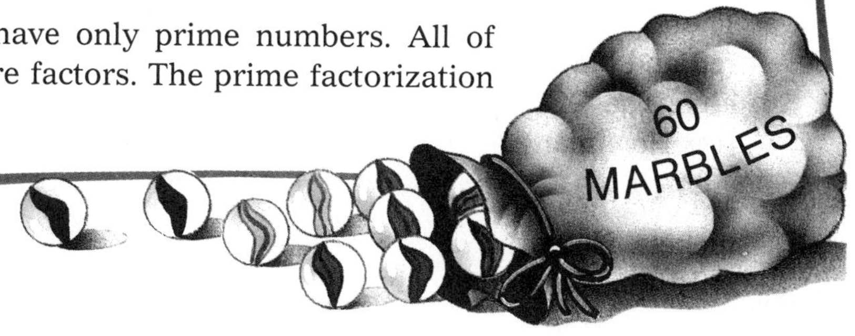

Factor Tree

60
6 10
2 3 2 5

60 MARBLES

Figure 7.1 HOUGHTON MIFFLIN MATHEMATICS, Grade 6, p. 218

Exercises 7.3

For Exercises 1–12, determine whether the given number is prime or composite. Find the prime factorization of each composite number.

1. 57	**2.** 149	**3.** 144
4. 133	**5.** 2261	**6.** 360
7. 18,900	**8.** 299	**9.** 701
10. 899	**11.** 511	**12.** 2047

13. (a) If you were to apply the Sieve of Eratosthenes to find all primes less than 1000, what would be the last prime number whose multiples would be crossed off the list?

(b) What would be the last (not the largest) number to be crossed off?

14. (a) Apply the Sieve of Eratosthenes to find all primes less than 300.

(b) Find a pair of twin primes between 101, 103 and 137, 139.

(c) How many pairs of twin primes are between 1 and 100? between 100 and 200? between 200 and 300?

15. What is the largest prime you must divide by to see if 5003 is a prime number?

16. (a) If n is a multiple of 3, what is the next multiple of 3?

(b) If $n - 1$ is a multiple of 3, what is the next multiple of 3?

(c) If $n + 1$ is a multiple of 3, what is the next multiple of 3?

17. Suppose that p is a prime greater than 3.

(a) Is p odd or even? Why? What about $p + 2$? What about $p + 4$?

(b) Can p be a multiple of 3? Why or why not?

(c) Explain why either $p - 1$ or $p + 1$ must be a multiple of 3.

(d) Explain why $p+2$ and $p+4$ cannot both be prime; that is, explain why {3, 5, 7} is the only set of prime triplets.

18. A teacher asks her sixth-grade class for the prime factorization of a certain large number. Alan says it is

$$2^4 \cdot 3^2 \cdot 7^3 \cdot 11 \cdot 13^2$$

Beth says it is

$$2^4 \cdot 3^3 \cdot 7^2 \cdot 11^2 \cdot 13$$

and Carl says it is

$$3^3 \cdot 2^2 \cdot 7^2 \cdot 2^2 \cdot 11^2 \cdot 13$$

(a) Can both Alan and Beth be right? Why or why not?

(b) Can both Beth and Carl be right? Why or why not?

(c) Can both Alan and Carl be right? Why or why not?

Christian Goldbach (1690–1764), a Prussian number theorist, proposed that every even number except 2 can be expressed as the sum of two prime numbers. This statement is known as **Goldbach's Conjecture** and is as yet unproved. For Exercises 19–27, express the given even number as the sum of two prime numbers (not necessarily different).

19. 6	**20.** 8	**21.** 10
22. 14	**23.** 16	**24.** 18
25. 20	**26.** 50	**27.** 100

28. (a) Which of the prime numbers less than 100 can be expressed in the form $2^p - 1$, where p is a prime? [Such numbers are called **Mersenne primes**; they are named after the French monk-mathematician-musician Marin Mersenne (1588–1648).]

(b) What is the numeral form of a Mersenne number in base two?

Exercises 29–31 show why there are infinitely many primes.

29. Determine whether each of these numbers is prime or composite. If composite, specify its prime factorization.

(a) $2 \cdot 3 - 1$

(b) $2 \cdot 3 \cdot 5 - 1$

(c) $2 \cdot 3 \cdot 5 \cdot 7 - 1$

(d) $2 \cdot 3 \cdot 5 \cdot 7 \cdot 11 - 1$

30. In general, if p is a prime greater than 2, explain why $(2 \cdot 3 \cdot 5 \ldots p) - 1$ is either a prime or else a composite number whose prime factorization contains a prime greater than p. (See Exercise 29.)

31. Use Exercise 30 to prove that, if p is a prime greater than 2, you can always find a prime greater than p. (This explains why there is no largest prime number.)

7.4 Common Factors and Multiples

While examining the factors of numbers in the previous section, you probably noticed that many numbers share some of the same factors. The factors that numbers have in common can tell us a lot about relationships among these numbers.

Definition A number d is a **common factor** of two (or more) whole numbers if d is a factor of both (or all) of the numbers.

Example 1 3 is a common factor of 6 and 15, but not of 6 and 8. □

Example 2 2 is a common factor of 6, 8, and 12, but 3 is not (because 3 is not a factor of 8). □

The Greatest Common Factor

It is easy to see that 1 is a common factor of every pair of whole numbers; in fact, it is the smallest common factor of every pair. Of more interest to us is the *largest* factor two numbers have in common.

Definition The **greatest common factor** of two numbers a and b is the largest factor of both a and b. It is denoted by

$$\operatorname{gcf}(a, b)$$

Note: Since *factor* and *divisor* mean the same thing, many people refer to this as the *greatest common divisor* and use the notation $\gcd(a, b)$. In fact, this used to be the standard terminology in school. However, nowadays mathematics educators prefer to use *factor*, and elementary textbooks are gradually changing over to this terminology. It appears that the reason for this change is the desire to avoid two sources of possible confusion for students:

- The term *divisor* is used in teaching division, where remainders are allowed, but *divisor* in this factoring sense implies that there is no remainder.
- In teaching fractions, the "least common denominator" of two or more fractions is abbreviated as "lcd," where the "d" in this abbreviation means something quite different than the "d" in "gcd."

In this book we use the more modern terminology. However, you are likely to run into both terms in your teaching, so you should be aware that "gcf" and "gcd" mean the same thing.

Example 3 The common factors of 12 and 30 are 1, 2, 3, and 6, so the greatest common factor of 12 and 30 is 6. We write

$$\text{gcf}(12, 30) = 6$$ □

Example 4

$$\text{gcf}(50, 70) = 10$$

because the common factors of 50 and 70 are 1, 2, 5, and 10. □

Example 5 The notion of greatest common factor is applied to more than two numbers in the obvious way. For instance, since the common factors of 12, 21, and 30 are 1 and 3, we say that

$$\text{gcf}(12, 21, 30) = 3$$ □

A special case occurs when two numbers have no common factors other than 1; that is, when 1 is the greatest common factor of the numbers. This case is important enough to deserve a name of its own.

Definition Two whole numbers a and b are **relatively prime** if the greatest common factor of a and b is 1.

It is important to realize that the phrase "relatively prime" describes a relationship between *pairs* of numbers, whereas the term "prime" by itself denotes a property of an individual number.

Example 6 4 and 15 are relatively prime because the only factor they have in common is 1. (Notice that neither number is itself a prime.) □

Example 7 2 and 3 are relatively prime, because $\text{gcf}(2, 3) = 1$. Also, 2 and 9 are relatively prime. In fact, 2 is relatively prime to every odd number. More generally, any prime number is relatively prime to every number of which it is not a factor. (Do you see why?) □

One way to find the greatest common factor of two or more numbers is to find the set of all factors of each number separately, then take the intersection of these sets. The greatest common factor will be the largest number in this intersection. For instance, to find gcf(28, 70), we can determine the set of all factors of 28—

$$\{1, 2, 4, 7, 14, 28\}$$

—and the set of all factors of 70—

$$\{1, 2, 5, 7, 10, 14, 35, 70\}$$

Then the set of all common factors is

$$\{1,2,4,7,14,28\} \cap \{1,2,5,7,10,14,35,70\} = \{1,2,7,14\}$$

so $\text{gcf}(28, 70) = 14$. □

This is not a very efficient procedure, however. Imagine, for example, finding gcf(144, 360) by this method. You would have to begin by finding all 15 factors of 144 and all 24 factors of 360! Fortunately, there is a better way—one that relies on prime factorizations. To find gcf(144, 360), we begin by finding the prime factorizations of the two numbers:

$$144 = 2^4 \cdot 3^2 \qquad \text{and} \qquad 360 = 2^3 \cdot 3^2 \cdot 5$$

Clearly, there are two prime numbers that are common factors of 144 and 360, namely 2 and 3, so the prime factorization of gcf(144, 360) must also contain powers of 2 and 3. Now, the highest power of 2 common to 144 and 360 is 2^3. (The highest power common to the two numbers is the smaller of the two powers. 2^4 is a factor of 144, but not of 360.) Similarly, the highest power of 3 common to 144 and 360 is 3^2, so the product $2^3 \cdot 3^2$ must also be a common factor of 144 and 360. The prime number 5 is not a factor of 144, so it cannot occur in the prime factorization of gcf(144, 360), and no other prime divides either number. Thus,

$$\text{gcf}(144, 360) = 2^3 \cdot 3^2 = 72$$

The example just given illustrates a general algorithm that can be used to find the greatest common factor of two or more whole numbers a, b, c, ..., k:

Step 1. Find the prime factorization of each number.

Step 2. Determine all the primes that are common factors of the numbers $a, b, c, \ldots, k$. If there are no common prime factors, then $\text{gcf}(a, b, c, \ldots, k) = 1$.

Step 3. Form the product of the highest common powers of each prime found in Step 2; that product is the greatest common factor of $a, b, c, \ldots, k$. (*Remember*: The highest *common* power of a prime is the *smallest* power of that prime that occurs *in every one* of the factorizations of $a, b, c, \ldots, k$.)

Example 8 Find $\mathrm{gcf}(84, 98)$.

Step 1. $$84 = 2^2 \cdot 3 \cdot 7 \quad \text{and} \quad 98 = 2 \cdot 7^2$$

Step 2. Observe that 2 and 7 are the only common primes.

Step 3. $$\mathrm{gcf}(84, 98) = 2 \cdot 7 = 14$$ □

Example 9 Find $\mathrm{gcf}(a, b)$ for

$$a = 2^3 \cdot 3^2 \cdot 5^4 \cdot 11 \cdot 13^3 \quad \text{and} \quad b = 2^2 \cdot 5^3 \cdot 11^4 \cdot 13^5 \cdot 17^2$$

Step 1. The prime factorizations of a and b are already given.

Step 2. The common factors are 2, 5, 11, and 13.

Step 3. $$\mathrm{gcf}(a, b) = 2^2 \cdot 5^3 \cdot 11 \cdot 13^3$$ □

Example 10 Find $\mathrm{gcf}(105, 286)$.

Step 1. $$105 = 3 \cdot 5 \cdot 7 \quad \text{and} \quad 286 = 2 \cdot 11 \cdot 13$$

Step 2. There are no common primes, so $\mathrm{gcf}(105, 286) = 1$. □

Example 11 Find $\mathrm{gcf}(36, 54, 90)$.

Step 1. $$36 = 2^2 \cdot 3^2, \quad 54 = 2 \cdot 3^3, \quad 90 = 2 \cdot 3^2 \cdot 5$$

Step 2. The common primes are 2 and 3.

Step 3. 2^1 and 3^2 are the highest powers of 2 and 3 common to all three numbers, so

$$\mathrm{gcf}(36, 54, 90) = 2 \cdot 3^2 = 18$$ □

The Least Common Multiple

As we have seen, to find the greatest common factor of two or more numbers, we take the product of the smallest powers of the common prime factors. If, instead, we took the product of the *highest* powers of all the prime factors, we would get another number that has an important relation to the original numbers.

For instance, if we consider

$$144 = 2^4 \cdot 3^2 \quad \text{and} \quad 360 = 2^3 \cdot 3^2 \cdot 5$$

and take the product of the higher powers of all the prime factors, we get

$$2^4 \cdot 3^2 \cdot 5 = 720$$

Now,

$$720 = 5 \cdot 144 \quad \text{and} \quad 720 = 2 \cdot 360$$

Because 720 is a multiple of both the original numbers it is called a *common multiple* of 144 and 360. In fact, 720 is the smallest number that is a multiple of both 144 and 360. ($1 \cdot 360 = 360$ is not a multiple of 144, and $4 \cdot 144 = 576$ is not a multiple of 360.) Hence, we call 720 the *least common multiple* of 144 and 360. In general:

Definition A whole number is a **common multiple** of two (or more) numbers if it is a multiple of each of the numbers. The **least common multiple** of two numbers a and b is the smallest nonzero whole number that is a multiple of both a and b. It is denoted by

$$\operatorname{lcm}(a, b)$$

More generally, the **least common multiple** of the numbers a, b, c, ..., k is the smallest nonzero whole number that is a multiple of each of these numbers. It is written

$$\operatorname{lcm}(a, b, c, \ldots, k)$$

Example 12 All multiples of 6 are in the set

$$\{6, 12, 18, 24, \ldots\}$$

and all multiples of 8 are in the set

$$\{8, 16, 24, 32, \ldots\}$$

The set of all common multiples of 6 and 8 is

$$\{24, 48, 72, \ldots\}$$

so the least common multiple of 6 and 8 is 24; that is,

$$\operatorname{lcm}(6, 8) = 24$$

□

Example 13 The least common multiple of 3 and 5 is 15; in fact, the set of all common multiples of 3 and 5 is

$$\{15, 30, 45, \ldots\}$$

Similarly, $\operatorname{lcm}(9, 10) = 90$; the set of common multiples of 9 and 10 is

$$\{90, 180, 270, \ldots\}$$

□

Notice that the numbers in each pair of Example 13 are relatively prime and their least common multiple is just the product of the two numbers.

This is true in general:

$$\text{Whenever } a \text{ and } b \text{ are relatively prime, } \operatorname{lcm}(a,b) = a \cdot b.$$

(Can you see why this is always true?)

There is also an efficient way to find the least common multiple of two or more whole numbers $a, b, c, \ldots, k$ by using the prime factorizations of the numbers:

Step 1. Find the prime factorization of each number.

Step 2. List all the primes that appear in any one (or more) of the prime factorizations.

Step 3. Form the product of the *highest* powers of each prime listed in Step 2; that product is the least common multiple of $a, b, c, \ldots, k$.

Example 14 Find $\operatorname{lcm}(14, 35)$.

Step 1. $$14 = 2 \cdot 7 \quad \text{and} \quad 35 = 5 \cdot 7$$

Step 2. The primes appearing in at least one factorization are 2, 5, and 7.

Step 3. Each prime listed in Step 2 occurs only to the first power; thus,

$$\operatorname{lcm}(14, 35) = 2^1 \cdot 5^1 \cdot 7^1 = 70$$ □

Example 15 Find $\operatorname{lcm}(36, 54, 90)$.

Step 1. $$36 = 2^2 \cdot 3^2, \quad 54 = 2 \cdot 3^3, \quad 90 = 2 \cdot 3^2 \cdot 5$$

Step 2. The primes appearing in at least one factorization are 2, 3, and 5.

Step 3. The highest powers of these primes are:

2^2 (in the factorization of 36)
3^3 (in the factorization of 54)
5^1 (in the factorization of 90)

Thus,

$$\operatorname{lcm}(36, 54, 90) = 2^2 \cdot 3^3 \cdot 5 = 540$$ □

Example 16 Find $\operatorname{lcm}(105, 286)$.

Step 1. $$105 = 3 \cdot 5 \cdot 7 \quad \text{and} \quad 286 = 2 \cdot 11 \cdot 13$$

Step 2. The primes appearing are 2, 3, 5, 7, 11, and 13.

Step 3. $$\operatorname{lcm}(105, 286) = 2 \cdot 3 \cdot 5 \cdot 7 \cdot 11 \cdot 13 = 30{,}030$$ □

Example 17 Find $\operatorname{lcm}(a, b)$ for

$$a = 2^3 \cdot 3^2 \cdot 5^4 \cdot 11 \cdot 13^3 \quad \text{and } b = 2^2 \cdot 5^3 \cdot 11^4 \cdot 13^5 \cdot 17^2$$

Step 1. The prime factorizations of a and b are already given.

Step 2. The primes that appear are 2, 3, 5, 11, 13, and 17.

Step 3. $$\operatorname{lcm}(a, b) = 2^3 \cdot 3^2 \cdot 5^4 \cdot 11^4 \cdot 13^5 \cdot 17^2$$ □

Example 18 Find both gcf (1500, 12,600) and lcm (1500, 12,600).

$$1500 = 2^2 \cdot 3 \cdot 5^3 \quad \text{and} \quad 12{,}600 = 2^3 \cdot 3^2 \cdot 5^2 \cdot 7$$

To find the gcf, we form the product of the *smallest* powers of the *common* prime factors. Thus,

$$\operatorname{gcf}(1500, 12{,}600) = 2^2 \cdot 3 \cdot 5^2 = 300$$

To find the lcm, we form the product of the *largest* powers of the primes that appear in *at least one* of the prime factorizations. Thus,

$$\operatorname{lcm}(1500, 12{,}600) = 2^3 \cdot 3^2 \cdot 5^3 \cdot 7 = 63{,}000$$ □

Example 19 Figure 7.2 shows a presentation of least common multiples to sixth graders. □

As we shall see in Chapter 10, the concepts of greatest common factor and least common multiple are very important in the discussion of fractions.

Exercises 7.4

1. (a) Find all the factors of 24.

(b) Find all the factors of 36.

(c) Find all the common factors of 24 and 36.

(d) Use the answer to part (c) to find gcf (24, 36).

(e) Use prime factorizations to find gcf (24, 36).

(f) Suppose you have just given your students the definition of gcf and you want them to "discover" how to find the gcf of two numbers using the technique outlined in this exercise. What helpful problem-solving questions might you ask them? Would the same questions help them discover how to find the lcm by the method outlined in Exercise 3?

2. (a) Find all the factors of 135.

(b) Find all the factors of 225.

(c) Find all the common factors of 135 and 225.

(d) Use the answer to part (c) to find gcf (135, 225).

(e) Use prime factorizations to find gcf (135, 225).

Multiples and Least Common Multiple

Wei's mother told him to box up part of his magazine collection and stack the boxes on shelves. The boxes are about 20 cm high and the holes for the shelf brackets are 6 cm apart. How far apart should Wei put the brackets and shelves if he wants to waste no space?

List the numbers that you can get by multiplying 20 and 6 by counting numbers. These are the *multiples* of 20 and of 6.

20 40 60 80 100 120

6 12 18 24 30 36 42 48 54 60 66

The number 60 is a *common multiple* of 20 and 6. So, if Wei puts the shelves 60 cm apart, he can stack three boxes and use all 60 cm of space.

Other common multiples of the two numbers 20 and 6 are 120, 180, and 360. The *least common multiple* of the two numbers is 60.

Zero and all multiples of 2 are called *even numbers*. Counting numbers that are not multiples of 2 are called *odd numbers*.

Try

a. List the first 6 multiples of 3.

b. Find 2 common multiples of 2 and 6.

c. Find the least common multiple of 3 and 5.

Figure 7.2 INVITATION TO MATHEMATICS, Grade 6, p. 54

3. (a) Find the first ten multiples of 12.

(b) Find the first ten multiples of 15.

(c) Find the first five common multiples of 12 and 15.

(d) Use the answer to part (c) to find lcm(12, 15).

(e) Use prime factorizations to find lcm(12, 15).

For Exercises 4–9, find two specific numbers a and b that satisfy the given conditions.

4. a and b are relatively prime, a is prime, and b is prime.

5. a and b are relatively prime, a is prime, and b is composite.

6. a and b are relatively prime, and both a and b are composite.

7. a and b are not relatively prime, and both a and b are prime.

8. a and b are not relatively prime, a is prime, and b is composite.

9. a and b are not relatively prime, and both a and b are composite.

For Exercises 10–15:

(a) Use prime factorizations to find both the gcf and the lcm of the given pair of numbers.

(b) Indicate which pairs are relatively prime and which are not.

10. 18 and 45

11. 162 and 60

12. 140 and 150

13. 6 and 35

14. 60 and 77

15. 550 and 567

For Exercises 16–21, use prime factorizations to find the gcf and the lcm of the given set of numbers.

16. 25, 35, and 45

17. 60, 162, and 360

18. 140, 150, and 160

19. 2100, 3960, and 11,880

20. 105, 165, and 1001

21. 24, 28, 42, and 48

In Exercises 22 and 23, the prime factorizations of two whole numbers m and n are given. Find $\text{gcf}(m,n)$ and $\text{lcm}(m,n)$. Leave your answers in prime factorization form.

22. $m = 2^5 \cdot 3^4 \cdot 5^8 \cdot 11^4$
$n = 2^2 \cdot 3^5 \cdot 5^6 \cdot 13$

23. $m = 3^{23} \cdot 5^7 \cdot 29^3 \cdot 31$
$n = 3^{21} \cdot 7^5 \cdot 31^3$

24. (a) For each of the following pairs of whole numbers a and b, calculate $\text{gcf}(a,b)$, $\text{lcm}(a,b)$, and $a \cdot b$.

i. $a = 2$, $b = 3$

ii. $a = 3$, $b = 4$

iii. $a = 6$, $b = 9$

iv. $a = 12$, $b = 13$

v. $a = 12$, $b = 10$

vi. $a = 100$, $b = 120$

(b) Use your answers to part (a) to form a conjecture about the relationship among $\text{gcf}(a,b)$, $\text{lcm}(a,b)$, and $a \cdot b$. Test your conjecture using $a = 24$ and $b = 36$.

In Exercises 25–31, let a, b, and n stand for arbitrary nonzero whole numbers. Make a conjecture about the answer to the given question; then use specific numbers to test your conjecture.

25. $\text{gcf}(a,a) = ?$

26. $\text{lcm}(a,a) = ?$

27. $\text{gcf}(1,a) = ?$

28. $\text{lcm}(1,a) = ?$

29. $\text{gcf}(a,an) = ?$

30. $\text{lcm}(a,an) = ?$

31. If $\text{lcm}(a,b) = ab$, then $\text{gcf}(a,b) = ?$

32. A restaurant cashier has 156 quarters and 180 half-dollars. If the cashier makes stacks of quarters and stacks of half-dollars so that each stack contains the same number of coins, what is the maximum number of coins each stack can contain?

33. (a) What would the term *least common factor* mean? Do you think it would be a useful concept? Why or why not?

(b) What would the term *greatest common multiple* mean? Do you think it would be a useful concept? Why or why not?

7.5 Clock Arithmetic

This section introduces a family of "small" number systems. Each one contains only finitely many numbers, but still has most of the same "nice" properties as the infinite number system **W** of whole numbers. In fact, some of these finite systems have useful properties that do not exist in **W**. These small number systems are used in the elementary grades to provide simple examples of arithmetic properties because they are easy to build and work with; yet they are unfamiliar enough to force students to focus on the definitions of the properties, rather than depend on their general notions about numbers. These systems make up what is called **clock arithmetic** or **remainder arithmetic** (for reasons that will become obvious), or, more formally, **modular arithmetic**.

Let us begin by considering a clock with a round face and the twelve numerals 1, 2, 3, ..., 12 representing the hours. (See Figure 7.3.) To simplify matters, we shall ignore minutes altogether and suppose that our clock has only an hour hand, which moves from one numeral to the next exactly on the hour and doesn't show the time between hours. If we start our clock at midnight on 12, then nine hours later the hand will point to 9. Seven hours after that, however, the hand will point to 4, because once the hand comes around to 12 (noon) it "starts over," so to speak. Thus, no matter how long the clock runs, it will show only a time between 1 and 12. If it runs for 27 hours, the hand will point to 3; if it runs for 31 hours, the hand will point to 7; if it runs for 48 hours, the hand will point to 12.

Modular Congruence

You have known all this, of course, ever since you first learned to tell time. Most children are familiar with these ideas by second grade or so. That familiarity makes clock arithmetic a useful tool for teaching arithmetic concepts in the upper elementary grades. If we make one small adjustment in the clock of Figure 7.3, then we can describe clock arithmetic concisely in a way that allows us to link up easily with number-system properties. In this way, a commonplace object familiar to elementary-school students becomes a vehicle for teaching them about the abstract properties of number systems. The adjustment is this: Replace the 12 on the clock by a 0, as in Figure 7.4. (You might say that the clock begins running at "zero hour.")

Now all hours that are multiples of 12 will show as "0" on the clock, but all the other times will appear just as before. The value of this change is that the clock will show the time of any number of hours as the remainder when that number is divided by 12. In other words, the clock ignores all

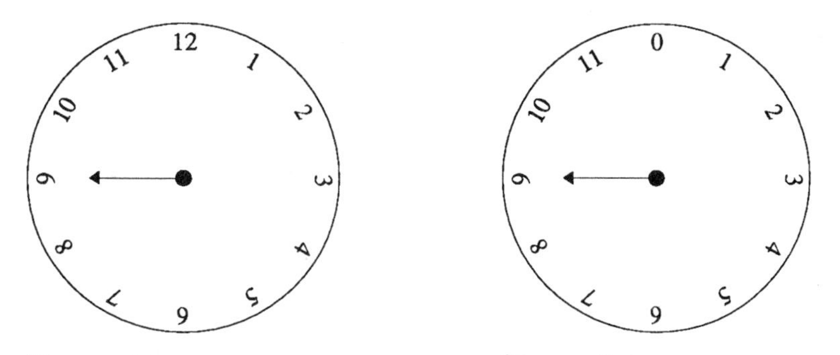

Figure 7.3
A twelve-hour clock.

Figure 7.4
Another twelve-hour clock.

multiples of 12 hours and shows only how many "extra" hours are in the time interval. Thus,

$$27 \text{ shows as } 3 \text{ because } 27 = 12 \cdot 2 + 3; \text{ that is, } 27 - 3 = 12 \cdot 2$$
$$43 \text{ shows as } 7 \text{ because } 43 = 12 \cdot 3 + 7; \text{ that is, } 31 - 7 = 12 \cdot 3$$
$$48 \text{ shows as } 0 \text{ because } 48 = 12 \cdot 4 + 0; \text{ that is, } 48 - 0 = 12 \cdot 4$$
$$9 \text{ shows as } 9 \text{ because } 9 = 12 \cdot 0 + 9; \text{ that is, } 9 - 9 = 12 \cdot 0$$

The clock face shows all the possible remainders—

$$0, 1, 2, \ldots, 11$$

—and all numbers that leave the same remainder when divided by 12 look the same on the clock. In the language of Chapter 3, our clock defines an equivalence relation on the set of whole numbers, with the twelve possible remainders representing the equivalence classes of the partition. The relation itself is usually called *congruence modulo 12*, a fancy title for a simple idea.

Definition Two whole numbers a and b are said to be **congruent modulo 12** if there is some whole number n such that either $a - b = 12n$ or $b - a = 12n$. We write this as

$$a \equiv b \pmod{12}$$

Example 1 $18 \equiv 6 \pmod{12}$ because $18 - 6 = 12 \cdot 1$. (That is, 18 equals $12 \cdot 1 + 6$.) 18 is also congruent modulo 12 to

$$30 \text{ because } 30 - 18 = 12 \cdot 1$$
$$42 \text{ because } 42 - 18 = 12 \cdot 2$$
$$54 \text{ because } 54 - 18 = 12 \cdot 3$$

□

Example 2 Any number is congruent modulo 12 to the remainder it leaves when divided by 12. For instance,

$$25 = 12 \cdot 2 + 1, \text{ so } 25 - 1 = 12 \cdot 2; \quad 25 \equiv 1 \pmod{12}$$
$$64 = 12 \cdot 5 + 4, \text{ so } 64 - 4 = 12 \cdot 5; \quad 64 \equiv 4 \pmod{12}$$
$$72 = 12 \cdot 6 + 0, \text{ so } 72 - 0 = 12 \cdot 6; \quad 72 \equiv 0 \pmod{12}$$

□

The mention of equivalence classes may remind you of the construction of the whole numbers (equivalence classes of sets). We are again constructing a number system from equivalence classes, this time classes of whole numbers. As we have just seen, the relation *congruence modulo 12* partitions the whole numbers into twelve classes, each one determined by a specific remainder under division by 12. We can name each of these classes by the remainder itself, which is the smallest number in its class. To distinguish between these two uses of the remainder numerals, we shall underline them when they are being used to represent the equivalence classes. Thus, the partition of **W** formed by *congruence modulo 12* looks like this:

$$\underline{0} = \{0, 12, 24, 36, 48, \dots\}$$
$$\underline{1} = \{1, 13, 25, 37, 49, \dots\}$$
$$\underline{2} = \{2, 14, 26, 38, 50, \dots\}$$
$$\underline{3} = \{3, 15, 27, 39, 51, \dots\}$$
$$\vdots$$
$$\underline{11} = \{11, 23, 35, 47, 59, \dots\}$$

The numerals appearing on the clock represent these equivalence classes. The numbers in each class represent the hours when the hand points to the numeral that represents the class. Thus, the hand points to "3" at 3 hours after its midnight (0 hour) start, and again 15 hours after its start, and again 27 hours after its start. From the clock's point of view, then, there are only twelve different times and, as the hours pass, those times just cycle around again and again.

Notation The set of equivalence classes of whole numbers formed by *congruence modulo 12* is denoted by $\mathbf{W}_{12}$. Thus,

$$\mathbf{W}_{12} = \{\underline{0}, \underline{1}, \underline{2}, \underline{3}, \dots, \underline{11}\}$$

Our interest in this set stems from the fact that, with the clock's help, we can turn it into a well-behaved number system with most of the properties of **W**. Its small size and easy construction, but slightly curious behavior (as we shall see), makes this system almost ideal for teaching children in the upper elementary grades about the properties of arithmetic.

Before we turn $\mathbf{W}_{12}$ into a number system (by defining arithmetic operations on it), let us observe that there is nothing "magic" about the number 12 here. $\mathbf{W}_{12}$ is only one of a large family of finite number systems; twelve was chosen because our everyday clocks are twelve-hour clocks. If we had used a five-hour clock or a twenty-four-hour clock we could have defined the relations *congruence modulo 5* or *congruence modulo 24* and the sets $\mathbf{W}_5$ or $\mathbf{W}_{24}$ in just the same way.

Definition Let m be any whole number greater than 1. Two whole numbers a and b are **congruent modulo** m if there is a whole number n such that either $a - b = m \cdot n$ or $b - a = m \cdot n$. In this case we write $a \equiv b \pmod{m}$. The number m is called the **modulus** of the congruence.

Example 3

$$17 \equiv 2 \pmod{5} \quad \text{because} \quad 17 - 2 = 5 \cdot 3$$

Put another way, $17 = 5 \cdot 3 + 2$, so 17 has a remainder of 2 when divided by 5. 17 is also congruent modulo 5 to

$$\begin{aligned} 7 &\text{ because } 17 - 7 = 5 \cdot 2 \\ 12 &\text{ because } 17 - 12 = 5 \cdot 1 \\ 17 &\text{ because } 17 - 17 = 5 \cdot 0 \\ 22 &\text{ because } 22 - 17 = 5 \cdot 1 \end{aligned}$$

and so forth. □

Example 4 $\mathbf{W}_5$ is $\{\underline{0}, \underline{1}, \underline{2}, \underline{3}, \underline{4}\}$. The equivalence classes are

$$\begin{aligned} \underline{0} &= \{0, 5, 10, 15, 20, \ldots\} \\ \underline{1} &= \{1, 6, 11, 16, 21, \ldots\} \\ \underline{2} &= \{2, 7, 12, 17, 22, \ldots\} \\ \underline{3} &= \{3, 8, 13, 18, 23, \ldots\} \\ \underline{4} &= \{4, 9, 14, 19, 24, \ldots\} \end{aligned}$$

This system shows up often in elementary-school texts. Its small size and its relation to "counting by fives" make it easy to work with. □

Example 5

$$\begin{aligned} 37 &\equiv 7 \pmod{10} && \text{because } 37 - 7 = 10 \cdot 3. && (37 = 10 \cdot 3 + 7) \\ 37 &\equiv 1 \pmod{6} && \text{because } 37 - 1 = 6 \cdot 6. && (37 = 6 \cdot 6 + 1) \\ 37 &\equiv 2 \pmod{7} && \text{because } 37 - 2 = 7 \cdot 5. && (37 = 7 \cdot 5 + 2) \\ 37 &\equiv 1 \pmod{3} && \text{because } 37 - 1 = 3 \cdot 12. && (37 = 3 \cdot 12 + 1) \\ 37 &\equiv 5 \pmod{8} && \text{because } 37 - 5 = 8 \cdot 4. && (37 = 8 \cdot 4 + 5) \end{aligned}$$

□

Example 6 $\mathbf{W}_4 = \{\underline{0}, \underline{1}, \underline{2}, \underline{3}\}$, where

$$\underline{0} = \{0, 4, 8, 12, 16, \ldots\}$$
$$\underline{1} = \{1, 5, 9, 13, 17, \ldots\}$$
$$\underline{2} = \{2, 6, 10, 14, 18, \ldots\}$$
$$\underline{3} = \{3, 7, 11, 15, 19, \ldots\}$$

Similarly, $\mathbf{W}_7 = \{\underline{0}, \underline{1}, \underline{2}, \underline{3}, \underline{4}, \underline{5}, \underline{6}\}$, where

$$\underline{0} = \{0, 7, 14, 21, 28, \ldots\}$$
$$\underline{1} = \{1, 8, 15, 22, 29, \ldots\}$$
$$\underline{2} = \{2, 9, 16, 23, 30, \ldots\}, \quad \text{etc.}$$ □

Example 7 The textbook page in Figure 7.5 illustrates how clock-arithmetic systems appear in the upper elementary-school curriculum as a form of enrichment. □

Modular Arithmetic

So far, $\mathbf{W}_m$ for any m is just a finite set (of equivalence classes of whole numbers). We call the elements "numbers" because they look like numbers; but to turn $\mathbf{W}_m$ into a number *system*, we must define addition and multiplication in such a way that they are operations on that finite set. That is, the sum and the product of any two elements of $\mathbf{W}_m$ must be back in $\mathbf{W}_m$. There is an easy, natural way to do this:

> Given any two elements of $\mathbf{W}_m$, add or multiply them as if they were ordinary whole numbers; then use as the answer the class that the resulting sum or product is in.

Notation We distinguish operations in $\mathbf{W}_m$ from the corresponding whole-number operations by putting circles around the symbols for the "new" ones.

Definitions Let a and b be whole numbers and let $\underline{a}$ and $\underline{b}$ be their respective equivalence classes in $\mathbf{W}_m$.

The **sum** of $\underline{a}$ and $\underline{b}$ is $\underline{r}$, where r is the remainder obtained from dividing $a + b$ by m. We write

$$\underline{a} \oplus \underline{b} = \underline{r}$$

The **product** of $\underline{a}$ and $\underline{b}$ is $\underline{s}$, where s is the remainder obtained from dividing $a \cdot b$ by m. We write

$$\underline{a} \odot \underline{b} = \underline{s}$$

the two equivalence classes will yield the same sum and product. Some examples of this follow.

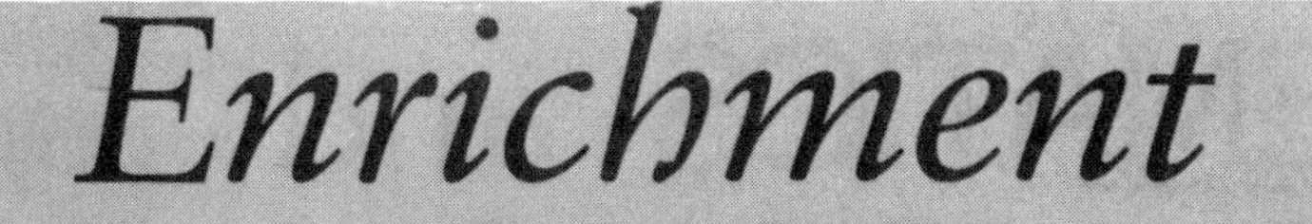

You are familiar with the 12-hour clock. Did you know that the numbers on the 12-hour clock form a system of numbers used in a special kind of arithmetic?

To find the sum of two clock numbers, think of the hour hand moving in a *clockwise* direction. For example, 5 h after 9 o'clock is 2 o'clock. We can express this sum using an equation. $9 + 5 \overset{12}{=} 2$

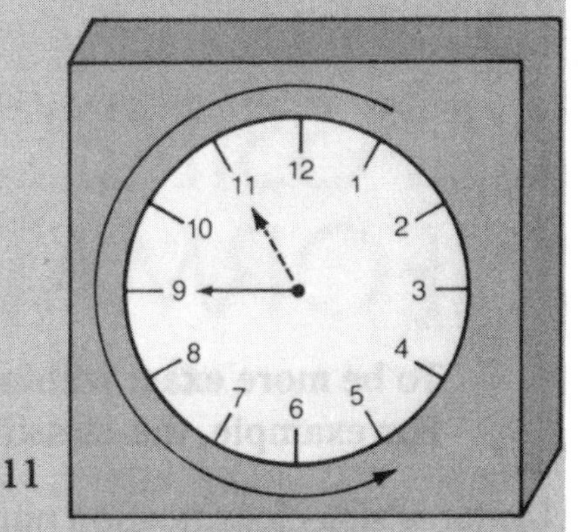

We show that this is a sum of clock numbers on a 12-hour clock by writing 12 over the equals sign.

To find the difference between two clock numbers, think of the hour hand moving in a *counterclockwise* direction. For example, 10 h before 9 o'clock is 11 o'clock. You can write an equation to show the difference. $9 - 10 \overset{12}{=} 11$

CLOCK ARITHMETIC

Complete.

1. 6 h after 2 o'clock is ■ o'clock. $2 + 6 \overset{12}{=}$ ■

2. 7 h before 10 o'clock is ■ o'clock. $10 - 7 \overset{12}{=}$ ■

3. 8 h before 2 o'clock is ■ o'clock. $2 - 8 \overset{12}{=}$ ■

4. 12 h after 5 o'clock is ■ o'clock. $5 + 12 \overset{12}{=}$ ■

Figure 7.5 HOUGHTON MIFFLIN MATHEMATICS, Grade 7, p. 56

Note: From here on, we confine our examples to the twelve-hour clock system $\mathbf{W}_{12}$ and the conveniently small system $\mathbf{W}_5$. Other systems are explored in the exercises.

Example 8 In $\mathbf{W}_5$:

$\underline{3} \oplus \underline{4} = \underline{2}$ because $3 + 4 = 7$ in $\mathbf{W}$, and $7 \equiv 2 \pmod 5$. Now, 8 is in $\underline{3}$ because $8 = 5 \cdot 1 + 3$, and 14 is in $\underline{4}$ because $14 = 5 \cdot 2 + 4$. Notice that, if we use 8 and 14 to represent the classes $\underline{3}$ and $\underline{4}$, we get the same sum:

$$8 + 14 = 22 = 4 \cdot 5 + 2 \equiv 2 \pmod 5$$

Other examples of sums:

$$\underline{3} \oplus \underline{3} = \underline{1} \text{ because } 3 + 3 = 6 \equiv 1 \pmod 5$$
$$\underline{3} \oplus \underline{2} = \underline{0} \text{ because } 3 + 2 = 5 \equiv 0 \pmod 5$$

$\underline{3} \odot \underline{4} = \underline{2}$ because $3 \cdot 4 = 12$ in $\mathbf{W}$, and $12 \equiv 2 \pmod 5$. Using 8 and 14 as representatives of $\underline{3}$ and $\underline{4}$, we still get $\underline{2}$:

$$8 \cdot 14 = 112 = 22 \cdot 5 + 2 \equiv 2 \pmod 5$$

Other examples of products:

$$\underline{3} \odot \underline{3} = \underline{4} \text{ because } 3 \cdot 3 = 9 \equiv 4 \pmod 5$$
$$\underline{3} \odot \underline{2} = \underline{1} \text{ because } 3 \cdot 2 = 6 \equiv 1 \pmod 5$$

□

Example 9 In $\mathbf{W}_{12}$:

$\underline{9} \oplus \underline{7} = \underline{4}$ because $9 + 7 = 16$ in $\mathbf{W}$, and $16 \equiv 4 \pmod{12}$. In clock-arithmetic language, this says that 7 hours after 9 o'clock the hour hand will point to 4. If we use 21 $(= 12 \cdot 1 + 9)$ and 67 $(= 12 \cdot 5 + 7)$ to represent the classes $\underline{9}$ and $\underline{7}$, we still get a sum of $\underline{4}$:

$$21 + 67 = 88 = 7 \cdot 12 + 4 \equiv 4 \pmod{12}$$

Other sums:

$$\underline{10} \oplus \underline{10} = \underline{8} \text{ because } 10 + 10 = 20 \equiv 8 \pmod{12}$$
$$\underline{5} \oplus \underline{6} = \underline{11} \text{ because } 5 + 6 = 11 \equiv 11 \pmod{12}$$
$$\underline{11} \oplus \underline{1} = \underline{0} \text{ because } 11 + 1 = 12 \equiv 0 \pmod{12}$$

$\underline{7} \odot \underline{5} = \underline{11}$ because $7 \cdot 5 = 35$, and $35 \equiv 11 \pmod{12}$. If we use 19 and 29 to represent the classes $\underline{7}$ and $\underline{5}$, respectively, we still get a product of $\underline{11}$:

$$19 \cdot 29 = 551 = 45 \cdot 12 + 11 \equiv 11 \pmod{12}$$

Other products:

$$\underline{2} \odot \underline{11} = \underline{10} \text{ because } 2 \cdot 11 = 22 \equiv 10 \pmod{12}$$
$$\underline{3} \odot \underline{2} = \underline{6} \text{ because } 3 \cdot 2 = 6 \equiv 6 \pmod{12}$$
$$\underline{8} \odot \underline{3} = \underline{0} \text{ because } 8 \cdot 3 = 24 \equiv 0 \pmod{12}$$

□

Let us examine these modular systems in relation to the properties of the whole-number system. A list of these properties appears at the end of Section 4.5. You might find it helpful to look back at that before going on.

The definitions of $\oplus$ and $\odot$ on $\mathbf{W}_m$ for any modulus m insure that these two processes are indeed operations; that is, the sum and product of any two elements of $\mathbf{W}_m$ are always elements of $\mathbf{W}_m$. Commutativity follows immediately from the commutativity of the corresponding operations on $\mathbf{W}$. It is also true (but not quite as obvious) that both operations are associative and that $\odot$ is distributive over $\oplus$. (See Exercises 28–30.) Moreover, it is easy to see that $\underline{0}$ is always the additive identity and $\underline{1}$ is always the multiplicative identity.

However, one aspect of the structure of $\mathbf{W}$ is missing in every $\mathbf{W}_m$—there is no useful order relation. That is, it is not possible to define a relation that is transitive, satisfies the Trichotomy Law, and preserves addition in $\mathbf{W}_m$. Consider the following example:

> Suppose there were such an order relation $\otimes$ on $\mathbf{W}_{12}$. Because $\underline{1}$ and $\underline{7}$ are unequal, one would have to be "less than" the other, by the Trichotomy Law. Assume $\underline{1} \otimes \underline{7}$. (The argument works either way.) Trichotomy implies that $\underline{7}$ is not less than $\underline{1}$. But $\otimes$ preserves addition, so we must have
>
> $$\underline{1} \oplus \underline{6} \quad \otimes \quad \underline{7} \oplus \underline{6}$$
>
> That is, $\underline{7} \otimes \underline{1}$, which is a contradiction.

Similar examples can be constructed for the other systems. (Transitivity is needed in the argument for systems of odd modulus; try constructing an example in $\mathbf{W}_5$ analogous to the previous one in $\mathbf{W}_{12}$.) Thus, it makes no sense to say that $\underline{1}$ is less than $\underline{2}$ (or $\underline{2}$ is less than $\underline{1}$) in any modular system, even though these numbers behave much like their namesakes in $\mathbf{W}$ when it comes to adding and multiplying. Of course, if you think of these systems in terms of time and clocks, this is not as strange as it first appears. A clock shows 2 o'clock after 1 o'clock, but then comes around to 1 o'clock again.

One particularly nice feature of the clock-arithmetic systems is that, even though there are no negative numbers, subtraction always works in every one of them. Moreover, despite the absence of fractions, there are many clock-arithmetic systems in which division always works. We postpone the discussion of these ideas until the next chapter, after we have introduced some useful terminology and notation.

As you work through the exercises for this section, notice how the computations, though unfamiliar, are easy once you get used to them. Moreover, they provide lots of practice in working with division and remainders. These features provide an interesting setting (if you present it well) for giving children arithmetic practice beyond mere repetition of routine drills. Moreover, oddities such as the order problem make the clock-arithmetic systems valuable teaching aids when you are trying to get (older) children to be careful in the ways they reason about numbers.

Exercises 7.5

For Exercises 1–16, complete the given congruence statement by filling in a whole number smaller than the modulus.

1. $19 \equiv$ ____ (mod 12)

2. $19 \equiv$ ____ (mod 5)

3. $68 \equiv$ ____ (mod 12)

4. $68 \equiv$ ____ (mod 5)

5. $35 \equiv$ ____ (mod 12)

6. $35 \equiv$ ____ (mod 5)

7. $72 \equiv$ ____ (mod 12)

8. $72 \equiv$ ____ (mod 5)

9. $25 \equiv$ ____ (mod 3)

10. $25 \equiv$ ____ (mod 4)

11. $25 \equiv$ ____ (mod 6)

12. $25 \equiv$ ____ (mod 7)

13. $25 \equiv$ ____ (mod 10)

14. $25 \equiv$ ____ (mod 15)

15. $25 \equiv$ ____ (mod 25)

16. $25 \equiv$ ____ (mod 30)

For Exercises 17–22, list the smallest five numbers in the given equivalence class.

17. $\underline{7}$ in $\mathbf{W}_{12}$ **18.** $\underline{10}$ in $\mathbf{W}_{12}$

19. $\underline{2}$ in $\mathbf{W}_3$ **20.** $\underline{1}$ in $\mathbf{W}_{10}$

21. $\underline{6}$ in $\mathbf{W}_7$ **22.** $\underline{0}$ in $\mathbf{W}_9$

23. List all the elements in $\mathbf{W}_3$ and in $\mathbf{W}_6$, describing each element as in Example 4 of this section.

For Exercises 24–27, compute the given sums and products. (See Examples 8 and 9.)

24. In $\mathbf{W}_5$: $\underline{2} \oplus \underline{4}$, $\underline{2} \odot \underline{4}$, $\underline{3} \oplus \underline{3}$, $\underline{3} \odot \underline{3}$

25. In $\mathbf{W}_{12}$: $\underline{5} \oplus \underline{10}$, $\underline{5} \odot \underline{10}$, $\underline{4} \oplus \underline{8}$, $\underline{4} \odot \underline{8}$

26. In $\mathbf{W}_7$: $\underline{6} \oplus \underline{3}$, $\underline{6} \odot \underline{3}$, $\underline{2} \oplus \underline{4}$, $\underline{2} \odot \underline{4}$

27. In $\mathbf{W}_3$: $\underline{1} \oplus \underline{2}$, $\underline{1} \odot \underline{2}$, $\underline{2} \oplus \underline{2}$, $\underline{2} \odot \underline{2}$

For Exercises 28–30, use the numbers $\underline{5}$, $\underline{8}$, and $\underline{10}$ in $\mathbf{W}_{12}$ to work out an example that illustrates the given fact.

28. $\oplus$ is associative. **29.** $\odot$ is associative.

30. $\odot$ is distributive over $\oplus$.

Review Exercises for Chapter 7

For Exercises 1–10, indicate whether the given statement is *true* or *false*.

1. The statements "a is a factor of b" and "a is a multiple of b" mean the same thing.
2. The Divisibility Rules for 2, 5, and 10 all depend only on the last digit of the numeral in base ten.
3. The Divisibility Rules for 4 and 9 depend only on the last two digits of the numeral in base ten.
4. A number is divisible by 12 if and only if it is divisible by 2 and 6.
5. 22, 222, 2222, and 22,222 are all divisible by 11.
6. Two distinct numbers are relatively prime if they are both prime.
7. Two distinct numbers are relatively prime only if they are both prime.
8. To be sure that a number n is prime it is necessary to test for divisibility by all numbers from 1 up to $\frac{n}{2}$.
9. To find the greatest common factor of two numbers from their prime factorizations, you multiply together the higher power of each prime found in either factorization.
10. In $\mathbf{W}_{12}$, the symbol $\underline{0}$ represents an infinite set of whole numbers.

For Exercises 11–15, fill in the blanks with *all* the appropriate phrases from among

is a factor of, is a multiple of
is divisible by, is a divisor of
divides

that make the sentence true. If none of the phrases are appropriate, write *none*.

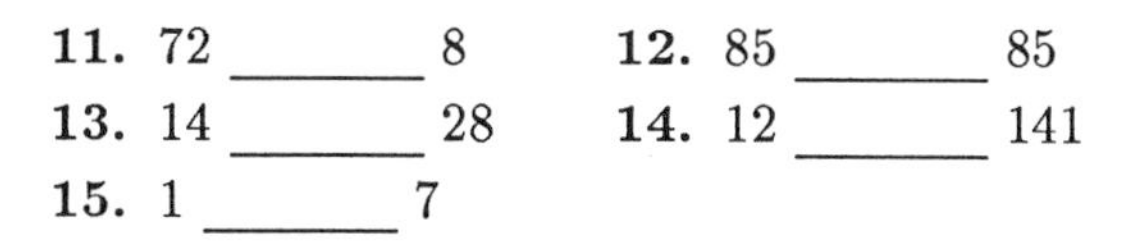

11. 72 ______ 8 **12.** 85 ______ 85
13. 14 ______ 28 **14.** 12 ______ 141
15. 1 ______ 7

16. Use the appropriate Divisibility Rules to determine whether 4620 is divisible by 2, 3, 4, 5, 6, 8, 9, 10, or 11.

For Exercises 17–21, classify the given number as *prime*, *composite*, or *neither*.

17. 9 **18.** 51 **19.** 101
20. 1 **21.** 143

22. Find all the factors of 126.

23. Find the first five multiples of 15.

For Exercises 24–29, use prime factorizations to calculate the required gcf or lcm.

24. $\text{gcf}(70, 80)$ **25.** $\text{lcm}(70, 80)$
26. $\text{gcf}(105, 242)$ **27.** $\text{lcm}(105, 242)$
28. $\text{gcf}(120, 252, 720)$ **29.** $\text{lcm}(120, 252, 720)$

30. Give an example of a composite number that is relatively prime to 108.

31. Yesterday you presented the Fundamental Theorem of Arithmetic to your sixth-grade class. Today two students raise their hands and...

> John says: "I think I've found a contradiction to the Fundamental Theorem of Arithmetic because $1547 = 17 \cdot 91$ and $1547 = 13 \cdot 119$, which are two different factorizations of 1547."
>
> Sarah says: "I think I've found a contradiction, too, because $3780 = 2^2 \cdot 3^3 \cdot 5 \cdot 7$ and $3780 = 2^2 \cdot 3^2 \cdot 5 \cdot 7 \cdot 3$, which are different factorizations of 3780."

How would you respond to these students?

For Exercises 32–37, complete the given congruence statement by filling in a whole number smaller than the modulus.

32. $38 \equiv$ ____ (mod 12)

33. $52 \equiv$ ____ (mod 5)

34. $112 \equiv$ ____ (mod 12)

35. $75 \equiv$ ____ (mod 5)

36. $112 \equiv$ ____ (mod 7)

37. $75 \equiv$ ____ (mod 4)

In Exercises 38 and 39, compute the given sums and products.

38. In $\mathbf{W}_5$: $\underline{3} \oplus \underline{4}$, $\underline{3} \odot \underline{4}$, $\underline{4} \oplus \underline{4}$, $\underline{4} \odot \underline{4}$

39. In $\mathbf{W}_{12}$: $\underline{7} \oplus \underline{11}$, $\underline{7} \odot \underline{11}$, $\underline{5} \oplus \underline{9}$, $\underline{5} \odot \underline{9}$

8.1 The Problem

When the whole-number system was developed in Chapter 4, a general definition of an *operation* on a set was given. That definition requires *each* ordered pair chosen from the set to be matched with a single element *of that set*. When a process defined on a set always yields results that are back in the same set we say the set is **closed under** that process. Of the four usual arithmetic processes defined on the set **W** of whole numbers, addition and multiplication are actually operations on **W** because **W** is closed under addition and multiplication. However, **W** is *not* closed under subtraction or division; there are many subtraction and division problems, such as $3-5=?$ and $3 \div 5 =?$, that do not have whole-number answers. Thus, subtraction and division are *not* operations on the set of whole numbers.

Since subtraction and division are fundamental arithmetic processes, it is important to have a number system in which they always make sense. Constructing such a number system is the task of this chapter and the next. In this chapter we deal with subtraction by constructing the system of *integers*, sometimes called "signed" (whole) numbers. The division problem is handled in Chapter 9 by constructing the *rational numbers*, a system of fractions. In many elementary textbooks, fractions are introduced before signed numbers, probably because it is easy to find concrete, familiar examples of fractional parts of things. That topical order also accurately reflects the historical development of numbers. Nevertheless, we have chosen to explain the integers first because we believe that beginning with addition and subtraction makes the logical structure of both "new" number systems easier to understand.

Before beginning the formal construction of the system of integers, perhaps we should pause to ask ourselves why anyone would care about subtracting a larger number from a smaller one. Putting the question less flippantly, is there any practical interpretation for $3 - 5 = ?$ or similar questions? If we interpret 3 and 5 as the cardinal numbers of two sets, then $3 - 5$ has no natural meaning, whereas $5 - 3$ does. "If Suzy has 5 marbles and loses 3, she has 2 left" is a reasonable instance of $5 - 3 = 2$, but a similar interpretation for $3 - 5$ makes no sense because Suzy can't lose 5 marbles if she only has 3 of them. Nevertheless, it is not too difficult to think of practical situations where $3 - 5$ does make sense, as the following examples show.

Example 1 The temperature is 3° above zero and drops 5°; the new temperature is 2° *below* zero. □

Example 2 A person has \$3 in his checking account and writes a check for \$5; he has *overdrawn* his account by \$2. □

Example 3 3 minutes after liftoff a course correction was made; 5 minutes earlier the support tower had been removed. The support tower was removed 2 minutes *before* liftoff. □

Example 4 National Widget Company's stock gained 3 points in the morning, then lost 5 points in the afternoon. The net change in National Widget's stock value for the day was a *loss* of 2 points. □

Example 5 In playing the game "Mother May I?," if a player is told to take 3 giant steps forward and then 5 giant steps backward, that player ends up 2 giant steps *in back of* the original position. □

Example 6 If a football team gains 3 yards on first down and loses 5 yards on second down, its position on third down is 2 yards *behind* the original line of scrimmage. □

The foregoing examples present reasonable interpretations for $3 - 5$. Although they describe different activities, these examples have common features for which the system of integers provides a useful model. After constructing the integers, we shall return (in Section 8.6) to these and other interpretations of integers that appear in the elementary-school curriculum.

PROBLEM-SOLVING COMMENT

As a first step in solving the main problem of this chapter, let us make sure we *understand the question.* We shall *restate the problem* in detail, *checking the definitions* and breaking the original problem down into *simpler problems* that will be handled one at a time (in Sections 8.2–8.5). ◇

Here is the main problem of this chapter: We want to extend the whole-number system to a larger system that is closed under subtraction. But what does "extend" mean in this context? It means something quite specific: We want a collection of things we shall call "numbers" of some sort, with the following properties:

- Some subset of this collection is a copy of the set **W** of whole numbers (and hence, we can say that our new collection "includes" the whole numbers).
- Operations of addition and multiplication can be defined on this new collection, and they will agree with the previous definitions of $+$ and $\cdot$ when they are applied to (the copy of) **W**.
- The new addition and multiplication "behave as nicely" as their counterparts in **W** (i. e., they are commutative, associative, etc.).
- This new collection *is closed under the subtraction process* (which is defined in terms of addition).
- An order relation can be defined on this new collection so that it agrees with $<$ on **W** and has similar nice properties (trichotomy, transitivity, etc.).

The situation we seek is diagrammed in Figure 8.1. The known set **W** must be matched with some subset of the new set (**?**) in such a way that the corresponding operations and relations agree, and the new set must also have subtraction as an operation.

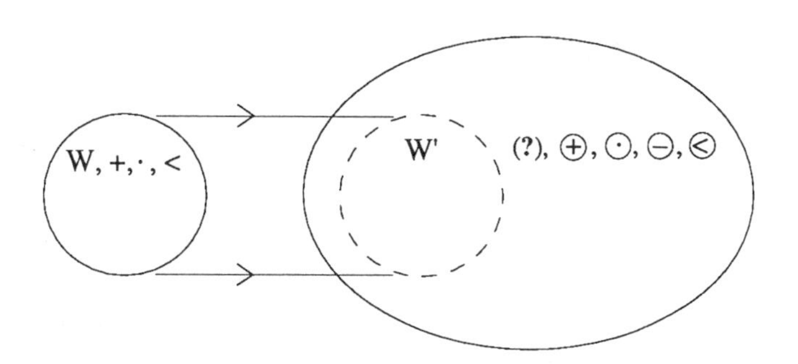

Figure 8.1 Extending the whole-number system.

HISTORICAL NOTE: NEGATIVE NUMBERS

Did you know that negative numbers were not generally recognized as legitimate mathematical objects, even among mathematicians, until about 400 years ago? It's true. Columbus discovered America almost a century before negatives were given first-class status as numbers.

Our mathematics, like much of Western culture, has most of its roots in the work of ancient Greek scholars. Despite the depth and subtlety of their mathematics and philosophy, however, the Greeks ignored negative numbers completely. Even the great number theorist Diophantus of Alexandria (c. A.D. 275) referred to negative solutions of equations as "absurd." Credit for first recognizing the existence of negative quantities belongs to the Indian mathematicians of the 7th, 8th, and 9th centuries. In some of their writings positive numbers are treated as possessions and negative numbers as debts, and the idea of positive and negative as opposite directions on a line also appears. Nevertheless, the Hindus still regarded negative quantities with suspicion. For instance, Bhaskara (c. 1150), after stating that the two roots of the equation $x^2 - 45x = 250$ are 50 and -5, says: "But the second value is in this case not to be taken, for it is inadequate; people do not approve of negative roots."[1]

Early European understanding of negatives was not influenced by the work of the Hindus. Indian mathematics first came to Europe through two books by the 9th-century Arab mathematician al-Khowarizmi, which were translated into Latin and circulated widely. However, al-Khowarizmi avoided using negative numbers, stating rules for negatives only in the context of differences in which a smaller number is taken from a larger one. Western mathematicians, left to deal with negative quantities on their own, proceeded much more slowly than their Eastern counterparts. Even as late as the mid-16th century, negative numbers were called "fictitious" and "absurd." Gradually, however, their utility became too obvious to ignore, and skepticism gave way to acceptance. By the middle of the 17th century, negative numbers had finally become an accepted part of mathematics. In the words of one eminent historian:

> The generalization of the conception of quantity so as to include the negative, was an exceedingly slow and difficult process in the development of algebra.[2] ◇

PROBLEM-SOLVING COMMENT

This seems like a huge job! Where shall we start? We might begin by examining the problem-solving tactics again. The first and most natural technique is often overlooked: *Ask questions.* The most natural question here seems to be "What do we have to work with?" and its answer is more helpful than it might first appear to be. Besides the general tools of logic, set language, and equivalence relations,

---

[1] Florian Cajori, *A History of Mathematics,* 2nd ed. New York: The Macmillan Co., 1961, p. 93.

[2] *Ibid.*, p. 141.

we have the whole numbers themselves. That is, we have the set **W** with the operations $+$ and $\cdot$ and the relation $<$ defined on it. In fact, it might be useful to review the properties of this system, summarized at the end of Chapter 4.

Now what? Another problem-solving tactic suggests that we consider what we want to end up with, then *reason backwards* from there to something we already know about. If done carefully, this process can provide a "blueprint" for the formal construction we need. The next section begins with this technique, and backwards reasoning will be used several times at key places in the development of the new number system. *However*, there is a subtle trap here. Many books, even at the college level, fall into this trap by taking a temptingly simple, but flawed, approach to the integers, reasoning backwards as follows:

We could do subtraction if only we had "additive inverses." That is, for each whole number n we need something that when added to n gives us 0. If we denote the additive inverse of n by $-n$, then the set of integers is just the set of all whole numbers and their additive inverses,

$$\{\ldots, -3, -2, -1, 0, 1, 2, 3, \ldots\}$$

There! Wasn't that easy? Sure, but *it doesn't solve the problem.* Defining something does *not* automatically assure that such a thing exists. Just describing a "hobbit" as a small, rotund, humanlike creature with furry feet does not guarantee that a hobbit actually exists, not even if we name one Bilbo Baggins.[3] Similarly, just defining additive inverses does not establish their existence, nor does choosing symbols for them. The point of this example is that, after reasoning backwards to get some idea of how to proceed, we must then *construct* these new numbers from what we have, so that we are sure they exist. This construction is accomplished in the next section. ◇

Exercises 8.1

1. List five whole-number subtraction problems that do not have answers in **W**; then list five that do.

For Exercises 2–5, list five different whole-number subtraction problems whose answer is the given number.

2. 1 **3.** 7 **4.** 100 **5.** 0

6. List three whole-number subtraction problems that have the same answer as $5-2=?$; then list three that have different answers.

7. Assuming that $a-b=?$ has the same solution as $5-2=?$, answer the question "$a+2=b+?$". (*Hint*: Try some examples from Exercise 6. Do you always get the same result? Why or why not?)

For Exercises 8 and 9, provide at least two different interpretations for the given difference.

8. $2-8$ **9.** $0-5$

[3]A complete definition may be found on p. 16 of J. R. R. Tolkien's *The Hobbit* (New York: Ballantine Books, 1966). A fascinating, but not compelling, argument for its existence is made in this book.

8.2 What Is an Integer?

We want a number system that contains an answer to every whole-number subtraction question, so it makes sense to begin by recalling the defining condition for subtraction:

$$a - b = x \quad \text{if} \quad b + x = a$$

Now, any question that seeks to subtract one whole number from another can be represented by the ordered pair of numbers themselves, whether or not the difference actually exists in $\mathbf{W}$:

$$5 - 3 = ? \text{ can be represented by the ordered pair } (5, 3)$$
$$2 - 9 = ? \text{ can be represented by the ordered pair } (2, 9)$$
$$87 - 64 = ? \text{ can be represented by the ordered pair } (87, 64)$$

and so forth. Thus, the set of all possible whole-number subtraction questions of the form $x - y = ?$ can be represented by the set of all ordered pairs (x, y) of whole numbers; that is, by $\mathbf{W} \times \mathbf{W}$.

We are looking for a system containing answers to all these subtraction questions, so it makes sense to consider two questions "the same" in some way if they have the same answer. $5 - 2 = ?$ and $7 - 4 = ?$ are related in this way, for instance. This observation suggests some sort of equivalence relation on the set of subtraction questions that have answers in $\mathbf{W}$, and hence on the set of ordered pairs that represent these questions. But what about the questions that do not have answers? Can we extend this notion of equivalence in a natural way to include them, too? The answer is *yes* (as you might expect), provided we look at "sameness" in just the right way.

The Equivalence Relation

Saying that $5 - 2 = ?$ and $7 - 4 = ?$ have the same answer means that

$$5 - 2 = 7 - 4$$

Adding 2 and 4 to both sides of this equation, we have

$$(*) \qquad (5 - 2) + 2 + 4 = (7 - 4) + 2 + 4 = 9$$

But the definition of subtraction tells us that

$$(5 - 2) + 2 = 5 \quad \text{and} \quad (7 - 4) + 4 = 7$$

so we can also write equation $(*)$ as

$$5 + 4 = 2 + 7 = 9$$

A general argument patterned after this example shows that $a - b = ?$ and $c - d = ?$ have the same answer whenever

$$a - b = c - d$$

and this is true if and only if

$$a + d = b + c$$

Now, this last expression does not involve subtraction at all, so it can be applied to *any* ordered pairs (a, b) and (c, d), whether or not the subtraction questions they represent have answers in **W**.

Definition Two ordered pairs (a, b) and (c, d) of whole numbers are **equivalent** if $a + d = b + c$. We shall denote this relation by the symbol "$\sim$". Thus,

$$(a, b) \sim (c, d) \quad \text{whenever} \quad a + d = b + c$$

Note: We have used the term *equivalent* and the symbol $\sim$ before to denote equivalence between sets (that is, the existence of a 1-1 correspondence). Rather than invent another name and another symbol for this type of equivalence, we use the same ones again in order to emphasize the analogy between the construction of this system and the construction of **W**. The context should establish which meaning is intended.

Example 1 $(5, 3) \sim (8, 6)$ because $5 + 6 = 3 + 8$. Notice that $5 - 3 = 2$ and $8 - 6 = 2$. □

Example 2 $(2, 7) \sim (5, 10)$ because $2 + 10 = 7 + 5$. Notice that $2 - 7 = ?$ and $5 - 10 = ?$ do not have answers in **W**. □

Example 3 $(7, 4) \not\sim (14, 8)$ because $7+8 \neq 4+14$. Notice that $7 - 4 = 3$ but $14 - 8 = 6$. □

The key fact, implied by the name given to $\sim$, is:

(8.1) $\boxed{\sim \text{ is an equivalence relation on } \mathbf{W} \times \mathbf{W}.}$

Before going any further, let us check to see why this is true. Recall that an equivalence relation must be reflexive, symmetric, and transitive. We give here an example to illustrate each property. A general proof can be constructed by mimicking these examples. (See Exercise 53.)

REFLEXIVE: If we choose any pair of whole numbers, such as $(13, 5)$, then $(13, 5) \sim (13, 5)$ because $13 + 5 = 5 + 13$.

SYMMETRIC: $(3, 7) \sim (5, 9)$ because $3 + 9 = 7 + 5 = 12$. But then $5 + 7 = 9 + 3 = 12$, so $(5, 9) \sim (3, 7)$.

TRANSITIVE: Since $(3, 7) \sim (5, 9)$ and $(5, 9) \sim (6, 10)$ [why?], we expect $(3, 7) \sim (6, 10)$, which is true because $3 + 10 = 7 + 6$.

The Set of Integers

Because $\sim$ is an equivalence relation, it partitions $\mathbf{W} \times \mathbf{W}$ into equivalence classes. That is, $\sim$ breaks $\mathbf{W} \times \mathbf{W}$ into nonoverlapping sets of ordered pairs of whole numbers, and if any pair in a class represents a subtraction question with a particular answer, then all the pairs in that class represent subtraction questions with that same answer. For instance, the class containing $(5,2)$ also contains $(7,4)$, $(3,0)$, and $(25,22)$; all these pairs represent subtraction questions with answer 3. Thus, *an entire equivalence class of ordered pairs represents a single subtraction-question answer.* This fact is the key to understanding the next definition.

Definition An **integer** is the set of all ordered pairs of whole numbers that are related to some particular pair (and hence to each other) by the relation $\sim$.

Notation The set of all ordered pairs that are equivalent to the pair (a,b) is denoted by $[a,b]$. In symbols,

$$[a,b] = \{(x,y) \mid (x,y) \in \mathbf{W} \times \mathbf{W} \text{ and } (x,y) \sim (a,b)\}$$

Furthermore, we denote the set of all integers by $\mathbf{I}$. Thus,

$$\mathbf{I} = \{[a,b] \mid a,b \in \mathbf{W}\}$$

Example 4 The integer $[5,3]$ contains the pairs $(5,3)$, $(8,6)$, $(75,73)$, $(91,89)$, and all the (infinitely many) others that represent all whole-number subtraction questions with the answer 2. □

Example 5 The integer $[2,9]$ contains the pairs $(2,9)$, $(1,8)$, $(6,13)$, $(0,7)$, $(53,60)$, and all the (infinitely many) others that represent all whole-number subtraction questions that ought to have the same answer as $2 - 9 = ?$. □

As you may recall from the discussion of equivalence classes in Section 3.4, two equivalence classes are equal whenever they have equivalent representatives. Thus, an integer may be represented by *any* ordered pair in it. For example, $[8,\ 6]$ is the same integer as $[5,\ 3]$ because the representative pairs are equivalent. That is,

$$[8,6] = [5,3] \text{ because } (8,6) \sim (5,3)$$

which is true because $8 + 3 = 6 + 5$. This is an example of a useful general statement about equality of integers:

$$[a,b] = [c,d] \text{ if and only if } a + d = b + c \tag{8.2}$$

Looking at a few other representatives of the integer $[5,3]$, we can find an interesting pattern:

$$(6,4) = (5+1, 3+1)$$
$$(7,5) = (5+2, 3+2)$$
$$(8,6) = (5+3, 3+3)$$

and so on. This suggests a general statement:

> *If we take an ordered pair of whole numbers and add the same whole number to each number in that pair, then the original pair and the new pair both represent the same integer.*

More formally,

(8.3)

> If a, b, and x are whole numbers, then
> $$[a+x, b+x] = [a, b]$$

It is easy to check that Statement (8.3) is always true. Applying (8.2), we see that

$$[a+x, b+x] = [a,b] \text{ if } (a+x)+b = (b+x)+a$$

and this second equation is true because addition of whole numbers is associative and commutative.

Example 6

$$[2,9] = [2+4, 9+4] = [6,13]$$

Also,

$$[2,9] = [0+2, 7+2] = [0,7]$$

Similarly,

$$[12,5] = [7+5, 0+5] = [7,0]$$ □

Standard Form

Example 6 illustrates an important general fact:

(8.4)

> Any integer can be represented by an ordered pair containing at least one zero.

For example, $[7,2]$ can be written as $[5,0]$ because $7 = 5+2$ and $2 = 0+2$. That is,

$$[7,2] = [5+2, 0+2] = [5,0]$$

Similarly,

$$[17,35] = [0+17, 18+17] = [0,18]$$

and

$$[6,6] = [0+6, 0+6] = [0,0]$$

In general, given any pair of unequal numbers, we just have to write them so that the larger number is expressed as the sum of the smaller number and something else; Statement (8.3) does the rest. (If the numbers in a pair are equal, then (0, 0) is always an equivalent representative.) Thus, any integer can be written in exactly one of three forms,

$$[n, 0],\ [0, 0],\ \text{or}\ [0, n]$$

where n is some natural number; this is called the **standard form** of the integer.

Because each standard-form pair uses at most one nonzero whole number, it is easy to abbreviate the way we write it. All we really need is that nonzero number and something to indicate whether it appears in the first place or in the second. There are many ways to do this. For instance, we might write $[5, 0]$ as 5_1 and $[0, 5]$ as 5_2. However, we choose instead a more traditional notation:

(8.5) For any natural number n, we write the integer $[n, 0]$ as ${}^{+}n$ or just as n and call it a **positive integer**; we write the integer $[0, n]$ as ${}^{-}n$ and call it a **negative integer**. The integer $[0, 0]$ is denoted by 0.

Note: The potential confusion between the *whole numbers* n and 0 and the *integers* n and 0 is usually eliminated by the context in which the symbols are used.

If you look back to the motivation for using classes of ordered pairs, this simplified, familiar notation should help to clarify what was done. For example, the integer $[9, 5]$, which can also be written as $[4, 0]$, represents the answer to the subtraction question $9 - 5 = ?$, and we write it as ${}^{+}4$. Similarly,

> $[2, 8]$ represents the answer to $2 - 8 = ?$, and we write that answer as ${}^{-}6$ because $[2, 8] = [0, 6]$.
>
> $[7, 7]$ represents the answer to $7 - 7 = ?$ and is written as 0 because $[7, 7] = [0, 0]$.

In general, since the difference $a - b$ of two whole numbers is represented by the integer $[a, b]$, and this, in turn, can be written in one of the forms $[n, 0]$, $[0, n]$, or $[0, 0]$ for some natural number n, *the answers to all whole-number subtraction questions can be written as* ${}^{+}n$, ${}^{-}n$, *or 0.*

Notation We denote the set of all positive integers by I^{+} and the set of all negative integers by I^{-}.

PROBLEM-SOLVING COMMENT

The choice of *appropriate notation* can be critical in helping students understand abstract ideas. In many parts of mathematics it is customary to use the same notation for different concepts, relying on the context for clarity. This practice may cause students unnecessary difficulty in learning the concepts. The difficulty may be eased by temporarily introducing alternative notation, until students understand the distinctions of usage well enough to infer for themselves which of the several meanings of a symbol is intended in a given instance.

The use of + and − signs in arithmetic is an example of this problem. The + sign represents addition and also denotes positive numbers; the − sign represents subtraction, denotes negative numbers, and (as you will soon see) indicates "additive inverses." Some elementary textbooks employ raised + and − signs to help students distinguish their use to denote positive and negative integers from their other uses (as we do here). Later, when students are better equipped to cope with the traditional ambiguity, the raised notation can be dropped. ◇

The Copy of W

Statement (8.5) establishes a 1-1 correspondence between the whole numbers and the nonnegative integers; that is, between $\mathbf{W}$ and $\mathbf{I}^+ \cup \{0\}$. Thus, we have solved the first part of our problem. We have *constructed* a familiar-looking set of things called "the integers," which contains a copy of the set of whole numbers and an answer to every whole-number subtraction question. The word "constructed" is italicized to emphasize that this set of things has been shown to exist! We *made* it from the whole-number system, and in so doing, we avoided the trap described in Section 8.1.

However, we are not yet finished. We cannot describe the set of integers as a number *system* until we define some operations on it and show that they behave as we think they should. This is done in the next two sections.

Exercises 8.2

1. Consider the following ordered pairs of whole numbers:

$(10,3)$, $(11,5)$, $(4,10)$, $(19,13)$, $(8,0)$, $(0,6)$, $(16,0)$, $(16,10)$, $(7,11)$, $(46,52)$

(a) Write the subtraction question that each pair represents.

(b) Which pairs in the list are equivalent to $(7,1)$?

(c) Which pairs in the list are equivalent to $(5,11)$?

For Exercises 2–21, fill in the missing numbers.

2. $(1,3) \sim (2,\quad)$

3. $(9,5) \sim (\quad,12)$

4. $(0,\quad) \sim (14,27)$

5. $(\quad,6) \sim (11,11)$

6. $[5,1] = [7,\quad]$

7. $[1,5] = [7,\quad]$

8. $[3,3] = [\quad,0]$

9. $[3,\quad] = [0,3]$

10. $^{+}5 = [\quad,0]$

11. $^{-}8 = [0,\quad]$

12. $^{+}7 = [7,\quad]$

13. $^{+}7 = [\quad,7]$

14. $^{-}4 = [4,\quad]$

15. $^{-}4 = [\quad,4]$

16. $0 = [7,\quad]$

17. $0 = [\quad,4]$

18. $^{+}2 = [\quad,3]$

19. $^{+}8 = [10,\quad]$

20. $^{-}3 = [5,\quad]$

21. $^{-}6 = [\quad,12]$

For Exercises 22–33, determine whether the given statement is *true* or *false*.

22. An integer is an ordered pair of whole numbers.

23. An integer is an infinite set of ordered pairs of whole numbers.

24. $(3,4) = {}^-1$

25. $[3,4] = {}^-1$

26. $(3,4) \in [5,6]$

27. $[3,4] \in (5,6)$

28. $(3,4) \subseteq [5,6]$

29. $(3,4) = [5,6]$

30. $(3,4) = (5,6)$

31. $[3,4] = [5,6]$

32. $(3,4) \sim (5,6)$

33. $[3,4] \sim [5,6]$

For Exercises 34–41, list five different representative pairs for the given integer.

34. $[2,7]$

35. $[10,1]$

36. $[3,3]$

37. ${}^+4$

38. ${}^-5$

39. 0

40. ${}^+17$

41. ${}^-25$

For Exercises 42–49, write the standard form for the given integer.

42. $[6,2]$

43. $[3,7]$

44. $[17,52]$

45. $[15,15]$

46. ${}^+8$

47. ${}^-9$

48. 0

49. ${}^+25$

50. Using the approach discussed at the beginning of this section, choose three more examples to illustrate that $a - b = c - d$ if and only if $a + d = b + c$.

51. Prove that $a - b = c - d$ if and only if $a + d = b + c$.

52. Work out examples to illustrate that $\sim$ is reflexive, symmetric, and transitive, using $(15,7)$ as one of the ordered pairs in each example.

53. Prove that the relation $\sim$ is reflexive, symmetric, and transitive. (*Hint*: If you have trouble proving transitivity, try constructing a few examples; then use them to reason backwards from the desired conclusion.)

8.3 Addition and Subtraction of Integers

As in the case of whole-number addition, the definition of integer addition flows easily from considering what integers "really" are and how we would like them to behave.

The Definition of Sum

What should it mean to add $[7,3]$ and $[5,2]$? These two integers are the classes of ordered pairs represented by $(7,3)$ and $(5,2)$, respectively, which, in turn, represent the subtraction problems $7 - 3 =?$ and $5 - 2 =?$. Using Statement (6.2), the Sum-Difference Property for **W**, we get

$$(7-3) + (5-2) = (7+5) - (3+2)$$

This suggests that the sum $[7,3] + [5,2]$ should be the integer $[7+5, 3+2]$. Moreover, if we actually do these subtraction problems in **W**, as we can in this case, we find that our expectation is justified—

$$7 - 3 = 4 \quad \text{and} \quad 5 - 2 = 3$$

and

$$(7+5)-(3+2)=12-5=7=4+3$$

In thinking about this example, notice that the symbol + is being used in two different senses. In the expressions $7+5$ and $3+2$ it denotes addition of whole numbers, as defined in Chapter 4. But in $[7,3]+[5,2]$ it stands for the still-to-be-defined addition of integers. To emphasize the distinction between what we are currently defining and what is already known, for the rest of this chapter the new + will be distinguished from the old one by putting a circle around the symbol when it represents the new addition. This convention will also be used for the other operation signs as they occur. (The Problem-Solving Comment on page 249 is also relevant here.)

We can define addition of integers, $\oplus$, by generalizing the preceding example.

Definition The **sum** of two integers $[a,b]$ and $[c,d]$ is the integer $[a+c,b+d]$. In symbols,

$$[a,b]\oplus[c,d]=[a+c,b+d]$$

Example 1 As we just saw, $[7,3]\oplus[5,2]=[12,5]$. Similarly,

$$[9,1]\oplus[4,8]=[9+4,1+8]=[13,9]$$ □

Before we go any further, a subtle, but crucial, problem must be addressed. The sum of two integers has been defined in terms of representative ordered pairs. However, each integer is an infinite set of ordered pairs, any one of which might have been used to represent it. Might our definition of *sum* give different answers for the sum of two integers, depending on which representative pairs are chosen? If the answer is *Yes*, then the operation is not "well-defined."

An operation on the integers (or on any set of things having different representatives) is **well-defined** if the result of combining two elements never depends on the particular representatives chosen for the elements. Thus, saying that addition of integers is well-defined means that, if you and I add the same two integers but use different representative pairs for them, our sums will be the same *integer* (even though we might end up with different representative pairs). Example 2 illustrates the fact that this is indeed the case for addition of integers; a general proof that $\oplus$ is always well-defined can be constructed along these same lines.

Example 2 We saw previously that $[7,3]\oplus[5,2]=[12,5]$. Now, $[7,3]=[8,4]$ (because the representative pairs are equivalent), so $[8,4]\oplus[5,2]$ should give us the same sum, and it does:

$$[8,4]\oplus[5,2]=[8+5,4+2]=[13,6]$$

and, by Statement (8.3),

$$[13, 6] = [12 + 1, 5 + 1] = [12, 5]$$

Even if we choose wildly different representative pairs for these two summands, we get the same sum:

$$[7, 3] = [20, 16] \quad \text{and} \quad [5, 2] = [98, 95] \qquad \text{(Why?)}$$

$$[20, 16] \oplus [98, 95] = [20 + 98, 16 + 95] = [118, 111]$$

By Statement (8.3),

$$[118, 111] = [12 + 106, 5 + 106] = [12, 5]$$

as required. □

Example 3 Using the shortened notation for integers, we can see that this addition works just the way we would expect it to:

$$\begin{aligned} {}^{+}6 \oplus {}^{+}8 &= [6, 0] \oplus [8, 0] \\ &= [6 + 8, 0 + 0] \\ &= [14, 0] \\ &= {}^{+}14 \end{aligned} \qquad \begin{aligned} {}^{-}3 \oplus {}^{+}2 &= [0, 3] \oplus [2, 0] \\ &= [0 + 2, 3 + 0] \\ &= [2, 3] \\ &= [0, 1] \\ &= {}^{-}1 \end{aligned}$$

□

Properties of Addition

It is not hard to see that $\oplus$ is commutative and associative. This new addition is defined just by applying the "old" + to the first and second components of representative pairs of whole numbers, so these properties of $\oplus$ follow immediately from the commutativity and associativity of + on **W**. Examples 4 and 5 illustrate these properties.

Example 4 *Addition of integers is commutative:*

$$\begin{aligned} [2, 5] \oplus [7, 1] &= [2 + 7, 5 + 1] && \text{Definition of } \oplus \\ &= [7 + 2, 1 + 5] && \text{Commutativity of + in } \mathbf{W} \\ &= [7, 1] \oplus [2, 5] && \text{Definition of } \oplus \end{aligned}$$

Similarly, in the shortened notation,

$${}^{+}8 \oplus {}^{-}3 = [8, 0] \oplus [0, 3] = [8, 3] = [5, 0] = {}^{+}5$$

$${}^{-}3 \oplus {}^{+}8 = [0, 3] \oplus [8, 0] = [8, 3] = [5, 0] = {}^{+}5$$

□

Example 5 *Addition of integers is associative:*

$$\begin{aligned}([5,1] \oplus [2,3]) \oplus [9,4] &= [5+2, 1+3] \oplus [9,4] \\ &= [(5+2)+9, (1+3)+4] \\ &= [5+(2+9), 1+(3+4)] \\ &= [5,1] \oplus [2+9, 3+4] \\ &= [5,1] \oplus ([2,3] \oplus [9,4])\end{aligned}$$

(Can you supply a reason for each of these steps?) □

As you might expect, the identity element for $\oplus$ is the integer 0 because for any integer $[a, b]$,

$$[a,b] \oplus [0,0] = [a+0, b+0] = [a,b]$$

Once again, the behavior of + on the separate parts of the representative pairs determines the behavior of $\oplus$.

Example 6

$$^{+}3 \oplus 0 = [3,0] \oplus [0,0] = [3+0, 0+0] = [3,0] = {}^{+}3$$ □

Example 7 The identity property of the integer 0 does not depend on the representative pair used. For instance, recalling that $[0,0] = [5,5]$, we can write

$$[3,0] \oplus [5,5] = [3+5, 0+5] = [8,5] = [3,0]$$ □

Inverses

So far, all these properties of $\oplus$ on **I** are just copies of the properties of + on **W**. However, there is one more property that not only distinguishes addition of integers from that of whole numbers, but also results in the integers being closed under subtraction. Essentially, this property says that, given any integer, there is an integer that can be added to it in order to get zero. For instance,

$$^{+}5 \oplus {}^{-}5 = [5,0] \oplus [0,5] = [5,5] = 0$$

Following the practice of Chapter 4, we first describe this property in the more general context of an arbitrary operation $*$ on a set, then apply it to addition of integers.

Definition Let $*$ be an operation on a set S with an identity element z, and let s be any element of S. If there is some element s' in S such that

$$s * s' = z \qquad \text{and} \qquad s' * s = z$$

then s' is called an **inverse** of s with respect to $*$.

The inverse of any integer $[a,b]$ with respect to $\oplus$ is the integer $[b,a]$ because

$$[a,b] \oplus [b,a] = [a+b, b+a] = [a+b, a+b] = 0$$

By the commutativity of $\oplus$,

$$[b,a] \oplus [a,b] = [a,b] \oplus [b,a] = 0$$

The integer $[b,a]$ is called the **additive inverse** of $[a,b]$; it is denoted by $-[a,b]$. In summary, we have shown:

(8.6) Every integer $[a,b]$ has an additive inverse, which, in fact, is $[b,a]$.

Moreover, every integer has *exactly one* such inverse. (See Exercises 17 and 18.)

Example 8 The additive inverse of $[7,5]$ is $[5,7]$ because

$$[7,5] \oplus [5,7] = [12,12] = [0,0] = 0$$

We write

$$-[7,5] = [5,7]$$

and read this: "The additive inverse of $[7,5]$ is $[5,7]$." □

Example 9 In the shortened notation for integers,

$$-({}^{+}6) = -[6,0] = [0,6] = {}^{-}6$$

$$-({}^{-}4) = -[0,4] = [4,0] = {}^{+}4$$

□

Example 9 illustrates an important general fact:

The additive inverse of a positive integer is negative;
the additive inverse of a negative integer is positive.

For this reason, many elementary mathematics books use the term **opposite** as a less imposing substitute for *additive inverse*, as in Figure 8.2. The additive inverse of an integer is also sometimes called the *negative of* that number. This terminology can be confusing, however, because the negative *of* a number may not be a negative number; for instance, the negative of ${}^{-}4$ is ${}^{+}4$. Regardless of the term chosen, the notation reflecting this distinction must be used with care. If the symbol "$-$" before a number is raised, it signals a negative integer; if it is not raised, it indicates the opposite of the integer. Thus, $-({}^{-}4)$ is ${}^{+}4$, the opposite (that is, the additive inverse) of the integer ${}^{-}4$.

Subtraction

Additive inverses are important because they enable us to do subtraction. The definition of difference given in Chapter 4 also applies here:

Definition Let p and q be integers. An integer x such that $q \oplus x = p$ is called the **difference** of p **minus** q; we write this number as $p \ominus q$. In symbols,

$$p \ominus q = x \quad \text{if} \quad q \oplus x = p$$

Example 10

$$^{+}8 \ominus {}^{+}3 = {}^{+}5 \text{ because } {}^{+}3 \oplus {}^{+}5 = {}^{+}8$$ □

You may recall that subtraction of signed numbers was described differently in elementary school. When faced with a problem such as $(^{+}3) \ominus (^{-}2)$, for instance, you were told to "change the sign" of the second number and then add. Thus, you were shown computations like

$$(^{+}3) \ominus (^{-}2) = (^{+}3) \oplus (^{+}2) = 5 \quad \text{and} \quad {}^{+}1 \ominus {}^{+}6 = {}^{+}1 \oplus {}^{-}6 = {}^{-}5$$

But $^{+}2$ is the opposite of $^{-}2$, and $^{-}6$ is the opposite of $^{+}6$, so this change-sign-and-add process is nothing more than adding the opposite of the second number to the first one.

Does this apparently different process always give the result required by the definition of difference? That is, for any integers p and q is it always true that $p \ominus q$ is the same as $p \oplus -q$? To prove that the answer is *Yes*, we verify that $p \oplus -q$ added to q gives us p, as required by the definition of difference:

$$\begin{aligned} q \oplus (p \oplus -q) &= q \oplus (-q \oplus p) && \text{Commutativity of } \oplus \\ &= (q \oplus -q) \oplus p && \text{Associativity of } \oplus \\ &= 0 \oplus p && \text{Definition of additive inverse} \\ &= p && \text{0 is the additive identity.} \end{aligned}$$

Thus, we have verified a useful alternative way of finding the difference of two integers, which we shall call the **Fundamental Subtraction Property**:

(8.7) The difference of two integers p and q is found by adding the additive inverse of the second integer to the first one. In symbols,

$$p \ominus q = p \oplus -q$$

Example 11

$$\begin{aligned} [2,5] \ominus [3,7] &= [2,5] \oplus -[3,7] \\ &= [2,5] \oplus [7,3] \\ &= [9,8] \end{aligned}$$ □

Example 12 $${}^{+}5 \ominus {}^{+}3 = {}^{+}5 \oplus -({}^{+}3) = {}^{+}5 \oplus {}^{-}3 = [5,0] \oplus [0,3] = [5,3] = [2,0] = {}^{+}2$$

Similarly,

$${}^{+}2 \ominus {}^{-}9 = {}^{+}2 \oplus -({}^{-}9) = {}^{+}2 \oplus {}^{+}9 = {}^{+}11$$ □

Example 13 Figure 8.2 shows a page from a sixth-grade text. Notice how the authors introduce the Fundamental Subtraction Property. □

Subtracting Integers

Study these pairs of equations.

$5 - 2 = 3$ $\quad 9 - 2 = 7$ $\quad 7 - 3 = 4$

${}^{+}5 + {}^{-}2 = {}^{+}3$ $\quad {}^{+}9 + {}^{-}2 = {}^{+}7$ $\quad {}^{+}7 + {}^{-}3 = {}^{+}4$

In each case, adding the opposite integer gives the same result as subtracting.

To subtract an integer, add its opposite.

A. The high temperature in Chicago during April was 21°C. The low temperature was ${}^{-}3$°C. How much warmer was the high temperature than the low?

Find ${}^{+}21 - {}^{-}3$.

$${}^{+}21 \quad - \quad {}^{-}3$$

Change to addition. ↓ ↓ Change to the opposite.

$${}^{+}21 \quad + \quad {}^{+}3 \quad = \quad {}^{+}24$$

The high temperature for April was 24 degrees warmer than the low.

B. Find ${}^{+}7 - {}^{-}5$.

${}^{+}7 - {}^{-}5$ Change ${}^{-}5$ to ${}^{+}5$ and add.

${}^{+}7 + {}^{+}5 = {}^{+}12$

C. Find ${}^{-}4 - {}^{-}9$.

${}^{-}4 - {}^{-}9$ Change ${}^{-}9$ to ${}^{+}9$ and add.

${}^{-}4 + {}^{+}9 = {}^{+}5$

Figure 8.2 INVITATION TO MATHEMATICS, Grade 6, p. 360

Example 14 Now we can return to the problem of finding $3 - 5$:

$$^{+}3 \ominus {}^{+}5 = {}^{+}3 \oplus {}^{-}5 = [3,0] \oplus [0,5] = [3,5] = [0,2] = {}^{-}2$$

Notice that this result is consistent with all the interpretations of $3-5$ given in Section 8.1. □

No matter what the integers p and q are, the additive inverse of q always exists in **I** because of Property (8.6), so the Fundamental Subtraction Property guarantees that the difference $p \ominus q$ *always exists in* **I** (because **I** is closed under $\oplus$). Thus, we have solved the main part of the problem that was stated at the beginning of the chapter:

(8.8) | The system of integers is closed under subtraction. |

Exercises 8.3

For Exercises 1–10, compute the given sum by using the definition of $\oplus$.

1. $[2,3] \oplus [7,4]$ **2.** $[3,2] \oplus [7,4]$

3. $[5,0] \oplus [1,6]$ **4.** $[8,8] \oplus [2,9]$

5. $^{+}5 \oplus {}^{+}3$ **6.** $^{+}5 \oplus {}^{-}3$

7. $^{-}5 \oplus {}^{+}3$ **8.** $^{-}5 \oplus {}^{-}3$

9. $0 \oplus {}^{-}4$ **10.** $^{-}4 \oplus {}^{+}4$

For Exercises 11–16, specify the additive inverse of the given integer.

11. $[4,1]$ **12.** $[3,8]$ **13.** $[6,6]$

14. $^{+}3$ **15.** $^{-}5$ **16.** 0

17. Suppose that x is an integer with the property that $^{+}5 \oplus x = 0$. Prove that x *must* equal $^{-}5$; that is, prove that x cannot be any other number.

18. Generalize Exercise 17 to prove that no integer can have more than one additive inverse.

For Exercises 19–32, compute the difference using the Fundamental Subtraction Property. Check your answers using the definition of subtraction.

19. $[7,2] \ominus [3,1]$ **20.** $[4,8] \ominus [5,2]$

21. $[3,10] \ominus [11,12]$ **22.** $[6,0] \ominus [7,7]$

23. $^{+}7 \ominus {}^{+}3$ **24.** $^{+}7 \ominus {}^{-}3$ **25.** $^{-}7 \ominus {}^{+}3$

26. $^{-}7 \ominus {}^{-}3$ **27.** $^{+}2 \ominus 0$ **28.** $0 \ominus {}^{+}2$

29. $^{+}2 \ominus {}^{+}2$ **30.** $^{+}2 \ominus {}^{-}2$ **31.** $^{-}2 \ominus {}^{+}2$

32. $^{-}2 \ominus {}^{-}2$

33. Give three specific numerical examples to illustrate that the Fundamental Subtraction Property agrees with the definition of difference.

34. (a) Show that
$$-({}^{+}5) \oplus -({}^{+}2) = -({}^{+}5 \oplus {}^{+}2)$$

(b) Prove: For any two integers p and q,
$$-p \oplus -q = -(p \oplus q)$$

(c) Use part (*b*) to determine
$$-({}^{+}5) \oplus -({}^{-}2)$$

35. Show by example that subtraction of integers is neither associative nor commutative. (Do the computations.)

For Exercises 36 and 37, use the material of this chapter to give reasons for each step in the computation shown.

36. $^+5 \ominus {}^+8$

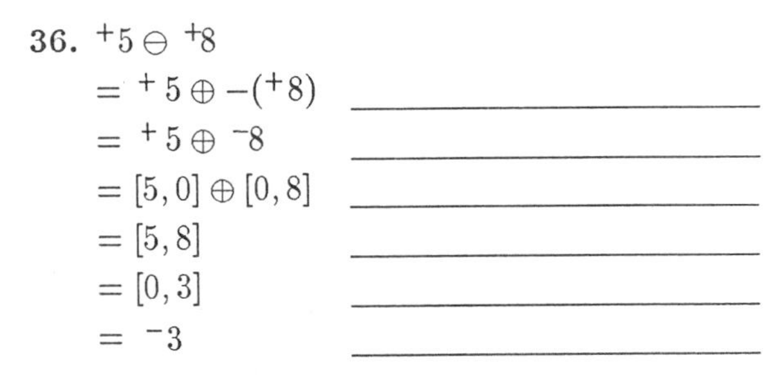

$$
\begin{aligned}
&= {}^+5 \oplus -({}^+8) \quad \_\_\_\_\_\_\_\_\_\_\_\_\\
&= {}^+5 \oplus {}^-8 \quad \_\_\_\_\_\_\_\_\_\_\_\_\\
&= [5,0] \oplus [0,8] \quad \_\_\_\_\_\_\_\_\_\_\_\_\\
&= [5,8] \quad \_\_\_\_\_\_\_\_\_\_\_\_\\
&= [0,3] \quad \_\_\_\_\_\_\_\_\_\_\_\_\\
&= {}^-3 \quad \_\_\_\_\_\_\_\_\_\_\_\_
\end{aligned}
$$

37. $^-4 \ominus {}^-7$

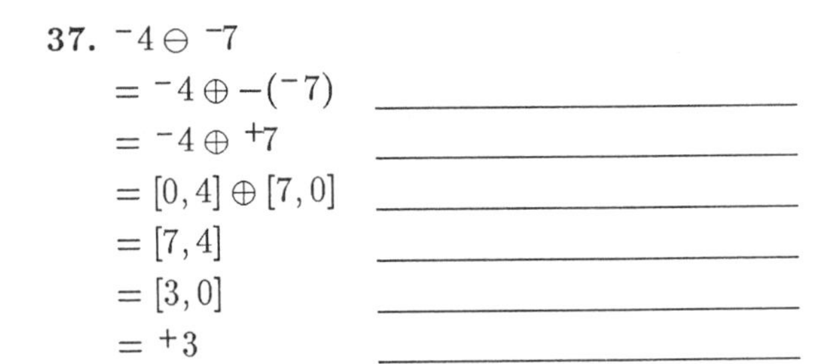

$$
\begin{aligned}
&= {}^-4 \oplus -({}^-7) \quad \_\_\_\_\_\_\_\_\_\_\_\_\\
&= {}^-4 \oplus {}^+7 \quad \_\_\_\_\_\_\_\_\_\_\_\_\\
&= [0,4] \oplus [7,0] \quad \_\_\_\_\_\_\_\_\_\_\_\_\\
&= [7,4] \quad \_\_\_\_\_\_\_\_\_\_\_\_\\
&= [3,0] \quad \_\_\_\_\_\_\_\_\_\_\_\_\\
&= {}^+3 \quad \_\_\_\_\_\_\_\_\_\_\_\_
\end{aligned}
$$

38. Prove that, for any integers p and q,

$$-(p \ominus q) = q \ominus p$$

39. Prove that $-(-p) = p$ for any integer p.

8.4 Multiplication of Integers

The definition of integer multiplication is a little more complicated than that of addition, but its motivational roots are the same. We approach the problem by deciding how we want multiplication of integers to behave, then reason backwards from there to construct a suitable definition.

> **PROBLEM-SOLVING COMMENT**
>
> Besides the technique of reasoning backwards from the desired conclusion, other problem-solving tactics are at work here, as they were in the previous section. We decide how we want the new operation to behave by *constructing an example* based on what we know from whole-number arithmetic, then *generalize* the form of that example by *introducing appropriate notation*. ◇

The Definition of Product

Consider multiplying the integers $[7,3]$ and $[5,2]$, for instance. If we were only interested in this particular product it would be easy, of course. Since these integers are just $^+4$ and $^+3$ and since we want positive-integer multiplication to agree with whole-number multiplication, we could use that "old" multiplication to compute the answer, 12. But this approach gives us no insight into how to extend the process to negative numbers. For that, we must take a closer look at how the integers in our example were formed and how they—typifying integers in general—"ought" to behave.

$[7,3]$ and $[5,2]$ represent answers to the whole-number subtraction problems $7 - 3 = ?$ and $5 - 2 = ?$, so it might help to look at how the product

$(7-3)\cdot(5-2)$ can be rearranged, using some basic arithmetic properties already developed. Since we are just looking for motivational guidance about how to define something, we shall allow ourselves to blur the distinction between nonzero whole numbers and positive integers as we explore this question.

By the distributivity of multiplication over subtraction, $(7-3)\cdot(5-2)$ can be rewritten as

$$(7-3)\cdot 5-(7-3)\cdot 2$$

and then as

$$(7\cdot 5-3\cdot 5)-(7\cdot 2-3\cdot 2)$$

Now, if we think of these individual products as positive integers, then the Fundamental Subtraction Property and Exercises 34 and 38 of Section 8.3 give us

$$(7\cdot 5+-(3\cdot 5))+(-(7\cdot 2)+3\cdot 2)$$

Associativity and commutativity of addition, along with the properties already cited, permit us (in several steps) to rearrange the previous expression as

$$(7\cdot 5+3\cdot 2)-(7\cdot 2+3\cdot 5)$$

Putting all this together, we have

$$(7-3)\cdot(5-2)=(7\cdot 5+3\cdot 2)-(7\cdot 2+3\cdot 5)$$

Some quick arithmetic verifies that this equation is true because both sides equal 12; but it is the *form* that is of special interest to us. If we replace the numbers by letters to help us focus on this form, we get

$$(a-b)\cdot(c-d)=(ac+bd)-(ad+bc)$$

This last equation tells us exactly how to multiply any two integers *because it describes the product of two subtraction problems as a single subtraction problem.*

Definition The **product** of two integers $[a,b]$ and $[c,d]$ is the integer $[ac+bd,ad+bc]$. In symbols,

$$[a,b]\odot[c,d]=[ac+bd,ad+bc]$$

Example 1

$$\begin{aligned}[7,3]\odot[5,2] &= [7\cdot 5+3\cdot 2,7\cdot 2+3\cdot 5]\\ &= [41,29]\\ &= [12,0]\end{aligned}$$

□

Because integers have different representative pairs, it is necessary to check that multiplication is well-defined. We omit the general proof of this fact, but Example 2 illustrates it with different representatives for the product in Example 1.

Example 2 $[7,3] = [10,6]$, so $[10,6] \odot [5,2]$ should be the same integer as $[7,3] \odot [5,2]$, which, as we saw in Example 1, is $[12,0]$:

$$\begin{aligned} [10,6] \odot [5,2] &= [10 \cdot 5 + 6 \cdot 2, 10 \cdot 2 + 6 \cdot 5] \\ &= [62,50] \\ &= [12,0] \quad \text{(Why?)} \end{aligned}$$

Even if we choose other representatives for these two integers, such as $[24,20]$ $(= [7,3])$ and $[11,8]$ $(= [5,2])$, we still get the same product:

$$\begin{aligned} [24,20] \odot [11,8] &= [24 \cdot 11 + 20 \cdot 8, 24 \cdot 8 + 20 \cdot 11] \\ &= [424,412] \\ &= [12,0] \end{aligned}$$

□

Example 3 Using the shortened notation for integers, we see that this definition of multiplication works exactly the way we expect it to:

$$\begin{aligned} {}^{+}6 \odot {}^{-}4 &= [6,0] \odot [0,4] \\ &= [6 \cdot 0 + 0 \cdot 4, 6 \cdot 4 + 0 \cdot 0] \\ &= [0,24] \\ &= {}^{-}24 \end{aligned}$$

Similarly,

$$\begin{aligned} {}^{-}11 \odot {}^{-}8 &= [0,11] \odot [0,8] \\ &= [0 \cdot 0 + 11 \cdot 8, 0 \cdot 8 + 11 \cdot 0] \\ &= [88,0] \\ &= {}^{+}88 \end{aligned}$$

□

Properties of Multiplication

This operation $\odot$ is commutative and associative, and it is also distributive over integer addition. General proofs of these properties are based on the corresponding properties of whole-number arithmetic. However, the details of those arguments are more distracting than useful as we pursue our goal of constructing the integer system, so we shall simply illustrate the properties by a few numerical examples.

Example 4 *Multiplication of integers is commutative*:

$$\begin{aligned} [7,3] \odot [5,2] &= [7 \cdot 5 + 3 \cdot 2, 7 \cdot 2 + 3 \cdot 5] \\ &= [5 \cdot 7 + 2 \cdot 3, 5 \cdot 3 + 2 \cdot 7] \\ &= [5,2] \odot [7,3] \end{aligned}$$

Similarly, in shortened notation,

$$^{+}6 \odot {}^{-}9 = [6,0] \odot [0,9] = [0,54] = {}^{-}54$$
$$^{-}9 \odot {}^{+}6 = [0,9] \odot [6,0] = [0,54] = {}^{-}54 \qquad \square$$

Example 5 *Multiplication of integers is associative*:

$$\begin{aligned}([5,1] \odot [2,3]) \odot [4,7] &= [10+3, 15+2] \odot [4,7] \\ &= [13,17] \odot [4,7] \\ &= [52+119, 91+68] \\ &= [171,159] \\ [5,1] \odot ([2,3] \odot [4,7]) &= [5,1] \odot [8+21, 14+12] \\ &= [5,1] \odot [29,26] \\ &= [145+26, 130+29] \\ &= [171,159] \qquad \square\end{aligned}$$

Example 6 *Integer multiplication is distributive over addition*:

$$\begin{aligned}{}^{-}3 \odot ({}^{-}2 + {}^{+}5) &= [0,3] \odot ([0,2] + [5,0]) \\ &= [0,3] \odot [5,2] \\ &= [6,15] \\ &= {}^{-}9 \\ ({}^{-}3 \odot {}^{-}2) \oplus ({}^{-}3 \odot {}^{+}5) &= ([0,3] \odot [0,2]) \oplus ([0,3] \odot [5,0]) \\ &= [6,0] \oplus [0,15] \\ &= [6,15] \\ &= {}^{-}9 \qquad \square\end{aligned}$$

The rules for multiplication of signed numbers sometimes seem mysterious to schoolchildren, and making them reasonable at the elementary-school level can be challenging at times. In Section 8.8 we discuss these rules and strategies for explaining them. Here we want you to see that the rules themselves are simply consequences of the way multiplication has been defined. As an example, we justify one rule that seems to be particularly bothersome in elementary school:

(8.9) | The product of two negative integers is positive. |

Let ${}^{-}a$ and ${}^{-}b$ denote negative integers. Then ${}^{-}a = [0, a]$ and ${}^{-}b = [0, b]$, so

$$^{-}a \odot {}^{-}b = [0,a] \odot [0,b] = [0 \cdot 0 + a \cdot b, 0 \cdot b + a \cdot 0] = [ab, 0]$$

which is the positive integer ab.

Example 7

$$\begin{aligned} {}^{-}2 \odot {}^{-}3 &= [0,2] \odot [0,3] \\ &= [0 \cdot 0 + 2 \cdot 3, 0 \cdot 3 + 2 \cdot 0] \\ &= [6,0] \\ &= {}^{+}6 \end{aligned}$$

□

The identity element for $\odot$ is $^{+}1$, as you would expect:

$$[a,b] \odot [1,0] = [a \cdot 1 + b \cdot 0, a \cdot 0 + b \cdot 1] = [a,b]$$

for any integer $[a,b]$. Also, by the commutativity of $\odot$,

$$[1,0] \odot [a,b] = [a,b]$$

Example 8

$$\begin{aligned} {}^{-}4 \odot {}^{+}1 &= [0,4] \odot [1,0] \\ &= [0 \cdot 1 + 4 \cdot 0, 0 \cdot 0 + 4 \cdot 1] \\ &= [0,4] \\ &= {}^{-}4 \end{aligned}$$

□

Division

The division process defined in Chapter 4 can be extended to the integers.

Definition Let p and q be integers. If there exists an integer x such that $q \odot x = p$, then we call x the **quotient** of p **divided by** q, and we write x as $p \ominus q$ or as p/q. In symbols,

$$p \ominus q = x \quad \text{if} \quad q \odot x = p$$

Example 9 $^{+}6 \ominus {}^{-}2 = {}^{-}3$ because $^{-}2 \odot {}^{-}3 = {}^{+}6$. (See Example 7.) □

Such quotients do not always exist in **I**. If we try to find the quotient $1 \ominus 2$, for example, we can see that such a number cannot exist in **I**. (See Exercise 15.) Thus, **I** is not closed under $\ominus$. This problem requires us to build a larger number system, which we shall do in Chapter 9.

Example 10 Figure 8.3 is an excerpt from a seventh-grade text. Observe how the definition of division is used to find the quotient of two integers. Also, notice that division problems involving zero are explained by relating them to multiplication. □

DIVISION OF INTEGERS

You divide positive integers just as you divide whole numbers. Remember how multiplication and division are related.

$$6 \times 3 = 18 \quad \text{so} \quad 18 \div 3 = 6$$

Multiplication and division of integers are related in the same way.

$$6 \times {}^-3 = {}^-18 \quad \text{so} \quad {}^-18 \div 3 = {}^-6$$
$${}^-6 \times 3 = {}^-18 \quad \text{so} \quad {}^-18 \div {}^-3 = 6$$
$${}^-6 \times {}^-3 = 18 \quad \text{so} \quad 18 \div {}^-3 = {}^-6$$

- The quotient of two positive or two negative integers is positive.
- The quotient of a positive integer and a negative integer is negative.

The quotient of zero divided by any other integer is zero.

$$0 \div 8 = 0$$
$$0 \div {}^-5 = 0$$

You cannot divide an integer by zero. You cannot divide 7 by 0, since no number multiplied by zero is 7.

$$7 \div 0 = ? \Rightarrow ? \times 0 = 7$$

If your calculator does not have a special key for entering negative numbers, divide integers as if they were whole numbers. Then write the quotient as a positive or negative number according to the rules above.

Figure 8.3 HOUGHTON MIFFLIN MATHEMATICS, Grade 7, p. 344

The Number Systems $\mathbf{W}_m$, Revisited

Now that inverses have been defined, we can complete our discussion of clock arithmetic. Recall from Section 7.5 that $\mathbf{W}_m$ represents the set

$$\{\underline{0}, \underline{1}, \ldots, \underline{m-1}\}$$

of equivalence classes of whole numbers under the relation "$\equiv \pmod{m}$." This relation says that two numbers are in the same class if and only if they leave the same remainder when divided by m; that class is represented by the remainder, underlined. Also recall that, in this context, the circled operation symbols refer to the remainder-arithmetic operations, where each answer is computed by doing the regular arithmetic operation and then finding the remainder modulo m. We again turn to $\mathbf{W}_{12}$ and $\mathbf{W}_5$ for typical examples.

Even though there are no negative numbers either in $\mathbf{W}$ or in $\mathbf{W}_{12}$, *every number in* $\mathbf{W}_{12}$ *has an additive inverse*, a number which when added to it yields $\underline{0}$. For instance,

$$\underline{7} \oplus \underline{5} = \underline{0} \text{ because } 7 + 5 = 12 \equiv 0 \pmod{12}$$

so 5 is the additive inverse of 7 (and vice versa). Using the usual arithmetic notation for the additive inverse of a number, we write $-\underline{7} = \underline{5}$ (and $-\underline{5} = \underline{7}$). Similarly, $-\underline{10} = \underline{2}$ because $\underline{10} \oplus \underline{2} = \underline{0}$. Thus,

$$\begin{array}{rcrcrcr} -\underline{1} & = & \underline{11} & \qquad & -\underline{11} & = & \underline{1} \\ -\underline{2} & = & \underline{10} & & -\underline{10} & = & \underline{2} \\ -\underline{3} & = & \underline{9} & & -\underline{9} & = & \underline{3} \\ -\underline{4} & = & \underline{8} & & -\underline{8} & = & \underline{4} \\ -\underline{6} & = & \underline{6} & & -\underline{0} & = & \underline{0} \end{array}$$

Since every number has an additive inverse, subtraction in $\mathbf{W}_{12}$ always makes sense. As the Fundamental Subtraction Property tells us, we can subtract by adding the additive inverse; for instance,

$$\underline{4} \ominus \underline{7} = \underline{4} \oplus (-\underline{7}) = \underline{4} \oplus \underline{5} = \underline{9}$$

In terms of our twelve-hour clock, this says that 7 hours *before* 4 o'clock, it was 9 o'clock.

An analogous question is: What about inverses for multiplication? That is, given a number $\underline{x}$ in $\mathbf{W}_{12}$, is there a number $\underline{y}$ in $\mathbf{W}_{12}$ such that $\underline{x} \odot \underline{y} = \underline{1}$? The answer is *Sometimes.* For instance, the multiplicative inverse of $\underline{7}$ is $\underline{7}$ itself, because $\underline{7} \odot \underline{7} = \underline{1}$. However, $\underline{2}$ does not have any multiplicative inverse. This can be checked by multiplying every element of $\mathbf{W}_{12}$ by $\underline{2}$ and seeing that $\underline{1}$ is never the answer; but you can avoid the nuisance of doing that by observing the following:

> 2 is an even number, so 2 times any number x must also be even, which implies that $2x$ must leave an even remainder when divided by the (even) number 12. In other words, $2x$ must always be congruent to an even number modulo 12, so $\underline{2} \odot \underline{x}$ cannot equal $\underline{1}$ for any x.

(This argument is worth looking at with some care; it can be adapted to tell us useful things about other numbers and other systems.)

In $\mathbf{W}_5$ there is also an additive inverse for every number:

$$-\underline{1} = \underline{4} \text{ because } 1 + 4 = 5 \text{ and } 5 \equiv 0 \pmod 5$$

Similarly,

$$-\underline{2} = \underline{3} \qquad -\underline{3} = \underline{2} \qquad -\underline{4} = \underline{1} \qquad -\underline{0} = \underline{0}$$

Therefore, subtraction is always possible in $\mathbf{W}_5$. For instance,

$$\underline{2} \ominus \underline{4} = \underline{2} \oplus -\underline{4} = \underline{2} \oplus \underline{1} = \underline{3}$$

Every number in $\mathbf{W}_5$ except $\underline{0}$ also has a multiplicative inverse:

$\underline{2}$ and $\underline{3}$ are inverses of each other, because $2 \cdot 3 \equiv 1 \pmod 5$;
$\underline{4}$ is its own inverse, because $4 \cdot 4 = 16 \equiv 1 \pmod 5$;
$\underline{1}$ is its own inverse, because $1 \cdot 1 = 1 \equiv 1 \pmod 5$.

Thus, if we think of dividing as multiplying by the multiplicative inverse, then division by nonzero numbers can always be done in $\mathbf{W}_5$. For instance,

$$\underline{1} \oslash \underline{3} = \underline{1} \odot \underline{2} = \underline{2}$$

Therefore, *all four arithmetic operations always work in* $\mathbf{W}_5$.

Two natural questions arise at this point:

- Does every modular system $\mathbf{W}_m$ have additive inverses for each of its numbers?
- For which modulus numbers m do all nonzero numbers have multiplicative inverses in $\mathbf{W}_m$?

The solutions are neat, elegant, and within the scope of the mathematics we have developed so far, provided it is used in just the right way. They are left as problem-solving challenges for you. (See Exercises 28 and 29.)

Exercises 8.4

For Exercises 1–10, compute the given product by using the definition of $\odot$.

1. $[8,3] \odot [5,1]$ **2.** $[3,8] \odot [5,1]$

3. $[3,8] \odot [1,5]$ **4.** $[2,2] \odot [9,6]$

5. $^+7 \odot {}^+4$ **6.** $^+7 \odot {}^-4$ **7.** $^-7 \odot {}^+4$

8. $^-7 \odot {}^-4$ **9.** $0 \odot {}^-5$ **10.** $^+1 \odot {}^+12$

11. Use the definition of $\odot$ to verify that $0 \odot {}^+7 = 0$.

12. Prove that $0 \odot x = 0$ for any integer x.

13. (a) Prove that, for any integers p and q,
$$p \odot -q = -p \odot q = -(p \odot q)$$
(b) Illustrate the general fact stated in part (a) by substituting specific integers for p and q and working out the computations.

14. Prove that $^-1 \odot p = -p$ for any integer p. (Notice the two different meanings for the "–" sign.)

15. Prove that there is no integer x such that $2 \odot x = 1$. (*Hint*: Reason backwards—If there were, then, for some integer $x = [a, b]$, we would have $2 \odot [a, b] = \ldots = [2a, 2b] = [1, 0]$, so ....)

16. Using the integers $^{+}2$ and $^{-}5$, work out an example to illustrate that $\odot$ is commutative.

17. Using the integers $^{+}2$, $^{-}5$, and $^{-}8$, work out an example to illustrate that $\odot$ is associative.

18. Using the integers $^{+}2$, $^{-}5$, and $^{-}8$, work out an example to illustrate that $\odot$ is distributive over $\oplus$.

19. Let $a = {}^{-}8$, $b = {}^{+}28$, $c = {}^{-}4$. Determine if each of the following statements is true:

i. $(a \oplus b) \odiv c = (a \odiv c) \oplus (b \odiv c)$
ii. $(a \ominus b) \odiv c = (a \odiv c) \ominus (b \odiv c)$

20. Let $a = {}^{-}210$, $b = {}^{-}2$, $c = {}^{+}4$. Determine if each of the following statements is true:

i. $a \odiv (b \oplus c) = (a \odiv b) \oplus (a \odiv c)$
ii. $a \odiv (b \ominus c) = (a \odiv b) \ominus (a \odiv c)$

21. Based on your answers to Exercises 19 and 20, state which of the following statements you believe to be true for integers (assuming they all refer to situations in which integer division is possible):

(a) Division is right distributive over addition.
(b) Division is left distributive over addition.
(c) Division is right distributive over subtraction.
(d) Division is left distributive over subtraction.

Exercises 22–29 refer to the clock-arithmetic systems $\mathbf{W}_m$.

22. Find the additive inverse of each number in $\mathbf{W}_6$.

23. Find (by checking all possible cases, if necessary) the multiplicative inverse of each number in $\mathbf{W}_6$ that has one.

24. Compute in $\mathbf{W}_6$:

$\underline{2} \ominus \underline{5}, \quad \underline{4} \ominus \underline{3}, \quad \underline{3} \ominus \underline{4}, \quad \underline{2} \ominus \underline{5}$

25. Find the additive inverse of each number in $\mathbf{W}_3$.

26. Find (by checking all possible cases, if necessary) the multiplicative inverse of each number in $\mathbf{W}_3$ that has one.

27. Compute in $\mathbf{W}_3$:

$\underline{1} \ominus \underline{2}, \quad \underline{2} \ominus \underline{1}, \quad \underline{0} \ominus \underline{2}, \quad \underline{2} \ominus \underline{2}$

28. (a) Looking back at the list of additive inverses in $\mathbf{W}_{12}$, describe, as specifically as you can, the connection between the modulus number 12 and the additive inverse of a number $\underline{x}$ in $\mathbf{W}_{12}$.
(b) Looking back at the list of additive inverses in $\mathbf{W}_5$, describe, as specifically as you can, the connection between the modulus number 5 and the additive inverse of a number $\underline{x}$ in $\mathbf{W}_5$.
(c) Generalize your results from parts (a) and (b) to give a specific description of the connection between any modulus m and the additive inverse of a number $\underline{x}$ in $\mathbf{W}_m$.
(d) Use your result from part (c) to justify the claim that every number in every system $\mathbf{W}_m$ has an additive inverse.

29. (a) Prove that $\underline{3}$ has no multiplicative inverse in $\mathbf{W}_{12}$. (*Hint*: Mimic the argument used to show that $\underline{2}$ has no multiplicative inverse in $\mathbf{W}_{12}$.)
(b) Which numbers in $\mathbf{W}_{12}$ have multiplicative inverses, and what are they?
(c) Prove that 5 has no multiplicative inverse in $\mathbf{W}_{15}$.
(d) Which numbers in $\mathbf{W}_{15}$ have multiplicative inverses? Justify your answer.
(e) Find a modulus m greater than 15 for which *every* nonzero number in $\mathbf{W}_m$ has a multiplicative inverse. Justify your answer.
(f) Characterize the modulus numbers m for which *every* nonzero number in $\mathbf{W}_m$ has a multiplicative inverse. Justify your answer.

8.5 Order and Absolute Value

The integers can be ordered in many ways, but we want to define an order relation that extends the relation $<$ already defined on the whole numbers. That is, we want to define a new order relation that arranges the positive integers in exactly the same order they have when they are considered as whole numbers and ordered by $<$.

One way to tackle this problem is simply to borrow the whole-number description of *less than* and see if it works for the integers. If we rephrase Statement (4.17) by substituting *integer* for *whole number*, we get

> An integer a is less than an integer b if there is a nonzero integer d such that $a \oplus d = b$.

This may seem tempting as a solution, but a moment's reflection should tell you that it is not right. For example, it says that $^+3$ is less than $^+5$ because $^+3 \oplus ^+2 = ^+5$ *and also* that $^+5$ is less than $^+3$ because $^+5 \oplus ^-2 = ^+3$. This means that the Trichotomy Law would no longer hold, even for positive integers, so this proposed definition is unacceptable.

Fortunately, a minor alteration fixes everything. As the preceding example shows, the difficulty arises when we reverse the the usual order of two positive integers by adding a *negative* number to the larger one. If this is prohibited, then everything will be fine.

Definition An integer p is **less than** an integer q if there is a positive integer d such that $p \oplus d = q$. In this case we also say q is **greater than** p, and write $p \otimes q$ and $q \otimes p$, respectively.

Notation We symbolize "is less than or equal to" by $\circledleq$, and "is greater than or equal to" by $\circledgeq$.

Example 1 $$^+3 \otimes ^+7 \text{ because } ^+3 \oplus ^+4 = ^+7$$ □

Example 2 $$^-5 \otimes ^-2 \text{ because } ^-5 \oplus ^+3 = ^-2$$ □

Example 3 Any negative integer is less than any positive integer. For instance, $^-2 \otimes ^+1$ because $^-2 \oplus (^+2 \oplus ^+1) = ^+1$. In general, if ^-p is a negative integer and q is a positive integer, then $p \oplus q$ is positive. Therefore, $^-p \otimes q$ because $^-p \oplus (p \oplus q) = q$. □

Properties of Order

The order relation $\otimes$ on the positive integers behaves exactly like $<$ on the whole numbers. For instance, $7 < 10$ in **W** and $^+7 \otimes ^+10$ in **I**. In general:

(8.10) $\boxed{\text{If } a, b \in \mathbf{W}, \text{ then } a < b \text{ if and only if } {}^{+}a \otimes {}^{+}b.}$

Moreover, the "nice" properties of $<$ on $\mathbf{W}$ carry over to $\otimes$ on $\mathbf{I}$ in an obvious way. In particular, $\otimes$ is transitive on $\mathbf{I}$; the argument is exactly like that for transitivity of $<$ on $\mathbf{W}$ (4.18) and is left as an exercise. This allows us to talk sensibly about an integer being *between* two others.

The Trichotomy Law (4.19) also holds; that is, if two integers are unequal, then one must be less than the other. We are so accustomed to this "obvious" fact that it may not seem worth mentioning. But there are some systems for which the Trichotomy Law does *not* hold, so we should check that it always works for integers. Its proof in $\mathbf{I}$ is based on the observation that, if p and q are unequal integers, then either $p \ominus q$ or $q \ominus p$ is positive; and when that positive integer is added to one of the original integers, we get the other. Example 4 illustrates this idea.

Example 4 To order ${}^{-}3$ and ${}^{+}2$, we observe that ${}^{+}2 \ominus {}^{-}3 = {}^{+}5$ and ${}^{-}3 \oplus {}^{+}5 = {}^{+}2$. Thus, ${}^{-}3 \otimes {}^{+}2$ by the definition of $\otimes$.

To order ${}^{+}3$ and ${}^{-}2$, we observe that ${}^{+}3 \ominus {}^{-}2 = {}^{+}5$ and ${}^{-}2 \oplus {}^{+}5 = {}^{+}3$. Thus, ${}^{-}2 \otimes {}^{+}3$ by the definition of $\otimes$. □

Integer addition preserves order, as we might expect. That is, for any integers p, q, and m,

(8.11) $\boxed{p \otimes q \text{ implies } p \oplus m \otimes q \oplus m.}$

Justification of this property is analogous to that of Property (4.20) in Section 4.5, and is left as an exercise.

We might also expect multiplication of integers to preserve order, but that is not always the case. Multiplication by a *positive* integer preserves order (and hence Property 4.21 is maintained by $\otimes$), but multiplication by a negative integer reverses the order: For any integers p, q, and m,

(8.12)

if m is positive, then $p \otimes q$ implies $p \odot m \otimes q \odot m$;
if m is negative, then $p \otimes q$ implies $p \odot m \circledast q \odot m$.

Example 5 Multiplying the inequality ${}^{+}2 \otimes {}^{+}5$ by ${}^{+}3$, we get ${}^{+}6 \otimes {}^{+}15$, which is true because ${}^{+}6 \oplus {}^{+}9 = {}^{+}15$ and ${}^{+}9$ is positive. On the other hand, if we multiply ${}^{+}2 \otimes {}^{+}5$ by ${}^{-}3$, we get ${}^{-}6$ on the left and ${}^{-}15$ on the right. But

$$^{-}15 \otimes {}^{-}6 \quad \text{because} \quad {}^{-}15 \oplus {}^{+}9 = {}^{-}6$$

Thus, the inequality has been reversed. □

The Number Line

The definition of ⊘ arranges the integers in an order pattern much like that of the whole numbers. As in **W**, each integer has a next one in size order, found by adding 1. The main difference between the order patterns of **W** and **I** is that **W** contains a smallest number, 0, whereas **I** does not. The order relation on these two sets can be pictured by extending the number line for **W** (introduced in Section 4.5) to include the negative integers. The number-line arrangements for **W** and **I** are pictured in Figure 8.4.

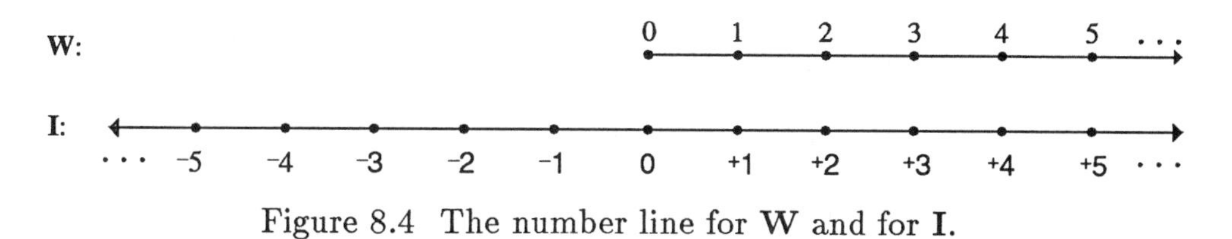

Figure 8.4 The number line for **W** and for **I**.

As you can see, **W** starts with 0, its smallest number, and marches off, step by step, into the sunset (if your book is headed south), whereas **I** comes from somewhere in the dawn of smaller and smaller negative numbers, steps up to 0, and then repeats the pattern of **W** with the positive integers. Notice, in particular, that the number line illustrates two elementary, but useful, observations:

(8.13) An integer is positive if it is greater than 0; an integer is negative if it is less than 0.

Example 6 Figure 8.5 shows how the number line is presented at the sixth-grade level. Notice how order is illustrated. □

Absolute Value

The foregoing observations support the claim that the system of integers serves as a useful model for the six situations described in Examples 1–6 of Section 8.1. Those examples (and many, many others) share three common features—a unit to measure quantity, a two-sided notion of direction, and a point of origin. These three concepts are reflected in the structure of the number line. The positive and negative directions lie on opposite sides of 0,

Integers on the Number Line

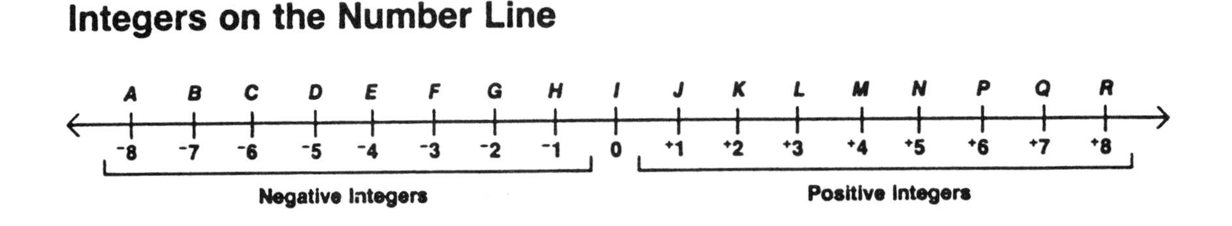

You can show integers on a number line. On a horizontal number line, positive integers are shown to the right of zero, and negative integers are shown to the left of zero. Zero is neither positive or negative.

A. Which is greater, $^{-}2$ or $^{-}5$?

The integer farther to the right on the number line is greater.

$^{-}2$ is the greater integer.

$$^{-}2 > {}^{-}5$$

B. Arrange the integers in order from least to greatest.

$^{-}1 \quad ^{+}4 \quad ^{-}3 \quad 0$

List the integers as they appear from left to right on the number line.

$^{-}3 \quad ^{-}1 \quad 0 \quad ^{+}4$

C. Name the integer for point *D* on the number line above.

$^{-}5$ is point *D*.

Figure 8.5 INVITATION TO MATHEMATICS, Grade 6, p. 354

the point of origin. The unit of measure is the distance between 0 and 1, so the quantity associated with each integer point is just the number of "unit steps" between that point and 0. Because this idea of distance "counts" unit steps, it always yields a positive number (or zero); this number is called the *absolute value* of the integer.

Definition The **absolute value** of an integer is the integer itself if it is positive or zero, and its additive inverse if it is negative. In symbols, we write $|x|$ for the absolute value of the integer x. Thus,

$$|x| = \begin{cases} x & \text{if } x \circledcirc\!\!\!\geq 0 \\ -x & \text{if } x \circledcirc\!\!\!< 0 \end{cases}$$

[*Remember*: If x is negative, then $-x$ (its "opposite") is positive.]

Example 7 The distance between $^+5$ and 0 is $|^+5|$; since $^+5 \geq 0$, $|^+5| = ^+5$.

The distance between $^-7$ and 0 is $|^-7|$; since $^-7 < 0$, $|^-7| = -(^-7) = ^+7$.

The distance between 0 and 0 is $|0|$; since $0 \geq 0$, $|0| = 0$. □

Absolute value can be used to define the more general idea of distance between any two integers. The occasional confusion in computing distances can usually be eased by drawing a diagram of part of the number line. For instance, consider the question: What is the distance (the number of units) between $^+2$ and $^-3$? Figure 8.6 makes it easy to see that the correct answer is $^+5$.

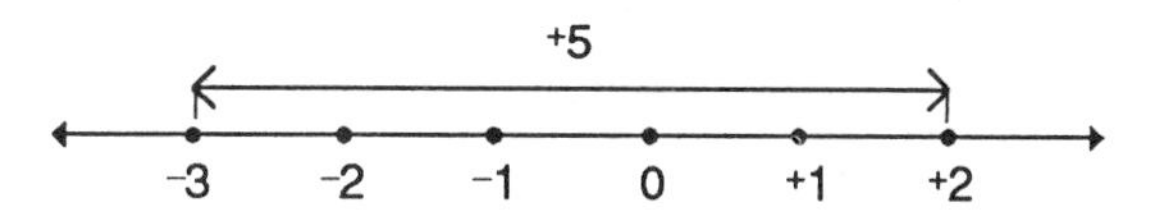

Figure 8.6 The distance between $^+2$ and $^-3$.

But how can we use the integers $^+2$ and $^-3$ without drawing a picture to compute the distance between them? Reasoning backwards, the fact that the distance must always be positive suggests that $^+5$ may be viewed as either $|^+5|$ or $|^-5|$, which are $|2 \ominus {}^-3|$ and $|^-3 \ominus {}^+2|$, respectively. This leads us to a general definition:

Definition The **distance** between two integers p and q is $|p \ominus q|$, the absolute value of their difference.

Example 8 The distance between $^+4$ and $^-5$ is $|^+4 \ominus {}^-5| = |^+9| = 9$. It is the same as the distance between $^-5$ and $^+4$, since $|^-5 \ominus {}^+4| = |^-9| = 9$. □

Example 9 In general, $|p \ominus q|$ is not the same as $|p| \ominus |q|$. For instance,

$$|^-4 \ominus {}^+1| = |^-4 \oplus {}^-1| = |^-5| = {}^+5$$

but

$$|^-4| \ominus |^+1| = {}^+4 \ominus {}^+1 = {}^+4 \oplus {}^-1 = {}^+3$$

□

A reasonable question to ask at this point is: How does absolute value behave with respect to addition and multiplication? Before giving a general answer, we look at some specific examples.

Example 10

$$|^-3 \odot ^+4| = |^-12| = ^+12$$
$$|^-3| \odot |^+4| = ^+3 \odot ^+4 = ^+12$$

Similarly,

$$|^-5 \odot ^-1| = |^+5| = ^+5$$
$$|^-5| \odot |^-1| = ^+5 \odot ^+1 = ^+5$$

□

Example 11

$$|^-3 \oplus ^+4| = |^+1| = ^+1 \quad \text{and} \quad |^-3| \oplus |^+4| = ^+3 \oplus ^+4 = ^+7$$

Therefore,

$$|^-3 \oplus ^+4| \circledless |^-3| \oplus |^+4|$$

$$|^-5 \oplus ^-1| = |^-6| = ^+6 \quad \text{and} \quad |^-5| \oplus |^-1| = ^+5 \oplus ^+1 = ^+6$$

Therefore,

$$|^-5 \oplus ^-1| = |^-5| \oplus |^-1|$$

□

Examples 10 and 11 illustrate the two fundamental rules governing the absolute values of sums and products: For any integers p and q,

(8.14)
$$\boxed{|p \odot q| = |p| \odot |q|}$$

(8.15)
$$\boxed{|p \oplus q| \;\; \circledleq \;\; |p| \oplus |q|}$$

Properties of the Integers

This completes the construction of the system of integers. If you refer to the problem described in Section 8.1, you will see that we have done all that we set out to do: We have extended the whole-number system to a larger system that is closed under subtraction. For review, we summarize the main properties of this larger system, the integers.

The set **I** of integers is the set of all equivalence classes of ordered pairs of whole numbers with respect to the relation $\sim$ defined by

$$(a, b) \sim (c, d) \quad \text{whenever } a + d = b + c$$

Each of these classes can be written in one of three standard forms:

$$\begin{aligned} [n, 0] &= \;^+n \quad \text{(positive integers)} \\ [0, 0] &= \;0 \quad \text{(zero)} \\ [0, n] &= \;^-n \quad \text{(negative integers)} \end{aligned}$$

The set $\mathbf{I}^+$ of positive integers may be identified with the set $\mathbf{W}$ of whole numbers by matching each integer ^+n in $\mathbf{I}$ with the whole number n in $\mathbf{W}$. Thus, the set of integers may be symbolized by

$$\mathbf{I} = \{\ldots, {}^-3, {}^-2, {}^-1, 0, {}^+1, {}^+2, {}^+3, \ldots\}$$

or by

$$\mathbf{I} = \{\ldots, {}^-3, {}^-2, {}^-1, 0, 1, 2, 3, \ldots\}$$

The operations of addition and multiplication are defined on $\mathbf{I}$ and have the following properties:

- Both operations are commutative: For all $p, q \in \mathbf{I}$,
$$p \oplus q = q \oplus p \qquad p \odot q = q \odot p$$
- Both operations are associative: For all $m, p, q \in \mathbf{I}$,
$$(m \oplus p) \oplus q = m \oplus (p \oplus q) \qquad (m \odot p) \odot q = m \odot (p \odot q)$$
- Both operations have identity elements: For all $p \in \mathbf{I}$,
$$0 \oplus p = p \oplus 0 = p \qquad 1 \odot p = p \odot 1 = p$$
- Multiplication is left and right distributive over addition: For all $m, p, q \in \mathbf{I}$,
$$m \odot (p \oplus q) = (m \odot p) \oplus (m \odot q)$$
$$(p \oplus q) \odot m = (p \odot m) \oplus (q \odot m)$$
- $\oplus$ and $\odot$ on $\mathbf{I}$ agree with $+$ and $\cdot$ on $\mathbf{W}$: For any $a, b \in \mathbf{W}$,
$$^+a \oplus {}^+b = [a, 0] \oplus [b, 0] = [a + b, 0] = {}^+(a + b)$$
$$^+a \odot {}^+b = [a, 0] \odot [b, 0] = [a \cdot b, 0] = {}^+(a \cdot b)$$

Thus, it is accurate to regard the integer operations as extensions of the corresponding whole-number operations.

Every integer $x = [a, b]$ has an additive inverse, $-x = [b, a]$. That is,

$$x \oplus -x = -x \oplus x = 0$$

This allows us to describe subtraction by the Fundamental Subtraction Property:

$$p \ominus q = p \oplus -q$$

The set $\mathbf{I}$ is closed under subtraction; that is, for every two integers p and q, $p \ominus q$ is always an integer.

An order relation $\otimes$ is defined on $\mathbf{I}$ by

$$p \otimes q \text{ if } p \oplus d = q \text{ for some } d \in \mathbf{I}^+$$

This relation has the following properties, where m, p, and q denote arbitrary integers:

- Transitivity: If $m \olessthan p$ and $p \olessthan q$, then $m \olessthan q$.
- Trichotomy: Exactly one of the following is true: $p = q$, $p \olessthan q$, $q \olessthan p$.
- Addition preserves order: $p \olessthan q$ implies $p \oplus m \olessthan q \oplus m$.
- Multiplication property:
  If m is positive, then $p \olessthan q$ implies $p \odot m \olessthan q \odot m$;
  if m is negative, then $p \olessthan q$ implies $p \odot m \ogreaterthan q \odot m$.
- $\olessthan$ on $\mathbf{I}^+$ agrees with $<$ on $\mathbf{W}$.

The division process can be defined on **I** just as it was on **W**, but it still is not always possible to find answers to division questions; so division is not an operation on **I**.

Now that the construction of the integers is complete, it will be useful to fix one notational inconvenience. In this chapter we have used circles around the signs $+$, $-$, $\cdot$, $<$, etc., to distinguish their integer usage from their whole-number usage. However, we have shown that the operations on **I** can be regarded as extensions of the operations on **W**, so there is no need to continue this distinction; the integer operations can be used for both systems. Thus, *after this section, the symbols for integer operations and order relations will no longer be encircled.*

Exercises 8.5

For Exercises 1–8, justify the given inequality by applying the definition of $\olessthan$.

1. ${}^+3 \olessthan {}^+8$ **2.** ${}^+4 \olessthan {}^+5$

3. ${}^-5 \olessthan {}^-1$ **4.** ${}^-2 \olessthan 0$

5. $0 \olessthan {}^+4$ **6.** ${}^-9 \olessthan {}^-6$

7. ${}^-6 \olessthan {}^+8$ **8.** ${}^-7 \olessthan {}^+3$

9. Give a specific numerical example to illustrate the Transitive Law for $\olessthan$ in **I**.

10. Prove the Transitive Law for $\olessthan$ in **I**. [*Hint*: Look back at the discussion of Statement (4.18).]

11. Give examples to show that $\olessthan$ is neither reflexive nor symmetric.

12. (a) Give a specific numerical example to illustrate the fact that integer addition preserves order. [This is Statement (8.11).]

(b) Prove Statement (8.11) for $\olessthan$. [*Hint*: Look at the analogous argument for Statement (4.20).]

For Exercises 13–34, compute the value of the given expression.

13. $|{}^+7|$ **14.** $|{}^-7|$

15. $|{}^+9 \oplus {}^+4|$ **16.** $|{}^+9 \ominus {}^+4|$

17. $|{}^+4 \ominus {}^+9|$ **18.** $|{}^-9 \oplus {}^+4|$

19. $|{}^+9 \odot {}^+4|$ **20.** $|{}^-9 \oplus {}^-4|$

21. $|{}^-9| \oplus |{}^-4|$ **22.** $|{}^-9| \ominus |{}^-4|$

23. $|{}^-9| \odot |{}^-4|$ **24.** $|-({}^+9 \ominus {}^+4)|$

25. $-|{}^+9 \oplus {}^+4|$ **26.** $-|{}^+4 \ominus {}^+9|$

27. $-(|{}^+4| \ominus |{}^+9|)$

28. the distance between ${}^+5$ and ${}^+8$

29. the distance between ${}^-5$ and ${}^+8$

30. the distance between ${}^-5$ and ${}^-8$

31. the distance between ${}^-5$ and ${}^+5$

32. the distance between integers p and $p \oplus {}^+5$

33. the distance between integers p and $p \ominus {}^{+}5$

34. the distance between integers p and ${}^{+}5 \ominus p$

For Exercises 35–50, state whether the given inequality is *true* or *false*.

35. ${}^{+}3 \textcircled{<} {}^{+}10$

36. ${}^{+}10 \textcircled{>} {}^{+}3$

37. ${}^{+}3 \textcircled{\le} {}^{+}10$

38. ${}^{+}3 \textcircled{>} {}^{+}10$

39. ${}^{+}3 \textcircled{\le} {}^{+}3$

40. ${}^{+}3 \textcircled{>} {}^{+}3$

41. ${}^{-}3 \textcircled{<} {}^{+}3$

42. ${}^{-}3 \textcircled{\le} {}^{+}3$

43. ${}^{-}3 \textcircled{<} {}^{-}10$

44. ${}^{-}3 \textcircled{>} {}^{-}10$

45. $|{}^{-}3| \textcircled{<} {}^{-}10$

46. $|{}^{-}10| \textcircled{<} {}^{-}3$

47. $|{}^{-}5 \oplus {}^{+}5| \textcircled{>} 0$

48. $|{}^{-}5| \oplus |{}^{+}5| \textcircled{>} 0$

49. $|{}^{-}2 \oplus {}^{+}3| \textcircled{<} |{}^{-}2 \ominus {}^{+}3|$

50. $|{}^{-}2 \odot {}^{+}3| \textcircled{<} |{}^{-}2| \odot |{}^{+}3|$

51. Use the definition of $\textcircled{<}$ to prove Property (8.13):

(a) 0 is less than every positive integer, and

(b) every negative integer is less than 0.

(*Hint*: Try some examples first.)

8.6 Interpretations of Integers

In the first section of this chapter we introduced some examples that typify how negative numbers occur in everyday experience. Now that we have developed a system containing these numbers, it is time to return to an examination of situations that use them. Such situations are called *interpretations* of the integers.

Look back at Examples 1–6 of Section 8.1, on page 240. Each of these six examples describes a reasonable, useful interpretation of the expression ${}^{+}3 + {}^{-}5$. In each case, the system of integers provides a model for the interpretation. These six situations have three common features—a unit of quantity, a sense of direction, and a point of origin, as in Table 8.1.

Interpretation	Quantity	Direction	Origin
Temperature	degrees	above or below 0	0
Money	dollars	to the good or overdrawn	no money
Space travel	minutes	before or after blast-off	blast-off
Stock market	points	gain or loss	original price
Mother May I?	giant steps	forward or back	starting position
Football	yards	gain or loss	line of scrimmage

Table 8.1 Six interpretations of **I**.

The system of integers provides a suitable approximate model for all these interpretations because it deals directly with the concepts of quantity, direction, and point of origin.[4] Each integer has a quantity (its absolute value), and each has a direction (its sign—positive or negative), except for zero, which serves as the point of origin. If we wanted diagrams to illustrate each of the interpretations, we could ask an artist to draw detailed pictures of each situation, or we could just make a diagram that stripped each situation of all but the essential numerical ideas. These diagrams would bear a striking resemblance to each other; in fact, they could all be drawn in the same way on a number line, as shown in Figure 8.7. Thus, besides providing a model for the separate situations, the integers serve to highlight the similarity of their underlying mathematical structure.

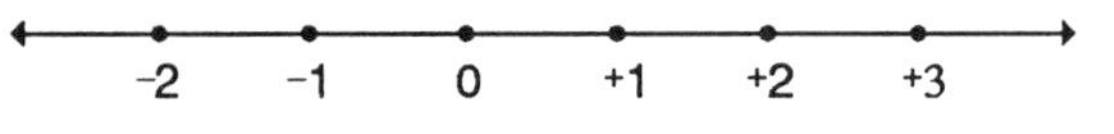

Figure 8.7 A diagram for all six examples.

Example 1 Look at Figure 8.8, a sixth-grade text page about the integers. Several interpretations are suggested there. Can you describe the quantity, direction, and point of origin for each one? □

Vectors

Integers can be illustrated by using directed arrows superimposed on or drawn above the number line, pointing to the right for positive integers and to the left for negative integers. The length of the arrow represents the absolute value of the integer. Zero is shown as a point, a "zero-length arrow" with no direction. Just as every integer can be represented by many different ordered pairs of whole numbers, with $(n, 0)$ or $(0, n)$ as the standard forms, so may the arrows be located anywhere on the number line. Such directed arrows are called **vectors**. Their standard position is that in which the tail corresponds to 0. Vectors are useful in physics and in other areas of study that analyze forces on objects. As a teacher, you may find vectors useful in explaining integer concepts. Note, in particular, that integers have two important characteristics—magnitude (absolute value) and direction (positive or negative)—whereas whole numbers have magnitude only. Thus, a vector is an ideal way to represent an integer visually.

[4]The accuracy of the model in several of these cases is somewhat limited by the absence of fractional parts for the unit of quantity, but this is not a serious drawback for our purposes.

Practice Write an integer to represent each situation.

1. Profit of $25
2. Loss of $85
3. Weight loss of 21 pounds
4. Weight gain of 13 pounds
5. Three flights of stairs down
6. Seven flights of stairs up
7. 10° below zero
8. 62° above zero
9. 1,012 feet above sea level
10. 635 feet below sea level
11. Forward 35 steps
12. Backward 26 steps
13. $65 savings deposit
14. $125 savings withdrawal
15. Eight minutes before the bell
16. Fifteen minutes after the bell
17. Sales decrease of $830
18. Sales increase of $250
19. Profit of $350
20. Loss of $625

*21. Altitude at sea level

*22. Time of liftoff

Give the opposite of each integer.

23. $^{+}1$	24. $^{-}3$	25. $^{+}9$	26. $^{-}8$	27. $^{-}15$	28. $^{+}20$
29. $^{-}62$	30. $^{+}34$	31. $^{+}160$	32. $^{-}183$	33. $^{-}211$	34. $^{+}235$

Apply Write an integer for each situation.

35. Eleven seconds after liftoff
36. Nine seconds before liftoff
37. At 117 seconds before liftoff, the flight crew begins pressurizing the liquid hydrogen tank.
38. At about 7 seconds before liftoff, the main engines are started.
39. In sunlight in space, the temperature can be as high as 200°C above zero.
40. On the night side of the orbit, the temperature is as cold as 100°C below zero.

Figure 8.8 INVITATION TO MATHEMATICS, Grade 6, p. 353

Example 2 In Figure 8.9, vectors for the integers $^{+}3$ and $^{-}5$ are drawn in standard position. □

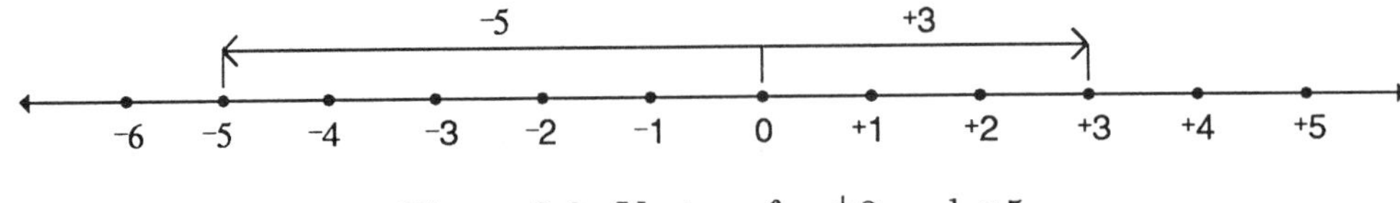

Figure 8.9 Vectors for $^{+}3$ and $^{-}5$.

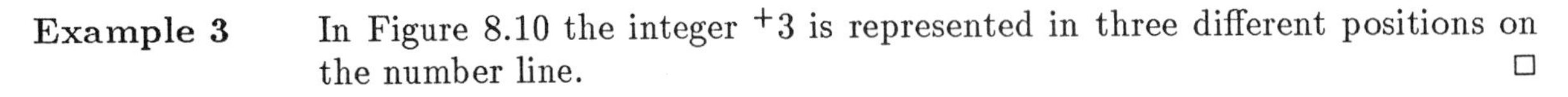

Example 3 In Figure 8.10 the integer $^{+}3$ is represented in three different positions on the number line. □

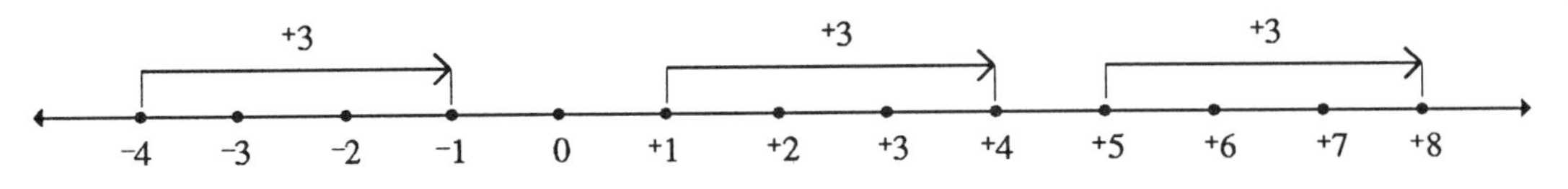

Figure 8.10 Three copies of the vector $^{+}3$.

Addition of integers can be illustrated using vectors, by a process called **vector addition**. To find $a + b$:

1. Draw the vector for the first integer a in standard position.
2. Draw the vector for the second integer b, placing the tail of b at the tip of a.
3. Draw a vector in standard position (its tail at zero) with its tip at the tip of the vector for b. This vector represents $a + b$.

In order that a, b, and $a + b$ do not overlap each other in your drawing, you should place each vector at a different level above the number line, using the first level for the first addend, the second level for the second addend, and the third level for their sum. There is another pedagogical reason for this arrangement, which we shall discuss later.

Example 4 The sum $^{+}2 + {}^{+}4$ can be determined by the diagram in Figure 8.11. □

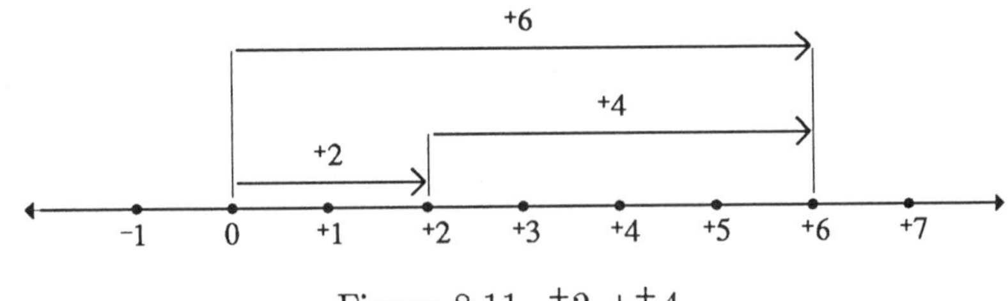

Figure 8.11 $^{+}2 + {}^{+}4$

Example 5 The sum $^{+}3 + {}^{-}5$ can be determined by the diagram in Figure 8.12. □

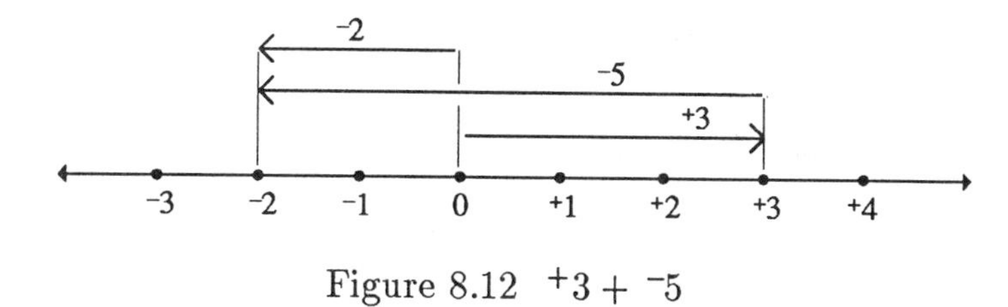

Figure 8.12 $^{+}3 + {}^{-}5$

Example 6 The sum $^{-}2 + {}^{-}3$ can be determined by the diagram in Figure 8.13. □

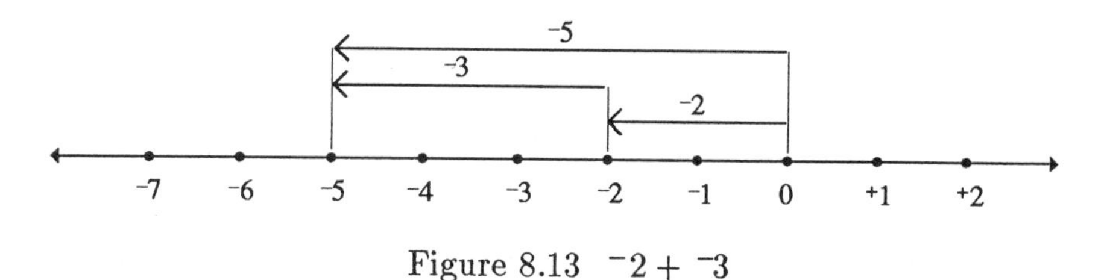

Figure 8.13 $^{-}2 + {}^{-}3$

Example 7 Figure 8.14 is a composite of pages from a sixth-grade text; it illustrates addition of integers. Observe how closely this represents vector addition. In the preceding examples each addition problem is represented by three vectors; the sixth-grade diagrams show only two vectors. Which one is missing? What is done to compensate for this? Is a third vector implied? □

Exercises 8.6

For Exercises 1–8, use vector addition to illustrate the given sum.

1. $^{+}3 + {}^{+}4$
2. $^{+}3 + {}^{-}7$
3. $^{-}2 + {}^{+}5$
4. $^{-}4 + {}^{-}2$
5. $^{+}5 + {}^{-}2$
6. $^{-}4 + {}^{+}1$
7. $^{+}2 + {}^{-}3 + {}^{+}4$
8. $^{-}2 + {}^{-}3 + {}^{+}4$

For Exercises 9–13, determine the integer-addition problem represented by the given interpretation. Also specify the point of origin in each case.

9. After a 15-yard penalty, the Redskins started a series of downs at their 30-yard line. On first down, they advanced to their 47-yard line. On second down, the quarterback was sacked and they ended up at their 38-yard line. What was the net gain (or loss) for these two plays?

10. The Rams began at their opponents' 9-yard line. On first down, they were thrown back to the 13-yard line. On second down, they made it to the 1-yard line. On third down, they made no gain. On fourth down, they were pushed back to the 3-yard line. What was the net gain (or loss)?

11. At midnight the thermometer read 1° below zero. At 5:00 a.m. it read 8° below. At 9:00 a.m. it read 7° above zero. At noon it read 27° above zero. What was the net increase in temperature for this 12-hour period?

Find $^{-}5 + {}^{-}7$.

Use a number line. Start at zero and move 5 units to the left for $^{-}5$. Then move 7 units farther left for $^{-}7$.

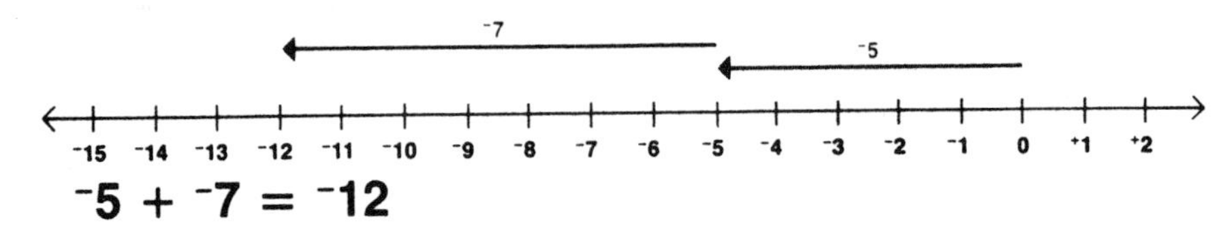

$$^{-}5 + {}^{-}7 = {}^{-}12$$

Erica's total withdrawals for March were $12.

B. Find $^{+}6 + {}^{+}3$.

Start at zero and move 6 units to the right for $^{+}6$. Then move 3 units farther right for $^{+}3$.

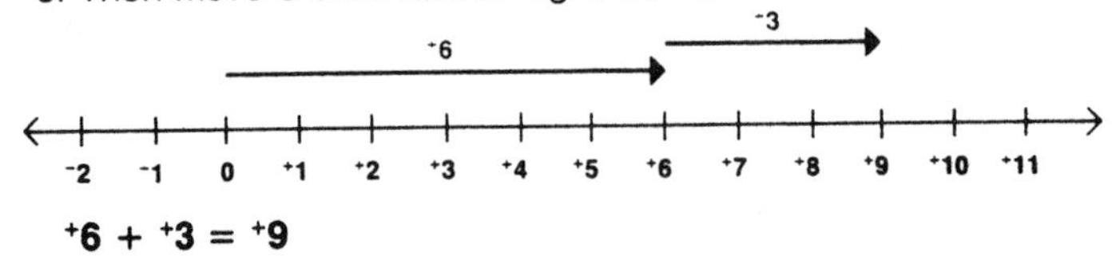

$$^{+}6 + {}^{+}3 = {}^{+}9$$

To add integers with the same sign, add without regard to the signs. Then use the sign of the numbers in your answer.

Adding Integers: Different Signs

A. Juan works at a Junior Achievement company. He earned $15 ($^{+}15$). Then he spent $9 ($^{-}9$) for a record album. How much money does he have left?

Find $^{+}15 + {}^{-}9$.

Use a number line. Start at zero and move 15 units to the right for $^{+}15$. Then move 9 units to the left for $^{-}9$.

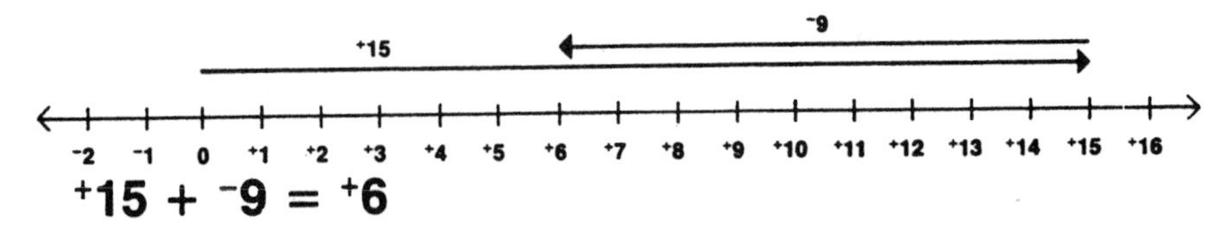

$$^{+}15 + {}^{-}9 = {}^{+}6$$

To add two integers with different signs, consider the distance each integer is from zero. Subtract the shorter distance from the longer distance. Then use the sign of the number farther from zero in your answer.

Figure 8.14 INVITATION TO MATHEMATICS, Grade 6, pp. 356 & 358

12. At 7:00 a.m. the thermometer read 70° above zero. By noon it had risen to 83°, but by 7:00 p.m. it had fallen to 68°. What was the net increase (or decrease) in temperature for this 12-hour period?

13. At 7:00 a.m. the thermometer read 65° above zero. By noon it had risen by 15°, but then by 10:00 p.m. it had dropped 20°. What was the net increase (or decrease) in temperature from 7:00 a.m. to 10:00 p.m.?

14. Describe two interpretations for each sum in Exercises 1–8.

15. Illustrate by vector addition that addition of integers is commutative, using the example $^{+}6 + {}^{-}2$.

16. Use two different vector-addition examples to show that the sum of an integer and its additive inverse is zero.

In Exercises 17–20 (Figures 8.15–8.18), the second addend (middle level) is missing from the diagram. The lowest level represents the first addend and the highest level represents the sum. Draw the vector for the missing addend.

17.

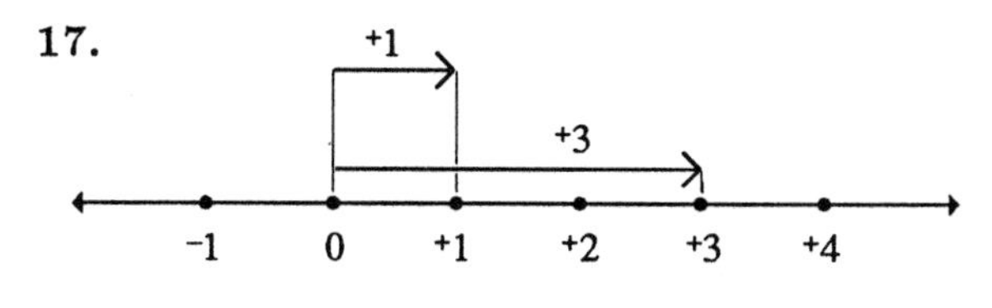

Figure 8.15 Exercise 17.

18. See Figure 8.16.

19. See Figure 8.17.

20. See Figure 8.18.

For Exercises 21–26, use vector addition to find the missing addend.

21. $^{+}5 + x = {}^{+}9$ **22.** $^{-}2 + x = {}^{+}5$

23. $^{+}3 + x = {}^{-}3$ **24.** $^{-}5 + x = {}^{-}7$

25. $^{-}8 + x = {}^{-}3$ **26.** $^{+}6 + x = {}^{+}2$

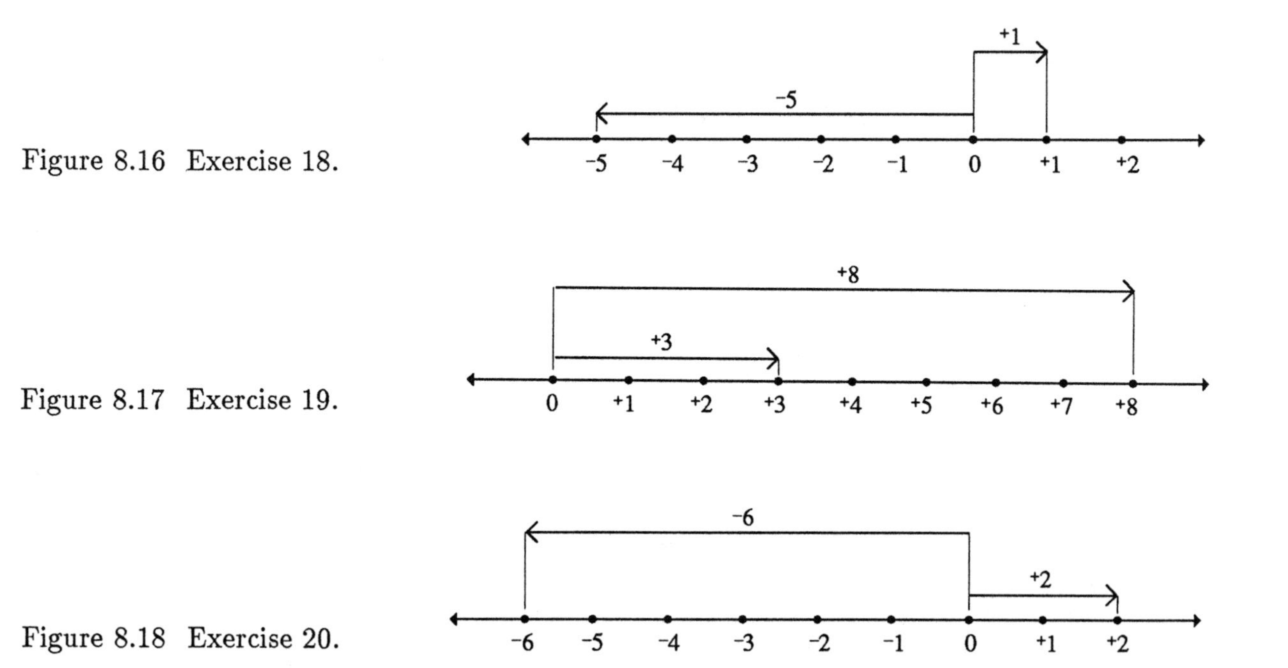

Figure 8.16 Exercise 18.

Figure 8.17 Exercise 19.

Figure 8.18 Exercise 20.

8.7 Algorithms for Adding and Subtracting Integers

Up to now, we have considered the operations on the integers strictly in terms of representative ordered pairs of whole numbers:

$$[a,b] + [c,d] = [a+c, b+d]$$
$$[a,b] - [c,d] = [a+d, b+c] \qquad \text{(Why?)}$$
$$[a,b] \cdot [c,d] = [ac+bd, ad+bc]$$

where $a, b, c, d \in \mathbf{W}$. This means that, before a problem in integer arithmetic can be solved, it must be reduced to a (longer) problem in whole-number arithmetic by means of these operation definitions. Using the definitions every time we want to add, subtract, or multiply two integers is not difficult, but it is time-consuming. As you know from previous experience, there are much more efficient methods for determining sums, differences, products, and quotients of signed numbers. In the remainder of this chapter we examine some of these methods—called *integer algorithms*—and the reasons why they work.

Our purpose in analyzing these algorithms is to help you see that the "rules for signed numbers" you learned in elementary school (and will soon teach to children) are direct, logical consequences of the way the integer operations have been defined. Our analyses in this chapter will not be as formal as those given in Chapter 6 for the whole-number algorithms, even though they certainly could be done that way. Rather, we shall rely more on an intuitive understanding of signed numbers, particularly as related to interpretations for which the integers serve as a model.

Note: In this section and the next, we use the previously established fact that the arithmetic of nonnegative integers is the same as the arithmetic of whole numbers. (See Sections 8.2–8.5.) Because of this, the absolute values of integers can be treated as if they were the corresponding whole numbers, and the rules of whole-number arithmetic can be applied to these absolute values. In the interest of efficiency, sometimes we shall blur the formal distinction between nonnegative integers and whole numbers, writing such expressions as $|{}^{-}5| = {}^{+}5 = 5$, even though technically the positive integer ${}^{+}5$ and the whole number 5 are in different number systems.

Addition

The algorithms for the sum of two integers depend on the signs of the integers being added. Specifically, what we do is determined by which of the following cases fits the particular sum $p + q$ we are trying to compute.

Case I: p and q are both positive.
Case II: p and q are both negative.

Case III: p is positive and q is negative.

These cases do not exhaust all possible combinations of integers. The situation where either p or q is 0 is not represented because 0 is neither positive nor negative; however, no algorithm is needed for this case because 0 is the additive identity. Sums such as $^-6 + ^+8$ are not explicitly represented either, because in none of these cases is the first addend negative and second addend positive. However, $^-6 + ^+8 = ^+8 + ^-6$, by commutativity of addition, so these sums are included in Case III.

The key idea to remember about summation algorithms, regardless of type or case, is this:

> *Since every nonzero integer has a direction and an absolute value, any algorithm to calculate the sum of two integers should be designed to determine both the direction and the absolute value of the sum.*

The algorithm for Case I is the simplest of the three because the arithmetic of the positive integers is a copy of the arithmetic of the whole numbers. Thus, $^+129 + ^+235$ can be treated as the whole-number sum $129 + 235 = 364$ (by the whole-number Addition Algorithm), which corresponds to the positive integer $^+364$. We can write this more compactly as

$$^+129 + ^+235 = 129 + 235 = 364 = ^+364$$

In Case I, then, the direction of the sum is positive and its absolute value can be found by adding the absolute values of the summands, treating them as whole numbers and using the whole number Addition Algorithm (if necessary). The vector diagram (Figure 8.19) shows a first vector pointing right, followed by a second vector that also points right, resulting in a vector sum that points right.

The algorithm for Case I simply says that the direction of the sum of two positive integers is positive, and its absolute value is found by adding the absolute values of the two integers (as if they were whole numbers).

(8.16) $\quad$ If $p > 0$ and $q > 0$, then $p + q = |p| + |q|$.

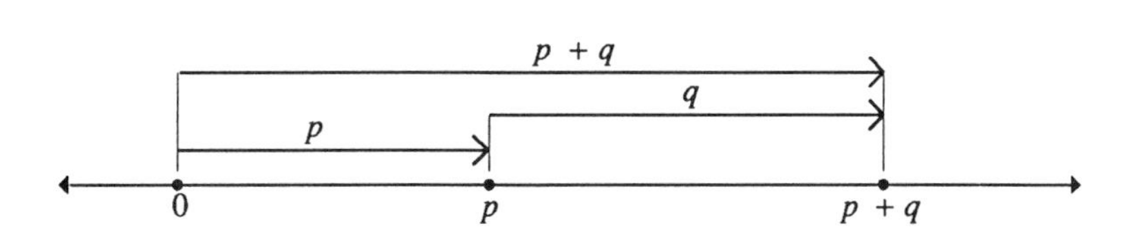

Figure 8.19 Case I.

Notice that "$p > 0$ and $q > 0$" says p and q are positive. Since $|p|$ and $|q|$ can be treated as whole numbers, the sum $|p| + |q|$ also corresponds to a whole number and hence must also be positive.

Example 1 If a child takes p giant steps forward and then q more giant steps forward, she will have taken $p + q$ giant steps forward. □

For the sum of two negative integers (Case II), the situation is analogous to Case I. The sum of two negative integers is negative, and its absolute value is the sum of the absolute values of the two integers. The vector representation is like that for Case I, except that all vectors point left, as in Figure 8.20.

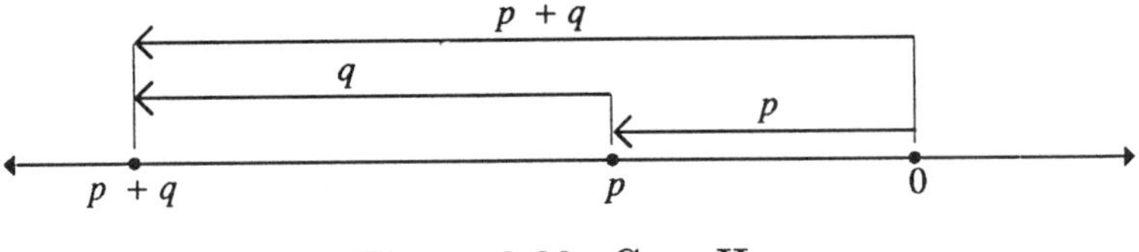

Figure 8.20 Case II.

Example 2 If the temperature drops p degrees and then drops q more degrees, it has dropped $p + q$ degrees altogether. □

The algorithm for the sum of two negative integers says: The direction of the sum of two negative integers is negative; to find its absolute value, add the absolute values of the two integers.

(8.17) If $p < 0$ and $q < 0$, then $p + q = -(|p| + |q|)$.

Note: "$p < 0$ and $q < 0$" says that both p and q are negative; $-(|p| + |q|)$ is negative because $|p| + |q|$ is positive.

Using vector addition in Case III, a positive and a negative integer, we can see that the first vector points right and the second one points left. The question is whether the tip of the second vector ends up to the right or the left of 0. Figures 8.21 and 8.22 show that $p + q$ can be either positive or

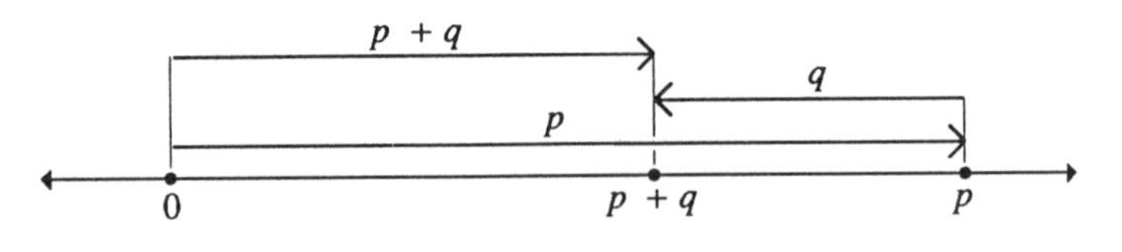

Figure 8.21 Case III(b).

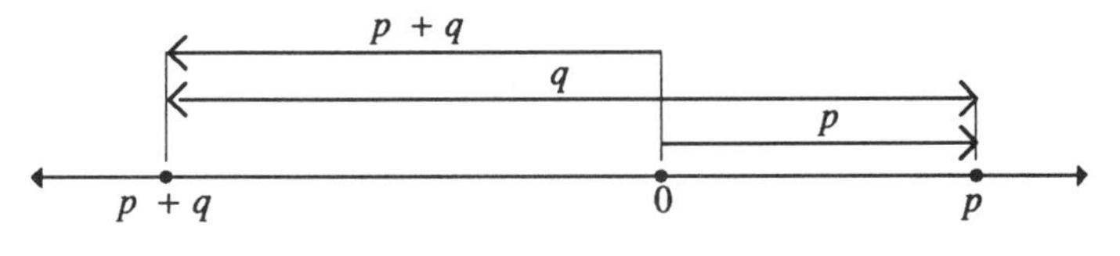

Figure 8.22 Case III(c).

negative, depending on the relative lengths of the two addends. If p is longer than q, then $p+q$ is positive; if q is longer than p, then $p+q$ is negative. Not pictured is the case where p and q are the same length. Thus, Case III should be separated into three subcases:

Case III: p is positive, q is negative, and

III(a). $|p| = |q|$ III(b). $|p| > |q|$ III(c). $|q| > |p|$

Example 3 $^{+}7 + ^{-}7$ is classified under III(a) because $|^{+}7| = |^{-}7|$.
$^{+}15 + ^{-}12$ comes under III(b) because $|^{+}15| > |^{-}12|$.
$^{+}8 + ^{-}9$ comes under III(c) because $|^{-}9| > |^{+}8|$. □

Example 4 $^{-}18 + ^{+}15$ is classified under III(c) because $^{-}18 + ^{+}15 = ^{+}15 + ^{-}18$ and $|^{-}18| > |^{+}15|$. □

Case III(a) describes the sum of an integer and its additive inverse, so no algorithm is needed; the sum of every such pair is zero.

In Case III(b), the vector to the right is longer than the vector to the left, implying that the sum points to the right, as in Figure 8.21. Since the second vector reverses the direction of the first, the net distance is that covered by the positive vector minus the negative one. Thus:

> If a negative integer is added to a positive integer of greater absolute value, then the direction of the sum is positive. To find its absolute value, subtract the absolute value of the negative integer from the absolute value of the positive one.

(8.18) $$\text{If } p > 0,\ q < 0, \text{ and } |p| > |q|, \text{ then } p + q = |p| - |q|$$

In Case III(c), the positive vector is shorter than the negative one, so we end up to the left of zero. How far to the left is determined by the difference between the lengths of the vectors, as shown in Figure 8.22. Thus:

> If a positive integer is added to a negative integer of greater absolute value, then the direction of the sum will be negative.

To find its absolute value, subtract the absolute value of the positive integer from the absolute value of the negative one.

In symbols:

$$\text{(8.19)} \quad \boxed{\text{If } p > 0,\ q < 0, \text{ and } |q| > |p|, \text{ then } p + q = -(|q| - |p|)}$$

Study the symbolic statements (8.18) and (8.19) carefully to make sure you understand how each one captures all the information in the verbal description preceding it. In particular, convince yourself that the expression $-(|q| - |p|)$ in (8.19) is always a negative integer. The following interpretations illustrate the reasonableness of these algorithms.

Example 5 If the temperature *rises* 8° and then *drops* 9°, it is 1° *below* the original temperature. [Case III(c)]

$$^{+}8 + ^{-}9 = -(9 - 8) = ^{-}1$$ □

Example 6 If a child takes 5 giant steps *forward* and then 2 giant steps *backward*, she ends up 3 giant steps *ahead* of her original position. [Case III(b)]

$$^{+}5 + ^{-}2 = 5 - 2 = ^{+}3$$ □

Example 7 If a football team *gains* 7 yards on the first down and *loses* 7 yards on the second down, it is back at the *original* line of scrimmage. [Case III(a)]

$$^{+}7 + ^{-}7 = 0$$ □

Example 8 If a cloud cover appears 18 minutes *before* blast-off and then dissipates 15 minutes *later*, it has cleared 3 minutes *before* blast-off. [Case III(c)]

$$^{-}18 + ^{+}15 = ^{+}15 + ^{-}18 = -(18 - 15) = ^{-}3$$ □

Thus, the choice of an algorithm for adding integers depends on the case a particular sum represents. Each case, summarized here in symbols, is a complete statement about how to determine both the direction and the absolute value of the sum. The absolute values are found by adding or subtracting whole numbers.

Case I: If $p > 0$ and $q > 0$, then $p + q = |p| + |q|$.
Case II: If $p < 0$ and $q < 0$, then $p + q = -(|p| + |q|)$.
Case III: If $p > 0$ and $q < 0$, and
[III(a)] if $|p| = |q|$, then $p + q = 0$.
[III(b)] if $|p| > |q|$, then $p + q = |p| - |q|$.
[III(c)] if $|q| > |p|$, then $p + q = -(|q| - |p|)$.

Subtraction

No new algorithms are needed for integer subtraction because

$$p - q = p + (-q)$$

by Statement (8.7), the Fundamental Subtraction Property. That is, the difference of two integers can be found by adding the first to the opposite of the second. Thus, any subtraction problem becomes an addition problem and can be solved by using the addition algorithms.

Sometimes students are surprised by their results when subtracting integers and doubt the answers they get. This often stems from their experience with subtraction in **W**, where the difference of two whole numbers is always less than the first one (if it exists at all). This may or may not be the case when integers are subtracted, as the next examples show.

Example 9 If the temperature was 2° above zero at noon and had been 8° below zero early in the morning, then the temperature *rose* 10°.

$$^{+}2 - {}^{-}8 = {}^{+}2 + {}^{+}8 = {}^{+}10$$ □

Example 10 If a person gets a notice that his checking account is overdrawn by \$5, but a week earlier he had had a balance of \$6, then he must have withdrawn \$11.

$$^{-}5 - {}^{+}6 = {}^{-}5 + {}^{-}6 = {}^{-}11$$ □

Example 11 If on second down the Giants are at their 17-yard line and they started on first down at their 19-yard line, then they have lost 2 yards.

$$^{+}17 - {}^{+}19 = {}^{+}17 + {}^{-}19 = {}^{-}2$$ □

Subtraction can be illustrated by vectors on the number line in two different ways. One, based on the definition of subtraction

$$p - q = x \quad \text{if} \quad q + x = p$$

is called the **missing-addend method**. The other, based on the Fundamental Subtraction Property

$$p - q = p + (-q)$$

is called the **additive-inverse method**. They are shown in the next two examples.

Example 12 Using the missing-addend method to find $^{+}5 - {}^{-}3$:

$$^{+}5 - {}^{-}3 = x \quad \text{if} \quad {}^{-}3 + x = {}^{+}5$$

To find the second addend we draw the vector-addition diagram shown in Figure 8.23. The second addend is represented by the middle arrow, which points to the right and is 8 units long. Therefore, $x = {}^{+}8$. □

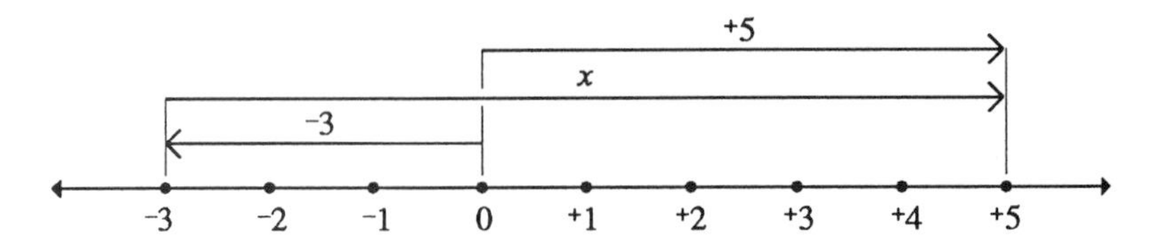

Figure 8.23 The missing-addend method.

Example 13 Using the additive-inverse method to find ${}^{+}5 - {}^{-}3$:

$$ {}^{+}5 - {}^{-}3 = {}^{+}5 + {}^{+}3 $$

We draw the vector sum, as shown in Figure 8.24. The sum is represented by the top arrow, which points to the right and is 8 units long. Therefore, $x = {}^{+}8$. □

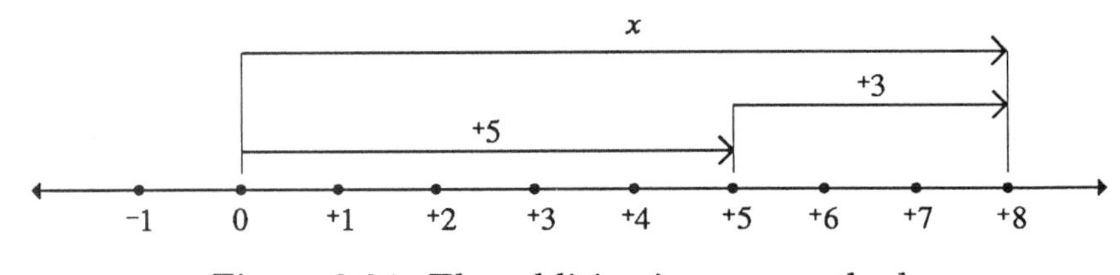

Figure 8.24 The additive-inverse method.

From a pedagogical standpoint there are advantages and disadvantages to each of these methods. The advantage of the missing-addend method is that the vectors representing all three numbers—p, q, and x—are visible in the diagram. A disadvantage of this method is that it is harder to do because one has to sketch in the first addend and the sum, then use these to find the position of the second addend. (This is why we suggested that the bottom, middle, and top vectors be used for first addend, second addend, and sum, respectively.) The advantage of the additive-inverse method is its simplicity; a disadvantage is that there is no vector to represent the integer q that is being "taken away."

PROBLEM-SOLVING COMMENT

A habit of *approximating answers* can help to eliminate calculation errors when you work with signed numbers. First decide whether your answer should be positive or negative, then estimate its absolute value. Useful estimation strategies include rounding off to multiples of a power of ten, drawing a vector sketch, and thinking of a natural interpretation. *Remember*: The purpose of making an estimate is to check on the reasonableness of your calculated answer. ◇

Example 14 Before calculating $^{+}79 + ^{-}32$, you might observe that the sum will be positive because the summand with the larger absolute value is positive; then estimate the answer by rounding the summands to the nearest multiple of 10:

$$^{+}79 + ^{-}32 \text{ is approximately } ^{+}80 + ^{-}30 = ^{+}50$$ □

Exercises 8.7

For Exercises 1–9, use an integer-addition algorithm to calculate the sum. Identify the case that applies.

1. $^{-}7 + ^{-}8$
2. $^{+}15 + ^{+}12$
3. $^{+}16 + ^{-}8$
4. $^{-}9 + ^{+}6$
5. $^{+}7 + ^{-}25$
6. $0 + ^{+}8$
7. $^{-}13 + ^{-}21$
8. $^{-}45 + 0$
9. $^{-}13 + ^{+}13$

For Exercises 10–18, *first* estimate the sum, *then* use an integer addition algorithm to calculate it. Identify the case that applies.

10. $^{+}21 + ^{+}32$
11. $^{-}81 + ^{+}22$
12. $^{+}38 + ^{-}19$
13. $^{+}53 + ^{-}53$
14. $^{-}138 + ^{+}57$
15. $^{-}135 + ^{-}28$
16. $^{-}129 + ^{-}33$
17. $^{+}48 + ^{-}235$
18. $^{-}582 + ^{+}78$

For Exercises 19–27, use the Fundamental Subtraction Property followed by an addition algorithm to calculate the difference.

19. $^{-}7 - ^{-}8$
20. $^{+}15 - ^{+}12$
21. $^{+}16 - ^{-}8$
22. $^{-}9 - ^{+}6$
23. $^{+}7 - ^{-}25$
24. $0 - ^{+}8$
25. $^{-}13 - ^{-}21$
26. $^{-}45 - 0$
27. $^{-}13 - ^{+}13$

For Exercises 28–36, *first* estimate the difference, *then* use the Fundamental Subtraction Property followed by an addition algorithm to calculate it.

28. $^{-}13 - ^{-}13$
29. $^{-}81 - ^{+}22$
30. $^{+}38 - ^{-}19$
31. $^{+}53 - ^{-}53$
32. $^{+}53 - ^{+}53$
33. $^{-}135 - ^{-}28$
34. $^{-}129 - ^{-}33$
35. $^{+}48 - ^{-}235$
36. $^{-}582 - ^{+}78$

For Exercises 37–42, use vectors on the number line to find the difference by *both* the missing-addend and the additive-inverse methods.

37. $^{+}3 - ^{+}1$
38. $^{+}5 - ^{+}6$
39. $^{+}4 - ^{-}2$
40. $^{-}6 - ^{+}3$
41. $^{-}5 - ^{-}4$
42. $^{-}7 - ^{-}10$

For Exercises 43–46, compute $|q| - |p|$ for the given values of p and q. Observe that in each case $|q| > |p|$.

43. $p = ^{-}2$, $q = ^{+}8$
44. $p = ^{+}7$, $q = ^{-}12$
45. $p = ^{-}13$, $q = ^{-}28$
46. $p = ^{-}36$, $q = ^{-}50$

47. If $|q| > |p|$, will $|q| - |p|$ be positive or negative? Always? (*Hint*: Consider your answers to Exercises 43–46.)

48. If $|q| > |p|$, will $-(|q| - |p|)$ be positive or negative? Always? (*Hint*: Consider your answers to Exercises 43–46.)

8.8 Algorithms for Multiplying and Dividing Integers

When considering algorithms for multiplication and division, we must classify the products and quotients of two integers according to whether

I. both integers are positive;

II. both integers are negative;

III. the first integer is positive and the second is negative.

These are the same three cases as for addition. Again, the algorithms must determine both the direction and the absolute value of the product in each case.

Multiplication

Before stating the algorithms for multiplication, let us consider some strategies leading to their development. Strategies for cases I and III center about the view of multiplication of whole numbers as repeated addition. Because the positive integers are "copies" of the whole numbers, a product involving at least one positive integer might reasonably be interpreted as repeated addition.

Example 1

$$({}^{+}3) \cdot ({}^{+}2) = {}^{+}2 + {}^{+}2 + {}^{+}2 = {}^{+}6$$ □

Example 2

$$({}^{+}4) \cdot ({}^{-}3) = {}^{-}3 + {}^{-}3 + {}^{-}3 + {}^{-}3 = {}^{-}12$$ □

Example 3

$$({}^{-}6) \cdot ({}^{+}3) = ({}^{+}3) \cdot ({}^{-}6) = {}^{-}6 + {}^{-}6 + {}^{-}6 = {}^{-}18$$ □

Vector addition is useful for problems like Examples 1–3. (Strictly speaking, to picture Example 3 by vector addition we must rewrite $({}^{-}6) \cdot ({}^{+}3)$ as $({}^{+}3) \cdot ({}^{-}6)$, by commutativity, so that three negative vectors of length 6 can be used.)

Example 4 A vector representation for $({}^{+}3) \cdot ({}^{+}2)$ is shown in Figure 8.25. □

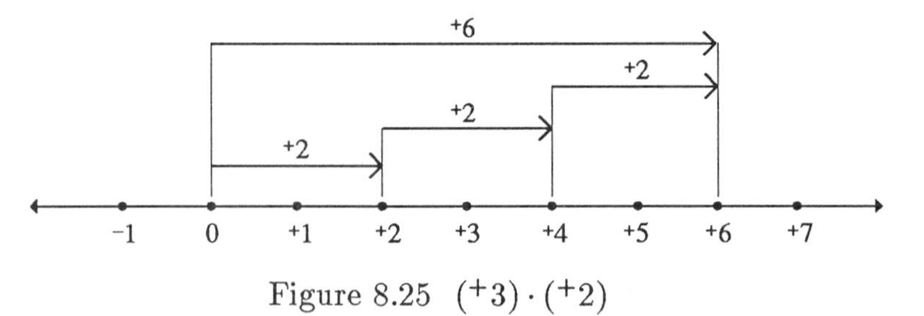

Figure 8.25 $({}^{+}3) \cdot ({}^{+}2)$

Example 5 A vector representation for $(^+4)\cdot(^-3)$ is shown in Figure 8.26. □

Figure 8.26 $(^+4)\cdot(^-3)$

When both factors are positive (Case I—Examples 1 and 4), the direction of the product is positive and its absolute value is the product of the absolute values of the two integers. When one integer is positive and the other is negative (Case III—Examples 2, 3, and 5), the product is negative and its absolute value is again the product of the absolute values of the two integers. However, neither repeated addition nor vector addition explains Case II, the product of two negative integers. This is often the most confusing case for elementary-school students, so let us take some time to look at three different approaches, or strategies, for explaining this type of product. They vary in abstractness, and their usefulness in your own classroom will depend on the level of abstraction your students can understand.

The first strategy relies on number patterns, prior knowledge of Case III, and the students' intuition. To determine $(^-2)\cdot(^-3)$, for instance, look at a series of products involving $^-2$ and successively smaller nonnegative factors:

$$\begin{aligned}(^-2)\cdot(^+4) &= {^-8}\\(^-2)\cdot(^+3) &= {^-6}\\(^-2)\cdot(^+2) &= {^-4}\\(^-2)\cdot(^+1) &= {^-2}\\(^-2)\cdot 0 &= 0\end{aligned}$$

Observe that each time the right factor is reduced by 1, the product is *increased* by 2. Continuing this pattern, when the right factor is reduced from 0 to $^-1$, the product should be increased from 0 to $^+2$. Reducing the right factor from $^-1$ to $^-2$ should increase the product from $^+2$ to $^+4$, etc.

$$\begin{aligned}(^-2)\cdot(^+1) &= {^-2}\\(^-2)\cdot 0 &= 0\\(^-2)\cdot(^-1) &= {^+2}\\(^-2)\cdot(^-2) &= {^+4}\\(^-2)\cdot(^-3) &= {^+6}\\ &\vdots\end{aligned}$$

Thus, the pattern of numbers makes it plausible that the product of two negative integers is positive. Of course, this is not a formal proof, but at least it makes the statement $(^-2) \cdot (^-3) = ^+6$ seem reasonable.

The second strategy for Case II is a bit more abstract than the first. It, too, is based on Case III, and again we use the example $(^-2) \cdot (^-3)$. We begin by observing that, whatever the value of $(^-2) \cdot (^-3)$ is, it must be an integer (by closure). Thus, we can add other integers to it. In particular, we can add $(^-2) \cdot (^+3)$ to it. Now let us apply some integer properties to this sum:

$$
\begin{aligned}
(^-2) \cdot (^-3) + (^-2) \cdot (^+3) &= -2 \cdot (^-3 + ^+3) && \text{Distributivity} \\
&= ^-2 \cdot 0 && \text{Sum of additive inverses} \\
&= 0 && \text{Zero multiplication}
\end{aligned}
$$

But we know that $(^-2) \cdot (^+3) = ^-6$ (by Case III), so the preceding equations tell us that

$$(^-2) \cdot (^-3) + ^-6 = 0$$

This means that $(^-2) \cdot (^-3)$ must be the additive inverse of $^-6$; that is,

$$(^-2) \cdot (^-3) = -(^-6) = ^+6$$

The third strategy uses interpretations for which the system of integers is a model. We lead up to the product of two negative integers by first looking at interpretations of products of two positive integers, a positive integer times a negative one, and a negative integer times a positive one. Examples 6 and 7 typify this approach.

Example 6 A "Mother, May I?" interpretation:

(a) $(^+2) \cdot (^+3) = ^+6$: Sally, who is calling out instructions to the players, has a good friend playing, too. Every time it's the friend's turn she is told to take 3 giant steps *forward* $(^+3)$. Two turns *from now* $(^+2)$ this friend will be 6 giant steps *ahead* of where she is now $(^+6)$.

(b) $(^-2) \cdot (^+3) = ^-6$: Two turns *ago* $(^-2)$ this friend was 6 giant steps *behind* where she is now $(^-6)$.

(c) $(^+2) \cdot (^-3) = ^-6$: Someone Sally doesn't like is also playing the game, and every time it's his turn he is told to take 3 giant steps *backward* $(^-3)$. Two turns *from now* he will be 6 giant steps *behind* where he is now $(^-6)$.

(d) $(^-2) \cdot (^-3) = ^+6$: Two turns *ago* $(^-2)$ he was 6 giant steps *ahead* of where he is now $(^+6)$. □

Example 7 A moving-picture interpretation:

(a) $(^{+}2) \cdot (^{+}3) =^{+} 6$: Pedro is a home-video buff, and his neighbor Sam owns a small swimming pool. In the spring, when Sam fills his pool, the water coming in *rises* at the rate of 2 inches per minute ($^{+}2$). Pedro videotapes the water as it is rising. When he runs the tape *forward* for 3 minutes ($^{+}3$), he sees the water *rising* 6 inches ($^{+}6$).

(b) $(^{+}2) \cdot (^{-}3) =^{-} 6$: When Pedro runs the tape *backwards* for 3 minutes ($^{-}3$), he sees the water *dropping* 6 inches ($^{-}6$).

(c) $(^{-}2) \cdot (^{+}3) =^{-} 6$: In the fall, Sam drains his pool, and the water *drops* at a rate of 2 inches per minute ($^{-}2$). Pedro also videotapes this event, and when he runs the tape *forward* for 3 minutes ($^{+}3$), he sees the water *dropping* 6 inches ($^{-}6$).

(d) $(^{-}2) \cdot (^{-}3) =^{+} 6$: When Pedro runs the tape *backwards* for 3 minutes ($^{-}3$), he sees the water *rising* 6 inches ($^{+}6$). □

Example 8 Figure 8.27 shows a seventh-grade text page that introduces multiplication algorithms for signed numbers. What strategies are used? How do they compare with the strategies of this section? □

The algorithms for integer multiplication can be summarized as follows:

(8.20) Case I: If $p > 0$ and $q > 0$, then $p \cdot q = |p| \cdot |q|$.

(8.21) Case II: If $p < 0$ and $q < 0$, then $p \cdot q = |p| \cdot |q|$.

(8.22) Case III: If p and q have opposite signs, then $p \cdot q = -(|p| \cdot |q|)$.

As with addition, the products involving 0 are not included in these three cases; they are covered by the fact (established earlier) that, for any integer p, $p \cdot 0 = 0 \cdot p = 0$.

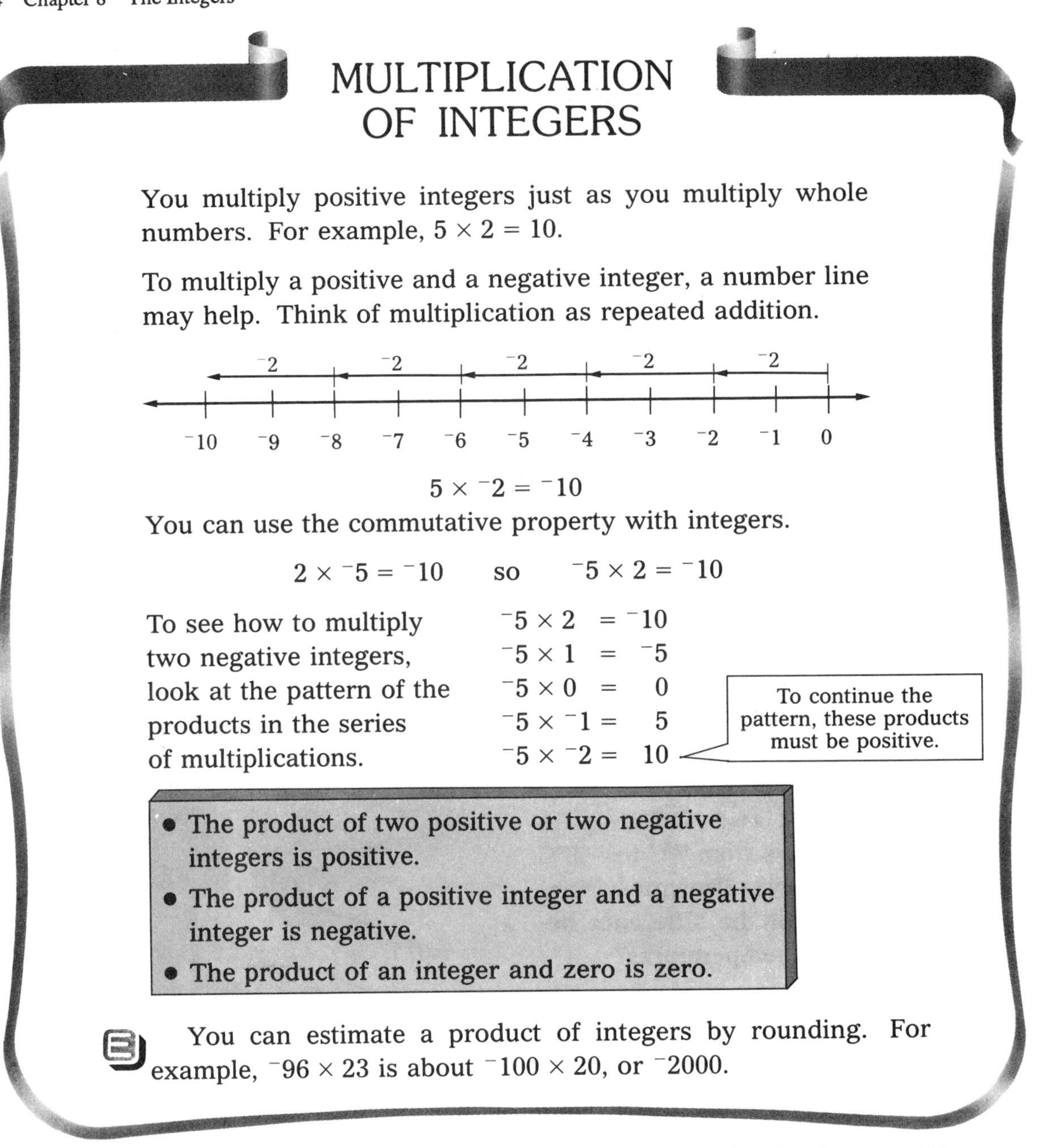

MULTIPLICATION OF INTEGERS

You multiply positive integers just as you multiply whole numbers. For example, $5 \times 2 = 10$.

To multiply a positive and a negative integer, a number line may help. Think of multiplication as repeated addition.

$$5 \times {}^-2 = {}^-10$$

You can use the commutative property with integers.

$$2 \times {}^-5 = {}^-10 \quad \text{so} \quad {}^-5 \times 2 = {}^-10$$

To see how to multiply two negative integers, look at the pattern of the products in the series of multiplications.

$$\begin{aligned} {}^-5 \times 2 &= {}^-10 \\ {}^-5 \times 1 &= {}^-5 \\ {}^-5 \times 0 &= 0 \\ {}^-5 \times {}^-1 &= 5 \\ {}^-5 \times {}^-2 &= 10 \end{aligned}$$

To continue the pattern, these products must be positive.

- The product of two positive or two negative integers is positive.
- The product of a positive integer and a negative integer is negative.
- The product of an integer and zero is zero.

You can estimate a product of integers by rounding. For example, ${}^-96 \times 23$ is about ${}^-100 \times 20$, or ${}^-2000$.

Figure 8.27 HOUGHTON MIFFLIN MATHEMATICS, Grade 7, p. 342

Division

Even though the integers are not closed under division, the process can be done sometimes, so we give algorithms for division here. These algorithms parallel the ones for multiplication, as you might expect from the way division is defined. In all three cases, these algorithms only make sense if the corresponding whole-number quotients exist.

(8.23) Case I: If $p > 0$ and $q > 0$, then $p \div q = |p| \div |q|$.

(8.24) Case II: If $p < 0$ and $q < 0$, then $p \div q = |p| \div |q|$.

(8.25) Case III: If p and q have opposite signs, then $p \div q = -(|p| \div |q|)$.

You might find it instructive to try to justify each of the algorithms for division by relating the corresponding multiplication algorithms to the definition of division,

$$p \div q = x \text{ if } q \cdot x = p$$

Example 9 Case I: ${}^{+}92 \div {}^{+}4 = 92 \div 4 = {}^{+}23$. This satisfies the definition of division, because $({}^{+}4) \cdot ({}^{+}23) = {}^{+}92$. □

Example 10 Case II: ${}^{-}84 \div {}^{-}14 = 84 \div 14 = {}^{+}6$ and $({}^{-}14) \cdot ({}^{+}6) = {}^{-}84$. □

Example 11 Case III: ${}^{+}27 \div {}^{-}3 = -(27 \div 3) = {}^{-}9$ and $({}^{-}3) \cdot ({}^{-}9) = {}^{+}27$. Similarly, ${}^{-}125 \div {}^{+}5 = -(125 \div 5) = {}^{-}25$ and $({}^{+}5) \cdot ({}^{-}25) = {}^{-}125$. □

Example 12 $0 \div {}^{-}3 = 0$ because $({}^{-}3) \cdot 0 = 0$. However, ${}^{-}8 \div 0$ is not possible; there is no integer x such that $0 \cdot x = {}^{-}8$. □

Exercises 8.8

For Exercises 1–9, use a multiplication algorithm to find the product. Indicate which case applies. For any absolute value that you cannot compute immediately "in your head," estimate the product before doing the explicit calculation.

1. $({}^{+}5) \cdot ({}^{+}8)$
2. $({}^{+}7) \cdot ({}^{-}3)$
3. $({}^{-}4) \cdot ({}^{-}8)$
4. $({}^{-}21) \cdot ({}^{+}6)$
5. $({}^{-}33) \cdot ({}^{-}7)$
6. $({}^{-}12) \cdot ({}^{+}9)$
7. $({}^{+}21) \cdot ({}^{-}32)$
8. $({}^{+}15) \cdot ({}^{+}10)$
9. $({}^{-}45) \cdot ({}^{-}52)$

For Exercises 10–15, use repeated addition to compute the product, if possible. If repeated addition cannot be used, state why not.

10. $(^+3) \cdot (^-4)$
11. $(^+5) \cdot (^+2)$
12. $(^-3) \cdot (^+4)$
13. $(^-2) \cdot (^-6)$
14. $(^+6) \cdot (^-1)$
15. $(^-3) \cdot 0$

For Exercises 16–18, use vector addition to illustrate the product.

16. $(^+3) \cdot (^-4)$
17. $(^+4) \cdot (^-2)$
18. $(^+3) \cdot (^+5)$

For Exercises 19–21, use the number-pattern strategy to illustrate the product.

19. $(^-3) \cdot (^-4)$
20. $(^-5) \cdot (^-2)$
21. $(^-9) \cdot (^-6)$

For Exercises 22–24, justify the given product by adding the appropriate product to it and applying the distributive property.

22. $(^-3) \cdot (^-4) = {}^+12$

23. $(^-5) \cdot (^-2) = {}^+10$

24. $(^-9) \cdot (^-6) = {}^+54$

For Exercises 25–28, use the "Mother, May I?" interpretation to illustrate each product.

25. $(^+3) \cdot (^+4)$
26. $(^+3) \cdot (^-4)$
27. $(^-3) \cdot (^+4)$
28. $(^-3) \cdot (^-4)$

For Exercises 29–32, use the moving-picture interpretation to illustrate each product.

29. $(^+3) \cdot (^+4)$
30. $(^+3) \cdot (^-4)$
31. $(^-3) \cdot (^+4)$
32. $(^-3) \cdot (^-4)$

For Exercises 33–36, develop a football interpretation to illustrate each product.

33. $(^+3) \cdot (^+4)$
34. $(^+3) \cdot (^-4)$
35. $(^-3) \cdot (^+4)$
36. $(^-3) \cdot (^-4)$

For Exercises 37–40, develop two new interpretations of your own to illustrate each product.

37. $(^+3) \cdot (^+4)$
38. $(^+3) \cdot (^-4)$
39. $(^-3) \cdot (^+4)$
40. $(^-3) \cdot (^-4)$

For Exercises 41–55, use the algorithms for division to determine each quotient that exists (in **I**). Indicate which case applies. For any quotient you cannot compute immediately "in your head," estimate the answer before doing the explicit calculation. Also show how the definition of division is satisfied.

41. ${}^+12 \div {}^+6$
42. ${}^-14 \div {}^+7$
43. ${}^+24 \div {}^-6$
44. ${}^-32 \div {}^-4$
45. ${}^-3 \div 0$
46. $0 \div {}^-9$
47. ${}^+48 \div {}^-16$
48. ${}^-50 \div {}^-10$
49. ${}^-225 \div {}^+5$
50. $0 \div {}^+8$
51. ${}^-81 \div {}^-27$
52. ${}^+25 \div 0$
53. ${}^+45 \div {}^-15$
54. ${}^-143 \div {}^+11$
55. ${}^+52 \div {}^-4$

Review Exercises for Chapter 8

For Exercises 1–10, indicate whether the given statement is *true* or *false*.

1. One motivation for the development of the integers is to construct a number system that is closed under subtraction.

2. There are no applications of negative numbers that elementary-school children are able to understand.

3. An integer is the set of all ordered pairs of whole numbers that are related to some par-

ticular pair by the equivalence relation $\sim$.

4. The operations of integer arithmetic are defined in terms of the operations of whole-number arithmetic.

5. $[a, b] = [c, d]$ if and only if $a + b = c + d$.

6. $[0, 2]$, $[4, 6]$, $[2, 0]$, and $^-2$ all represent the same integer.

7. There is no understandable way to explain to elementary-school children why the product of two negative numbers is positive.

8. To subtract two integers, you can add the first integer to the additive inverse of the second.

9. If $x \neq 0$, then $-x$ always represents a negative number.

10. If $x \neq 0$, then $|x|$ always represents a positive number.

For Exercises 11–15, complete the following table so that all the entries in the same row represent the same integer.

	Integer Form	Standard [a, b] Form	Other [a, b] Form
11.	$^+2$	[,]	[5,]
12.		[50, 0]	[,]
13.		[,]	[3, 7]
14.	$^-14$	[,]	[, 17]
15.		[,]	[2, 2]

For Exercises 16–19, use the definition of addition or multiplication to compute the given sum or product. You need not convert your answer to standard form.

16. $[5, 8] \oplus [9, 2]$ **17.** $[15, 2] \oplus [3, 10]$

18. $[2, 3] \odot [7, 9]$ **19.** $[12, 4] \odot [3, 8]$

For Exercises 20 and 21, use the definition of subtraction to compute the difference. Leave your answer in the form of the given problem.

20. $[4, 9] \ominus [9, 2]$ **21.** $^-3 - {}^+8$

For Exercises 22 and 23, use the Fundamental Subtraction Property to compute the difference. Leave your answer in the form of the given problem.

22. $[3, 7] \ominus [8, 4]$ **23.** $^+3 - {}^+9$

For Exercises 24–27, use vectors to find the sum.

24. $^+2 + {}^-5$ **25.** $^-3 + {}^-4$

26. $^-4 + {}^+9$ **27.** $^-2 + {}^-5 + {}^+10$

For Exercises 28 and 29, use vectors to find the difference by the missing-addend method.

28. $^-4 - {}^-6$ **29.** $^+3 - {}^+7$

For Exercises 30 and 31, use vectors to find the difference by the additive-inverse method.

30. $^+2 - {}^-3$ **31.** $^-5 - {}^+3$

For Exercises 32 and 33, use an appropriate interpretation to illustrate the given sum.

32. $^+7 + {}^-9$ **33.** $^-3 + {}^-7$

For Exercises 34–36, illustrate the product $^-2 \odot {}^-4$ by the method specified.

34. Use number patterns.

35. Add an appropriate product to it and use integer properties.

36. Use an interpretation.

Exercises 37–43 refer to the cases for addition of integers as described in Section 8.7. For each exercise, identify which case applies and find the sum. (In some situations you may have to use commutativity before a case applies.)

37. $^{+}8 + {}^{-}2$

38. $^{-}5 + {}^{-}9$

39. $^{-}5 + {}^{+}15$

40. $^{-}19 + {}^{+}3$

41 $^{+}21 + {}^{-}38$

42. $^{+}7 + {}^{+}12$

43. $^{-}9 + {}^{+}9$

For Exercises 44–47, compute the given product or quotient by the methods discussed in this chapter.

44. $({}^{+}2) \cdot ({}^{-}5)$

45. $({}^{-}6) \cdot ({}^{-}3)$

46. $^{-}24 \div {}^{-}3$

47. $^{+}49 \div {}^{-}7$

48. State the definition of *less than* for integers. How does it differ from the definition of *less than* for whole numbers?

For Exercises 49–51, use the definition of *less than* for integers to identify the smaller number of the given pair.

49. $^{-}13$ and $^{-}3$

50. $^{+}15$ and $^{+}6$

51. $^{+}12$ and $^{-}23$

9.1 The Problem

We noted in Chapter 8 that division does not always make sense in the system of integers. The process can be defined, as it was in the whole-number system, but there are many pairs of integers for which it just does not work. For instance, the question $6 \div 2 = ?$ has the answer 3 in the integers because $2 \cdot 3 = 6$; however, $5 \div 2 = ?$ cannot be answered within the integers, because there is no integer that can be multiplied by 2 to get 5. In other words, the integers—which are closed under addition, subtraction, and multiplication—are not closed under division. In this chapter we construct a number system that it is closed under all four elementary arithmetic operations.

Note: From here on, following common practice, unsigned symbols are used for both whole numbers and positive integers. Thus, 3 represents the positive integer $^{+}3$ as well as the whole number 3, and so forth.

In Chapter 8, before we formally extended **W** to a system that allows us to subtract 5 from 3, we asked why anyone would want to do such computations. In answering that question we identified a variety of interpretations for which $3 - 5$ made sense. Important characteristics of the system of integers—quantity, direction, and point of origin—were common to all those interpretations. In this chapter we start by asking the analogous question: Why would anyone want to divide 3 by 5? In fact, there are many everyday situations in which $3 \div 5$ does make sense, as the following examples show.

Example 1 If there are 3 pizzas to be equally divided among five people, how much pizza does each person get? □

Example 2 If you can buy 5 loaves of bread for \$3, how much does each loaf cost? □

Example 3 If you want to make 5 shelves of equal length from one board 3 meters long, how long will each shelf be? □

Example 4 If a plant grows from a height of 5 inches to a height of 8 inches, what is the ratio of the increase in height to the original height? □

Example 5 At a school's field-day event there was a 5-lap race. The winner's total time was 3 minutes; what was her average lap time? □

Example 6 If the temperature drops at a constant rate of 5° in 3 hours, how long did it take to drop 1°? □

Each of the preceding examples describes a situation in which $3 \div 5$ has a reasonable interpretation. Although they describe different activities, these examples have common features for which the rational-number system provides a useful model. After constructing the rational numbers, we shall return (in Chapter 10) to these and other interpretations of rational numbers that appear in the elementary-school curriculum.

PROBLEM-SOLVING COMMENT

This entire chapter is an example of the mathematical use of *analogy*. As you will see, the construction of the rational-number system is virtually identical to that of the integers. In fact, Chapter 8 provides a step-by-step guide to Chapter 9, which is emphasized by the similarity of the text in the two chapters.

The key to this parallel between the integers and the rational numbers is the fact that rational-number multiplication is the precise analogue of integer addition in the constructions of the two systems. This situation arises from the similarity between the definitions of difference and quotient. Let us compare these two definitions, originally stated in Chapter 4:

If there exists a number x such that $\left\{ \begin{array}{c} b + x = a \\ b \cdot x = a \end{array} \right\}$, then we call x the $\left\{ \begin{array}{c} \text{difference} \\ \text{quotient} \end{array} \right\}$ of $a \left\{ \begin{array}{c} \text{minus} \\ \text{divided by} \end{array} \right\} b$.

The construction of the integers, a system in which differences can always be found, was based squarely on this definition of difference. Thus, we might reasonably expect to be able to use the analogous definition of quotient to construct a system in which quotients can always be found. Moreover, the pattern of constructing the integers should provide a "blueprint" for constructing this new system, provided we replace addition with its analogue, multiplication. If you refer to Chapter 8 occasionally as you study this chapter, you should be able to see the parallelism in the structures of the two systems as they develop step by step, starting with the outline of the problem. ◇

Here is the main problem of this chapter. We want to extend the system of integers to a larger system that is closed under division. Specifically, we want a collection of things we shall call "numbers" of some sort, with the following properties:

- Some subset of this collection is a copy of the set **I** of integers (and hence we can say that our new collection "includes" the integers);
- Operations of addition, subtraction, and multiplication can be defined on this new collection, and they will agree with the previous definitions of $+$, $-$, and $\cdot$ when they are applied to (the copy of) **I**;
- These new operations "behave as nicely" as their counterparts in **I** (i.e., they are commutative, associative, etc.);
- This new collection *is closed under the division process* (which is defined in terms of multiplication);
- An order relation can be defined on this new collection so that it agrees with $<$ on **I** and has similar nice properties (trichotomy, transitivity, etc.).

The situation we seek is diagrammed in Figure 9.1. The known set **I** must be matched with some subset of the new set **(?)** in such a way that the corresponding operations and relations agree, and the new set must also have division as an operation.

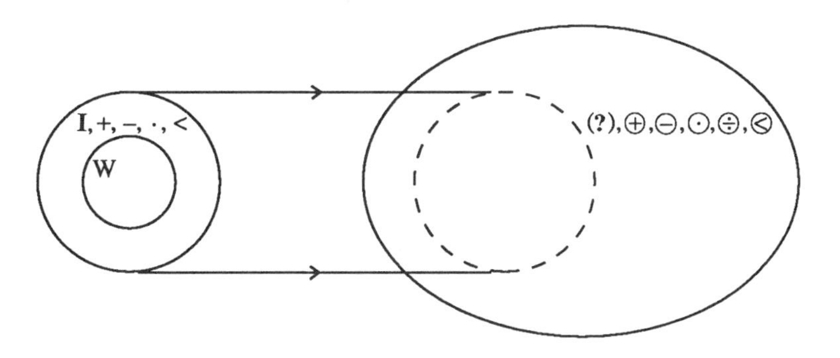

Figure 9.1 Extending the system of integers.

This problem is just as big as its counterpart in Chapter 8, but our task now is much easier. Instead of having to grope for a starting place and appropriate questions to ask, we have been handed a road map to the solution; we just imitate the construction of the integers.

HISTORICAL NOTE: UNITS AND PARTS

It seems that mankind's need for numbers was satisified by the natural (or "counting") numbers until sometime around the Bronze Age. In earlier times, when it was necessary to account for portions of objects, the objects were broken down (sometimes literally) into smaller pieces and then the pieces were counted. This evolved into primitive systems of weights and measures, where the basic units of measure were made smaller as more precision was desired. In modern terms, we might say that ounces would be counted instead of pounds, inches instead of feet, cents instead of dollars, and so forth. Of course, these particular units of measure were not used in early times, but their predecessors were. Some measurement systems still in use today reflect the desire to count smaller units, rather than deal with fractional parts. Consider, for example, the following list of liquid measures, where each unit is half the size of its predecessor:

gallon, half-gallon, quart, pint, cup, gill

In fact, each of these can be expressed in terms of an even smaller unit, the *fluid ounce*. (A *gill* is 4 fluid ounces.)

The idea of subdividing and then counting smaller units may well have been the reason that the first treatment of "partial quantities" as numbers came in the form of *unit fractions*, fractions with numerator 1. (Even our word "fraction," which has the same root as "fracture" and "fragment," suggests breaking something up.) The Rhind Papyrus, part of an Egyptian mathematical treatise by the scribe Ahmes (about 1650 B.C.), contains an extensive treatment of unit fractions, including ways to express partial quantities as sums of unit fractions.

The practice of dealing with fractional values as sums of unit fractions dominated the practical arithmetic of fractions from the time of Ahmes, through the Greek and Roman periods, well into the Middle Ages. Fibonacci's *Liber Abaci*, an influential 13th-century European mathematics text, made extensive use of unit fractions and discussed various ways of converting other fractions to sums of unit fractions. Long after such methods had been replaced by more efficient ways of expressing fractions, unit fractions were still being taught in schools and transcribed in books. As recently as the 17th century, Russian manuscripts on surveying referred to one ninety-sixth of a particular measure as "half-half-half-half-half-third,"[1] expecting the reader to think in terms of successive subdivisions:

$$\frac{1}{2}\cdot\frac{1}{2}\cdot\frac{1}{2}\cdot\frac{1}{2}\cdot\frac{1}{2}\cdot\frac{1}{3}=\frac{1}{96} \qquad \diamond$$

[1]David Eugene Smith, *History of Mathematics*, Vol. II, New York: Dover Publications, Inc., 1958, p. 213.

Exercises 9.1

1. Using integers, list five division problems that do not have answers in **I**; then list five that do.

For Exercises 2–7, list five different integer division problems whose answer is the given number.

2. 2 **3.** $^-5$ **4.** $^-60$

5. 1 **6.** 0 **7.** $^-1$

8. List three integer division problems that have the same answer as $6 \div 2 = ?$, then list three that have different answers.

9. Assuming that $a \div b = ?$ has the same solution as $6 \div 2 = ?$, answer the question "$a \cdot 2 = b \cdot ?$". (*Hint*: Try some examples from Exercise 8. Do you always get the same result? Why or why not?)

For Exercises 10–12, provide at least two different interpretations for the given quotient.

10. $2 \div 8$ **11.** $7 \div 3$ **12.** $0 \div 5$

9.2 What Is a Rational Number?

We want a number system that contains an answer to every integer-division question, so it makes sense to begin by recalling the defining condition for division:

$$a \div b = x \text{ if } b \cdot x = a$$

Now, any question that seeks to divide one integer by another can be represented by the ordered pair of numbers themselves, whether or not the quotient actually exists in **I**:

$$6 \div 3 = ? \text{ can be represented by the ordered pair } (6, 3)$$
$$2 \div 7 = ? \text{ can be represented by the ordered pair } (2, 7)$$
$$^-70 \div 14 = ? \text{ can be represented by the ordered pair } (^-70, 14)$$

Thus, the set of all possible integer-division questions of the form $x \div y = ?$ can be represented by the set of all ordered pairs (x, y) of integers; that is, by $\mathbf{I} \times \mathbf{I}$.

Some of these division questions, however, cannot have reasonable answers even under the best of conditions. For instance, claiming that $2 \div 0 = ?$ has some number n as its answer would require that $0 \cdot n = 2$. But we would expect 0 multiplied by anything to yield 0 in any number system, as happens in **I**, so there is no hope of finding a suitable answer in this case. In general, the question $n \div 0 = ?$ is not answerable for any nonzero integer n. Moreover, the question $0 \div 0 = ?$ is useless because the definition of division allows too many possible answers. Any integer q would fit because $0 \cdot q = 0$.

Thus, *all* division questions with divisor 0 will be excluded from consideration. That is:

$n \div 0 = ?$ is not a permissible question for any integer n.

PROBLEM-SOLVING COMMENT

Although division *by* zero is not permitted, dividing *into* zero (by nonzero numbers) is permitted. Many times students (and occasionally even their teachers) remember that there is some difficulty about zero and division, but forget what is allowed and what is not. This problem is easily solved by *constructing division examples* and then *checking them with the definition of division.* What is allowed and what is not will virtually scream out at you! ◇

We shall denote the set of nonzero integers by $\mathbf{I}^*$. In this notation, the set of all division questions for which we need answers is represented by the set of ordered pairs $\mathbf{I} \times \mathbf{I}^*$. Since we are looking for a system containing answers to all these division questions, it makes sense to consider two questions "the same" in some way if they have the same answer. $6 \div 3 = ?$ and $10 \div 5 = ?$ are related in this way, for instance. This observation suggests some sort of equivalence relation on the set of division questions that have answers in $\mathbf{I}$, and hence on the set of ordered pairs representing these questions. Moreover, as in the construction of the integers, if we define that relation cleverly enough, we might be able to extend it to all the pairs in $\mathbf{I} \times \mathbf{I}^*$.

The Equivalence Relation

Saying that $6 \div 3 = ?$ and $10 \div 5 = ?$ have the same answer means that

$$6 \div 3 = 10 \div 5$$

Multiplying both sides of this equation by 3 and 5, we see that

$$(6 \div 3) \cdot 3 \cdot 5 = (10 \div 5) \cdot 3 \cdot 5 = 30$$

Observe that we can also write the preceding line as

$$6 \cdot 5 = 10 \cdot 3 = 30$$

In general, $a \div b = ?$ and $c \div d = ?$ have the same answer whenever

$$a \div b = c \div d$$

and this is true if and only if

$$a \cdot d = b \cdot c$$

Now, this last expression does not involve division at all, so it can be applied to *any* ordered pairs (a, b) and (c, d), whether or not the division questions they represent have answers in **I**.

Definition Two ordered pairs (a, b) and (c, d) of integers are **equivalent** if $a \cdot d = b \cdot c$. We shall denote this relation by the symbol "$\sim$". Thus,

$$(a, b) \sim (c, d) \quad \text{whenever} \quad a \cdot d = b \cdot c$$

(**Note:** We have used the term "equivalent" and the symbol $\sim$ before, in constructing the whole numbers from sets and in constructing the integers from pairs of whole numbers. Rather than invent another name and another symbol for this type of equivalence, we use the same ones again in order to emphasize the analogy between the construction of this system and the constructions of **W** and **I**. The context should establish which meaning is intended.)

Example 1 $(9, 3) \sim (12, 4)$ because $9 \cdot 4 = 3 \cdot 12$. Notice that $9 \div 3 = 3$ and $12 \div 4 = 3$. □

Example 2 $(2, 7) \sim (6, 21)$ because $2 \cdot 21 = 7 \cdot 6$. Notice that $2 \div 7 = ?$ and $6 \div 21 = ?$ do not have answers in **I**. □

Example 3 $(8, 2) \not\sim (12, 6)$ because $8 \cdot 6 \neq 2 \cdot 12$. Notice that $8 \div 2 = 4$, but $12 \div 6 = 2$. □

The key fact, implied by the name given to $\sim$, is:

(9.1) $\sim$ is an equivalence relation on $\mathbf{I} \times \mathbf{I}^*$.

That is, $\sim$ is a reflexive, symmetric, and transitive relation on the set of ordered pairs of integers with nonzero second elements. These properties are easily verified by arguments exactly analogous to those for $\sim$ on $\mathbf{W} \times \mathbf{W}$; they are left as exercises for you. (Exercises 2 and 55.)

The Set of Rational Numbers

Because $\sim$ is an equivalence relation, it partitions $\mathbf{I} \times \mathbf{I}^*$ into equivalence classes. That is, $\sim$ breaks $\mathbf{I} \times \mathbf{I}^*$ into nonoverlapping sets of ordered pairs of integers, and if some pair in a class represents a division question with a particular answer, then all the pairs in that class represent division questions with that same answer. For instance, the class containing $(^-6, 3)$ also contains $(^-8, 4)$, $(^-2, 1)$, and $(100, ^-50)$; all these pairs represent division questions with answer $^-2$. Thus, *an entire equivalence class of ordered pairs* *represents a single division-question answer.* This fact is the key to understanding the next definition.

Definition A **rational number** is the set of all ordered pairs of integers that are related to some particular pair (and hence to each other) by the relation $\sim$.

Notation The set of all ordered pairs that are equivalent to the pair (p, q) is denoted by $[p, q]$. In symbols,

$$[p, q] = \{(x, y) \mid (x, y) \in \mathbf{I} \times \mathbf{I}^* \text{ and } (x, y) \sim (p, q)\}$$

Furthermore, we denote the set of all rational numbers by $\mathbf{Q}$ (for "quotients"). Thus,

$$\mathbf{Q} = \{[p, q] \mid p, q \in \mathbf{I},\ q \neq 0\}$$

Example 4 The rational number $[12, 4]$ contains the pairs $(12, 4)$, $(6, 2)$, $(75, 25)$, $(^-3, ^-1)$, and all the (infinitely many) others that represent all integer division questions with the answer 3. □

Example 5 The rational number $[^-3, 7]$ contains the pairs $(^-3, 7)$, $(^-6, 14)$, $(30, ^-70)$, $(^-21, 49)$, and all the (infinitely many) others that represent all integer division questions that ought to have the same answer as $^-3 \div 7 = ?$. □

As you may recall from the discussion of equivalence classes in Section 3.4, two equivalence classes are equal whenever they have equivalent representatives. Thus, a rational number may be represented by *any* ordered pair in it. For example, $[15, 6]$ is the same rational number as $[5, 2]$ because the representative pairs are equivalent. That is,

$$[15, 6] = [5, 2] \quad \text{because} \quad (15, 6) \sim (5, 2)$$

which is true because $15 \cdot 2 = 6 \cdot 5$. This is an example of a useful general statement about equality of rational numbers:

(9.2) $$\boxed{[a, b] = [c, d] \text{ if and only if } a \cdot d = b \cdot c.}$$

Looking at a few other representatives of the rational number $[5, 2]$, we can find an interesting pattern:

$$\begin{aligned}(15, 6) &= (5 \cdot 3, 2 \cdot 3)\\ (10, 4) &= (5 \cdot 2, 2 \cdot 2)\\ (20, 8) &= (5 \cdot 4, 2 \cdot 4)\\ (^-15, ^-6) &= (5 \cdot (^-3), 2 \cdot (^-3))\end{aligned}$$

and so on. This suggests a general statement:

> *If we take an ordered pair of integers and multiply each number in that pair by the same integer, then the original pair and the new pair both represent the same rational number.*

There is an obvious exception to this rule. If we multiply any pair of numbers by zero, the result will be $(0,0)$, so that case will have to be excluded. However, in all other cases the pattern holds. More formally,

(9.3) If p, q, and x are integers, with $q \neq 0$ and $x \neq 0$, then

$$[p \cdot x, q \cdot x] = [p, q]$$

It is easy to check that Statement (9.3) is always true. Applying (9.2), we see that

$$[p \cdot x, q \cdot x] = [p, q] \quad \text{if} \quad (p \cdot x) \cdot q = (q \cdot x) \cdot p$$

and that second equation is true because multiplication of integers is associative and commutative.

Example 6

$$[^-7, 2] = [^-21, 6] = [^-7 \cdot 3, 2 \cdot 3]$$ □

Example 7 $[6, 10] = [9, 15]$ because $6 \cdot 15 = 10 \cdot 9$. Observe that

$$[6, 10] = [3 \cdot 2, 5 \cdot 2] \quad \text{and} \quad [9, 15] = [3 \cdot 3, 5 \cdot 3]$$

so this rational number can also be written as $[3, 5]$. □

Using Statement (9.3), it is easy to generate as many representative pairs for a rational number as we want, just by multiplying both numbers in the original pair by a nonzero integer. For instance,

$$[2, 3] = [4, 6] = [6, 9] = [8, 12] = \ldots = [^-2, ^-3] = [^-4, ^-6] = [^-6, ^-9] = \ldots$$

Traditional Notation

Once again, we have manufactured a new set of "numbers," and each number in this set is an infinite collection of ordered pairs. But this time the structure should not seem so strange; you have been working with rational numbers in this way, more or less, since elementary school. If you have not already recognized the connection between our formal construction and your previous arithmetic experience, one small notational adjustment should do the trick:

$$\text{Instead of writing } (2,3) \text{ to represent } 2 \div 3 = ?, \text{ write } \frac{2}{3}.$$

There! Doesn't that feel better? Both $(2,3)$ and $\frac{2}{3}$ are pairs of integers, but we are more accustomed to pairs in the latter form as division expressions. In elementary school you called such a pair a **fraction** and called its first and second integers the **numerator** and the **denominator**, respectively.

You were taught to check if two fractions are "equal" by "cross multiplying." For instance, you were told that

$$\frac{2}{3} = \frac{4}{6} \quad \text{because} \quad 2 \cdot 6 = 3 \cdot 4$$

This is exactly the definition of $\sim$ and Property (9.2)! Moreover, you were told that equal fractions really represent the same quantity; that is, $\frac{2}{3}, \frac{4}{6}, \frac{6}{9}, \frac{8}{12}, \ldots$ are all names for the same number. You were also taught that you could multiply both the numerator and the denominator of a fraction by the same nonzero number without changing its "value"; sometimes this process, used in reverse, was known as "canceling" a common factor.

All of this is exactly what we have constructed in developing the set $\mathbf{Q}$! The "fractions" are the ordered pairs; "cross multiplying" is the basis for the definition of the equivalence relation $\sim$; and the representation of the same quantity by different fractions is just the assertion that a rational number $[p, q]$ is a class containing infinitely many equivalent pairs, any one of which can be used to represent that class. Finally, the "canceling" concept is no more nor less than Property (9.3). Thus, in elementary-school notation, with all its ambiguities, the rational number $\frac{2}{3}$ *is* the set of fractions

$$\left\{\ldots, \frac{^-4}{^-6}, \frac{^-2}{^-3}, \frac{2}{3}, \frac{4}{6}, \frac{6}{9}, \frac{8}{12}, \ldots\right\}$$

and all the fractions in this set are equivalent. The only significant difference between that approach and ours is that we explicitly distinguish the *number* $[2,3]$—a class of ordered pairs—from the *pair* $(2,3)$ that represents the class; this distinction is rarely made in elementary school. When it's not, a fourth- or fifth-grader must distinguish between the number $\frac{2}{3}$ and the representative fraction $\frac{2}{3}$ *solely from context*! ("How come $\frac{2}{3}$ and $\frac{4}{6}$ are the same? They sure look different!")

Example 8 Figure 9.2 is a page from a fourth-grade text. Notice how pictures and number lines are used to illustrate concretely the abstract idea of equivalence classes of fractions. □

Equal Fractions

A. The same figure is used in these three pictures.

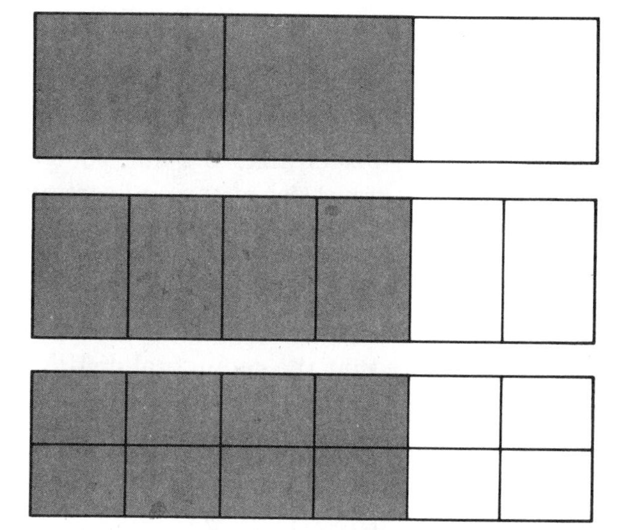

2 of the 3 equal parts are shaded.
$\frac{2}{3}$ of the figure is shaded.

4 of the 6 equal parts are shaded.
$\frac{4}{6}$ of the figure is shaded.

8 of the 12 equal parts are shaded.
$\frac{8}{12}$ of the figure is shaded.

$\frac{2}{3}$, $\frac{4}{6}$, and $\frac{8}{12}$ describe the same amount. They are ***equal fractions.***

$\frac{2}{3} = \frac{4}{6} = \frac{8}{12}$

B. Equal fractions can also be shown on number lines.

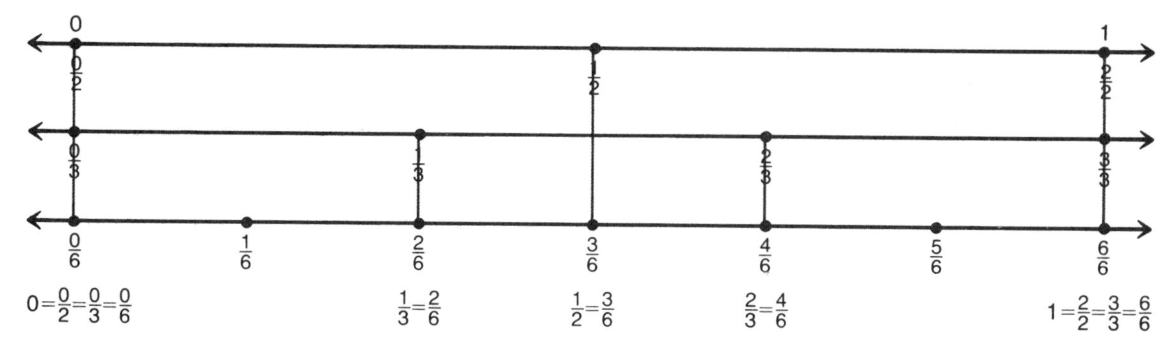

Figure 9.2 INVITATION TO MATHEMATICS, Grade 4, p. 318

The less formal elementary-school approach to numbers, while adequate for doing arithmetic, glosses over some important structural features of the number system. As this chapter illustrates, the construction of the rationals parallels that of the integers, with multiplication replacing addition. In elementary school the integers are not developed using pairs, so this parallel is hidden and much of the underlying unity of the number systems is obscured. The formal construction is neither necessary nor recommended for children, but you—as their teacher—should be familiar with it so that you will be better able to guide your students through the pathways of arithmetic at any level of their understanding. That is why we take the classes-of-pairs approach to the integers and the rationals.

Having said all this, we shall (grudgingly) begin to convert back to the traditional fraction notation, simply because it is more familiar. Throughout the rest of this chapter, examples are given in both notations, in the hope that the contextual distinction between class and pair will become clear by the time you reach the end. In the remaining chapters we use the traditional notation exclusively.

Standard Form

This connection with the traditional fraction notation should clarify what has been done in the class-of-pairs construction of the rationals. Take as an example the rational number $[6,3]$, which can also be written as $[2,1]$. This represents the answer to the division question $6 \div 3 = ?$, and we write it as $\frac{6}{3}$, or as $\frac{2}{1}$. Similarly, $[4,6]$ represents the answer to $4 \div 6 = ?$ and, since $[4,6] = [2,3]$, we can write that answer either as $\frac{4}{6}$ or as $\frac{2}{3}$.

We can always represent a rational number by an ordered pair of integers whose second number (denominator) is positive. For instance, $[4,{}^-3]$ can be rewritten as $[{}^-4,3]$ using Property (9.3) because

$$[4,{}^-3] = [4 \cdot ({}^-1), 3 \cdot ({}^-1)] = [{}^-4,3]$$

Similarly,

$$[{}^-2,{}^-5] = [{}^-2 \cdot ({}^-1), {}^-5 \cdot ({}^-1)] = [2,5]$$

Thus, we do not lose any equivalence classes if we restrict the set of representative integer pairs to $\mathbf{I} \times \mathbf{I}^+$. *We shall make this restriction from now on.* This allows us to call a rational number **positive** or **negative** according to whether its first integer (its numerator) is positive or negative.

Notation We denote the set of positive rational numbers by Q^+ and the set of negative rationals by Q^-.

Property (9.3) is also convenient in that it allows us to represent any rational number by a pair of integers that have no common factors (other than 1 and $^-1$); that is, by a pair of *relatively prime* integers. For instance, the rational number $[^-75, 50]$ can be written as $[^-3, 2]$ because

$$[^-75, 50] = [^-3 \cdot 25, 2 \cdot 25] = [^-3, 2]$$

Property (9.3) guarantees that such a reduction can always be done; it says that any nonzero common factor can be inserted or removed without changing the number represented. A rational number represented by a relatively prime pair is said to be **in lowest terms** (or **in simplest form**). Hence, we have established that:

(9.4) | Any rational number can be put in lowest terms.

Moreover, *each rational number contains only one lowest-terms pair*, which can be used as the **standard form** of that number.

Example 9 $[32, 12]$ can be put in lowest terms by factoring out the gcf of 32 and 12, which is 4. Thus,

$$[32, 12] = [8 \cdot 4, 3 \cdot 4] = [8, 3]$$

In fraction notation,

$$\frac{32}{12} = \frac{8 \cdot 4}{3 \cdot 4} = \frac{8}{3}$$

Similarly, since the gcf of 1001 and 154 is 77,

$$[^-1001,\ 154] = [^-13 \cdot 77,\ 2 \cdot 77] = [^-13,\ 2]$$

$$\frac{^-1001}{154} = \frac{^-13 \cdot 77}{2 \cdot 77} = \frac{^-13}{2}$$

□

The Copy of I

The rational numbers contain a copy of the integers in just the way you would expect. For any integer p, the answer to the division question $p \div 1 = ?$ obviously is p itself because $1 \cdot p = p$. This means:

(9.5) | Each integer p corresponds in a natural way to the rational number $[p, 1]$ (or $\frac{p}{1}$, if you prefer).

Thus, the set $\mathbf{I}$ of integers corresponds to the subset

$$\{[p, 1] \mid p \in \mathbf{I}\}$$

of $\mathbf{Q}$. (In formal terms, we are saying that the function $f : \mathbf{I} \to \mathbf{Q}$ defined by $f(p) = [p, 1]$ is a 1-1 correspondence.) Thus, in the context of rational numbers, we shall regard 5 as a "shorthand" form of $\frac{5}{1}$ or $[5, 1]$, and so forth.

Example 10 $[30, 5]$ represents an integer because 5 is a factor of 30, so that

$$[30, 5] = [6 \cdot 5, 1 \cdot 5] = [6, 1]$$

$$\frac{30}{5} = \frac{6 \cdot 5}{1 \cdot 5} = \frac{6}{1} = 6$$ □

If you review the results of this section in relation to the problem posed in Section 9.1, you will see that the first part has been solved. We have *constructed* a familiar-looking set of things called "the rational numbers," which contains a copy of the set of integers and an answer to every integer-division question (except for division by 0). Again, the word "constructed" means that we have *made* the the set $\mathbf{Q}$ of rational numbers from the (known) system $\mathbf{I}$ of integers. However, $\mathbf{Q}$ itself is not yet a number *system* because we have not defined any operations on it. This is done in the next two sections.

Exercises 9.2

1. Consider the following ordered pairs of integers:

$$(5,3),\ (^-7,^-1),\ (12,20),\ (14,^-2),$$
$$(^-8,2),\ (1,3),\ (0,7),\ (^-70,10),$$
$$(^-21,^-35),\ (30,50),\ (15,^-7),\ (15,25)$$

(a) Write the division question that each pair represents.

(b) Which pairs in the list are equivalent to $(^-7,1)$?

(c) Which pairs in the list are equivalent to $(3,5)$?

2. Work out examples to illustrate that $\sim$ is reflexive, symmetric, and transitive, using the ordered pair $(^-1,5)$ in each example.

For Exercises 3–18, fill in the missing numbers.

3. $(1,3) \sim (2,\quad)$ **4.** $(^-12,4) \sim (\quad,2)$

5. $(\quad,15) \sim (0,27)$ **6.** $(6,\quad) \sim (11,11)$

7. $[4,1] = [\quad,4]$ **8.** $[15,9] = [10,\quad]$

9. $[^-20,\quad] = [4,1]$ **10.** $[\quad,5] = [1,^-5]$

11. $\frac{3}{4} = [12,\quad]$ **12.** $^-17 = [\quad,2]$

13. $5 = [\quad,5]$ **14.** $^-5 = [^-5,\quad]$

15. $\frac{16}{6} = \frac{8}{\ }$ **16.** $0 = \frac{\ }{7}$

17. $\frac{10}{\ } = \frac{^-5}{6}$ **18.** $\frac{\ }{3} = 4$

For Exercises 19–30, determine whether the given statement is *true* or *false*.

19. A rational number is an ordered pair of integers.

20. A rational number is an infinite set of ordered pairs of integers.

21. $[3,4] = \frac{3}{4}$ **22.** $(3,4) \in \frac{3}{4}$

23. $[3,4] \in \frac{3}{4}$ **24.** $[3,4] = [5,6]$

25. $[3,4]=[6,8]$ **26.** $(3,4)=(6,8)$
27. $(3,4)\sim(5,6)$ **28.** $(3,4)\sim(6,8)$
29. $[3,4]\sim[6,8]$ **30.** $(6,8)\in[3,4]$

For Exercises 31–42, list five different representative pairs for the given rational number.

31. $[7,5]$ **32.** $[^-1,2]$ **33.** $[6,1]$
34. $[^-6,3]$ **35.** $\frac{8}{3}$ **36.** $\frac{3}{9}$
37. $\frac{20}{4}$ **38.** $\frac{^-7}{4}$ **39.** 3
40. 0 **41.** 1 **42.** $^-10$

For Exercises 43–54, express the given rational number in lowest terms.

43. $[8,10]$ **44.** $[^-45,15]$ **45.** $[140,16]$
46. $[22,^-11]$ **47.** $\frac{25}{40}$ **48.** $\frac{^-84}{17}$
49. $\frac{1001}{13}$ **50.** $\frac{^-26}{52}$ **51.** $\frac{26}{9}$
52. $\frac{48}{^-72}$ **53.** $\frac{^-132}{144}$ **54.** $\frac{^-320}{^-288}$

55. Prove that the relation $\sim$ defined in this chapter is an equivalence relation on $\mathbf{I}\times\mathbf{I}^*$.

9.3 Multiplication and Division of Rational Numbers

(**Note:** Although addition is usually discussed before multiplication, we have reversed the order in this chapter to emphasize the structural parallel between addition in **I** and multiplication in **Q**. Addition and subtraction of rational numbers are discussed in Section 9.4.)

The Definition of Product

Figure 9.3 shows how multiplication is introduced to fifth graders. Two features of the approach on that page are significant for us. First, the word "of" is used for the multiplication process. This is a natural extension of the idea (introduced in Chapter 4) of multiplication as repeated addition, where $3\cdot 4$ is represented by 3 sets *of* 4 things. It's an easy step from there to $\frac{1}{2}\cdot 4$ as "half *of* 4," another easy step to $\frac{1}{2}\cdot\frac{3}{5}$ as "half *of* $\frac{3}{5}$," and so forth.

The second significant feature of this approach is its reliance on a visual, geometric argument. It begins with the number 1 represented as a rectangle of area 1. (It might be better to use a square.) To represent the product $\frac{a}{b}\cdot\frac{c}{d}$, a rectangle with one dimension $\frac{a}{b}$ and the other $\frac{c}{d}$ is constructed within the square; the area of that rectangle represents the product of the two fractions.

Let us use this approach to find the product $\frac{2}{3}\cdot\frac{4}{5}$. Start with a square 1 unit on a side [Figure 9.4(a)]. Divide one side into thirds and identify a length of $\frac{2}{3}$ [Figure 9.4(b)]. Next, divide an adjacent side of the square into fifths and identify a length of $\frac{4}{5}$ [Figure 9.4(c)]. The unit square has now been divided into $3\cdot 5$ or 15 cells of equal size and the same shape. Now look at the rectangular part of the square with dimensions $\frac{2}{3}$ by $\frac{4}{5}$ [Figure 9.4(d)]. Its area represents the product $\frac{2}{3}\cdot\frac{4}{5}$. It is composed of $2\cdot 4$ or 8 of the 15 cells, showing that

Using Pictures to Multiply Fractions

Forests once covered about $\frac{3}{5}$ of the earth's land surface. Today, only about $\frac{1}{2}$ of those forest areas remain. What fraction of the earth's land surface is now covered by forests?

You can draw a picture to help you find $\frac{1}{2}$ of $\frac{3}{5}$.

First shade $\frac{3}{5}$.

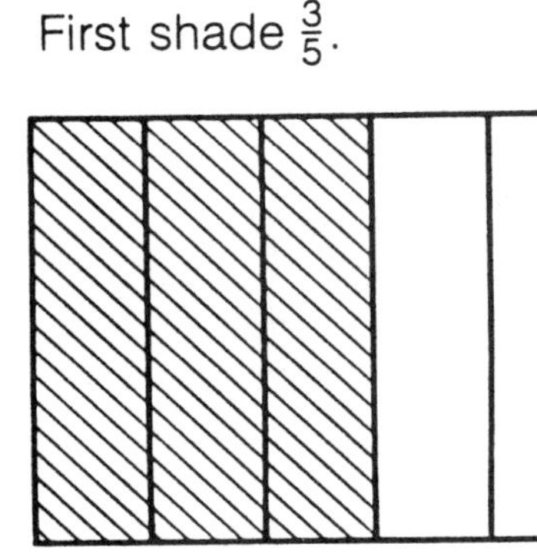

Next shade $\frac{1}{2}$ of that.

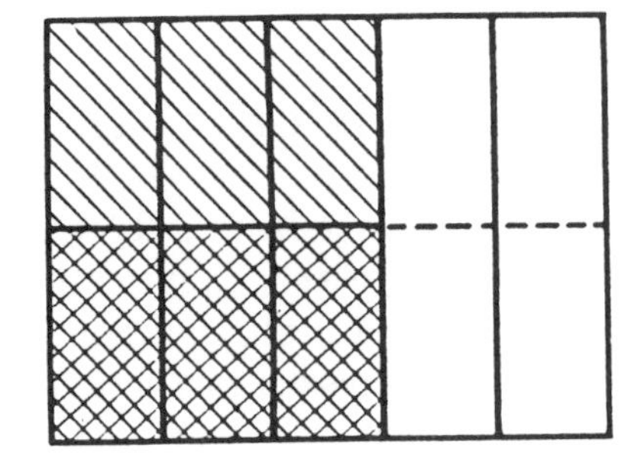

You can see that $\frac{1}{2}$ of $\frac{3}{5}$ is $\frac{3}{10}$.

$$\mathbf{\frac{1}{2} \times \frac{3}{5} = \frac{3}{10}}$$

About $\frac{3}{10}$ of the earth's land surface is covered by forests.

Figure 9.3 INVITATION TO MATHEMATICS, Grade 5, p. 250

$$\frac{2}{3} \cdot \frac{4}{5} = \frac{2 \cdot 4}{3 \cdot 5} = \frac{8}{15}$$

In thinking about this example, notice that the symbol $\cdot$ is being used in two different senses. In the expressions $2 \cdot 4$ and $3 \cdot 5$ it denotes multiplication of integers, as defined in Chapter 8. However, in $\frac{2}{3} \cdot \frac{4}{5}$ it stands

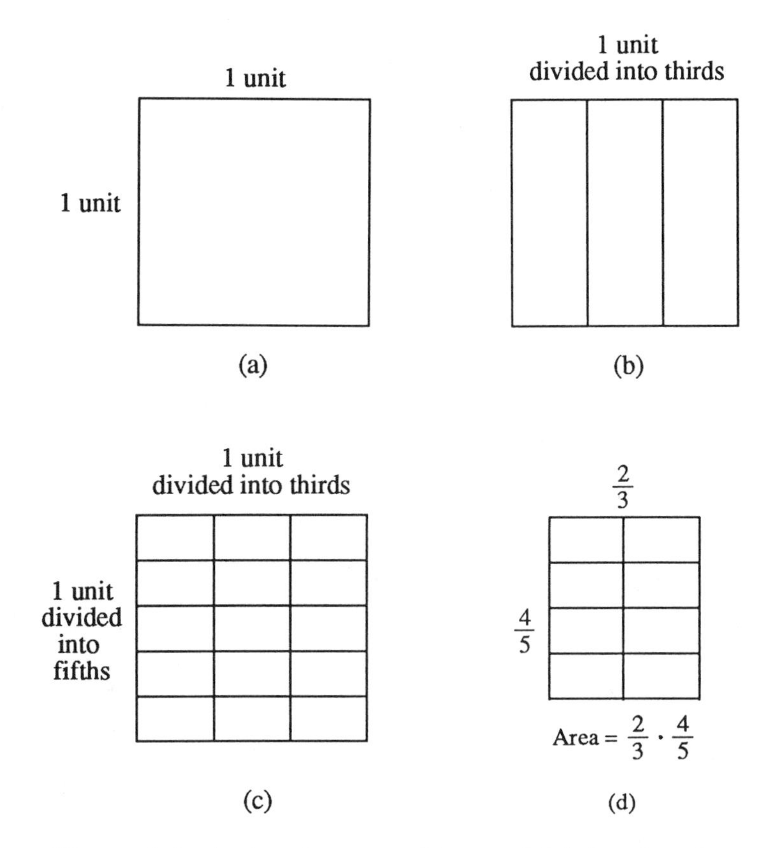

Figure 9.4 A visual approach to multiplying fractions.

for the still-to-be-defined multiplication of rational numbers. To emphasize the distinction between what we are currently defining and what is already known, for the rest of this chapter the new symbol will be distinguished from the old one by putting a circle around the $\cdot$ when it represents the new multiplication, as was done in Chapter 8. This convention will also be used for the other operation signs as they occur.

We can define multiplication of rational numbers, $\odot$, by generalizing the preceding example:

Definition The **product** of two rational numbers $[p, q]$ and $[m, n]$ is the rational number $[p \cdot m, q \cdot n]$. In symbols,

$$[p, q] \odot [m, n] = [p \cdot m, q \cdot n]$$

In fraction notation,

$$\frac{p}{q} \odot \frac{m}{n} = \frac{p \cdot m}{q \cdot n}$$

Example 1 $[12,3] \odot [10,2] = [12 \cdot 10, 3 \cdot 2] = [120,6]$, which equals [20, 1] by Property (9.3). Which division questions do these two rational numbers represent? Does the answer seem reasonable from that viewpoint? □

Example 2

$$[4,5] \odot [7,6] = [4 \cdot 7, 5 \cdot 6] = [28,30] = [14,15]$$

In fraction notation,

$$\frac{4}{5} \odot \frac{7}{6} = \frac{14}{15}$$ □

Example 3

$$[^-1,2] \odot [^-1,2] = [(^-1) \cdot (^-1),\ 2 \cdot 2] = [1,4]$$

That is,

$$\frac{^-1}{2} \odot \frac{^-1}{2} = \frac{1}{4}$$ □

Multiplication of rational numbers is well-defined; that is, the definition of *product* does not depend on the pairs chosen to represent the factors. This fact, which is illustrated in Example 4, is not immediately obvious, but it can be proved fairly easily from the definition of $\odot$.

Example 4

$$[4,9] \odot [3,2] = [4 \cdot 3, 9 \cdot 2] = [12,18] = [2,3]$$

Let us redo this product using different pairs to represent the numbers:

$$[8,18] \odot [9,6] = [8 \cdot 9, 18 \cdot 6] = [72,108] = [2,3]$$

In fraction notation,

$$\frac{4}{9} \odot \frac{3}{2} = \frac{8}{18} \odot \frac{9}{6} = \frac{2}{3}$$ □

PROBLEM-SOLVING COMMENT

Look back at the definition of the sum of two integers (in Section 8.3) and notice how the *analogy* between addition in **I** and multiplication in **Q** is preserved by this definition of product. The two definitions have precisely the same form, except that + in one is replaced by · in the other. ◇

Properties of Multiplication

It is not hard to see that $\odot$ is commutative and associative. This new multiplication is defined just by applying the "old" · to the first and second components of representative pairs of integers, so these properties of $\odot$ follow immediately from the commutativity and associativity of · on **I**. Examples 5 and 6 illustrate these properties.

Example 5 *Multiplication of rational numbers is commutative:*

$$\begin{aligned}[6,5] \odot [2,7] &= [6 \cdot 2, 5 \cdot 7] && \text{Definition of } \odot \\ &= [2 \cdot 6, 7 \cdot 5] && \text{Commutativity of } \cdot \text{ in } \mathbf{I} \\ &= [2,7] \odot [6,5] && \text{Definition of } \odot\end{aligned}$$

Similarly, in fraction notation,

$$\frac{{}^-3}{4} \odot \frac{5}{8} = \frac{{}^-3 \cdot 5}{4 \cdot 8} = \frac{5 \cdot ({}^-3)}{8 \cdot 4} = \frac{5}{8} \odot \frac{{}^-3}{4}$$ □

Example 6 *Multiplication of rational numbers is associative:*

$$\begin{aligned}([1,2] \odot [3,4]) \odot [5,6] &= [1 \cdot 3, 2 \cdot 4] \odot [5,6] \\ &= [(1 \cdot 3) \cdot 5, (2 \cdot 4) \cdot 6] \\ &= [1 \cdot (3 \cdot 5), 2 \cdot (4 \cdot 6)] \\ &= [1,2] \odot [3 \cdot 5, 4 \cdot 6] \\ &= [1,2] \odot ([3,4] \odot [5,6])\end{aligned}$$

(Can you supply a reason for each of these steps? Can you rewrite this example in fraction notation?) □

As you might expect, the identity element for $\odot$ is the rational number $[1,1]$ (or $\frac{1}{1}$) because, for any rational number $[p,q]$,

$$[p,q] \odot [1,1] = [p \cdot 1, q \cdot 1] = [p,q]$$

Once again, the behavior of $\cdot$ on the separate parts of the representative pairs determines the behavior of $\odot$. As noted in Section 9.2, $[1,1]$ is identified with the integer 1.

It is useful to bear in mind that the identity property of 1 does not depend on the representative pair used. By Property (9.3), *any* ordered pair with both (nonzero) integers the same is in the equivalence class $[1,1]$, and hence can be used to represent it.

Example 7

$$[{}^-4,3] \odot [1,1] = [{}^-4 \cdot 1, 3 \cdot 1] = [{}^-4,3]$$

In fraction notation,

$$\frac{{}^-4}{3} \odot 1 = \frac{{}^-4}{3} \odot \frac{1}{1} = \frac{{}^-4 \cdot 1}{3 \cdot 1} = \frac{{}^-4}{3}$$

Recalling that $[1,1] = [7,7]$, we can write

$$[{}^-4,3] \odot [7,7] = [{}^-4 \cdot 7, 3 \cdot 7] = [{}^-4,3]$$

This last equality is true by Property (9.3). □

Inverses

So far, all these properties of $\odot$ on **Q** are just copies of the properties of $\cdot$ on **I**. However, there is one more property that not only distinguishes multiplication of rational numbers from multiplication of integers, but also is the key to the nonzero rationals being closed under division. Essentially, this property says that, given any nonzero rational number, there is a rational number that can be multiplied by it in order to get 1. In other words, every nonzero rational number has an *inverse* with respect to multiplication.

You know, of course, that half of 2 is 1, one-third of 3 is 1, and so forth. If we write these facts as symbols, a pattern emerges:

$$\frac{2}{1} \odot \frac{1}{2} = 1$$
$$\frac{3}{1} \odot \frac{1}{3} = 1$$
$$\frac{4}{1} \odot \frac{1}{4} = 1$$
$$\vdots$$

In general, the inverse with respect to $\odot$ of any nonzero rational number $[p, q]$ is the rational number $[q, p]$ because

$$[p, q] \odot [q, p] = [p \cdot q, q \cdot p] = [p \cdot q, p \cdot q] = [1, 1] = 1$$

(Can you justify these steps?) The number $[q, p]$ is called the **multiplicative inverse** or **reciprocal** of $[p, q]$. In summary, we have shown:

(9.6) Every nonzero rational number $[p, q]$ has a multiplicative inverse, which is $[q, p]$.

Moreover, each rational number has *exactly one* such inverse. (See Exercises 19 and 20.) The multiplicative inverse of $[p, q]$ is denoted by $[p, q]^{-1}$. In fraction notation,

$$\left(\frac{p}{q}\right)^{-1} = \frac{q}{p}$$

Note: The superscript "-1" is just a symbol for now; its connection with negative exponents will be made later.

Example 8 The multiplicative inverse of $[3, 2]$ is $[2, 3]$ because

$$[3, 2] \odot [2, 3] = [6, 6] = [1, 1] = 1$$

(and $[2,3] \odot [3,2] = 1$ by commutativity.) We write $[3,2]^{-1} = [2,3]$. This is the same as saying that the reciprocal of $\frac{3}{2}$ is $\frac{2}{3}$ because

$$\frac{3}{2} \odot \frac{2}{3} = \frac{3 \cdot 2}{2 \cdot 3} = \frac{6}{6} = 1 \qquad \square$$

Example 9 The multiplicative inverse, or reciprocal, of $[^-5,7]$ is $[7,^-5]$. It is usually written with a positive denominator, as $[^-7,5]$. Thus,

$$[^-5,7] \odot [^-7,5] = [(^-5) \cdot (^-7),\ 7 \cdot 5] = [35,35] = [1,1]$$

In fraction notation,

$$\frac{^-5}{7} \odot \frac{^-7}{5} = \frac{(^-5) \cdot (^-7)}{7 \cdot 5} = \frac{35}{35} = 1 \qquad \square$$

Division

Multiplicative inverses are important because they enable us to do division. The definition of a quotient given in Chapter 4 also applies here:

Definition Let r and s be rational numbers. A rational number x such that $s \odot x = r$ is called the **quotient** of r **divided by** s; we write this number as $r \oplus s$. In symbols,

$$r \oplus s = x \quad \text{if} \quad s \odot x = r$$

(*Remember*: As discussed in Section 9.2, s cannot be zero.)

Example 10

$$[15,8] \oplus [3,2] = [5,4]$$

because

$$[3,2] \odot [5,4] = [3 \cdot 5, 2 \cdot 4] = [15,8]$$

In fraction notation,

$$\frac{15}{8} \oplus \frac{3}{2} = \frac{5}{4} \text{ because } \frac{3}{2} \odot \frac{5}{4} = \frac{15}{8} \qquad \square$$

You may recall that division of fractions was described differently in elementary school. When faced with a problem such as $\frac{3}{7} \oplus \frac{2}{5}$, for instance, you were told to "invert and multiply." Thus, you were shown computations like

$$\frac{3}{7} \oplus \frac{2}{5} = \frac{3}{7} \odot \frac{5}{2} = \frac{3 \cdot 5}{7 \cdot 2} = \frac{15}{14}$$

But $\frac{5}{2}$ is the reciprocal of $\frac{2}{5}$, so this invert-and-multiply process is nothing more than multiplying the first number by the reciprocal of the second.

Does this apparently different process always give the result required by the definition of a quotient? That is, for any rational numbers r and s, where $s \neq 0$, is it always true that $r \ominus s$ is the same as $r \odot s^{-1}$? To prove that the answer is *Yes*, we verify that s multiplied by $r \odot s^{-1}$ gives us r, as required by the definition of a quotient:

$$\begin{aligned} s \odot (r \odot s^{-1}) &= s \odot (s^{-1} \odot r) && \text{Commutativity of } \odot \\ &= (s \odot s^{-1}) \cdot r && \text{Associativity of } \odot \\ &= 1 \odot r && \text{Definition of reciprocal} \\ &= r && \text{1 is the multiplicative identity.} \end{aligned}$$

Thus, we have verified a useful alternative way of finding the quotient of two rational numbers, which we shall call the **Fundamental Division Property**:

(9.7) The quotient of two rational number r and s, where $s \neq 0$, is found by multiplying the first by the multiplicative inverse of the second.

In symbols,

$$r \ominus s = r \odot s^{-1}$$

In fraction notation, if $r = \frac{p}{q}$ and $s = \frac{m}{n}$, then

$$\frac{p}{q} \ominus \frac{m}{n} = \frac{p}{q} \odot \left(\frac{m}{n}\right)^{-1} = \frac{p}{q} \odot \frac{n}{m}$$

(For each step in the following examples, you should be able to supply a reason from the material in this section.)

Example 11 $[7,8] \ominus [3,5] = [7,8] \odot [3,5]^{-1} = [7,8] \odot [5,3] = [35,24]$ □

Example 12 $$\frac{2}{3} \ominus \frac{4}{5} = \frac{2}{3} \odot \left(\frac{4}{5}\right)^{-1} = \frac{2}{3} \odot \frac{5}{4} = \frac{10}{12} = \frac{5}{6}$$ □

Example 13 $$\frac{7}{2} \ominus {}^{-}4 = \frac{7}{2} \odot \left(\frac{{}^{-}4}{1}\right)^{-1} = \frac{7}{2} \odot \frac{{}^{-}1}{4} = \frac{{}^{-}7}{8}$$ □

Example 14 The division question $3 \div 5 = ?$, for which various interpretations were described in Section 9.1, can now be answered formally:

$$3 \div 5 = \frac{3}{1} \ominus \frac{5}{1} = \frac{3}{1} \odot \left(\frac{5}{1}\right)^{-1} = \frac{3}{1} \odot \frac{1}{5} = \frac{3}{5}$$

Thus, the definition of division given in this section agrees with our expectations. □

No matter which rational numbers r and s are considered (provided that $s \neq 0$), the reciprocal of s always exists in **Q** because of Property (9.6), so the Fundamental Division Property guarantees that the quotient $r \ominus s$ *always exists in* **Q** (because **Q** is closed under $\odot$). Thus, we have solved the main part of the problem that was stated at the beginning of this chapter:

(9.8) The rational number system is closed under division, except for division by zero.

Exercises 9.3

For Exercises 1–10, compute the given product by using the definition of $\odot$. Express all answers in lowest terms.

1. $[3, 5] \odot [5, 8]$
2. $[4, 7] \odot [2, 5]$
3. $[^-4, 9] \odot [12, 5]$
4. $[0, 7] \odot [21, 19]$
5. $\frac{4}{3} \odot \frac{5}{8}$
6. $\frac{^-13}{7} \odot \frac{1}{2}$
7. $\frac{5}{17} \odot 4$
8. $\frac{^-2}{3} \odot \frac{^-3}{2}$
9. $\frac{5}{16} \odot \frac{^-20}{7}$
10. $\frac{40}{21} \odot \frac{35}{12}$

For Exercises 11–16, specify the multiplicative inverse of the given rational numbers.

11. $[3, 5]$
12. $[^-7, 2]$
13. $\frac{5}{7}$
14. $\frac{^-9}{5}$
15. 3
16. $^-8$

17. Suppose that x is a rational number with the property $\frac{3}{5} \odot x = \frac{3}{5}$. Prove that x *must* equal 1; that is, prove that x cannot be any other number.

18. Generalize Exercise 17 to prove that **Q** can have no more than one identity element for multiplication.

19. Suppose that x is a rational number with the property $\frac{3}{5} \odot x = 1$. Prove that x *must* equal $\frac{5}{3}$; that is, prove that x cannot be any other number.

20. Generalize Exercise 19 to prove that no rational number can have more than one multiplicative inverse.

For Exercises 21–32, compute the quotient by using the Fundamental Division Property. Express your answers in lowest terms and check them with the definition of division.

21. $[7, 2] \ominus [5, 3]$
22. $[12, 7] \ominus [3, 8]$
23. $[15, 4] \ominus [6, 1]$
24. $[12, 9] \ominus [^-5, 5]$
25. $\frac{4}{5} \ominus \frac{3}{2}$
26. $\frac{18}{7} \ominus \frac{8}{9}$
27. $\frac{^-13}{6} \ominus 4$
28. $5 \ominus 3$
29. $5 \ominus \frac{1}{3}$
30. $\frac{7}{2} \ominus 1$
31. $1 \ominus \frac{7}{2}$
32. $\frac{^-10}{7} \ominus {^-6}$

33. Give three specific numerical examples to illustrate that the Fundamental Division Property agrees with the definition of quotient.

34. (a) Prove: For any two nonzero rational numbers r and s,
$$r^{-1} \odot s^{-1} = (r \odot s)^{-1}$$
(b) Using $r = \frac{2}{3}$ and $s = \frac{^-5}{4}$ as a specific example, work out the computation to verify part (a).

35. Show by example that division of rational numbers is neither associative nor commutative. (Do the computations.)

36. Using the material of this chapter, give reasons to justify each step in the following computation.

$\frac{6}{5} \ominus {}^-3$
$= [6,5] \ominus [{}^-3,1]$ ________________
$= [6,5] \odot [{}^-3,1]^{-1}$ ________________
$= [6,5] \odot [1,{}^-3]$ ________________
$= [6,5] \odot [{}^-1,3]$ ________________
$= [{}^-6,15]$ ________________
$= [{}^-2,5]$ ________________
$= \frac{{}^-2}{5}$ ________________

37. Prove that, for any nonzero rational numbers r and s,

$$(r \ominus s)^{-1} = s \ominus r$$

38. Prove that $(r^{-1})^{-1} = r$ for any nonzero rational number r.

39. A comment in Chapter 8 says that Example 9 of Section 8.3 "illustrates an important general fact:

> The additive inverse of a positive integer is negative; the additive inverse of a negative integer is positive."

What is the analogous statement for multiplicative inverses in **Q**? Explain your answer.

9.4 Addition and Subtraction of Rational Numbers

The development of addition and subtraction for the rationals does not exactly parallel the development of any integer operations. Nevertheless, because subtraction is defined in terms of addition, only the definition of sum is actually new.

The Definition of Sum

The general definition of addition of rational numbers flows naturally from an important special case—adding rationals with the same denominator. An elementary-school teacher (using appropriate teaching aids) might explain $\frac{2}{12} + \frac{5}{12}$ like this:

> You have a rectangle divided into 12 equal parts. Shade in 2 of them; this is $\frac{2}{12}$. To add $\frac{5}{12}$, shade in 5 more parts. Altogether, you have shaded in $2+5$, or 7, parts, showing that $\frac{2}{12} + \frac{5}{12} = \frac{7}{12}$. (See Figure 9.5.)

This suggests an appropriate definition for the sum of two rational numbers that have the same denominator:

Definition The **sum** of two rational numbers $[p,q]$ and $[m,q]$ *with the same denominator*

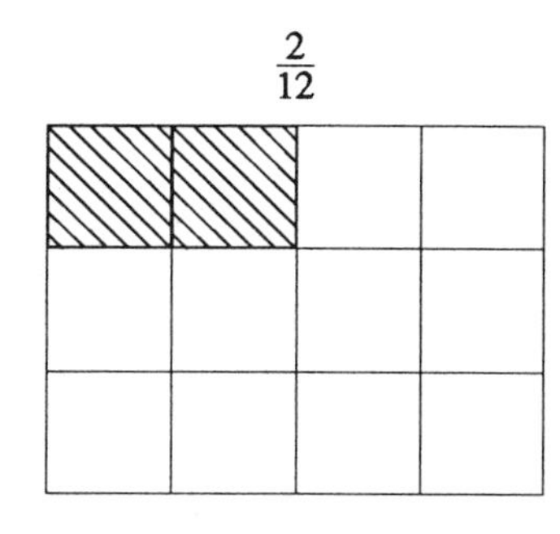

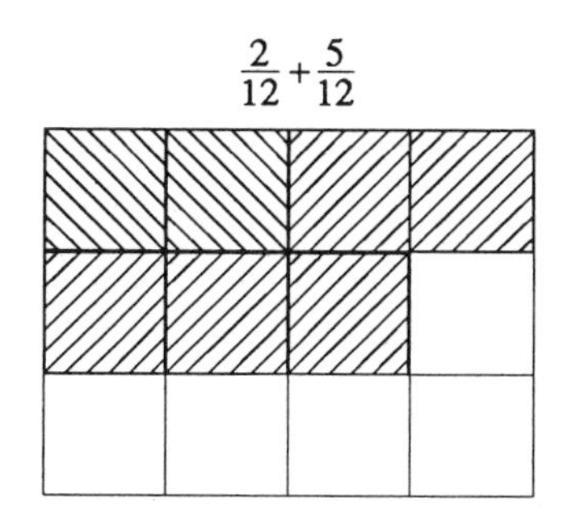

Figure 9.5 $\frac{2}{12} + \frac{5}{12} = \frac{7}{12}$

is the rational number $[p+m, q]$. In symbols,

$$[p, q] \oplus [m, q] = [p+m, q]$$

In fraction notation,

$$\frac{p}{q} \oplus \frac{m}{q} = \frac{p+m}{q}$$

Example 1

$$[3, 7] \oplus [5, 7] = [3+5, 7] = [8, 7]$$

That is,

$$\frac{3}{7} \oplus \frac{5}{7} = \frac{3+5}{7} = \frac{8}{7}$$

□

The general definition of the sum of two rational numbers is an easy and obvious extension of this special case. Given any two rational numbers, we choose representative pairs for each having the same denominator, then use the previous definition. Representatives with the same denominator (sometimes called a *common denominator*) can always be found, using Property (9.3). For example, if we want to add $[2, 3]$ and $[4, 5]$, we can write

$$[2, 3] = [2 \cdot 5, 3 \cdot 5] \qquad \text{and} \qquad [4, 5] = [3 \cdot 4, 3 \cdot 5]$$

so that, by the previous definition,

$$[2 \cdot 5, 3 \cdot 5] \oplus [3 \cdot 4, 3 \cdot 5] = [2 \cdot 5 + 3 \cdot 4, 3 \cdot 5]$$

Thus,

$$[2, 3] + [4, 5] = [22, 15]$$

A common denominator that *always* works is the product of the two denominators of the original representative pairs. This leads us to:

Definition The **sum** of two rational numbers $[p, q]$ and $[m, n]$ is the rational number $[pn + qm, qn]$. In symbols,

$$[p, q] \oplus [m, n] = [pn + qm, qn]$$

In fraction notation,

$$\frac{p}{q} \oplus \frac{m}{n} = \frac{pn + qm}{qn}$$

Example 2

$$[5,6] \oplus [^-3,10] = [5 \cdot 10 + 6 \cdot (^-3), 6 \cdot 10] = [32,60] = [8,15]$$

In fraction notation,

$$\frac{5}{6} \oplus \frac{^-3}{10} = \frac{5 \cdot 10}{6 \cdot 10} \oplus \frac{6 \cdot (^-3)}{6 \cdot 10} = \frac{5 \cdot 10 + 6 \cdot (^-3)}{6 \cdot 10} = \frac{32}{60} = \frac{8}{15}$$

A smaller common denominator—30—can be found in this case, but it does not have to be used. □

This definition of sum gives us a well-defined operation of addition for rational numbers. A proof of this fact is not hard, but it requires keeping track of some tedious notation in the process, so we omit it. The idea is illustrated by Example 3.

Example 3 We have already seen that $[2,3] \oplus [4,5] = [22,15]$. Now, $[2,3] = [6,9]$ and $[4,5] = [8,10]$. We should get the same sum if we add these same two rational numbers using the other representative pairs:

$$[6,9] \oplus [8,10] = [6 \cdot 10 + 9 \cdot 8, 9 \cdot 10] = [132,90]$$

and $[22,15] = [132,90]$ because the representative pairs are equivalent. You can also see this equality by applying Property (9.3) to remove a common factor from the terms of [132, 90]:

$$[132,90] = [22 \cdot 6, 15 \cdot 6] = [22,15]$$

(To get an idea of how the general proof works, think about the connection between this common factor, 6, in the sum and the common factors used to form the other representatives of the original summands.) □

Example 4 When schoolchildren first try to add fractions, they sometimes make the mistake of adding the numerators and adding the denominators. Clearly, this is wrong because it doesn't reflect the idea of adding up parts of a whole thing; $\frac{1}{2} \oplus \frac{1}{2}$ should equal 1, not $\frac{2}{4}$. But, apart from that, could this process be an operation "(+)" on **Q** that just gives strange results? The answer is *No* because *it's not well-defined.* That is, using different representative pairs for the same rational numbers will yield answers that are *different rational numbers*! Consider this: We know that $\frac{1}{2} = \frac{2}{4}$ and $\frac{2}{3} = \frac{6}{9}$, so combining $\frac{1}{2}$ and $\frac{2}{3}$ by an operation should give the same answer as when we combine $\frac{2}{4}$ and $\frac{6}{9}$ by that same operation. However,

$$\frac{1}{2}(+)\frac{2}{3}=\frac{3}{5} \quad \text{but} \quad \frac{2}{4}(+)\frac{6}{9}=\frac{8}{13}$$

and $\frac{8}{13}$ is *not* another way of writing $\frac{3}{5}$; they are different rational numbers. □

Properties of Addition

The operation $\oplus$ is commutative and associative, and $\odot$ is distributive over $\oplus$. General proofs of these facts can be given, based on the corresponding properties of integer arithmetic. However, the detail required in such arguments would be more distracting than useful as we pursue our goal of constructing the rational-number system, so we shall simply illustrate the properties with a few numerical examples.

Example 5 *Addition of rational numbers is commutative:*

$$\begin{aligned}[2,3]\oplus[4,5] &= [2\cdot 5+3\cdot 4, 3\cdot 5]\\ &= [4\cdot 3+5\cdot 2, 5\cdot 3]\\ &= [4,5]\oplus[2,3]\end{aligned}$$

because both $\cdot$ and $+$ are commutative in **I**. Similarly,

$$\frac{1}{4}\oplus\frac{3}{7}=\frac{1\cdot 7+4\cdot 3}{4\cdot 7}=\frac{3\cdot 4+7\cdot 1}{7\cdot 4}=\frac{3}{7}\oplus\frac{1}{4}$$ □

Example 6 *Addition of rational numbers is associative:*

$$\left(\frac{7}{2}\oplus\frac{4}{9}\right)\oplus\frac{3}{5}=\frac{71}{18}\oplus\frac{3}{5}=\frac{409}{90}$$

$$\frac{7}{2}\oplus\left(\frac{4}{9}\oplus\frac{3}{5}\right)=\frac{7}{2}\oplus\frac{47}{45}=\frac{409}{90}$$ □

Example 7 *Rational-number multiplication is distributive over addition:*

$$\frac{3}{4}\odot\left(\frac{1}{2}\oplus\frac{7}{5}\right)=\frac{3}{4}\odot\frac{19}{10}=\frac{57}{40}$$

and

$$\left(\frac{3}{4}\odot\frac{1}{2}\right)\oplus\left(\frac{3}{4}\odot\frac{7}{5}\right)=\frac{3}{8}\oplus\frac{21}{20}=\frac{228}{160}=\frac{57}{40}$$

(Can you justify the steps of these computations?) □

The identity element for $\oplus$ is 0, as you would expect, because

$$[p,q]\oplus[0,1]=[p\cdot 1+q\cdot 0, q\cdot 1]=[p,q]$$

for any rational number $[p,q]$. Moreover, every rational number $[p,q]$ has an additive inverse, $[-p, q]$, represented by the opposite of its numerator with the same denominator:

$$\begin{aligned}[p,q] \oplus [-p,q] &= [p \cdot q + q \cdot (-p), q \cdot q] \\ &= [pq + -(pq), q^2] \\ &= [0, q^2] \\ &= [0,1]\end{aligned}$$

That is:

(9.9) The additive inverse of a rational number $[p, q]$ is $[-p, q]$.

In symbols,

$$-[p,q] = [-p,q]$$

In fraction notation,

$$-\left(\frac{p}{q}\right) = \frac{-p}{q}$$

(which is also $\frac{p}{-q}$).

Example 8 Zero $(= [0,1] = \frac{0}{1})$ is the additive identity:

$$[3,8] \oplus [0,1] = [3 \cdot 1 + 8 \cdot 0,\ 8 \cdot 1] = [3,8]$$

Also,

$$\frac{0}{1} \oplus \frac{^{-}7}{9} = \frac{0 \cdot 9 + 1 \cdot (^{-}7)}{1 \cdot 9} = \frac{^{-}7}{9}$$ □

Example 9 The additive inverse of $[3,2]$ is $[^{-}3,2]$:

$$[3,2] \oplus [^{-}3,2] = [3 \cdot 2 + 2 \cdot (^{-}3),\ 2 \cdot 2] = [0,4] = [0,1]$$

Similarly, $-\left(\frac{^{-}5}{8}\right) = \frac{-(^{-}5)}{8} = \frac{5}{8}$:

$$\frac{^{-}5}{8} \oplus \frac{5}{8} = \frac{^{-}40 + 40}{64} = \frac{0}{64} = 0$$ □

Subtraction

As in the integers, additive inverses can be used to perform the subtraction operation, because the Fundamental Subtraction Property holds in **Q**:

(9.10) For any rational numbers $[p, q]$ and $[m, n]$,

$$[p,q] \ominus [m,n] = [p,q] \oplus [-m,n]$$

In fraction notation,

$$\frac{p}{q} \ominus \frac{m}{n} = \frac{p}{q} \oplus \frac{-m}{n}$$

That is, if we define subtraction as usual—

$$r \ominus s = x \quad \text{if} \quad s \oplus x = r$$

—then, for any rational numbers $r = [p, q]$ and $s = [m, n]$, we have

$$[m, n] \oplus ([p, q] \oplus [-m, n]) = [p, q]$$

so Property (9.10) must be true. (Can you fill in the missing steps?)

Example 10 By the Fundamental Subtraction Property,

$$[3,4] \ominus [1,2] = [3,4] \oplus [^-1,2] = [3 \cdot 2 + 4 \cdot (^-1), 4 \cdot 2] = [2,8] = [1,4]$$

Similarly,

$$\frac{4}{3} \ominus \frac{3}{5} = \frac{4}{3} \oplus \frac{^-3}{5} = \frac{4 \cdot 5 + 3 \cdot (^-3)}{3 \cdot 5} = \frac{11}{15}$$

□

Exercises 9.4

For Exercises 1–10, compute the given sum by using the definition of $\oplus$. Express your answers in lowest terms.

1. $[3,4] \oplus [7,4]$ **2.** $[4,3] \oplus [^-1,3]$

3. $[^-1,6] \oplus [3,4]$ **4.** $[2,3] \oplus [8,5]$

5. $\frac{2}{3} \oplus \frac{9}{10}$ **6.** $\frac{1}{3} \oplus \frac{1}{4}$ **7.** $\frac{^-1}{2} \oplus \frac{2}{7}$

8. $\frac{1}{4} \oplus 3$ **9.** $3 \oplus 2$ **10.** $\frac{^-2}{3} \oplus \frac{^-2}{3}$

For Exercises 11–16, specify the additive inverse of the given rational number.

11. $[3,4]$ **12.** $[5,2]$ **13.** $[^-1,6]$

14. $\frac{7}{8}$ **15.** $\frac{9}{4}$ **16.** $\frac{^-3}{10}$

For Exercises 17–26, compute the given difference by using the Fundamental Subtraction Property. Express your answers in lowest terms and check them by applying the definition of subtraction.

17. $[2,3] \ominus [1,3]$ **18.** $[4,5] \ominus [3,2]$

19. $[3,8] \ominus [^-1,4]$ **20.** $[^-2,7] \ominus [^-7,2]$

21. $\frac{6}{11} \ominus \frac{1}{5}$ **22.** $\frac{7}{3} \ominus \frac{3}{7}$ **23.** $\frac{7}{10} \ominus \frac{^-2}{3}$

24. $\frac{^-5}{6} \ominus \frac{^-1}{3}$ **25.** $2 \ominus \frac{3}{8}$ **26.** $\frac{4}{5} \ominus 1$

27. (a) Give an example of specific rational numbers r and s that satisfy
$$-r \oplus -s = -(r \oplus s)$$
(b) Prove that the equation in part (a) holds for *any* two rational numbers.

28. (a) Give an example of specific rational numbers r and s that satisfy
$$-(r \ominus s) = s \ominus r$$
(b) Prove that the equation in part (a) holds for *any* two rational numbers.

29. Show by example that subtraction of rational numbers is neither commutative nor associative.

30. (a) Give an example to show that $\odot$ is distributive over $\ominus$ in **Q**.

(b) Prove that $\odot$ is (always) distributive over $\ominus$ in **Q**.

9.5 Ordering the Rational Numbers

The rational numbers can be ordered in many ways, but we want to define an order relation that extends the relation $<$ already defined on **I**. That is, we want to define a new order relation that arranges the rationals with denominator 1 in exactly the same order they have when considered as integers and ordered by $<$. In this case the obvious approach works; we just borrow the integer definition for $<$, substituting *rational number* for *integer.*

Definition A rational number r is **less than** a rational number s if there is a positive rational number d such that $r \oplus d = s$. In this case we also say that s is **greater than** r, and we write $r \circledless s$ and $s \circledgreater r$, respectively.

Notation We symbolize "is less than or equal to" by $\circledleq$, and "is greater than or equal to" by $\circledgeq$.

This definition works just fine in theory, but sometimes there is a problem with its application to particular rational numbers. For instance, while it is easy to show

$$\frac{1}{2} \circledless \frac{3}{4} \quad \text{because} \quad \frac{1}{2} \oplus \frac{1}{4} = \frac{3}{4}$$

it is not at all obvious from this definition whether or not $\frac{5}{7}$ is smaller than $\frac{8}{11}$. To determine this from the definition, we must know which of the questions

$$\frac{5}{7} \oplus (?) = \frac{8}{11} \qquad \text{and} \qquad \frac{8}{11} \oplus (?) = \frac{5}{7}$$

can be answered with a *positive* number. Since these two addition questions involve rational numbers with different denominators, it is useful to begin by converting the numbers to fractions with the same denominator. In this case, $7 \cdot 11$ will work as a common denominator, so we can write

$$\frac{5}{7} = \frac{5 \cdot 11}{7 \cdot 11} = \frac{55}{77}$$

and

$$\frac{8}{11} = \frac{7 \cdot 8}{7 \cdot 11} = \frac{56}{77}$$

Now the answer is obvious: $55 < 56$ (because $55 + 1 = 56$ in **I**), so we know that $\frac{55}{77} \circledless \frac{56}{77}$; that is,

$$\frac{5}{7} \circledless \frac{8}{11} \quad \text{because} \quad \frac{5}{7} \oplus \frac{1}{77} = \frac{8}{11}$$

This solution is based on converting the two fractions to equivalent fractions with the same (positive) denominator; then the fraction with the smaller numerator represents the smaller rational number. Since a common denominator for *any* pair of fractions is the product of the two given denominators, we can rewrite two fractions $\frac{p}{q}$ and $\frac{m}{n}$ as $\frac{pn}{qn}$ and $\frac{qm}{qn}$. Thus, we have a useful general principle:

(9.11) For any two rational numbers $[p, q]$ and $[m, n]$, (with q and n positive),

$$[p, q] \otimes [m, n] \quad \text{if and only if} \quad pn < qm$$

In fraction notation,

$$\frac{p}{q} \otimes \frac{m}{n} \quad \text{if and only if} \quad pn < qm$$

Example 1 Using the definition of *less than*,

$$[3, 5] \otimes [4, 5] \quad \text{because} \quad [3, 5] \oplus [1, 5] = [4, 5]$$

Similarly,

$$\frac{^-3}{7} \otimes \frac{2}{7} \quad \text{because} \quad \frac{^-3}{7} \oplus \frac{5}{7} = \frac{2}{7}$$

□

Example 2 By Statement (9.11), $[7, 9] \otimes [4, 5]$ because $7 \cdot 5 < 9 \cdot 4$ in **I**. Similarly,

$$\frac{8}{10} \otimes \frac{6}{7} \quad \text{because} \quad 56 = 8 \cdot 7 < 10 \cdot 6 = 60$$

□

Example 3 Any negative rational number is less than any positive rational number. For instance,

$$\frac{^-12}{5} \otimes \frac{1}{2} \quad \text{because} \quad ^-12 \cdot 2 = ^-24 < 5 = 5 \cdot 1 \text{ in } \mathbf{I}.$$

In general, for any *positive* rationals $\frac{p}{q}$ and $\frac{m}{n}$,

$$\frac{^-p}{q} \otimes \frac{m}{n} \quad \text{because} \quad ^-(pn) < qm$$

and any negative integer is less than any positive integer. (See Example 3 of Section 8.5.)

□

Properties of Order

The order relation $\otimes$ on the fractions with denominator 1 behaves exactly like $<$ on the integers. For instance, $\frac{3}{1} \otimes \frac{8}{1}$ in **Q** precisely because $3 < 8$ in **I**. In general:

(9.12)

$$\text{If } p, m \in \mathbf{I}, \text{ then } p < m \text{ if and only if } \frac{p}{1} \circledless \frac{m}{1}.$$

Moreover, the "nice" properties of $<$ on **I** carry over to $\circledless$ on **Q** in an obvious way. In fact, the parallel between order on **I** and order on **Q** is so exact that we just list here the main properties of $\circledless$, leaving to you the task of looking back to Section 8.5 to see why they work:

- $\circledless$ is transitive on **Q**, so we may talk sensibly about a rational number being *between* two others.
- The Trichotomy Law holds; that is, if two rationals are unequal, then one must be less than the other.
- Rational-number addition preserves order: If r, s, and t are rational numbers and $r \circledless s$, then $r \oplus t \circledless s \oplus t$.
- Multiplication by a positive rational number preserves the order relation; multiplication by a negative rational reverses it.

The Number Line

Because of transitivity and trichotomy, the ordering of **Q** can be represented on a number line. We can picture the rationals in much the same way as we did the integers in Figure 8.4, with all the positive rationals stretching out from 0 to the right in increasing size order and all the negatives aligned in decreasing size order from 0 to the left. (See Figure 9.6.) In particular, the number line illustrates that:

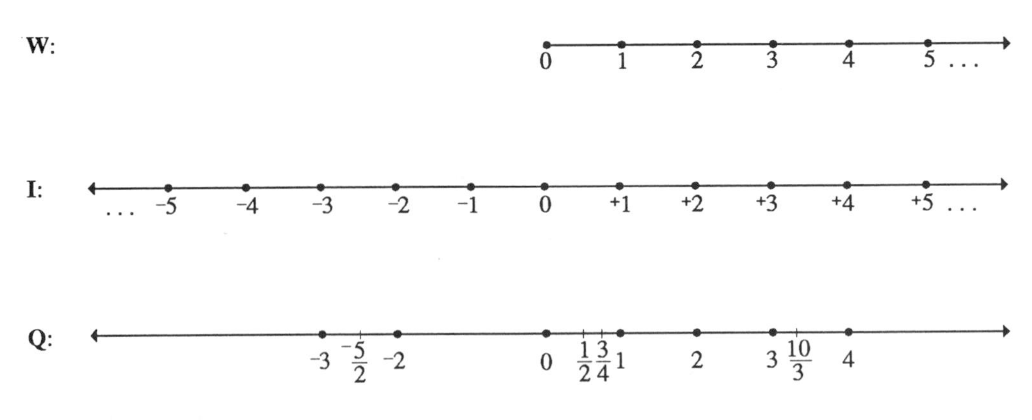

Figure 9.6 The number line for **W**, for **I**, and for **Q**.

(9.13)

> A rational number is positive if it is greater than 0; it is negative if it is less than 0.

There is, however, a fundamental difference between the number line for **Q** and the number line for **I**. Each integer is separated from its next larger or next smaller one by a space 1 unit long, but in the rationals the concept "next" doesn't even make sense! For instance, $\frac{3}{4}$ is greater than $\frac{1}{2}$, but it certainly isn't the *next* larger number, because $\frac{5}{8}$ lies between them; nor is $\frac{5}{8}$ next to $\frac{1}{2}$, because $\frac{9}{16}$ lies between them; and so on. In general, given any two rational numbers r and s, there are always other rationals between them. An obvious one is their "average,"

$$(r \oplus s) \ominus 2$$

which is halfway between r and s. (See Exercise 47.) Thus, between any two rational numbers there is a third one; between this third one and one of the first two there is a fourth one; between the third and fourth there is a fifth one; and so on. In other words, between any two rational numbers there are *infinitely* many others. This fact can be used to prove that

every segment of the number line (no matter how short) contains rational-number points.

We refer to this property of **Q** by saying that the rationals are **dense** on the number line.

Example 4 The rational number halfway between $\frac{9}{7}$ and $\frac{10}{7}$ can be found by adding the numbers and dividing by 2:

$$\left(\frac{9}{7} \oplus \frac{10}{7}\right) \ominus 2 = \frac{19}{7} \odot \frac{1}{2} = \frac{19}{14}$$

and

$$\frac{9}{7} = \frac{18}{14} \otimes \frac{19}{14} \otimes \frac{20}{14} = \frac{10}{7}$$

□

Example 5 There are many easy ways of specifying infinitely many rational numbers between 0 and 1; for instance, we might look at successive "halfway" points—

$$0 \otimes \frac{1}{2} \otimes \frac{3}{4} \otimes \frac{7}{8} \otimes \frac{15}{16} \otimes \ldots \otimes 1$$

—or add 1 to successive numerators and denominators—

$$0 \otimes \frac{1}{2} \otimes \frac{2}{3} \otimes \frac{3}{4} \otimes \frac{4}{5} \otimes \ldots \otimes 1$$

Notice that these increasing chains of rationals are getting closer and closer to 1 but will never actually reach 1. Similarly, we can specify infinite decreasing chains of rationals between 0 and 1:

$$1 \circledgt \frac{1}{2} \circledgt \frac{1}{4} \circledgt \frac{1}{8} \circledgt \cdots \circledgt \frac{1}{2^{n-1}} \circledgt \cdots \circledgt 0$$

$$1 \circledgt \frac{1}{2} \circledgt \frac{1}{3} \circledgt \frac{1}{4} \circledgt \cdots \circledgt \frac{1}{n} \circledgt \cdots \circledgt 0 \qquad \square$$

Example 6 To find four rational numbers between $\frac{1}{3}$ and $\frac{2}{3}$, we can proceed like this: $\frac{1}{3} = \frac{5}{15}$ and $\frac{2}{3} = \frac{10}{15}$, so

$$\frac{1}{3} \circledless \frac{6}{15} \circledless \frac{7}{15} \circledless \frac{8}{15} \circledless \frac{9}{15} \circledless \frac{2}{3}$$

Or, we could pick a larger denominator and accomplish the same thing: $\frac{1}{3} = \frac{10}{30}$ and $\frac{2}{3} = \frac{20}{30}$, so

$$\frac{1}{3} \circledless \frac{11}{30} \circledless \frac{12}{30} \circledless \frac{13}{30} \circledless \frac{14}{30} \circledless \frac{2}{3} \qquad \square$$

Finally, we note that the concepts of absolute value and distance are defined in exactly the same way for rational numbers as they were for integers, and the resulting computational rules are the same.

Example 7 The distance between $\frac{7}{8}$ and $\frac{5}{3}$ is

$$\left|\frac{7}{8} \ominus \frac{5}{3}\right| = \left|\frac{21-40}{24}\right| = \left|\frac{{}^{-}19}{24}\right| = \frac{19}{24} \qquad \square$$

Properties of the Rational Numbers

This completes the construction of the system of rational numbers. If you refer to the problem described in Section 9.1, you will see that we have done all that we set out to do: We have extended the system of integers to a larger system that is closed under division. For review, we summarize the main properties of this larger system, the rational numbers.

The set $\mathbf{Q}$ of rational numbers is the set of all equivalence classes of ordered pairs of integers with respect to the relation $\sim$ defined by

$$(a,b) \sim (c,d) \quad \text{whenever } a \cdot d = b \cdot c$$

Each of these classes can be written as $[p,q]$ or as $\frac{p}{q}$, where $q > 0$, and the integers p and q can be chosen so that they are relatively prime; in this case the rational number is said to be *in lowest terms.* Thus, the set of rational numbers may be represented as

$$\mathbf{Q} = \left\{ \frac{p}{q} \mid p, q \in \mathbf{I},\ q > 0,\ p \text{ and } q \text{ are relatively prime} \right\}$$

The operations of addition and multiplication are defined on **Q** as follows:

$$\frac{p}{q} \oplus \frac{m}{n} = \frac{pn + qm}{qn} \qquad \text{and} \qquad \frac{p}{q} \odot \frac{m}{n} = \frac{pm}{qn}$$

These operations have the following properties:

- Both operations are commutative: For all $r, s \in \mathbf{Q}$,

$$r \oplus s = s \oplus r \qquad\qquad r \odot s = s \odot r$$

- Both operations are associative: For all $r, s, t \in \mathbf{Q}$,

$$(r \oplus s) \oplus t = r \oplus (s \oplus t) \qquad\qquad (r \odot s) \odot t = r \odot (s \odot t)$$

- Both operations have identity elements: For all $r \in \mathbf{Q}$,

$$0 \oplus r = r \oplus 0 = r \qquad\qquad 1 \odot r = r \odot 1 = r$$

- Multiplication is left and right distributive over addition: For all $r, s, t \in \mathbf{Q}$,

$$r \odot (s \oplus t) = (r \odot s) \oplus (r \odot t)$$
$$(s \oplus t) \odot r = (s \odot r) \oplus (t \odot r)$$

Subtraction is also an operation on **Q**, but it is neither commutative nor associative, and it does not have an identity element. However, multiplication is left and right distributive over subtraction.

$\oplus$, $\odot$, and $\ominus$ on the rationals with denominator 1 agree with $+$, $\cdot$, and $-$ on **I**: For any $p, m \in \mathbf{I}$,

$$\frac{p}{1} \oplus \frac{m}{1} = \frac{p \cdot 1 + 1 \cdot m}{1 \cdot 1} = \frac{p + m}{1} = p + m$$
$$\frac{p}{1} \odot \frac{m}{1} = \frac{p \cdot m}{1 \cdot 1} = \frac{pm}{1} = pm$$
$$\frac{p}{1} \ominus \frac{m}{1} = \frac{p}{1} \oplus \frac{{}^{-}m}{1} = \frac{p + ({}^{-}m)}{1} = p - m$$

Thus, it is accurate to regard the rational-number operations as extensions of the corresponding integer operations.

Every nonzero rational number $r = \frac{p}{q}$ has a multiplicative inverse, which is $\frac{q}{p}$. That is,

$$r \odot r^{-1} = r^{-1} \odot r = 1$$

This allows us to describe division on the nonzero rationals by the Fundamental Division Property:

$$r \odiv s = r \odot s^{-1}$$

The set of nonzero rationals is closed under division.

An order relation $\olessthan$ is defined on $\mathbf{Q}$ by

$$r \olessthan s \text{ if } r \oplus d = s \text{ for some } d \in \mathbf{Q}^+$$

This relation has the following properties, where r, s, and t denote arbitrary rational numbers:

- Transitivity: If $r \olessthan s$ and $s \olessthan t$, then $r \olessthan t$.
- Trichotomy: Exactly one of the following is true: $r = s$, $r \olessthan s$, $s \olessthan r$.
- Addition preserves order: $r \olessthan s$ implies $r \oplus t \olessthan s \oplus t$.
- Multiplication property:
  If t is positive, then $r \olessthan s$ implies $r \odot t \olessthan s \odot t$;
  if t is negative, then $r \olessthan s$ implies $r \odot t \ogreaterthan s \odot t$.
- $\olessthan$ on the rationals with denominator 1 agrees with $<$ on $\mathbf{I}$.

The rational numbers are dense on the number line; that is, every segment, regardless of length, contains rational numbers.

*   *   *   *   *

Now that the construction of the rational numbers is complete, it will be useful to fix one notational inconvenience. In this chapter we have used circles around the signs $+$, $-$, $\cdot$, $\div$, and $<$ to distinguish their rational-number usage from their integer usage. However, we have shown that the operations on $\mathbf{Q}$ can be regarded as extensions of the operations on $\mathbf{I}$, so there is no need to continue this distinction; the rational-number operations can be used for both systems. Thus, *after this section, the symbols for rational-number operations and order relations will no longer be encircled.* Moreover, we shall write the rationals with denominator 1 and the integers interchangeably, no longer distinguishing between 5 and $\frac{5}{1}$, etc.

Exercises 9.5

For Exercises 1–10, justify the given inequality by applying the definition of $\olessthan$ and/or Property (9.11).

1. $[2,3] \olessthan [4,5]$ **2.** $[3,2] \ogreaterthan [2,3]$

3. $\frac{6}{7} \olessthan \frac{8}{7}$ **4.** $\frac{^-1}{5} \olessthan \frac{2}{5}$ **5.** $\frac{5}{6} \olessthan \frac{6}{5}$

6. $\frac{13}{8} \olessthan \frac{20}{12}$ **7.** $\frac{^-9}{4} \olessthan \frac{2}{3}$ **8.** $\frac{^-5}{3} \olessthan \frac{3}{5}$

9. $\frac{11}{7} \ogreaterthan 1$ **10.** $\frac{^-14}{5} \ogreaterthan {}^-3$

11. Give a specific numerical example to illustrate the Transitive Law for $\olessthan$ in $\mathbf{Q}$.

12. Prove the Transitive Law for $\circledless$ in **Q**. [*Hint*: Look back at the discussion of Statement (4.18).]

13. Give examples to show that $\circledless$ is neither reflexive nor symmetric.

14. Figure 9.7 shows a third-grade text page that deals with inequality of fractions. How does it "prove" that $\frac{3}{6} < \frac{5}{6}$? How can you prove this fact using the material in this section? How might you use the pizza theme to illustrate the formal proof? How might you use this theme to show that $\frac{1}{8} < \frac{1}{4}$?

For Exercises 15–22, decide which of the two given rational numbers is larger; then find another rational number between them.

15. $\frac{7}{8}, \frac{4}{5}$ **16.** $\frac{10}{7}, \frac{6}{4}$ **17.** $\frac{9}{11}, \frac{13}{16}$

18. $\frac{23}{15}, \frac{8}{3}$ **19.** $\frac{1}{100}, \frac{1}{1000}$ **20.** $\frac{^{-}23}{12}, {}^{-}2$

21. $\frac{1}{2}, {}^{-}1$ **22.** $\frac{^{-}2}{21}, \frac{^{-}3}{31}$

For Exercises 23–26, find the distance between the two given rational numbers.

23. $\frac{7}{2}$ and $\frac{4}{9}$ **24.** $\frac{3}{4}$ and $\frac{4}{3}$

25. $\frac{3}{7}$ and $\frac{^{-}5}{7}$ **26.** $\frac{^{-}4}{5}$ and $\frac{5}{8}$

For Exercises 27–40, state whether the given inequality is *true* or *false*.

27. $\frac{3}{8} \circledless \frac{3}{9}$ **28.** $\frac{5}{4} \circledless \frac{6}{4}$ **29.** $\frac{^{-}7}{3} \circledless \frac{^{-}6}{4}$

30. $\frac{1}{6} \circledless \frac{1}{7}$ **31.** $2 \circledgtr \frac{11}{5}$ **32.** $2 \circledgtr \frac{11}{6}$

33. $\frac{9}{13} \circledgtr 1$ **34.** $\frac{13}{9} \circledgtr 1$

35. $\frac{^{-}2}{3} \circledless {}^{-}1$ **36.** $\frac{^{-}3}{2} \circledless {}^{-}1$

37. $\left|\frac{2}{3} \oplus \frac{^{-}3}{4}\right| \circledgtr 0$ **38.** $\left|\frac{2}{3} \oplus \frac{^{-}3}{4}\right| \circledgtr 1$

39. $\left|\frac{2}{3}\right| \oplus \left|\frac{^{-}3}{4}\right| \circledgtr 0$ **40.** $\left|\frac{2}{3}\right| \oplus \left|\frac{^{-}3}{4}\right| \circledgtr 1$

For Exercises 41–44, find five rational numbers between the two given numbers.

41. $\frac{1}{3}$ and $\frac{1}{4}$ **42.** $\frac{7}{8}$ and $\frac{8}{7}$

43. $\frac{33}{16}$ and 2 **44.** $\frac{^{-}4}{5}$ and $\frac{^{-}7}{12}$

45. Arrange from smallest to largest:

$$\frac{1}{4}, \frac{2}{7}, \frac{2}{9}, \frac{3}{11}, \frac{3}{13}$$

46. Arrange from smallest to largest:

$$\frac{5}{3}, \frac{^{-}4}{3}, \frac{3}{5}, \frac{^{-}3}{5}, \frac{3}{4}, \frac{^{-}3}{4}, \frac{4}{3}, \frac{^{-}5}{3}$$

47. Use the definition of *less than* to prove:

If $r, s \in \mathbf{Q}$ and $r \circledless s$, then
$$r \circledless [(r \oplus s) \odiv 2] \circledless s$$

48. Prove that there is no smallest positive rational number.

49. Let r be a rational number. Prove:

(a) If $0 \circledless r \circledless 1$, then $r^{-1} \circledgtr 1$. (*Hint*: Try some examples first.)

(b) If $r \circledgtr 1$, then $0 \circledless r^{-1} \circledless 1$.

9.6 Negative Exponents

When we first encountered whole-number exponents (in Chapter 4), we needed two definitions—one for b^n, where n is a natural number, and one for b^0. Now that we have rational numbers, we can extend the exponent concept to negative integers, giving meaning to such expressions as $5^{^{-}2}$, $(^{-}4)^{^{-}3}$, and $(\frac{2}{3})^{^{-}1}$. In Chapter 8 two different strategies were used to explain why the product of two negative integers is positive. One used number patterns and

Comparing Fractions

After a party, $\frac{5}{6}$ of a large sausage pizza and $\frac{3}{6}$ of a large cheese pizza were left over. Was there more sausage pizza or cheese pizza left over? Compare $\frac{5}{6}$ and $\frac{3}{6}$.

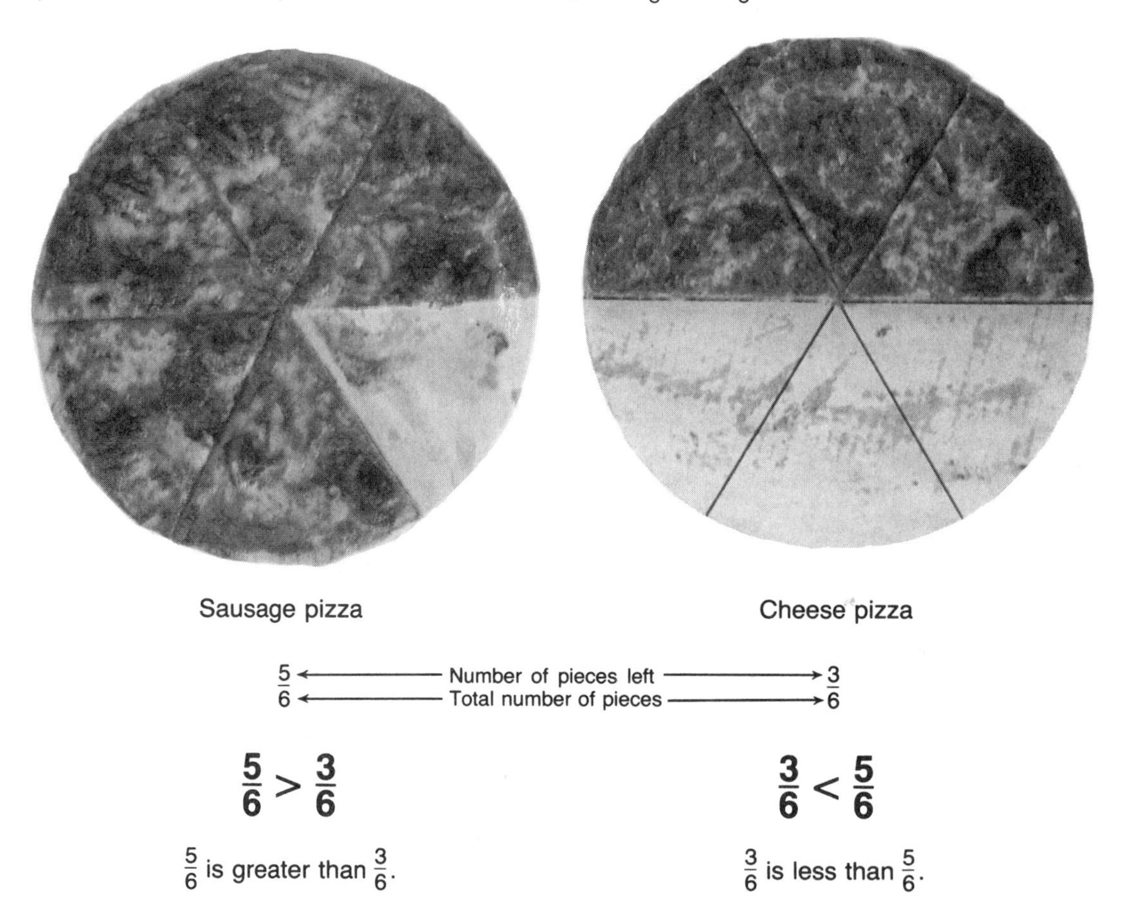

$5 \leftarrow$ Number of pieces left $\rightarrow 3$
$6 \leftarrow$ Total number of pieces $\rightarrow 6$

$$\frac{5}{6} > \frac{3}{6} \qquad \frac{3}{6} < \frac{5}{6}$$

$\frac{5}{6}$ is greater than $\frac{3}{6}$. $\frac{3}{6}$ is less than $\frac{5}{6}$.

There was more sausage pizza left over.

Try Replace each ● with $<$, $>$, or $=$.

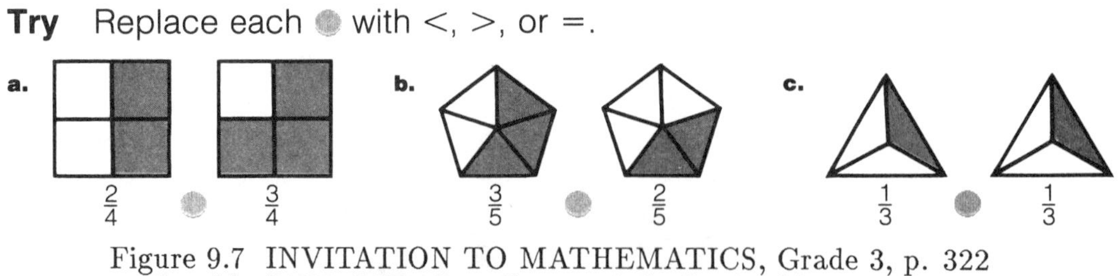

Figure 9.7 INVITATION TO MATHEMATICS, Grade 3, p. 322

the other used the formal properties of the integers. In a similar way we describe in this section two different approaches to negative exponents.

Suppose we want to give some meaning to the expression $2^{{}^-3}$. We could begin to generate a pattern by listing some known powers of 2:

$$\begin{aligned} 2^3 &= 8 \\ 2^2 &= 4 \\ 2^1 &= 2 \\ 2^0 &= 1 \end{aligned}$$

Observe that, as the exponent in each of these expressions is decreased by 1, the corresponding result is divided by 2. Taking the pattern one step further, as the exponent changes from 0 to ${}^-1$, the corresponding result should change from 1 to $\frac{1}{2}$. Continuing the same pattern, as the exponent decreases further, from ${}^-1$ to ${}^-2$, then to ${}^-3$, the results should change from $\frac{1}{2}$ to $\frac{1}{4}$, then to $\frac{1}{8}$, as in Figure 9.8.

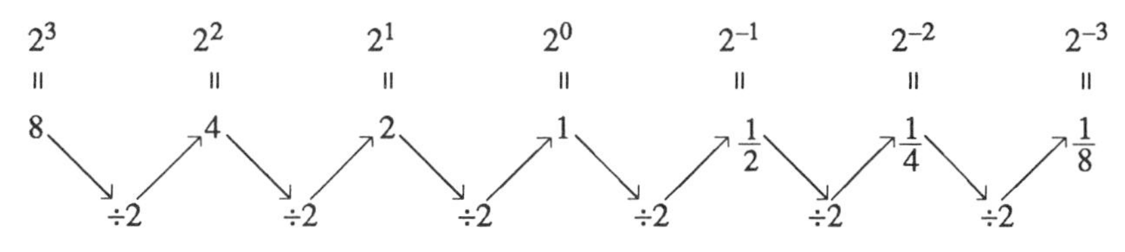

Figure 9.8 A decreasing-exponent pattern.

The denominator of each fraction in Figure 9.8 is a power of 2:

$$\frac{1}{2} = \frac{1}{2^1} \qquad \frac{1}{4} = \frac{1}{2^2} \qquad \frac{1}{8} = \frac{1}{2^3}$$

The general pattern illustrated here can be expressed by a formula: For any negative integer ${}^-n$,

$$2^{{}^-n} = \frac{1}{2^n}$$

This pattern gives a meaning to any nonzero rational raised to any negative integral power.

The same idea can be explained using an argument based on the formal properties of integers and exponents. If we want to define negative exponents at all, then surely we want to define them so that the addition property of exponents holds. That is, if b is a nonzero rational number and if p and q are any integers, then we want the equality

$$b^p \cdot b^q = b^{p+q}$$

For instance, this property should work for $2^{{}^-3} \cdot 2^4$; that is,

$$2^{{}^-3} \cdot 2^4 = 2^{{}^-3+4} = 2^1 = 2$$

Thus, 2^{-3} must be the number such that $2^{-3} \cdot 16 = 2$, implying that 2^{-3} must be $\frac{1}{8}$.

By a similar argument we can determine the value of any exponential expression involving a nonzero rational number as base and a negative integer as exponent. Therefore, the following definition is the way we *must* treat negative powers if we expect to maintain some consistency in the laws of exponents.

Definition If b is any nonzero rational number and if ^{-}n is any negative integer, then

$$b^{-n} = \frac{1}{b^n}$$

Example 1

$$5^{-2} = \frac{1}{5^2} = \frac{1}{25}$$ □

Example 2

$$(^{-}4)^{-3} = \frac{1}{(^{-}4)^3} = \frac{1}{^{-}64} = \frac{^{-}1}{64}$$ □

Example 3

$$\left(\frac{2}{3}\right)^{-1} = \frac{1}{(\frac{2}{3})^1} = \frac{1}{\frac{2}{3}} = \frac{3}{2}$$ □

PROBLEM-SOLVING COMMENT

You have just seen two different approaches to the definition of negative exponents. The first is based on *looking for a pattern*; the second is based on *reasoning backwards from the desired conclusion.* Either approach might be more appropriate than the other in a particular elementary-school class, depending on the background and ability level of the students. ◇

The properties stated in Chapter 4 for whole-number exponents are true for negative integer exponents, as well:

If a and b are nonzero rational numbers and if p and q are any integers, then

$$b^p \cdot b^q = b^{p+q}$$
$$a^p \cdot b^p = (a \cdot b)^p$$
$$(b^p)^q = b^{pq}$$
$$b^p \div b^q = b^{p-q}$$

Notice that this last property no longer needs the restriction "$p \geq q$" that appeared in Chapter 4; now that we have negative exponents, b^{p-q} has meaning even when $p < q$.

Example 4

$$5^{-4} \cdot 5^3 = 5^{-4+3} = 5^{-1} = \frac{1}{5}$$

Similarly,

$$(^-3)^2 \cdot (^-3)^{-6} = (^-3)^{2+(^-6)} = (^-3)^{-4} = \frac{1}{(^-3)^4} = \frac{1}{81}$$

□

Example 5

$$(^-2)^{-3} \cdot 3^{-3} = [(^-2) \cdot 3]^{-3} = (^-6)^{-3} = \frac{1}{(^-6)^3} = \frac{1}{^-216} = \frac{^-1}{216}$$

This example can be done another way:

$$(^-2)^{-3} \cdot 3^{-3} = \frac{1}{(^-2)^3} \cdot \frac{1}{3^3} = \frac{1}{^-8} \cdot \frac{1}{27} = \frac{1}{^-216} = \frac{^-1}{216}$$

□

Example 6

$$\left[(^-3)^2\right]^{-4} = (^-3)^{2 \cdot (^-4)} = (^-3)^{-8} = \frac{1}{(^-3)^8} = \frac{1}{6561}$$

□

Example 7

$$3^3 \div 3^5 = 3^{3-5} = 3^{-2} = \frac{1}{9}$$

Similarly,

$$2^5 \div 2^{-3} = 2^{5-(^-3)} = 2^8 = 256$$

□

Electronic calculators and computers use the exponential notation we have been discussing. A calculator or computer has a maximum number of digits it can handle; in order to deal with large numbers it must resort to some kind of "shorthand" to overcome this limitation. For example, if you multiply 50,000 by 50,000 on an eight-digit calculator, the answer—2,500,000,000—cannot be displayed in that form. Some calculators will give you an error message in this situation, but others will express the answer as "`2.5 09`" or "`2.5 E 09`". In either case, the answer means

$$(2.5) \cdot 10^9 = (2.5) \cdot 1,000,000,000 = 2,500,000,000$$

Suppose you divide 2.46 by 200,000,000 on an eight-digit calculator. The correct answer is .0000000123, but again the calculator cannot display it in this form. If you don't get an error message, you probably will get "`1.23 -08`" or "`1.23 E -08`". In either case, this means

$$1.23 \cdot 10^{-8} = 1.23 \cdot \frac{1}{10^8} = .0000000123$$

Example 8 Both "`5.3 05`" and "`5.3 E 05`" mean $5.3 \cdot 10^5$, which equals $5.3 \cdot 100{,}000$, or 530,000. □

Example 9 Both "`7.59 -03`" and "`7.59 E -03`" mean

$$7.59 \cdot 10^{^-3} = 7.59 \cdot \frac{1}{10^3} = \frac{7.59}{1000} = .00759$$

Similarly, both "`-3.5 -02`" and "`-3.5 E -02`" mean

$$(^-3.5) \cdot 10^{^-2} = (^-3.5) \cdot \frac{1}{10^2} = \frac{^-3.5}{100} = {}^-.035$$ □

The fact that the negative sign is not raised in the calculator expressions of Example 9 reflects common custom. Machines make no distinction between "$^-3.5$," the negative number 3.5, and "-3.5," the additive inverse of the (positive) number 3.5. We have used the raised negative sign to emphasize the difference between these two concepts, as is customary in most elementary- and secondary-school texts. However, this notational distinction is rarely used beyond high school; people (and machines) use "-3.5" for both meanings. There is no danger of computational confusion because $^-3.5$ *is* the additive inverse of 3.5; that is,

$$^-3.5 = -3.5$$

We shall continue to use the raised-sign notation in the next chapter, where we discuss algorithms for the rational numbers as they are taught in elementary school. From Section 10.7 on, we shall follow the common custom and use the in-line negative sign for both meanings.

Exercises 9.6

For Examples 1–18, evaluate the given expression.

1. $8^{^-2}$ **2.** $5^{^-4}$ **3.** $(^-5)^4$

4. $(^-5)^{^-4}$ **5.** $-(5^4)$ **6.** $-(5^{^-4})$

7. $(^-13)^0$ **8.** $7^{^-2}$ **9.** $(^-7)^{^-2}$

10. $(\frac{1}{2})^{^-3}$ **11.** $(^-\frac{1}{2})^3$ **12.** $(\frac{3}{2})^4$

13. $(\frac{3}{2})^{^-4}$ **14.** $2^{^-4} \cdot 3^2$

15. $(\frac{1}{2})^{^-4} \cdot 3^2$ **16.** $(\frac{1}{2})^4 \cdot 3^{^-2}$

17. $\left[(\frac{1}{2})^{^-4}\right]^{^-2}$ **18.** $[5^8 \cdot (^-8)^{^-10}]^0$

19. Use both of the strategies explained in this section to illustrate why $3^{^-2} = \frac{1}{9}$.

20. In the definition of b^{-n} (for b a nonzero rational number), the number $-n$ was said to be a negative integer, implying that n represents a positive integer.

(a) By substituting appropriate numbers for n, give examples to illustrate that the equation $b^{-n} = \frac{1}{b^n}$ is true for *any* integer n, not just for positive values.

(b) Prove that $b^{-n} = \frac{1}{b^n}$ is true for any integer n.

For Exercises 21–34, use the appropriate properties of integer exponents to evaluate each expression.

21. $3^{-2} \cdot 3^{-3}$
22. $[({}^{-}4)^{-3}]^2$
23. $({}^{-}2)^{-4} \cdot ({}^{-}2)^2$
24. $({}^{-}5)^{-3} \cdot 2^{-3}$
25. $(3^2)^{-1}$
26. $7^9 \div 7^7$
27. $5^{-2} \cdot 5^3$
28. $({}^{-}8)^2 \div ({}^{-}8)^{-1}$
29. $2^3 \div 2^5$
30. $(2^3 \cdot 2^{-4})^{-5} \div (2^{-5})^2$
31. $\left(\frac{5}{7}\right)^{-3} \cdot \left(\frac{5}{7}\right)^4$
32. $\left(\frac{17}{11}\right)^{-9} \cdot \left(\frac{11}{17}\right)^{-9}$
33. $\left(\frac{1}{2}\right)^{-3} + ({}^{-}2)^3$
34. $\left[\left(\frac{2}{3}\right)^{-2}\right]^{-1}$

For Exercises 35–40, evaluate the given calculator expression.

35. `5.73 E 05`
36. `-2.9 03`
37. `4.723 E -08`
38. `-4.82391 -07`
39. `5.67 E 00`
40. `1.0001 -03`

Review Exercises for Chapter 9

For Exercises 1–10, indicate whether the given statement is *true* or *false*.

1. One motivation for the development of the rational numbers is to obtain a number system that is closed under division.
2. The set of integers is to addition and subtraction as the set of nonzero rational numbers is to multiplication and division.
3. The expression $0 \div 6$ is undefined, but $6 \div 0 = 0$.
4. Every rational number r is an ordered pair of integers (a, b), where b is not zero.
5. $[3, 4]$, $[{}^{-}3, {}^{-}4]$, $[6, 8]$, and $\frac{3}{4}$ all represent the same rational number.
6. $-\frac{5}{6}$, $\frac{{}^{-}5}{6}$, $\frac{5}{{}^{-}6}$, and $\frac{{}^{-}5}{{}^{-}6}$ all represent the same rational number.
7. To divide one rational number by another, you can multiply the first number by the additive inverse of the second.
8. In $\mathbf{Q}$, $[p, q] \oplus [r, s] = [p + q, r + s]$.
9. A rational number r is less than a rational number s whenever there is a rational number d such that $r \oplus d = s$.
10. In the rational-number system, $b^{-n} = -b^n$.

In Exercises 11–15, the given statement describes some property or definition from the system of integers. Write a true *analogous* statement from the system of rational numbers.

11. $[a, b] = [c, d]$ if and only if $a + d = b + c$.
12. $[a, b] \oplus [c, d] = [a + c,\ b + d]$.
13. For every $a \in \mathbf{I}$, $a + 0 = 0 + a = a$.
14. For each $a \in \mathbf{I}$, there is an element $-a \in \mathbf{I}$ such that $a + (-a) = (-a) + a = 0$.
15. For all $a, b \in \mathbf{I}$, $a - b = a + (-b)$.

16. State two structural properties of the rational numbers that are *not* true for the integers.

In Exercises 17–20, use the appropriate definition to perform the indicated operation. Show your work in the $[a, b]$ form, and put your answer in simplest $[a, b]$ form.

17. $[3, 10] \oplus [2, 15]$
18. $[{}^{-}6, 2] \oplus [3, 5]$
19. $[4, 7] \odot [{}^{-}5, 6]$
20. $[{}^{-}4, 5] \odot [{}^{-}5, 7]$

21. Use a diagram to illustrate the product $\frac{3}{4} \cdot \frac{2}{5}$. Be sure to explain how the diagram is related to the problem and its answer.

For Exercises 22 and 23, determine the smaller of the two numbers.

22. $\frac{4}{7}, \frac{5}{9}$

23. $\frac{^-3}{8}, \frac{^-2}{7}$

24. Insert three distinct rational numbers between $\frac{4}{7}$ and $\frac{3}{5}$.

25. Use number patterns to explain why $3^{^-2}$ is defined to be $\frac{1}{9}$.

26. Use properties of integers and exponents to explain why $3^{^-2}$ is defined to be $\frac{1}{9}$.

For Exercises 27–29, evaluate the given expression.

27. $2^{^-3} \cdot 5^{^-1}$ **28.** $(^-3)^{^-4}$ **29.** $(^-2)^{^-3} \cdot 3^2$

For Exercises 30–33, evaluate the given calculator expression.

30. `5.743 E 08` **31.** `4.3276 -05`

32. `-2.358 06` **33.** `-7.89 E -07`

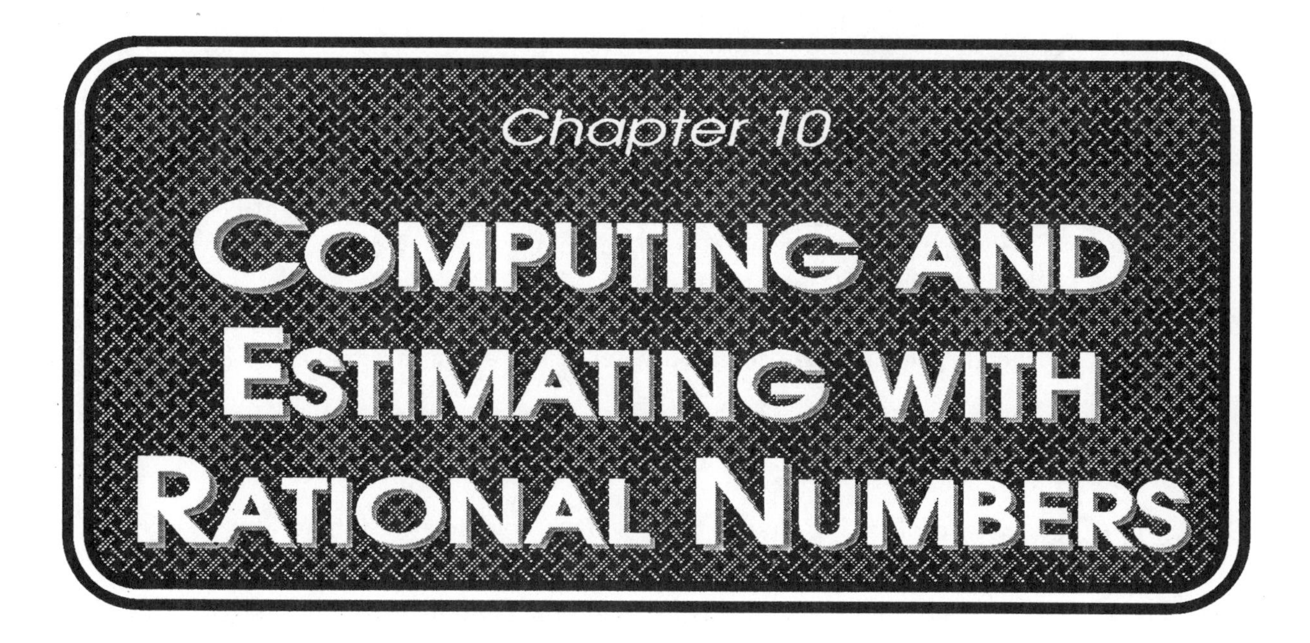

10.1 The Problem

If you look at the treatment of rational numbers in an elementary-school text, you most likely will find that the sequence of topics does not parallel our development in Chapter 9. The main reason for this is that the structural concepts in Chapter 9 begin with and depend on the premise that the rational number $[a, b]$ represents the solution to the division problem $a \div b = ?$, whereas fractions are first introduced to elementary-school children in a much more intuitive way, as representing parts of some whole thing. The link between fractional parts and division is taught only after the initial fraction idea has gradually been made more abstract. This chapter relates our formal structural development with the techniques taught in the elementary grades.

At the beginning of Chapter 9 we saw six different interpretations for the division question $3 \div 5 = ?$. Although $\frac{3}{5}$ clearly represents the correct answer for all six questions, that particular fraction form may not always be the best way to express it in every case.

Example 1 If each of the 3 pizzas had been cut into 10 pieces, then the 5 people sharing them would each get $\frac{6}{10}$ of a pizza, or 6 pieces. □

Example 2 Rather than saying that a loaf of bread costs $\frac{3}{5}$ of a dollar, it might be more appropriate to say it costs $\frac{60}{100}$ of a dollar, or \$.60. □

Example 3 Instead of saying that each shelf is $\frac{3}{5}$ of a meter long, it might be more useful to say it is $\frac{60}{100}$ of a meter, or 60 cm long. □

Example 4 If $\frac{3}{5}$ is the ratio of the increase in a plant's height relative to its original height, this is more commonly expressed as an increase of 60%. □

Example 5 A runner's average lap time stated as $\frac{3}{5}$ of a minute is harder to understand than the statement that it is $\frac{36}{60}$ of a minute, or 36 seconds. □

Example 6 Instead of saying that it took $\frac{3}{5}$ of an hour for the temperature to drop 1°, it might be more informative to say that it took 36 minutes. □

These examples illustrate that, depending on the application, one numeral form may be more suitable than another to represent a particular number. Three different numeration forms are commonly used to express rational numbers: fractions, decimals, and percents. For each of these forms there are special techniques, or algorithms, for handling rational numbers written in that form. We shall review many of those techniques in the course of this chapter. However,

> *The main purpose of this chapter is not to reteach you how to use the algorithms, but rather to provide logical explanations for those formal processes, so that you will be able to teach them to your future students in ways that make them seem reasonable.*

The decimal (base ten) numeration system and its algorithms for whole-number arithmetic provide the foundation for the three rational numeration forms and their associated algorithms. Hence, it is important for elementary-school teachers to provide their students with the conceptual links between the whole numbers and the rationals in each of the three numeration forms. Most elementary textbooks link the whole numbers to fractions as we did, through the idea

$$\frac{a}{b} = a \div b$$

Fraction form is then linked with decimal form by using powers of 10 as denominators. Finally, percent—the link between the fraction and decimal forms—is established by multiplying or dividing by 100. In this chapter we present the material in much the same order as it often appears in the elementary-school curriculum. Thus, we first discuss numeration and computation for fractions, then for decimals, and finally for percents.

Note: Up to now we have used the dot symbol · to denote multiplication of numbers, reserving the symbol $\times$ primarily for the Cartesian product of sets. This is standard practice in the mathematical literature above the secondary-school level. However, elementary-school books often use $\times$ for number multiplication (as you have seen in some of the page excerpts in Chapter 9). As a teacher, you will have to be familiar with both symbols, so we shall use them interchangeably from here on, letting our choice be guided purely by our sense of visual clarity in the expressions we write. Your reading should make no distinction in their meaning.

Historical Note: Writing Fractions

The fraction concept has been part of mathematics for 4000 years or so, but the way we write fractions appeared much later. Not until the Middle Ages did the Hindu custom of writing fractions as one number over another become common in the West. The earliest evidence of this notation in Indian writings dates back to Brahmagupta (c. A.D. 650), who wrote fractions in this form, but without a bar to separate the numbers. The horizontal bar was inserted by the Arabs and appeared in most Latin manuscripts of the Middle Ages; but it was omitted in the early days of printing, probably because of typesetting problems. It gradually came back into common use in the 16th and 17th centuries. Although 3/4 is easier to print than $\frac{3}{4}$, the "slash" notation did not appear until about 1850.

The form in which fractions were written directly affected the arithmetic that developed. For instance, the "invert and multiply" rule for dividing one fraction by another was used by the Hindu mathematician Mahavira (c. A.D. 850). However, it was not part of Western arithmetic until the 16th century, probably because it made no sense unless fractions were written as one number over another. Latin writers of the Middle Ages were the first to use the terms *numerator* ("numberer"—how many) and *denominator* ("namer"—of what size) as a convenient way of distinguishing the top number of a fraction from the bottom one. Despite the occasional introduction of simpler terms, including *top* and *base*, the clumsier, but more suggestive, Latin terms have survived for 600 years.

The first use of decimals as a computational device occurred in the 16th century. The power of decimal fractions was described in a book by the Flemish mathematician Simon Stevin in 1585. Within a generation, the use of decimal fractions by scientists such as Johannes Kepler and John Napier, including the use of a dot (the *decimal point*) to separate the whole and fractional parts of a number, paved the way for popular acceptance of decimal arithmetic. The term *per cent* ("for every hundred") as a special name for fractions with denominator 100 emerged in the commercial arithmetic of the 15th and 16th centuries, when it was common to quote interest rates in hundredths. The continuation of such customs in business, reinforced in this country by a monetary system based on dollars and *cents* (hundredths of dollars), has insured the continued use of percents as a special branch of decimal arithmetic.[1] ◇

[1]Much of the material for this Historical Note is from David Eugene Smith, *History of Mathematics*, Vol. II, New York: Dover Publications, Inc., 1958. Pages 208–250 of this book contain a wealth of material on the historical development of rational numbers and their various notational forms.

10.2 Multiplying and Reducing Fractions

Two ideas are central to understanding the fraction forms of rational numbers. The first is that different fractions can be used to represent the same rational number. (This is the equivalence-class concept explained in Chapter 9.) The second is that the fraction forms can be classified as being "proper" or "improper," or in "mixed numeral" form. As you may recall from your early schooling, a fraction is **proper** if its absolute value is less than 1, it is **improper** if its absolute value is greater than or equal to 1, and it is in **mixed numeral** form if it is written as an integer followed by a proper fraction. For example, $\frac{4}{5}$ is proper, $\frac{6}{5}$ is improper, and $1\frac{1}{5}$ is a mixed numeral.

Related to each of these concepts are specific algorithms that are used within the algorithms for addition, subtraction, multiplication, and division. One changes a fraction to a different, but equivalent, form, especially to the form in lowest terms; another expresses an improper fraction as a mixed numeral, and vice versa. Associated with these algorithms is the idea that, for any integer a, the fractions $\frac{a}{a}$ and $\frac{a}{1}$ can be replaced by the integers 1 and a, respectively. As Examples 1 and 2 show, all these concepts are so interwoven that the logical sequence for teaching them is not at all obvious.

Example 1 The procedure for reducing $\frac{18}{30}$ to lowest terms works like this:

$$\frac{18}{30} = \frac{6 \cdot 3}{6 \cdot 5} = \frac{6}{6} \times \frac{3}{5} = 1 \times \frac{3}{5} = \frac{3}{5}$$ □

Example 2 Adding a mixed numeral to a fraction works as follows:

$$\begin{aligned} 2\frac{2}{3} + \frac{4}{5} &= 2 + \frac{2}{3} + \frac{4}{5} = \frac{2}{1} + \frac{2}{3} + \frac{4}{5} \\ &= \left(\frac{2}{1} \times \frac{15}{15}\right) + \left(\frac{2}{3} \times \frac{5}{5}\right) + \left(\frac{4}{5} \times \frac{3}{3}\right) \\ &= \frac{30}{15} + \frac{10}{15} + \frac{12}{15} = \frac{52}{15} = 3\frac{7}{15} \end{aligned}$$ □

Most elementary textbooks begin the presentation of these concepts with an intuitive introduction to fractions by using geometric figures divided into regions of equal size and the same shape. The students are told that the original figure represents the whole number 1. The denominator of the fraction indicates how many regions the "1" figure is to be divided into, and the numerator indicates how many of these subdivisions are to be chosen.

The text excerpt in Figure 10.1 shows a typical example of such a presentation at the third-grade level. This approach is useful primarily for

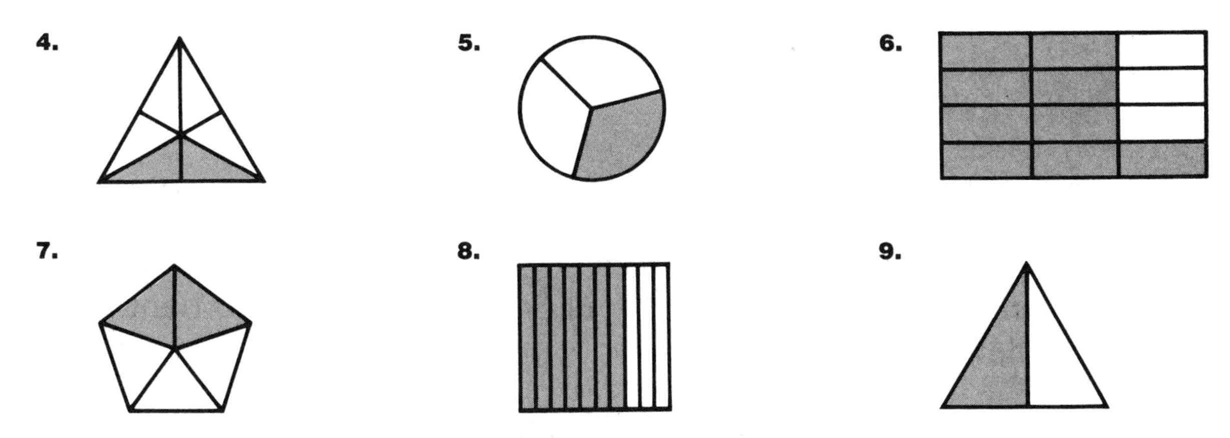

Figure 10.1 INVITATION TO MATHEMATICS, Grade 3, p. 315

introducing proper fractions, such as $\frac{1}{2}$, $\frac{2}{3}$, $\frac{3}{5}$, etc. Of course, it is possible to draw a diagram that represents an improper fraction, such as $\frac{8}{3}$, but children often have a difficult time understanding how you can divide the "1" shape into 3 regions and shade 8 of them! Also, this geometric approach to fractions does not readily suggest that $\frac{a}{b} = a \div b$, an idea essential to a thorough treatment of improper fractions and mixed numerals.

In spite of these shortcomings, this pictorial approach does permit the teacher to introduce the concepts of equivalent fractions and addition and subtraction of fractions, as well as the idea that a fraction of the form $\frac{a}{a}$ can be replaced by the whole number 1. However, the related algorithms cannot easily be explained by pictures, except for special cases of addition and subtraction (of proper fractions with like denominators, sum not exceeding 1, and nonnegative difference). Before these algorithms can be extended to include general sums and differences of fractions, the algorithms for equivalent fractions and multiplication of fractions must be developed, and the students need to learn that $\frac{a}{1} = a$ and $\frac{a}{b} = a \div b$.

Of all these ideas, the next logical choice is the multiplication algorithm. It, too, can be approached geometrically, by beginning with the number 1 represented as a square 1 unit on a side, as was shown in Section 9.3. (See, in particular, Figures 9.3 and 9.4.) To find the product $\frac{a}{b} \times \frac{c}{d}$, a rectangle with one dimension $\frac{a}{b}$ and the other $\frac{c}{d}$ is constructed within the square; the area of this rectangle represents the product of the two fractions. The pictures also illustrate that the area is found by multiplying the numerators and multiplying the denominators. Thus, in general,

(10.1)
$$\frac{a}{b} \times \frac{c}{d} = \frac{ac}{bd}$$

We call this the **Multiplication Algorithm** (for fractions). It is the definition of rational-number multiplication that was introduced in Section 9.3.

Once the Multiplication Algorithm has been established, several other properties and algorithms can be introduced. The repeated-addition approach to multiplication suggests a reasonable interpretation for the product of a whole number and a fraction, so we can establish the idea that the fraction $\frac{a}{1}$ and the whole number a represent the same quantity. For instance, it is easy to see that

$$3 \times \frac{2}{7} = \frac{2}{7} + \frac{2}{7} + \frac{2}{7} = \frac{(2+2+2)}{7} = \frac{6}{7}$$

By the Multiplication Algorithm,

$$\frac{3}{1} \times \frac{2}{7} = \frac{3 \cdot 2}{1 \cdot 7} = \frac{6}{7}$$

Therefore, it makes sense to regard the fraction $\frac{3}{1}$ as interchangeable with the whole number 3. Examples like this can be used to lead students to the general conclusion that

(10.2)
$$\frac{a}{1} = a$$

This, in turn, can be used to establish the more general property

(10.3)
$$\frac{a}{b} = a \div b$$

The argument that justifies Statement (10.3) is as follows: By the definition of division, $a \div b = x$ if $b \cdot x = a$. Hence, to show that $a \div b = \frac{a}{b}$, we must show that

$$b \cdot \left(\frac{a}{b}\right) = a$$

Using Equation (10.2) and the Multiplication Algorithm, we have

$$b \cdot \frac{a}{b} = \frac{b}{1} \cdot \frac{a}{b} = \frac{b \times a}{1 \times b}$$

By the commutativity of multiplication and the two properties we just used,

$$\frac{b \times a}{1 \times b} = \frac{a \times b}{1 \times b} = \frac{a}{1} \times \frac{b}{b} = \frac{a}{1} \times 1 = \frac{a}{1} = a$$

The Multiplication Algorithm also enables us to develop algorithms for writing equivalent fractions. These algorithms are useful for reducing fractions, adding fractions with unlike denominators, and even for simplifying the Multiplication Algorithm itself, as the following examples show. In particular, Example 3 shows how to justify the process of reducing fractions by "cancelling" common factors from the numerator and the denominator. Recall from Section 9.2 that a rational number $\frac{a}{b}$, where $b > 0$, is in lowest terms (or in simplest form) if a and b are relatively prime; that is, if $\gcd(a, b) = 1$. Using Equation (10.1), the common factors are separated from the others in the form of a fraction with the same numerator and denominator, which is just a representative of 1.

Example 3 To reduce $\frac{10}{15}$ to lowest terms:

$$\frac{10}{15} = \frac{2 \times 5}{3 \times 5} = \frac{2}{3} \times \frac{5}{5} = \frac{2}{3} \times 1 = \frac{2}{3} \qquad \square$$

Example 4 This example illustrates the algorithm for expressing a fraction as an equivalent fraction by multiplying the numerator and the denominator by the same number. It is the same process as in Example 3, done in reverse. To express $\frac{4}{9}$ as a fraction with denominator 18:

$$\frac{4}{9} = \frac{4}{9} \times 1 = \frac{4}{9} \times \frac{2}{2} = \frac{4 \times 2}{9 \times 2} = \frac{8}{18} \qquad \square$$

Example 5 The Multiplication Algorithm can be made more efficient if factors that are common to *any* numerator and *any* denominator are eliminated *before* the actual multiplication is done. We show here how a typical example might look in practice, then write out in detail the steps that justify it.

$$\frac{\overset{2}{\cancel{10}}}{\underset{9}{\cancel{27}}} \times \frac{\overset{\overset{1}{\cancel{7}}}{\cancel{21}}}{\underset{\underset{11}{\cancel{77}}}{\cancel{385}}} = \frac{2}{99}$$

Justification:

$$\begin{aligned} \frac{10}{27} \times \frac{21}{385} &= \frac{10 \times 21}{27 \times 385} = \frac{2 \times 3 \times 5 \times 7}{3 \times 9 \times 5 \times 7 \times 11} = \frac{2 \times 3 \times 5 \times 7}{99 \times 3 \times 5 \times 7} \\ &= \frac{2}{99} \times \frac{3}{3} \times \frac{5}{5} \times \frac{7}{7} = \frac{2}{99} \times 1 \times 1 \times 1 = \frac{2}{99} \qquad \square \end{aligned}$$

The greatest common factor of the numerator and the denominator of a fraction plays an important role in reducing a fraction to its lowest terms.

(10.4) Any fraction $\frac{a}{b}$ can be reduced to lowest terms in one step by dividing both a and b by $\gcd(a, b)$.

Example 6 $\frac{210}{270}$ can be reduced to lowest terms in one step by dividing 210 and 270 by $\gcd(210, 270) = 30$. Thus,

$$\frac{210}{270} = \frac{7}{9}$$ □

Example 7 $\frac{10}{21}$ is already reduced to lowest terms because $\gcd(10, 21) = 1$. □

Example 8 The elementary-text page in Figure 10.2 shows how reducing fractions to lowest terms is handled in the fifth grade. Notice that the shorter process described there is our Statement (10.4). □

Exercises 10.2

1. Use appropriate diagrams to show that $\frac{2}{3}$ and $\frac{4}{6}$ are equivalent fractions.

2. Draw a diagram to illustrate $\frac{4}{7} \times \frac{5}{9}$. (*Hint*: Refer to Figure 9.3 on page 314.)

For Exercises 3–8, compute the product by using the Multiplication Algorithm.

3. $\frac{3}{4} \times \frac{7}{6}$
4. $\frac{1}{3} \times \frac{7}{8}$
5. $\frac{15}{2} \times \frac{10}{9}$
6. $\frac{7}{3} \times \frac{14}{15}$
7. $\frac{21}{4} \times \frac{8}{27}$
8. $\frac{23}{25} \times \frac{12}{32}$

9. Use a method similar to the one given in this section to show that $\frac{4}{1}$ is equivalent to 4.

10. Consider the following interchange between a sixth-grade teacher and a student:

S: Why does $\frac{a}{1} = a$?

T: Because $\frac{a}{1} = a \div 1 = a$.

S: Why does $\frac{a}{b} = a \div b$?

T: By the definition of division, that's true if $b \times \frac{a}{b} = a$, so we compute:

$$b \times \frac{a}{b} = \frac{b}{1} \times \frac{a}{b} = \frac{b \cdot a}{b \cdot 1} = \frac{a}{1} \times \frac{b}{b} = \frac{a}{1} \times 1 = \frac{a}{1} = a$$

Are the teacher's responses logically correct? Explain.

11. Using the method shown in Example 3, explain how $\frac{9}{12}$ can be reduced to $\frac{3}{4}$.

12. Using the method shown in Example 3, explain how $\frac{30}{42}$ can be reduced to $\frac{5}{7}$.

13. Using the method shown in Example 4, explain how $\frac{7}{8}$ can be expressed as $\frac{84}{96}$.

Fractions in Lowest Terms

A fraction is in ***lowest terms*** when the only number that will divide both the numerator and the denominator is 1.

12 members of the orchestra play the violin. 8 of these members are women. $\frac{8}{12}$ of the violin players are women.

Write $\frac{8}{12}$ in lowest terms.

You could do the work like this.

$$\frac{8}{12} = \frac{4}{6} \quad (8 \div 2,\ 12 \div 2)$$

You can divide both 8 and 12 by 2 to find a fraction equal to $\frac{8}{12}$.

$$\frac{4}{6} = \frac{2}{3} \quad (4 \div 2,\ 6 \div 2)$$

$\frac{4}{6}$ is not in lowest terms because you can divide both 4 and 6 by 2.

To shorten your work, you could divide both the numerator and the denominator by the ***greatest common factor***, 4.

$$\frac{8}{12} = \frac{2}{3} \quad (8 \div 4,\ 12 \div 4)$$

$\frac{2}{3}$ is in lowest terms because the greatest number that divides both 2 and 3 is 1.

Try Is each fraction in lowest terms? Write *yes* or *no*.

a. $\frac{3}{5}$ **b.** $\frac{6}{10}$ **c.** $\frac{9}{15}$

Write each fraction in lowest terms.

d. $\frac{3}{9}$ **e.** $\frac{8}{10}$ **f.** $\frac{12}{18}$

Figure 10.2 INVITATION TO MATHEMATICS, Grade 5, p. 220

14. Using the method shown in Example 4, explain how $\frac{5}{9}$ can be expressed as $\frac{75}{135}$.

15. Using the method shown in Example 5, explain why $\frac{14}{30} \times \frac{99}{154} = \frac{3}{10}$.

16. Using the method shown in Example 5, explain why $\frac{180}{77} \times \frac{143}{1350} = \frac{26}{105}$.

For Exercises 17–22, show how Statement (10.4) can be used to reduce the fraction to lowest terms.

17. $\dfrac{420}{480}$ **18.** $\dfrac{35}{105}$ **19.** $\dfrac{104}{286}$

20. $\dfrac{1050}{1500}$ **21.** $\dfrac{207}{209}$ **22.** $\dfrac{^-77}{1001}$

23. Describe how you would accurately divide up each of the following figures to illustrate the fraction $\frac{3}{5}$. What instruments would you need?

(a) a 5-inch by 1-inch rectangle

(b) a square 5 inches on a side

(c) a circle 5 inches in diameter

In general, do you think it is easier for students to use rectangles, squares, or circles to represent fractions? Explain.

10.3 Adding, Subtracting, and Dividing Fractions

PROBLEM-SOLVING COMMENT

The development of algorithms for adding fractions provides an excellent opportunity to apply many of the problem-solving tactics discussed in Chapter 1. The problem of adding two fractions is attacked by first *solving the simpler problem* of adding fractions with the same denominator. This can be done by *drawing diagrams* to represent the fractions, then *arguing by analogy*: Since addition of whole numbers is based on the union of disjoint sets, perhaps the addition of fractions could be based on the "union" of disjoint fractional diagrams. (Elementary-school teachers often refer to this as "forming trains.") By judiciously *constructing examples* and *reasoning backwards from the desired conclusion*, we solve the simpler problem:

> The numerator of the sum of two fractions with the same denominator is the sum of the numerators, and the denominator of the sum is the denominator that the two fractions have in common.

From there, it is an easy step for students to *generalize the solution* by realizing that they can add *any* two fractions by expressing them as equivalent ones with the same denominator. To keep the numbers simple, it is often desirable to choose the denominator as small as possible. Thus, we can begin by finding the "least common denominator"; then we add the two fractions by the "LCD Addition Algorithm," which follows next. ◇

Definition The **least common denominator (LCD)** of two fractions is the least common multiple of their denominators; that is,

$$\text{LCD of } \frac{a}{b} \text{ and } \frac{c}{d} = \text{lcm}\,(b, d)$$

The **LCD Addition Algorithm** works like this: To add any two fractions $\frac{a}{b}$ and $\frac{c}{d}$:

Step 1. Find the LCD of $\frac{a}{b}$ and $\frac{c}{d}$; that is, find $\operatorname{lcm}(b, d)$. (An algorithm for the lcm is given in Chapter 7.) This is the denominator of the sum.

Step 2. Express each fraction as an equivalent fraction with the LCD as denominator.

Step 3. Add the numerators of these two fractions. This is the numerator of the sum.

Example 1 Add $\dfrac{25}{21} + \dfrac{31}{77}$.

Step 1. Find LCD of $\frac{25}{21}$ and $\frac{31}{77} = \operatorname{lcm}(21, 77)$:

$$21 = 3 \times 7 \quad \text{and} \quad 77 = 7 \times 11$$

so

$$\operatorname{lcm}(21, 77) = 3 \times 7 \times 11 = 231$$

Step 2. Express both fractions as equivalent fractions with denominator 231:

$$\frac{25}{21} = \frac{25}{21} \times \frac{11}{11} = \frac{275}{231} \quad \text{and} \quad \frac{31}{77} = \frac{31}{77} \times \frac{3}{3} = \frac{93}{231}$$

Step 3. Add the new numerators:

$$\frac{25}{21} + \frac{31}{77} = \frac{275}{231} + \frac{93}{231} = \frac{368}{231}$$

□

Example 2 Add $\dfrac{2}{5} + \dfrac{3}{7}$:

Step 1. LCD $= \operatorname{lcm}(5, 7) = 35$ (5 and 7 are relatively prime.)

Step 2. $$\frac{2}{5} = \frac{2}{5} \times \frac{7}{7} = \frac{14}{35} \quad \text{and} \quad \frac{3}{7} = \frac{3}{5} \times \frac{5}{5} = \frac{15}{35}$$

Step 3. $$\frac{2}{5} + \frac{3}{7} = \frac{14}{35} + \frac{15}{35} = \frac{29}{35}$$

□

Example 2 illustrates the general fact that, if $\frac{a}{b}$ and $\frac{c}{d}$ are two fractions with $\operatorname{gcf}(b, d) = 1$, then their sum can be found by using

$$\frac{a}{b} + \frac{c}{d} = \frac{ad + bc}{bd} \tag{10.5}$$

This formula can be used for the sum of any two fractions, whether or not their denominators are relatively prime, because it is just the definition of *sum* in the rational-number system. For elementary-school students, it can be justified directly from the previous algorithms, as follows:

$$\frac{a}{b} + \frac{c}{d} = \left(\frac{a}{b} \times \frac{d}{d}\right) + \left(\frac{c}{d} \times \frac{b}{b}\right) = \frac{ad}{bd} + \frac{bc}{bd} = \frac{ad + bc}{bd}$$

Of course, bd will be the least common denominator (LCD) only if b and d are relatively prime, but the sum will always be correct. Thus, this formula provides another common algorithm for adding fractions, which we shall call the **Addition-by-Formula Algorithm**. In some cases, this algorithm is easier to use than the LCD method. You might find it interesting to try to express the LCD Addition Algorithm as a formula and then compare the two formulas.

Example 3 $\frac{7}{15} + \frac{2}{9} = \frac{(7 \times 9) + (15 \times 2)}{15 \times 9} = \frac{93}{135}$ (The LCD is 45.) □

Example 4 $\frac{3}{4} + \frac{4}{5} = \frac{(3 \times 5) + (4 \times 4)}{4 \times 5} = \frac{31}{20}$ (The LCD is 20.) □

Example 5 Figure 10.3 shows an excerpt from a fifth-grade text. Which of the addition algorithms is presented there? □

Find $\frac{2}{3} + \frac{1}{4}$.

Write the fractions with a common denominator.

The least common denominator is 12.

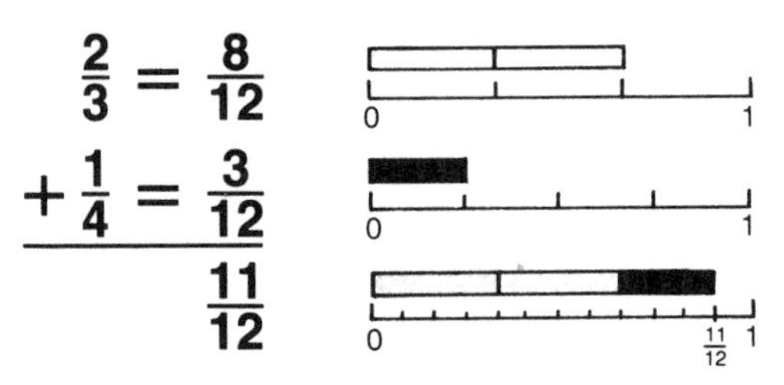

Figure 10.3 INVITATION TO MATHEMATICS, Grade 5, p. 278

Now that the addition algorithms have been established, we can discuss the notation for improper fractions and mixed numerals. For example, consider the improper fraction $\frac{9}{5}$, which represents $9 \div 5$. By the Division Algorithm for whole numbers, we know that

$$9 \div 5 = 1\ R\ 4$$

(That is, the quotient is 1 and the remainder is 4.) But we can show that $1\ R\ 4$ can be written as $1 + \frac{4}{5}$, using the definition of division to verify that $9 \div 5 = 1 + \frac{4}{5}$:

$$5 \times \left(1 + \frac{4}{5}\right) = 5 \times \left(\frac{5}{5} + \frac{4}{5}\right) = 5 \times \frac{9}{5} = 9$$

The same approach can be used to verify the corresponding general statement for any fraction $\frac{a}{b}$:

(10.6) If $a \div b = c\ R\ r$, then $a \div b = c + \frac{r}{b}$.

The details of the justification are left as an exercise. This fact confirms the accuracy of the way we usually think of mixed numerals—

$$1\tfrac{4}{5} = 1 + \tfrac{4}{5}$$
$$7\tfrac{2}{3} = 7 + \tfrac{2}{3}$$

and so forth. In general:

(10.7) The numeral "$c\,\frac{r}{b}$" represents the sum $c + \frac{r}{b}$.

The subtraction algorithms are so similar to those for addition that there is very little to say about them:

- The LCD Algorithm for subtraction is identical to the one for addition, except that in Step 3 the second numerator is subtracted from, rather than added to, the first.

- The Subtraction-by-Formula Algorithm is obtained from the corresponding addition algorithm by changing the "+" in the numerator of the formula to "−". Thus,

$$\frac{a}{b} - \frac{c}{d} = \frac{ad - bc}{bd}$$

Example 6 Subtract $\frac{7}{15} - \frac{2}{9}$ by the LCD algorithm:

Step 1. LCD of $\frac{7}{15}$ and $\frac{2}{9}$ = lcm$(15, 9) = 3^2 \times 5 = 45$

Step 2. $\dfrac{7}{15} = \dfrac{7}{15} \times \dfrac{3}{3} = \dfrac{21}{45}$ and $\dfrac{2}{9} = \dfrac{2}{9} \times \dfrac{5}{5} = \dfrac{10}{45}$

Step 3. $$\frac{7}{15} - \frac{2}{9} = \frac{21}{45} - \frac{10}{45} = \frac{21-10}{45} = \frac{11}{45}$$ □

Example 7 Using the Subtraction-by-Formula Algorithm,

$$\frac{7}{15} - \frac{2}{9} = \frac{(7 \times 9) - (15 \times 2)}{15 \times 9} = \frac{63-30}{135} = \frac{33}{135}$$ □

Although it is easy to prove, the Division Algorithm is a bit more difficult to make plausible to young students. You may remember being taught: "To divide, you invert and multiply." This idea can be explained by using the concept of multiplicative inverses (reciprocals), which was discussed in Section 9.3. It is easy to demonstrate by example that dividing one whole number by another is equivalent to multiplying the first number by the reciprocal of the second. (In fact, the general statement

$$a \div b = a \times \frac{1}{b}$$

is easy to prove; it is left as an exercise.) Once this principle has been established, it becomes reasonable to extend it to division of rational numbers. Thus, the Division Algorithm can be stated as follows:

(10.8) Dividing one rational number by another is equivalent to multiplying the first number by the reciprocal of the second. In symbols,

$$\frac{a}{b} \div \frac{c}{d} = \frac{a}{b} \times \frac{d}{c} = \frac{ad}{bc}$$

PROBLEM-SOLVING COMMENT

Statement (10.8) has an *analogue* in the integers; it is the Fundamental Subtraction Property (of Section 8.3):

> Subtracting one integer from another is equivalent to adding the first number to the opposite of the second. In symbols,
>
> $$a - b = a + (-b)$$

This is an extension of the analogy between multiplication in **Q** and addition in **I**. ◇

Example 8

$$\frac{2}{3} \div \frac{4}{5} = \frac{2}{3} \times \frac{5}{4} = \frac{2 \times 5}{3 \times 4} = \frac{10}{12} = \frac{5}{6} \qquad \square$$

Example 9

$$5 \div 7 = \frac{5}{1} \div \frac{7}{1} = \frac{5}{1} \times \frac{1}{7} = \frac{5 \times 1}{1 \times 7} = \frac{5}{7} \qquad \square$$

Example 10 Figure 10.4 shows how a sixth-grade text presents division of fractions. $\square$

Two final notes are in order before we leave algorithms for fractions:

- First, although the examples provided so far have been restricted to nonnegative rationals, these algorithms apply to *all* rationals. However, because most texts in elementary school treat fractions before signed numbers, we have provided examples consistent with that order.
- Second, recall from Chapter 9 that the additive inverse of a fraction $\frac{a}{b}$ is equivalent to the fractions $\frac{-a}{b}$ and $\frac{a}{-b}$; that is:

$$-\left(\frac{a}{b}\right) = \frac{(-a)}{b} = \frac{a}{(-b)} \tag{10.9}$$

B. Here is a way to find $1\frac{1}{2} \div \frac{1}{6}$.

THINK How many sixths in $1\frac{1}{2}$?

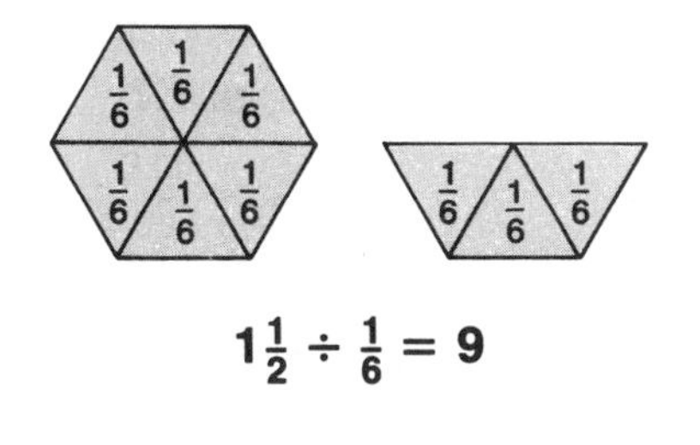

Try Give the reciprocal for each number.

a. $\frac{1}{3}$ **b.** 2 **c.** $1\frac{3}{5}$

Use the picture to find the answer to Exercise d. Multiply to find the answer to Exercise e.

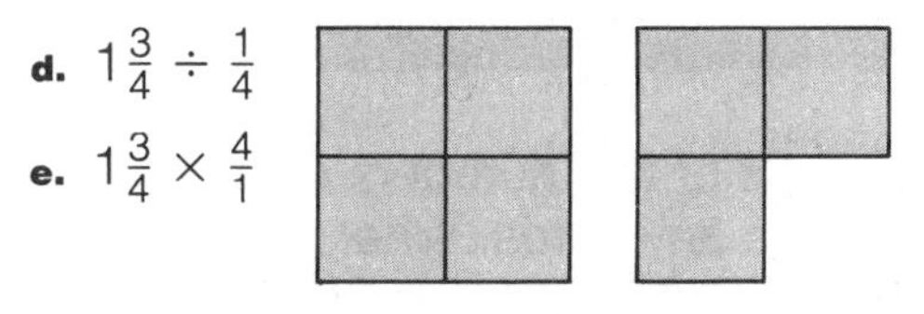

Figure 10.4 INVITATION TO MATHEMATICS, Grade 6, p. 230

Example 11

$$-\left(\frac{21}{37}\right) = \frac{^-21}{37} = \frac{21}{^-37}$$ □

Example 12

$$\frac{^-3}{4} \times \frac{5}{^-7} = \frac{^-15}{^-28} = \frac{15}{28}$$ □

Example 13

$$\frac{^-2}{9} + \frac{^-6}{12} = \frac{(^-2)\cdot 12 + 9\cdot(^-6)}{9\cdot 12} = \frac{^-78}{108} = \frac{^-13}{18}$$ □

Example 14

$$\frac{^-8}{15} \div \frac{3}{^-20} = \frac{^-8}{15} \times \frac{^-20}{3} = \frac{160}{45} = \frac{32}{9}$$ □

Example 15

$$\frac{^-5}{14} + \frac{5}{14} = \frac{^-5+5}{14} = \frac{0}{14} = 0$$ □

Exercises 10.3

For Exercises 1–4, illustrate the sum with an appropriate diagram.

1. $\frac{2}{7} + \frac{3}{7}$ **2.** $\frac{2}{5} + \frac{3}{5}$

3. $\frac{1}{4} + \frac{2}{3}$ **4.** $\frac{1}{6} + \frac{4}{5}$

For Exercises 5–8, use *both* the LCD Addition Algorithm and the Addition-by-Formula Algorithm to compute the sum.

5. $\frac{3}{14} + \frac{6}{35}$ **6.** $\frac{8}{15} + \frac{11}{18}$

7. $\frac{7}{8} + \frac{11}{15}$ **8.** $\frac{6}{5} + \frac{14}{9}$

9. Prove that both the LCD Addition Algorithm and Addition-by-Formula Algorithm work for adding two fractions with the same denominator. Prove also that the results are equivalent to the sum obtained by just adding the numerators.

10. Use the LCD Addition Algorithm to prove that if $\frac{a}{b}$ and $\frac{c}{d}$ are two fractions with $\gcd(b, d) = 1$, then

$$\frac{a}{b} + \frac{c}{d} = \frac{ad+bc}{bd}$$

11. Express the LCD Addition Algorithm as a formula and compare it with the Addition-by-Formula Algorithm.

12. Use the algorithm for division of whole numbers to show that $17 \div 5 = 3\ R\ 2$.

13. Use the definition of division to show that $17 \div 5 = 3 + \frac{2}{5}$

14. Prove: If $a \div b = c$ with remainder r, then $a \div b = c + \frac{r}{b}$. (*Hint*: Generalize Exercises 12 and 13.)

For Exercises 15–18, use both the LCD Subtraction Algorithm and the Subtraction-by-Formula Algorithm to compute the difference.

15. $\frac{3}{14} - \frac{6}{35}$ **16.** $\frac{8}{15} - \frac{11}{18}$

17. $\frac{7}{8} - \frac{11}{15}$ **18.** $\frac{6}{5} - \frac{14}{9}$

For Exercises 19–22, use Statement (10.8), the Division Algorithm, to compute the quotient.

19. $\frac{3}{14} \div \frac{6}{35}$ **20.** $\frac{8}{15} \div \frac{11}{18}$

21. $\frac{7}{8} \div \frac{11}{15}$ **22.** $\frac{6}{5} \div \frac{14}{9}$

23. Prove: If $a \neq 0$ and $b \neq 0$, then $\frac{1}{a}$ is the multiplicative inverse of a and $\frac{b}{a}$ is the multiplicative inverse of $\frac{a}{b}$.

24. Explain why 0 does not have a multiplicative inverse.

For Exercises 25–30, use appropriate algorithms to add, subtract, multiply, and divide the pair of fractions (in the order given). Express your answer in lowest terms with positive denominator.

25. $\frac{^-2}{5}, \frac{^-6}{^-7}$ **26.** $\frac{3}{^-10}, \frac{^-7}{8}$

27. $\frac{^-9}{11}, \frac{^-5}{6}$ **28.** $\frac{^-12}{^-13}, \frac{17}{^-15}$

29. $\frac{6}{^-11}, \frac{13}{^-12}$ **30.** $\frac{^-4}{3}, \frac{7}{12}$

31. Which of these fractions are equivalent to $-\frac{3}{5}$?

$$\frac{^-3}{5}, \frac{^-3}{^-5}, \frac{3}{^-5}, -\frac{3}{^-5}, -\frac{^-3}{^-5}, -\frac{^-3}{5}$$

32. Which of these fractions are equivalent to $\frac{4}{7}$?

$$\frac{^-4}{7}, \frac{^-4}{^-7}, \frac{4}{^-7}, -\frac{4}{^-7}, -\frac{^-4}{^-7}, -\frac{^-4}{7}$$

33. A "borrowing" algorithm sometimes taught for subtracting mixed numerals is shown here. Analyze in detail all steps of this algorithm.

$$\begin{array}{rr} \overset{5}{\cancel{6}} & \overset{4}{\frac{\cancel{1}}{3}} \\ -\ 1 & \frac{2}{3} \\ \hline 4 & \frac{2}{3} \end{array}$$

10.4 Rational Numbers as Decimals

The base-ten numeration system for whole numbers was presented in Chapter 5. It was shown there that a whole number can be expressed either in standard form, such as 1379, or in expanded form, such as

$$1 \cdot 10^3 + 3 \cdot 10^2 + 7 \cdot 10^1 + 9 \cdot 10^0$$

This numeration system can be extended so that rational numbers may be expressed in a similar way by using a decimal point and negative exponents. For instance,

$$12.3 \text{ stands for } 1 \cdot 10^1 + 2 \cdot 10^0 + 3 \cdot 10^{-1}$$

$$4.56 \text{ stands for } 4 \cdot 10^0 + 5 \cdot 10^{-1} + 6 \cdot 10^{-2}$$

$$.7089 \text{ stands for } 7 \cdot 10^{-1} + 0 \cdot 10^{-2} + 8 \cdot 10^{-3} + 9 \cdot 10^{-4}$$

In general:

Definitions If $a_n, a_{n-1}, \ldots, a_1, a_0$ and $b_1, b_2, \ldots, b_m$ are digits, then the numeral

$$a_n a_{n-1} \ldots a_1 a_0 \,.\, b_1 b_2 \ldots b_m$$

is called the **standard decimal form** of the number represented, and the numeral

$$a_n 10^n + a_{n-1} 10^{n-1} + \ldots + a_1 10^1 + a_0 10^0 + b_1 10^{-1} + b_2 10^{-2} + \ldots + b_m 10^{-m}$$

is called its **expanded decimal form.** Here, a_0 is the **units digit,** a_1 the **tens digit,** a_2 the **hundreds digit,** and so on; b_1 is the **tenths digit,** b_2 the **hundredths digit,** b_3 the **thousandths digit,** and so on. $a_n a_{n-1} \ldots a_1 a_0$ is called the **integral part** and $.b_1 b_2 \ldots b_m$ is called the **decimal part.** The number of digits in the decimal part is m; it is called the number of **decimal places** of the numeral.

Example 1 The numeral 23.497 is in standard decimal form. Its expanded decimal form is

$$2 \cdot 10^1 + 3 \cdot 10^0 + 4 \cdot 10^{-1} + 9 \cdot 10^{-2} + 7 \cdot 10^{-3}$$

2 is the tens digit and 3 is the units digit; 4 is the tenths digit, 9 is the hundredths digit, and 7 is the thousandths digit. 23 is the integral part and .497 is the decimal part. □

Example 2

$$5 \cdot 10^1 + 3 \cdot 10^0 + 4 \cdot 10^{-1} + 8 \cdot 10^{-2} + 6 \cdot 10^{-3} + 1 \cdot 10^{-4}$$

is in expanded decimal form. Its standard decimal form is 53.4861. Using

$$10^1 = 10, \quad 10^0 = 1, \quad 10^{-1} = \tfrac{1}{10}, \quad \ldots, \quad 10^{-n} = \tfrac{1}{10^n}$$

the expanded decimal form can be expressed as

$$50 + 3 + \frac{4}{10} + \frac{8}{100} + \frac{6}{1000} + \frac{1}{10{,}000}$$

This called the **expanded fractional form** of the numeral. □

Example 3 The expanded fractional form of the numeral 236.705 is

$$200 + 30 + 6 + \frac{7}{10} + \frac{0}{100} + \frac{5}{1000}$$

or simply

$$200 + 30 + 6 + \frac{7}{10} + \frac{5}{1000}$$ □

The expanded fractional form for decimal numerals allows us to convert from the decimal representation of a rational number to an equivalent fractional form. For instance, consider the numeral in Example 3:

$$\begin{aligned} 236.705 &= 200 + 30 + 6 + \frac{7}{10} + \frac{5}{1000} \\ &= 236 + \frac{7}{10} + \frac{5}{1000} \end{aligned}$$

By the Addition Algorithm for fractions, this sum equals

$$\frac{236{,}000}{1000} + \frac{700}{1000} + \frac{5}{1000} = \frac{236{,}705}{1000}$$

That is,

$$236.705 = \frac{236{,}705}{10^3}$$

This computation exemplifies the analysis of the familiar **algorithm for converting standard decimal form to fractional form:**

(10.10) If n is the number of decimal places, then the denominator of the equivalent fraction is 10^n and the numerator is formed by removing the decimal point from the original numeral.

Example 4 To convert 346.78 to a fraction, write 34, 678 for the numerator and 10^2 for the denominator (because there are 2 decimal places). Thus, the fractional equivalent of 346.78 is $\frac{34{,}678}{100}$.

The analysis of the algorithm in this case is:

346.78

$= 3 \cdot 10^2 + 4 \cdot 10 + 6 + 7 \cdot 10^{-1} + 8 \cdot 10^{-2}$ Expanded decimal form

$= 3 \cdot 100 + 4 \cdot 10 + 6 + \dfrac{7}{10} + \dfrac{8}{100}$ Expanded fractional form

$= 346 + \dfrac{7}{10} + \dfrac{8}{100}$ Standard form for whole numbers

$= \dfrac{34{,}678}{100}$ Addition Algorithm for fractions □

Example 5 The decimal numeral .236 converts to the fraction $\frac{236}{1000}$.

Analysis:

.236

$= \dfrac{2}{10} + \dfrac{3}{100} + \dfrac{6}{1000}$ Expanded fractional form

$= \dfrac{236}{1000}$ Addition Algorithm for fractions □

The next question is: How does one convert a rational number from its fractional form to its decimal form? Computationally, the answer is simple—divide.

Example 6 To convert $\frac{5}{8}$ to a decimal, divide 5 by 8:

$$\begin{array}{r} .625 \\ 8\overline{)5.000} \\ \underline{4\,8} \\ 20 \\ \underline{16} \\ 40 \\ \underline{40} \\ 0 \end{array}$$

□

But *why* does this process, with its seemingly casual placement of the decimal point, give us an answer that makes sense relative to what decimal representation means? The answer comes from the Division Algorithm for whole numbers, extended to include rationals by means of the familiar statement $\frac{a}{b} = a \div b$. For instance, recall (from Chapter 6) Division Algorithm II, applied here to 23,224 ÷ 64:

$$\begin{array}{rcl} 23224 & & \\ \underline{-\ 19200} & & 300 \\ 4024 & & \\ \underline{-\ 3840} & & 60 \\ 184 & & \\ \underline{-\ 128} & & 2 \\ 56 & \leftarrow & \text{(remainder)} \end{array}$$

At each stage of this process we are finding the greatest power-of-ten multiple of the divisor that is less than the remaining number. When we are left with 56, which is less than 64, the process ends *in* **W** and the result is stated as 362 *R* 56. However, now that we have rational numbers to work with, the process does not have to end at that step. Recognizing that $56 = \frac{560}{10}$, we can ask how many *tenths* of 64 are contained in $\frac{560}{10}$. Once this number of tenths is subtracted, we can ask how many *hundredths* of 64 are contained in the remainder (expressed in hundredths), subtract that amount, then go on to *thousandths*, etc. The extended process can be displayed as follows:

$$\begin{array}{rr} 23224 & \\ \underline{-\ 19200} & 300 \\ 4024 & \\ \underline{-\ 3840} & 60 \\ 184 & \\ \underline{-\ 128} & 2 \end{array}$$

$$\begin{array}{rrrr}
56 & = & 560/10 & \\
 & & -\ 512/10 & 8/10 \\ \cline{3-3}
 & & 48/10 & \\
 & & = 480/100 & \\
 & & -\ 448/100 & 7/100 \\ \cline{3-3}
 & & 32/100 & \\
 & & = 320/1000 & \\
 & & -\ 320/1000 & 5/1000 \\ \cline{3-3}
 & & 0 &
\end{array}$$

Therefore,

$$\begin{aligned}
23{,}224 \div 64 &= 362 + \frac{8}{10} + \frac{7}{100} + \frac{5}{1000} \\
&= 362 + 8 \cdot 10^{-1} + 7 \cdot 10^{-2} + 5 \cdot 10^{-3} \\
&= 362.875
\end{aligned}$$

and the conversion is complete. Using the format of Division Algorithm III, the process looks like this:

$$\begin{array}{rl}
 & \quad\;\; 5/1000 \\
 & \quad\; 7/100 \\
 & \; 8/10 \\
 & 2 \\
 & 60 \\
 & 300 \qquad\qquad = \quad 362.875 \\
64 & \overline{)\,23224} \\
 & -\ 19200 \\ \hline
 & 4024 \\
 & -\ 3840 \\ \hline
 & 184 \\
 & -\ 128 \\ \hline
 & \;\; 560/10 \\
 & \;\; -\ 512/10 \\ \hline
 & \quad 480/100 \\
 & \quad -\ 448/100 \\ \hline
 & \qquad 320/1000 \\
 & \qquad -\ 320/1000 \\ \hline
 & \qquad 0
\end{array}$$

Finally, when Division Algorithm IV is extended to the rationals, the format should look quite familiar:

```
        362.875
   64 ⟌23224.000
       192
        402
        384
         184
         128
          56 0
          51 2
           4 80
           4 48
             320
             320
               0
```

Thus, to convert a fraction $\frac{a}{b}$ to decimal form, we express the fraction as $a \div b$ and apply (the extended) Division Algorithm IV.

Example 7 To convert $\frac{23}{99}$ to a decimal, divide 23 by 99:

```
         .2323...
   99 ⟌23.0000...
       19 8
        3 20
        2 97
          230
          198
           320
           297
             ...
```

□

As Example 7 shows, sometimes the Division Algorithm will not terminate in a finite number of steps, so the decimal numeral will not be expressible in a finite number of decimal places. The numeral in Example 7 is an **infinite repeating decimal**—a decimal in which, from some specific digit on, a finite sequence of digits repeats again and again in the same order, without interruption. The decimal in Example 6 is called a **terminating decimal** because the division process yields a zero remainder after some finite number of steps. The decimal form for any rational number will be one of these two types. The reason for this can be seen by examining the extended Division Algorithm.

Consider, for example, the algorithm for dividing any whole number less than 7 by 7. According to that algorithm, there are only a finite number of possible remainders at each subtraction step—0, 1, 2, 3, 4, 5, or 6. If 0 ever occurs as a remainder, then the algorithm is finished and the result will be a terminating decimal. If 0 has not appeared after six subtraction steps, then the seventh remainder *must* be either 0 or a remainder appearing for a second time. If it is 0, then the decimal terminates. If a repeated remainder occurs, then the pattern of digits in the quotient will repeat from that point on, resulting in an infinite repeating decimal.

If we were to divide a number greater than 7 by 7, we would arrive at a similar result because such a rational number could be expressed as a mixed numeral—the sum of a whole number and a proper fraction—and the decimal equivalent of the proper fraction would be terminating or repeating, as before. Thus, the sum of the whole number and the decimal would be terminating or repeating. Similar arguments can be used for *any* divisor, establishing that:

> *If any rational number is expressed in decimal form, the resulting decimal will either terminate or repeat.*

The block of digits forming the repeating part of the decimal is called the **repetend**, and the number of digits in the repetend is called the **period** of the decimal.

Notation A bar is placed over the repetend to indicate how the decimal repeats. Thus, the decimal equivalent of $\frac{23}{99}$ (as in Example 7) is

$$.\overline{23} = .23232323\ldots$$

Example 8 The repetend of the (infinite) decimal $27.13\overline{5768}$ is the block of digits 5768, and its period is 4. Thus,

$$27.13\overline{5768} = 27.135768576857685768\ldots$$ □

Example 9 The repetend of the infinite decimal .2139139139...is 139, and its period is 3. This decimal can be written as $.2\overline{139}$. □

At this point you might be wondering if there is an algorithm for changing repeating decimals to fractional form. The answer is *Yes.* A discussion of that process appears in Chapter 13, as part of a more comprehensive treatment of infinite decimals. (See Section 13.5.)

Exercises 10.4

For Exercises 1–9, express the decimal numeral in expanded decimal form and in expanded fractional form.

1. 357.9 **2.** 35.79 **3.** 3.579
4. .3579 **5.** 2.708 **6.** 30.03
7. 137.5089 **8.** .000076 **9.** 1007.00908

For Exercises 10–18, write the numeral in standard decimal form.

10. $2 \cdot 10^1 + 4 \cdot 10^0 + 6 \cdot 10^{-1} + 8 \cdot 10^{-2}$

11. $2 \cdot 10^{-1} + 3 \cdot 10^{-2} + 7 \cdot 10^{-3} + 3 \cdot 10^{-4}$

12. $4 \cdot 10^2 + 2 \cdot 10^1 + 7 \cdot 10^0 + 3 \cdot 10^{-1} + 6 \cdot 10^{-2} + 7 \cdot 10^{-3}$

13. $5 \cdot 10^4 + 3 \cdot 10^2 + 2 \cdot 10^{-3} + 6 \cdot 10^{-5}$

14. $7 \cdot 100 + 6 \cdot 10 + 3 + \frac{2}{10} + \frac{5}{100} + \frac{4}{1000} + \frac{8}{10{,}000}$

15. $5 \cdot 1000 + 5 \cdot 10 + \frac{5}{10} + \frac{5}{1000}$

16. $3000 + 100 + 50 + \frac{3}{100} + \frac{2}{1000}$

17. $9000 + \frac{9}{1000}$

18. $\frac{7}{1{,}000{,}000}$

For Exercises 19–26, convert the decimal numeral to equivalent fractional form.

19. .34 **20.** .446 **21.** 25.78
22. 135.089 **23.** 235.21 **24.** 10.00307
25. 500.0009 **26.** .0000005

27. Analyze the conversion algorithm (10.10) for Exercise 19. (See Example 4.)

28. Analyze the conversion algorithm for Exercise 20. (See Example 4.)

29. Analyze the conversion algorithm for Exercise 23. (See Example 4.)

30. Analyze the conversion algorithm for Exercise 24. (See Example 4.)

31. Convert $\frac{3917}{40}$ to an equivalent decimal using the formats of Division Algorithms II, III, and IV.

32. Convert $\frac{754}{250}$ to an equivalent decimal using the formats of Division Algorithms II, III, and IV.

For Exercises 33–40, convert the given fraction to an equivalent decimal. In the case of an infinite repeating decimal, indicate its repetend and period.

33. $\frac{3}{16}$ **34.** $\frac{578}{32}$ **35.** $\frac{48}{11}$
36. $\frac{17}{25}$ **37.** $\frac{21}{15}$ **38.** $\frac{5}{7}$
39. $\frac{13}{999}$ **40.** $\frac{51}{1111}$

10.5 Algorithms for Decimals

The algorithms for computing with decimals are very similar to those for whole numbers, except that one has to worry about the location of the decimal points. Analysis of these algorithms depends on the algorithms for fraction arithmetic, which, in turn, rely upon the whole-number algorithms. We present here algorithms for terminating decimals only. Infinite repeating decimals may be treated by rounding them off to appropriate terminating decimals (discussed in Section 10.7) or by converting them to fractions (as shown in Chapter 13).

As you may recall from elementary school, the algorithms for adding or subtracting decimals begin by "lining up" the decimal points and annexing appropriately many zeros to each decimal so that they are all "the same length." The numbers are then added or subtracted as if they were whole numbers. The analyses in the following examples explain why such a process works.

Example 1 To add 2.3, 13.456, and .48, we follow the method described in the preceding paragraph:

$$\begin{array}{r} 2.300 \\ 13.456 \\ +\ \ .480 \\ \hline 16.236 \end{array}$$

Analysis:

$2.3 + 13.456 + .48$

$= \frac{23}{10} + \frac{13{,}456}{1000} + \frac{48}{100}$	Conversion to fractions
$= \frac{2300}{1000} + \frac{13{,}456}{1000} + \frac{480}{1000}$	Equivalent fractions
$= \frac{16{,}236}{1000}$	Addition Algorithm for fractions
$= 16.236$	Conversion to decimal

□

Example 2 To subtract .7865 from 156.4, we write

$$\begin{array}{r} 156.4000 \\ -\ \ .7865 \\ \hline 155.6135 \end{array}$$

Analysis:

$156.4 - .7865$

$= \frac{1564}{10} - \frac{7865}{10{,}000}$	Conversion to fractions
$= \frac{1{,}564{,}000}{10{,}000} - \frac{7865}{10{,}000}$	Equivalent fractions
$= \frac{1{,}556{,}135}{10{,}000}$	Subtraction Algorithm for fractions
$= 155.6135$	Conversion to decimal

□

> The algorithmic process of "lining up the decimal points and annexing zeros" is equivalent to expressing the fractions with a common denominator that is a power of ten. We then "add as if they were whole numbers" because the decimal digits become the digits of the numerators, which are whole numbers.

The Multiplication Algorithm for decimals differs from the Addition and Subtraction Algorithms in that we don't have to "line up" the decimal points. Instead, we multiply the two numbers "as if they were whole numbers," and then place the decimal point so that the number of decimal places of the product equals the sum of the numbers of decimal places of the two factors. The next two examples show why this works.

Example 3 The algorithm for multiplying 2.35×67.789 is

$$\begin{array}{rl} 67.789 & \text{(3 decimal places)} \\ \times\ \ 2.35 & \text{(2 decimal places)} \\ \hline 338945 & \\ 203367 & \\ 135578 & \\ \hline 159.30415 & \text{(5 decimal places)} \end{array}$$

Analysis:

2.35×67.789

$$= \frac{235}{100} \times \frac{67{,}789}{1000} \qquad \text{Conversion to fractions}$$

$$= \frac{15{,}930{,}415}{100{,}000} \qquad \text{Multiplication Algorithm for fractions}$$

$$= 159.30415 \qquad \text{Conversion to decimal}$$

□

Example 4 To multiply $.6 \times .00035$, we write

$$\begin{array}{r} .00035 \\ \times\ \ .6 \\ \hline .000210 \end{array}$$

Analysis:

$.6 \times .00035$

$$= \frac{6}{10} \times \frac{35}{100{,}000} \qquad \text{Conversion to fractions}$$

$$= \frac{210}{1{,}000{,}000} \qquad \text{Multiplication algorithm for fractions}$$

$$= .000210 \qquad \text{Conversion to decimal}$$

□

As you can see from these examples, multiplication of fractions does not require the fractions to have a common denominator, so the decimal points do not have to be "lined up" as they were for addition and subtraction. Furthermore, the digits of the two decimals can be treated as digits of whole numbers because they are the numerators of the corresponding fractions. Finally:

> The number of decimal places of the product is the *sum* of the corresponding numbers of decimal places of the factors because the denominators all are powers of ten.

This means that the power of the product denominator is found by *adding* the exponents in the denominators of the factors. For instance, in Example 3 above, the denominator of the product is

$$10^2 \times 10^3 = 10^{2+3} = 10^5$$

Next, recall the Division Algorithm for decimals, as you learned it in school:

> Move the decimal point all the way to the right end of the divisor numeral, move the decimal point the same number of places to the right in the dividend, then divide as with whole numbers.

To analyze this algorithm, it is best to consider first a special case—dividing a number in decimal form by a nonzero whole number, as in Example 5. In this case we divide using the whole-number Division Algorithm, ignoring the decimal point. When this division process is finished, the decimal point in the quotient is placed directly above the decimal point in the dividend. Once the analysis of this algorithm is understood, it is easy to analyze the general Division Algorithm for any two decimals, as shown in Example 6.

Example 5 Divide 7.851 by 6:

```
     1.3085
  6 |7.8510
     6
     1 8
     1 8
       051
        48
         30
         30
          0
```

Analysis:

$7.851 \div 6$

$= \frac{7851}{1000} \div \frac{6}{1}$ Conversion to fractions

$= \frac{7851}{1000} \times \frac{1}{6}$ Fundamental Division Property

$= \frac{7851 \cdot 1}{1000 \cdot 6}$ Multiplication Algorithm for fractions

$= \frac{7851 \cdot 1}{6 \cdot 1000}$ Commutativity of multiplication

$= \frac{7851}{6} \times \frac{1}{1000}$ Multiplication Algorithm for fractions

$= 1308.5 \times \frac{1}{1000}$ Extended Division Algorithm (whole numbers)

$= 1308.5 \times .001$ Conversion to decimal

$= 1.3085$ Multiplication Algorithm for decimals □

Example 6 Divide 2.356 by .25:

```
            9.4 2 4
  .2 5. |2.3 5.6 0 0
          2 2 5
            1 0 6
            1 0 0
                6 0
                5 0
                1 0 0
                1 0 0
                    0
```

Analysis:

$2.356 \div .25$	
$= \dfrac{2356}{1000} \div \dfrac{25}{100}$	Conversion to fractions
$= \dfrac{2356}{1000} \times \dfrac{100}{25}$	Fundamental Division Property
$= \dfrac{235{,}600}{1000 \cdot 25}$	Multiplication Algorithm for fractions
$= \dfrac{235{,}600}{1000} \times \dfrac{1}{25}$	Multiplication Algorithm for fractions
$= \dfrac{235{,}600}{1000} \div \dfrac{25}{1}$	Fundamental Division Property
$= 235.600 \div 25$	Conversion to decimals
$= 9.424$	Division Algorithm (decimal by whole number) □

So far, our treatment of the fraction form for rational numbers has been restricted to fractions with integer numerators and denominators. These are called **simple fractions**. The fraction form may also be used when the numerators and/or denominators are themselves fractions; such numerals are called *complex fractions*. That is, if either a or b is itself a fraction, then $\frac{a}{b}$ is a **complex fraction**, and it means $a \div b$. It is often useful to express a complex fraction as an equivalent simple fraction. This is easily done using the algorithms for fractions, as in the following example.

Example 7 To express $\frac{2/3}{5/8}$ as an equivalent fraction with integer numerator and denominator, we use the fact that $\frac{a}{b} = a \div b$:

$$\frac{2/3}{5/8} = \frac{2}{3} \div \frac{5}{8} = \frac{2}{3} \times \frac{8}{5} = \frac{16}{15}$$ □

One note of caution is in order here. Observe that

$$\frac{\frac{2}{3}}{5} = \frac{2}{3} \div \frac{5}{1} = \frac{2}{3} \times \frac{1}{5} = \frac{2}{15}$$

whereas

$$\frac{2}{\frac{3}{5}} = \frac{2}{1} \div \frac{3}{5} = \frac{2}{1} \times \frac{5}{3} = \frac{10}{3}$$

In general, except for specially chosen values for a, b, and c,

$$\frac{\frac{a}{b}}{c} \neq \frac{a}{\frac{b}{c}}$$

Exercises 10.5

For Exercises 1–18, use the appropriate algorithm to perform the computation; then analyze the process used.

1. $1.3 + .579$

2. $2.34 + 567.468$

3. $32.6 + 1.085 + .4044$

4. $15.367 + 4.7432 + 138.89$

5. $25.4 - 1.67$

6. $45 - .0051$

7. $236.78 - 1.8654$

8. $78.0007 - 2.3$

9. 32.135×1.23

10. 4.08×5.11

11. $.00067 \times .72$

12. 7.0002×7000.2

13. $2.45 \div 8$

14. $87.03 \div 4$

15. $94.7 \div 25$

16. $.372 \div 150$

17. $2.85 \div 1.2$

18. $.0049 \div .014$

For Exercises 19 and 20, simplify the complex fraction by expressing it as an equivalent simple fraction.

19. $\dfrac{4/7}{8/9}$

20. $\dfrac{2/5}{3}$

10.6 Percent

Another numeration form used to express rational numbers is *percent*, which (as we saw in the Historical Note of Section 10.1) literally means "by hundreds." For any number n, the term n **percent** means "n hundredths." In symbols,

$$n\% = \frac{n}{100}$$

Because $\frac{n}{100} = n \cdot \frac{1}{100} = n \times .01 = .01n$, we can also say that

$$n\% = .01n$$

Both forms are useful interpretations of $n\%$. The first is usually more convenient when n is expressed as a fraction; the second is often better when n is expressed as a decimal. When n is an integer, both forms are commonly used.

Example 1

$$23\% = \frac{23}{100} = .23 \qquad \square$$

Example 2

$$3.7\% = .01 \times 3.7 = .037 \qquad \square$$

Example 3

$$\frac{3}{5}\% = \frac{3/5}{100} = \frac{3}{5} \times \frac{1}{100} = \frac{3}{500} \qquad \square$$

Dividing a decimal numeral by 100 is accomplished by just moving the decimal point two places to the left; dividing a fraction by 100 is equivalent to multiplying its denominator by 100. These two observations permit quick conversion of a rational number in percent form to an equivalent fraction or decimal, as shown in the following examples.

Example 4

$$124\% = 1.24 \qquad 45.67\% = .4567$$

$$3.6\% = .036 \qquad .043\% = .00043$$

□

Example 5

$$\frac{25}{3}\% = \frac{25}{300} \qquad \frac{1}{10}\% = \frac{1}{1000}$$

$$200\% = \frac{200}{100} = 2 \qquad 5\tfrac{1}{4}\% = \frac{21}{4}\% = \frac{21}{400}$$

□

Since we can convert from percent form to either fraction or decimal form by dividing by 100, we can also convert from fractions or decimals to percents by multiplying by 100. For decimals this means moving the decimal point two places to the right; for fractions it means multiplying the numerator by 100.

Example 6

$$.76 = 76\% \qquad .234 = 23.4\% \qquad 56.321 = 5632.1\%$$

$$1.2 = 120\% \qquad .00432 = .432\%$$

□

Example 7

$$\frac{3}{4} = \frac{300}{4}\% = 75\% \qquad \frac{2}{3} = \frac{200}{3}\% = 66\tfrac{2}{3}\% \qquad \frac{1}{10} = \frac{100}{10}\% = 10\%$$

□

There are several common verbal expressions involving percents. One is "Take $n\%$ of a number." Now, $n\%$ means "n hundredths"; so taking $n\%$ of a number means multiplying that number by $\frac{n}{100}$, or by $.01n$. This product is called the **percentage**, and the number you are multiplying by $\frac{n}{100}$ is called the **base**. In symbols, if b represents the base and p represents the percentage, then the statement "the percentage is $n\%$ of the base" can be written as

(10.11)
$$\boxed{p = \left(\frac{n}{100}\right)b \quad \text{or} \quad p = .01nb}$$

An expression such as "He gives 100% of himself" means that the person makes a complete effort, because

$$100\% = \frac{100}{100} = 1$$

The familiar sportscaster expression that an athlete is "out there giving 110%" is a way of saying the athlete is giving *more* than he or she has, because

$$110\% = \frac{110}{100} = 1.1$$

The expression "a thousand percent" used in this context is very nearly nonsense; it implies an athlete is playing at *ten times* capacity:

$$1000\% = \frac{1000}{100} = 10$$

Example 8 7.5% of 20 is

$$.01 \times 7.5 \times 20 = .075 \times 20 = 1.5$$

The base is 20 and the percentage is 1.5. □

Example 9 $\frac{1}{2}\%$ of 6 is

$$\frac{1}{200} \cdot 6 = \frac{6}{200} = \frac{3}{100}$$

The base is 6 and the percentage is $\frac{3}{100}$. □

The formula that relates the percentage and the base can also be expressed as a **proportion**—that is, as an equality of two ratios:

$$\boxed{\frac{n}{100} = \frac{p}{b}} \qquad (10.12)$$

Many educators prefer to teach percentages by using this proportion because they believe that it demonstrates clearly the relationships among the numbers n, b, and p. If any two of these three numbers are known, the other one is easily found by substituting the two known values into the proportion and solving for the third one. Of course, the same thing can be done using the formula $p = .01nb$; the choice of which formula to use is largely a matter of individual preference.

Example 10 Problem: 8 is 32% of what number?

Here the percentage is known and the base is sought. Substituting into

Equation (10.12), we get $\frac{32}{100} = \frac{8}{b}$. By the "cross product" comparison of equal fractions, this becomes $32b = 800$, so $b = 25$.

Example 11 Problem: 54 is what percent of 60?

In this case the percentage and the base are both known and the percent number n is sought. Substitution into Equation (10.12) yields

$$\frac{n}{100} = \frac{54}{60}$$

By cross multiplication we have $60n = 5400$, or

$$n = \frac{5400}{60} = 90$$

Thus, 54 is 90% of 60. The same result can also be obtained by substituting the known values into Equation (10.11), $p = .01nb$:

$$\begin{aligned} 54 &= .01 \cdot n \cdot 60 \\ \frac{54}{60} &= .01 \cdot n \\ \frac{5400}{60} &= n \\ 90 &= n \end{aligned}$$

□

Sometimes the $n\%$-term in these expressions is called the **rate of percentage** or simply the **rate**, and in that case it is often denoted by r. Thus,

$$r = n\% = \frac{n}{100} = .01n$$

and Equation (10.12) becomes $r = \frac{p}{b}$, implying

(10.13) $$\boxed{p = br}$$

These formulas are fairly easy to remember, so they are useful for solving simple percentage problems. However, they are *not* new additions to the theory of rational-number algorithms; the computations involved are just special cases of the fraction and decimal algorithms already presented.

Example 12 To find the base if the percentage is 15 and the rate is 3%, we can substitute into the formula $p = br$:

$$15 = b \cdot 3\% \qquad \text{or} \qquad 15 = .03b$$

Thus, $b = \frac{15}{.03} = 500$. □

3 To find the percentage when the rate is 12% and the base is 23, the equation $p = br$ is convenient:

$$p = 23 \cdot 12\% = 23 \times .12 = 2.76 \qquad \square$$

e 14 For a percentage of 125 and a base of 60, the rate is found from the same formula (10.13) by using $125 = 60r$. Its solution is given by

$$r = \frac{125}{60} = 2.083 = 208.3\% \qquad \square$$

nple 15 Figure 10.5 shows how sixth-grade students are encouraged to write an equation to solve a percent problem. Do you see how this is related to our formula (10.13)? $\square$

xercises 10.6

For Exercises 1–12, express each percent as a decimal.

1. 26%
2. 55%
3. 57.2%
4. 37.5%
5. 254%
6. 300%
7. .32%
8. .71%
9. .005%
10. .0001%
11. $\frac{1}{2}\%$
12. $\frac{3}{4}\%$

For Exercises 13–24, express each percent as a simple fraction.

13. 47%
14. 20%
15. 150%
16. 235%
17. $\frac{1}{2}\%$
18. $\frac{3}{8}\%$
19. $3\frac{4}{5}\%$
20. $87\frac{1}{2}\%$
21. $16\frac{1}{3}\%$
22. .1%
23. .01%
24. .007%

For Exercises 25–36, express each number as a percent.

25. .59
26. .043
27. 25
28. 54.7
29. 3
30. $\frac{4}{5}$
31. $\frac{5}{6}$
32. $\frac{1}{100}$
33. $34\frac{1}{2}$
34. $\frac{7}{8}$
35. $\frac{3}{4}$
36. $\frac{4}{3}$

For Exercises 37–44, answer the question by making appropriate substitutions into either the proportion $\frac{n}{100} = \frac{p}{b}$ or the formula $p = .01nb$.

37. What is 24% of 51?
38. 38 is what percent of 133?
39. 426 is 75% of what number?
40. What percent is 25 of 200?
41. 85% of what number is 4.097?
42. What is $\frac{1}{2}\%$ of 10?
43. $\frac{3}{4}$ is what percent of 100?
44. $\frac{1}{2}$ is what percent of $\frac{2}{3}$?

Use the formula $p = br$ to answer Exercises 45–50.

45. Find the base if the percentage is 216 and the rate is 30%.
46. Find the base if the percentage is 6 and the rate is 12%.
47. Find the percentage if the rate is 5.5% and the base is 500.
48. Find the percentage if the rate is 32.6% and the base is 187.

Problem Solving **Write an Equation**

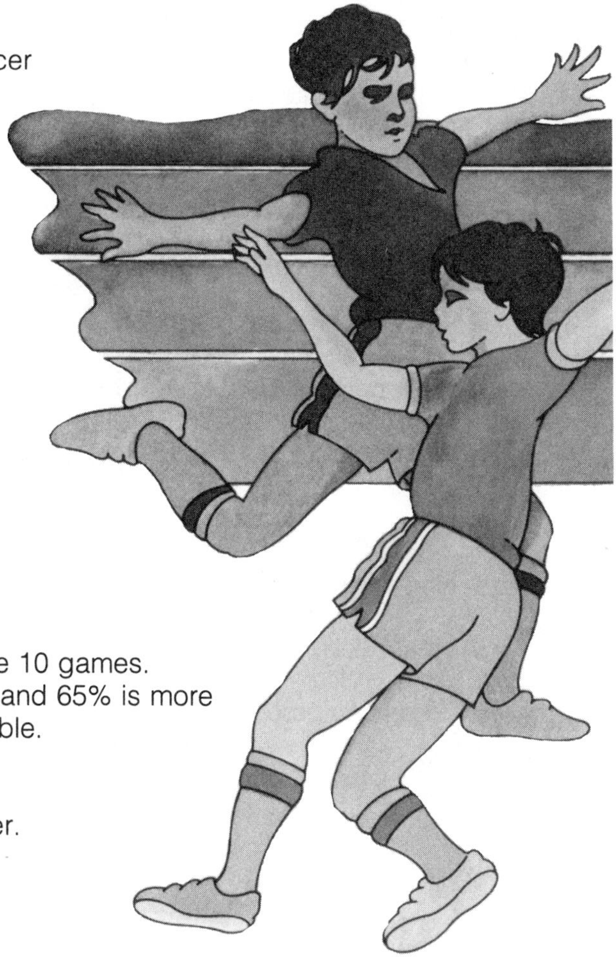

Read The Jets won 13 out of their 20 soccer games this year. What percent of the games played did they win?

Plan Write an equation to show what percent of the games they played were won by the Jets. What percent of 20 is 13?

Percent	Games played	Games won
n $\times$	20 =	13

Solve
$n \times 20 = 13$
$n = 13 \div 20$
$n = 0.65$ or 65%

Answer The Jets won 65% of their games.

Look Back Half, or 50%, of the games would be 10 games. The Jets won more than 10 games, and 65% is more than 50%, so the answer is reasonable.

Try Write an equation. Then give the answer.

a. The Jets played 60% of the 20 games at home. How many games did they play at home?

Figure 10.5 INVITATION TO MATHEMATICS, Grade 6, p. 330

49. Find the rate if the base is 8.3 and the percentage is .6225.

50. Find the rate if the base is 250 and the percentage is 5.

For Exercises 51–59, identify the percentage, the base, and the rate; then solve for the unknown quantity.

51. In a special sale, a coat that regularly sells for $254 was sold at a 15% discount. What was the amount of the discount? What was the selling price?

52. A saleswoman works on commission, receiving 12% of her total sales. One week she earned $421.20. What was the total amount of her sales for that week?

53. In one month gasoline prices increased by $.05 per gallon. If the original price at the beginning of the month was $1.25 per gallon, what was the rate of increase?

54. If last year's sales were $16,000 and this year's sales are $17,000, what percent are this year's sales of last year's?

55. An agent receives $1394 commission for selling $16,400 worth of goods. What is the commission rate?

56. A family has an annual income of $32,000, of which it spends 32% for food and 15% for medical expenses? How much does it spend on both food and medical expenses?

57. A family spent $9240 on rent last year, which was 28% of its total income for the year. What was the total income?

58. A company spent 20% of its total sales revenue on advertising. If the company spent $226,000 on advertising, what was its total sales revenue?

59. Connecticut's sales tax is 7.5% of the selling price. How much sales tax must be paid on an item that sells for $78.95? (Round to the nearest cent.)

10.7 Approximating Rational Numbers by Decimals

As we saw at the end of Chapter 9, every calculator and computer has a maximum number of digits for each number that it handles. Numbers whose decimal representations require more decimal places than the maximum cannot be handled with complete accuracy. Even though devices are being designed to allow an ever-increasing number of decimal places, the problem will never be completely resolved because, as you know, some numbers have infinite decimal representations. Also, as we shall see in Chapter 12, all measuring instruments are similarly limited in their capacity for accuracy. It is necessary, then, to have systematic ways of approximating rational numbers by finite decimals of a fixed length. It is also necessary to understand "the margin of error" of these approximations. These two topics are the focal points of this section.

PROBLEM-SOLVING COMMENT

Besides being a useful tactic for your problem solving, *approximating the answer* is used by scientists and engineers to accomodate their theories to reality. The application of abstract science to the real world is a continuous exercise in approximation, and the techniques for estimating and coping with the quantitative errors resulting from approximation is an entire field of study in and of itself. In this section you will see a glimpse of that field, perhaps just enough to convince you that approximation is a valuable tool. Remember, however, that it is always necessary and sometimes tricky to deal with the margin of error that occurs as part of any approximation process. ◇

Notation After this remark, we shall conform to the common practice of using the in-line negative sign to denote negative numbers as well as additive inverses. There is no danger of computational confusion because the additive inverse of any positive number p is the negative number ^-p; that is, $-p = {}^-p$. (See the end of Section 9.6 for a related comment.)

In Section 9.6 we saw that when the usual form of a number is too long for a calculator or computer, the machine may express that number in exponential form. For instance, a six-digit calculator might treat 25,000,000 as 2.5×10^7 and .0000000034521 as 3.4521×10^{-9}. In fact, any nonzero number can be written as a number with absolute value between 1 and 10 multiplied by some integral power of 10.

Definition A nonzero number N is said to be in **scientific notation** if $N = n \times 10^p$, where p is an integer and $1 \leq |n| < 10$.

Example 1

$$1.023478 \times 10^{-2} \text{ represents } .01023478$$

□

Example 2

number	*scientific notation*
2356	2.356×10^3
-67.321	-6.7321×10^1
.0005891	5.891×10^{-4}

□

Notation Throughout this section we shall refer to a number's "true value" and an "approximate value" for it. If the true value is represented by N, then we shall represent an approximate value by $\overline{N}$. We write the statement "N is approximately equal to $\overline{N}$" as "$N \approx \overline{N}$."

Definition If one number is used to approximate another, then the **absolute error**, E, of this approximation is the difference found by subtracting the approximate value from the true value. In symbols,

$$E = N - \overline{N}$$

A positive absolute error shows that the true value exceeds the approximate value; a negative absolute error shows that the true value is less than the approximation.

Example 3 If $N = 2345.6247$ and $\overline{N} = 2345.62$, then $E = .0047$. □

Example 4 For $N = 34{,}537$ and $\overline{N} = 35{,}000$, $E = -463 = -4.63 \times 10^2$. □

In many cases the absolute error does not convey enough information about the accuracy of an approximation. For example, a 24-foot measurement that is short by a foot is far less accurate (in some sense) than a 23,584-foot measurement that is short by a foot, yet they have the same absolute error. To reflect this type of situation better, we relate the error to the original number:

Definition If N is approximated by $\overline{N}$, then the **relative error**, $R.E.$, of this approximation is the ratio of the absolute error to the true value. In symbols,

$$R.E. = \frac{N - \overline{N}}{N} = \frac{E}{N}$$

Example 5 If $N = 4$ and $\overline{N} = 3.8$, then

$$R.E. = \frac{4 - 3.8}{4} = \frac{.2}{4} = .05 = 5\%$$

Relative error is often expressed as a percent. A relative error of 5% means that the absolute error is 5% of the true value. □

Example 6 If $N = 400$ and $\overline{N} = 399.8$, then

$$R.E. = \frac{400 - 399.8}{400} = \frac{.2}{400} = .0005 = .05\%$$

The approximation in Example 5 has the same *absolute* error as this one, but its *relative* error is one hundred times as large! □

Definition In the decimal representation of a number, each nonzero digit and each zero that does not serve only to fix the position of the decimal point is called a **significant digit**.

Example 7 0.0358 has 3 significant digits. The two zeros serve only to fix the position of the decimal point, so they are not significant. In contrast, the zeros in 30,058 are significant because they do not fix the decimal point; this number has 5 significant digits. □

Example 8 On the right side of the decimal point, ending zeros are always significant. *If* there are no nonzero digits to the left of the decimal point, then the zeros immediately to the right of the decimal point are not significant. Thus, .6700, 67.00, .006700, and 6.007 all have 4 significant digits. In particular, notice that the last two zeros in .006700 are significant; the first two are not. □

Example 9 It is impossible to tell (without more information) whether or not the zeros in 4600 are significant. Scientific notation is often used to clarify this:

- If 4600 represents some unknown number between 4550 and 4650, then neither zero is significant. In scientific notation, we write 4.6×10^3.
- If 4600 represents some unknown number between 4595 and 4605, then the zero in the tens place is significant, but the zero in the units place is not. In scientific notation, we write 4.60×10^3, signifying that the zero in the hundredths place conveys information.
- If 4600 represents some unknown number between 4599.5 and 4600.5, or if it represents the true value, then both zeros are significant. In scientific notation, we write 4.600×10^3. A decimal point immediately after the second zero (4600.) also shows that both zeros are significant.
- 4600.0 is unambiguous; it has 5 significant digits. In scientific notation, we write 4.6000×10^3. □

Rounding (or **rounding off**) a number to n significant digits is a way to approximate the original number by a (simpler) number with n significant digits. This is done by discarding all digits after the nth significant digit, according to some rule, and adding zeros, if necessary, to fix the position of the decimal point.

Don't be put off by that formal description! This is something we do all the time as shoppers. A $9.95 turkey is essentially a $10.00 item, a price with 2 or 1 significant digits, depending on whether you think of spending money in terms of dollars or tens. If you haven't brought much money along, as you put things into the shopping cart you might keep track of the approximate total cost by rounding off the prices to the nearest dollar, or even to the nearest dime.

To minimize the error in the round-off process and avoid embarrassment at the checkout counter, you round some prices up and others down. If you are thinking in dollar amounts, you usually round prices more than halfway between dollar amounts to the next higher dollar and prices less than halfway to the next lower dollar amount. A price that is exactly halfway between, such as $2.50, is a bit of a puzzle, but most people round those up to the next dollar, too. However, if you happen to choose a lot of items with prices exactly on the half-dollar amount, your total estimate will end up far higher than your actual cost. This may not be a major shopping problem, but there are situations where the error introduced in this way becomes unacceptably large. Part (3) of the round-off process, which follows, eases this difficulty; parts (1) and (2) are just general restatements of the dollar example.

When rounding a number to n significant digits, the unit amount is determined by the place of the last significant digit to be retained, called here the "last digit."

(1) If the value of the discarded digits is less than half a unit, keep the last digit unchanged.

(2) If the value of the discarded digits is more than half a unit, add 1 to the last digit (carrying, if necessary).

(3) If the value of the discarded digits is equal to half a unit, make the last digit even by keeping it unchanged if it is already even and adding 1 if it is odd.

Part (3) balances the overestimates with the underestimates "in the long run." The easier (but somewhat less accurate) rule of always increasing the last digit in the half-unit case is often used in the elementary grades. Either procedure assures us of a key fact:

> If a number N is rounded to n significant digits, the size of the absolute error $|E|$ will always be less than or equal to half the unit determined by the last significant digit.

We can write this more compactly in symbols, assuming that the last significant digit has place value 10^k:

$$|E| \leq .5 \times 10^k$$

Example 10 Each of the numbers in the first column of Table 10.1 has been rounded to 4 significant digits. The maximum margin of error of each approximation is listed in the right column. □

N		$\overline{N}$	maximum error for $\lvert E\rvert$
234,789	$\approx$	234,800	$.5 \times 10^2 = 50$
47.73499	$\approx$	47.73	$.5 \times 10^{-2} = .005$
0.00398251	$\approx$	0.003983	$.5 \times 10^{-6} = .0000005$
47,005,000	$\approx$	47,000,000	$.5 \times 10^4 = 5000$
513.9500	$\approx$	514.0	$.5 \times 10^{-1} = .05$
513.8500	$\approx$	513.8	$.5 \times 10^{-1} = .05$

Table 10.1 Rounding to 4 significant digits.

Rounding can be used in a natural way to measure the accuracy of an approximation. We say that an approximation $\overline{N}$ of a number N is **correct to n significant digits** if $\overline{N}$ and N both round to the same n significant digits but round differently to $n+1$ significant digits.

Example 11 .7782347 is approximately $\frac{7}{9} = .7777\ldots$. This approximation is correct to 3 significant digits because both numbers round to .778, but it is not correct to 4 significant digits because they round to .7782 and .7778, respectively. □

More often than not, when an approximate number $\overline{N}$ is used in calculations, the true value N is unknown. However, if the number of significant digits of accuracy is known, we can estimate N by finding upper and lower limits for it. In this case, the set of numbers between these two limits is called an **interval estimate** of N. If a and b are lower and upper limits for an interval estimate of N, it is common to write the interval estimate as $[a, b]$.

Example 12 If 2.3 has been rounded to 2 significant digits, then

$$|E| \leq .5 \times 10^{-1} = .05$$

so N is between $2.3 - .05$ and $2.3 + .05$. In other words, an interval estimate for N is $[2.25, 2.35]$. □

Example 13 If 4.0725 is an approximate value that is correct to 3 significant digits, then the true value N rounds to 4.07. This means that

$$|E| \leq .5 \times 10^{-2} = .005$$

so an interval estimate for N is $[4.065, 4.075]$. □

An important area of interest about approximations centers about what happens when approximations are used in calculations. In particular, how do the results of such calculations compare with the true results, and how can an accurate interval estimate for the true result be determined from the result of a calculation that involves approximate numbers? Specific answers to these questions can be derived from examining the standard arithmetic algorithms to see their (sometimes magnifying) effects on the error values of the initial numbers. However, the details would take us beyond the scope of this introduction to approximations, so we forego the temptation to present the arguments here. From a practical point of view, the main lesson to be learned about this topic is:

> When two or more numbers rounded to n significant digits are used in a calculation, the answer is usually not correct to n significant digits.

Depending on the purpose of the calculation, then, it may be useful or even necessary to improve its accuracy by increasing the number of significant digits of the original numbers. If your calculations must be correct to 4 significant digits, you might need to begin with numbers that are correct to

5 or 6 significant digits. If your initial numbers are measurements of some sort, you may have to use a more precise measuring instrument. If you are using a calculator or computer to do the computations, you will have to make sure it allows for enough significant digits.

Actually, two calculators that allow the same number of significant digits may differ in the accuracy of their outputs. This happens because some machines do the *internal* computations with more significant digits than they display. If a calculator does all its computations using 10 significant digits internally but only displays 6 digits, all 6 of the displayed digits are probably correct. This is also the case with computers, but often the number of significant digits can be controlled to some extent. Many computers give the user the option of "single-precision" or "double-precision" modes for calculation. In single-precision mode, computations usually are done using 8 significant digits, whereas double-precision mode may use 15 or 16 significant digits. (As might be expected, double-precision calculations are quite a bit slower than single-precision ones.)

Sometimes the numbers in a calculation do not have equally many significant digits. Approximate numbers can come from a variety of sources, some producing more significant digits than others. When this happens, be sure to use all of the significant digits of each number throughout the calculation; then round the answer to the smallest number of significant digits of any of the original numbers. When you are doing a sequence of calculations and the result of one of them is used in a later calculation, carry forward as many significant digits as possible (That's why there's a memory button!) and round only the final answer. Rounding at intermediate stages introduces a round-off error at each stage, which may unnecessarily increase the round-off error of the final answer. This effect is called **accumulated round-off error**.

Example 14 The computations of this example and the next were done on a calculator that computes internally to 12 significant digits and can display up to 10 digits. Ordinarily, many of the intermediate numbers would not be displayed, but we show them here to illustrate the process.

Calculate $3.4241 + (578.325 \times .000754)$.

* * * * * * *

First multiply: $578.325 \times .000754 \approx .43605705$

Then add: $3.4241 + .43605705 \approx 3.86015705$

The smallest number of significant digits in the original numbers is 3 (in .000754), so we round the final answer to 3 significant digits: 3.86. □

Example 15 Calculate $43.5785 \cdot \sqrt{23.21007 - 4.573^2}$.

* * * * * * *

First square: $4.573^2 \approx 20.912329$

Then subtract: $23.21007 - 20.912329 \approx 2.297741$

Next, take the square root: $\sqrt{2.297741} \approx 1.515830136$

Finally, multiply: $43.5785 \times 1.515830136 \approx 66.05760358$

Since 4 is the smallest number of significant digits in the original numbers, we round the final answer to 4 significant digits: 66.06 □

Exercises 10.7

For Exercises 1–8, use scientific notation to express each number.

1. 768.21 **2.** 76,821
3. 7.6821 **4.** .0008654
5. 9,854,000 **6.** .32187
7. .000000321 **8.** 3,215,600,000,000

For Exercises 9–16, use standard decimal form to express each number.

9. 1.237×10^4 **10.** 1.037×10^{-2}
11. 7.332×10^{-1} **12.** 7.332×10^1
13. 4.975×10^{-6} **14.** 4.975×10^6
15. 5.78×10^{-7} **16.** 3.561×10^0

For Exercises 17–20, find E and $R.E.$

17. $N = 45.2$, $\overline{N} = 45$

18. $N = 4120$, $\overline{N} = 4100$

19. $N = .325$, $\overline{N} = .32$

20. $N = 1.247 \times 10^{-5}$, $\overline{N} = 1.2 \times 10^{-5}$

For Exercises 21–26, state the number of significant digits of the given number.

21. 203 **22.** 2.30
23. 0.00675870 **24.** 3.40×10^3
25. 3.400×10^3 **26.** 3.4×10^3

For Exercises 27–35, round the given number to:
(a) 2 significant digits
(b) 3 significant digits
(c) 4 significant digits

Write your answers in standard form and in scientific notation.

27. 43,500 **28.** 26.500 **29.** 4.6789
30. .003273 **31.** 2,378,900 **32.** .7765001
33. 2200.001 **34.** .0050607 **35.** .000054399

For Exercises 36–41, find the best upper limit for $|E|$ and also find the resulting interval estimate.

36. 63,000 is rounded to 2 significant digits.

37. 63,000 is rounded to 3 significant digits.

38. 63,000 is rounded to 4 significant digits.

39. 3.14 is rounded to 3 significant digits.

40. 3.142 is rounded to 3 significant digits.

41. 3.1416 is rounded to 3 significant digits.

In Exercises 42–47, find the true value N rounded to the given number of significant digits; then find a good interval estimate for N. (*Hint*: See Example 14.)

42. 1357 is correct to 2 significant digits.

43. 1357 is correct to 3 significant digits.

44. 1357 is correct to 4 significant digits.

45. .002468 is correct to 2 significant digits.

46. .002468 is correct to 3 significant digits.

47. .002468 is correct to 4 significant digits.

For Exercises 48–52, compute using a calculator, and round your answer appropriately. (Assume that the original numbers are rounded to the significant digits given.)

48. $2.354 + (34.074 \times 7.653)$

49. $(1.579 \times 10^{-5}) + \sqrt{4.34 \times 10^{-11}}$

50. $578.34 - (32.43 + 21.78^2)$

51. $\dfrac{.5642 + (.034752 \cdot 27.65^2)}{.0046787}$

52. $\dfrac{(6.80 \times 10^{-7}) \cdot (5.321 \times 10^3)}{(5.76 \times 10^2) + (2.17 \times 10^{-1})}$

Review Exercises for Chapter 10

For Exercises 1–10, indicate whether the given statement is *true* or *false.*

1. $\frac{3}{5}$, $\frac{6}{10}$, .6, and 60% are four different numbers used to represent the same numeral.

2. $2\frac{5}{8}$ means $2 + \frac{5}{8}$.

3. The fraction $\frac{a}{b}$ is reduced to lowest terms if and only if $\gcd(a, b) = 1$.

4. The least common denominator of two fractions is the least common multiple of the denominators.

5. If b and d are relatively prime, then the least common denominator of $\frac{a}{b}$ and $\frac{c}{d}$ is bd.

6. To convert $\frac{a}{b}$ to a decimal, you divide b by a.

7. If any rational number is expressed in decimal form, the resulting decimal will either terminate or repeat.

8. In multiplying decimals, the number of decimal places in the product equals the sum of the numbers of decimal places of the factors because of a property of exponents.

9. To convert a percent to a decimal, you move the decimal point two places to the right, which is equivalent to multiplying by 100.

10. When two or more numbers rounded to n significant digits are used in a calculation, the answer usually is correct to n significant digits.

11. Provide a step-by-step illustration of how to express $\frac{4}{5}$ as an equivalent fraction with denominator 35. Give a reason to justify each step.

12. Provide a step-by-step illustration of how to reduce $\frac{12}{18}$ to lowest terms. Give a reason to justify each step.

In Exercises 13–16, use prime factorizations to find the least common denominator; then use the LCD Algorithm to find each sum or difference.

13. $\dfrac{11}{84} + \dfrac{13}{126}$ **14.** $\dfrac{^-21}{20} + \dfrac{5}{14}$

15. $\dfrac{3}{35} - \dfrac{^-2}{15}$ **16.** $\dfrac{11}{36} + \dfrac{7}{120}$

In Exercises 17–20, use the Formula Algorithm to find each sum or difference.

17. $\dfrac{4}{15} - \dfrac{6}{5}$ **18.** $\dfrac{7}{24} + \dfrac{5}{6}$

19. $\dfrac{^-3}{25} + \dfrac{2}{10}$ **20.** $\dfrac{5}{6} - \dfrac{7}{9}$

In Exercises 21–26, express each rational number as a fraction with integer numerator and denominator.

21. 12.4 **22.** 3.295 **23.** 5.2%

24. .004 **25.** 32.75 **26.** .78%

For Exercises 27–34, do the indicated calculation using an appropriate algorithm.

27. $32.45 + 9.234$ **28.** 5.156×1.57

29. $35.4 - 39.852$ **30.** $.71346 \div 2.3$

31. $21.37 + 9.181$ **32.** $2.359 \times .35$

33. $83.1 - 39.237$ **34.** $7.94592 \div 2.48$

For Exercises 35–37, express the given rational number in decimal form. If the decimal is repeating, indicate the repetend and the period.

35. $\frac{3}{8}$ **36.** $\frac{5}{111}$ **37.** $\frac{6}{7}$

38. One sixth grader told her teacher that she had been trying to convert $\frac{5}{17}$ to a decimal and that after 35 divisions the decimal still had not started to repeat. Without doing any calculations, the teacher knew immediately that the student had made a mistake. How did the teacher know this?

For Exercises 39–44, set up a proportion that can be used to solve the problem with appropriate substitutions; then solve the problem.

39. Sarah receives a commission of 8.3% of her total sales. During one week she received a commission of $37.35. What was the total of her sales for that week?

40. What percent is 25 of 15?

41. What is 32% of 75?

42. 45 is 1.5% of some number. What is the number?

43. A teacher's salary rose from $22,000 to $25,000. What was the percent of increase?

44. If a radio originally priced at $30 is to be sold at a 25% discount, what will the discount amount to?

In Exercises 45–47, express the given complex fraction as an equivalent simple fraction.

45. $\frac{3/5}{4/7}$ **46.** $\frac{8}{2/5}$ **47.** $\frac{3.2}{.7589}$

In Exercises 48–50, compute using a calculator, and round your answer appropriately. (Assume that the original numbers are rounded to the significant digits given.)

48. $.075 + (3.3456 \times 456.1)$

49. $(5.76 \times 10^{-6}) + \sqrt{5.789 \times 10^4}$

50. $\frac{.5329 + (78.976 \cdot 43.789^3)}{.0000543}$

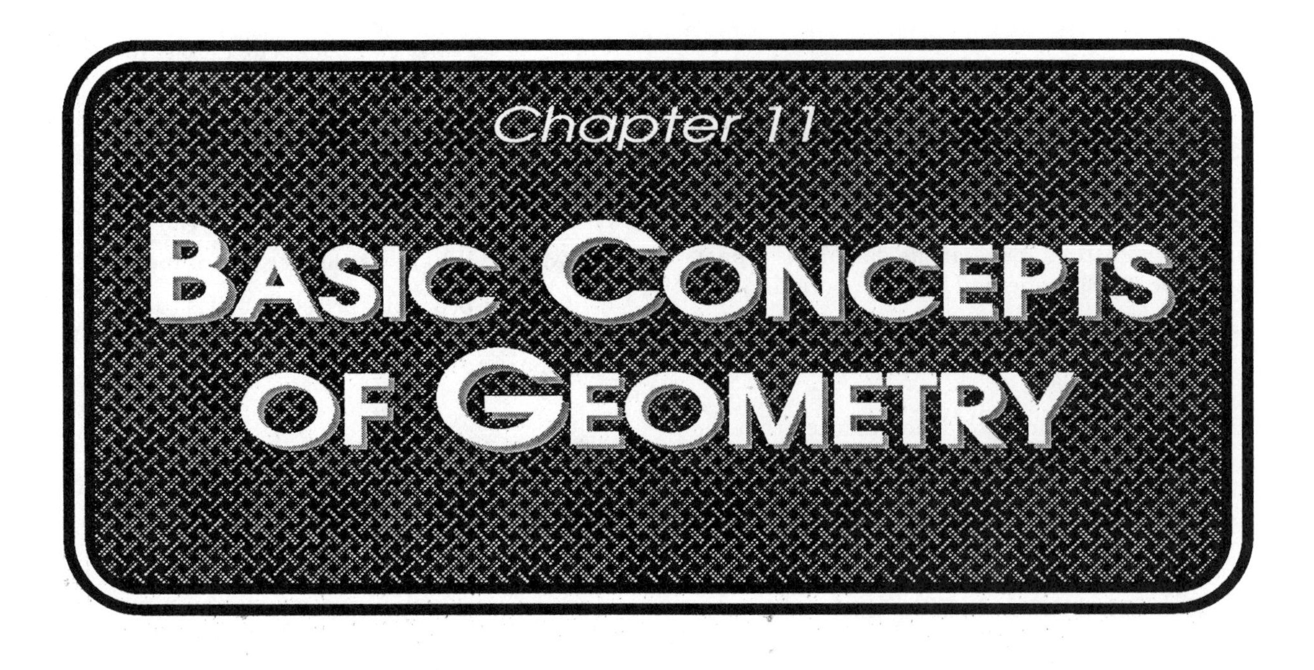

11.1 The Geometry of Euclid

The plane geometry that you learned in high school is a system that was developed more than 2000 years ago by the Greek mathematician Euclid. Sometime around 300 B.C., he wrote a 13-volume work called the *Elements*, which organized and extended all the mathematics known to the Greeks at that time. The first six of these volumes contained the system of plane geometry, presented in the form of theorems proved from five basic assumptions called *postulates.*

Euclid's five postulates (paraphrased a little) are:

1. Two points determine a line segment.
2. Any line segment can be extended continuously to form a line.
3. Given any point and any line segment, there exists a circle with that point as center and the length of that line segment as radius.
4. All right angles are congruent.
5. If two straight lines in a plane are cut by a transversal making the sum of the measures of two interior angles on the same side of the transversal less than 180°, then the two straight lines will meet on that side of the transversal.

HISTORICAL NOTE: THE ORIGINS OF GEOMETRY

Our knowledge of early geometry comes from two primary sources—the Rhind Papyrus, written more than 2000 years before Christ by an Egyptian priest named Ahmes, and a summary of early geometry written by the Greek historian Proclus (A.D. 412–485). Proclus attributes the early development of geometry to the ancient Egyptians who were trying to restore landmarks periodically destroyed by floodings of the Nile. The Rhind Papyrus gives a fairly detailed account of the geometrical knowledge attained by the Egyptians. It also shows that the Egyptians did not convey geometric information by statements of general properties; instead, they just provided examples of solutions to specific problems.

Egyptian geometry was introduced to the Greeks by Thales of Miletus (c. 640–546 B.C.), who visited Egypt and brought back many of the geometric ideas he studied there. Proclus asserted that, although Thales presented these ideas as a series of isolated propositions, not arranged in any logical order, many of the propositions were accompanied by deductive proofs. Subsequently, other Greek mathematicians sought to improve on Thales' work by linking the propositions in a logical sequence. Significant contributions to this effort were made by Pythagoras (c. 580 B.C.), Archytus of Tarentum (428–347 B.C.), Plato (429–348 B.C.), Hippocrates of Chios (c. 470 B.C.), and Menaechmus (c. 375–325 B.C.). However, it was Euclid who organized all of these ideas into a single unified system, thereby shaping geometry as it was to be learned and taught for the next two thousand years. His work still forms the basis for most of the geometry commonly studied in today's elementary and secondary schools. In recognition of his monumental contribution, this geometry is called **Euclidean Geometry**. ◇

It seems that the ink was hardly dry on the *Elements* before it was subjected to harsh criticism from Euclid's contemporaries. Much of this criticism centered around Euclid's fifth postulate. They regarded it as too complex to be considered a postulate and thought it could be proved from the other four. This controversy commanded the attention of mathematicians for nearly two thousand years before it was resolved. (We shall have more to say about this matter in the next chapter.)

Besides the criticism of the fifth postulate, Euclid's work was also criticized for being based on tacit assumptions in addition to the five explicit postulates. The modern viewpoint of this criticism was best articulated by David Hilbert (1862–1943). He showed that assumptions about *incidence*, *separation*, and *betweenness* were contained in many of Euclid's proofs, but were nowhere acknowledged. As a result of Hilbert's work, the geometry of Euclid, although not substantially changed in content, was made into a more carefully developed system of logical deduction. Most modern high-school geometry texts reflect the influence of Hilbert.

In addition to the changes in Euclid's postulational system, a subtler change had evolved in geometry by the beginning of this century. In Euclid's time the basic geometric components—*point*, *line*, and *plane*—were defined as concrete physical entities. Furthermore, the postulates were regarded as "obvious truths" about these components. The theorems, on the other hand, were less obvious truths, and that was the reason for proving them from the postulates. This view is one of the primary reasons why Euclid's fifth postulate was subjected to so much criticism. It did not appear to be an "obvious truth," and for that reason it seemed to require proof. The modern view of geometry is that *point*, *line*, and *plane* are undefined abstractions, and that the postulates are assumptions about these terms that are not necessarily "truths" about the real world. Theorems are just statements that follow logically from the postulates, again with no necessary connection to the real world. This viewpoint regards the basic terms *point*, *line*, and *plane* merely as formal symbols, thereby allowing them to have many different concrete interpretations. Any interpretation that satisfies the postulates is a *model* of the geometry, and the theorems become necessarily true statements about that model. (We shall have more to say about interpretations and models in the next section.)

Thus, geometry evolved from a specific, concrete, problem-oriented body of knowledge of the ancient Egyptians to a deductive treatment of that same body of knowledge by the Greeks, and then to an abstract deductive system representing not just one concrete body of knowledge, but a wide variety of concrete applications. Our purpose in the rest of this chapter and the next is to discuss some basic geometric terms and properties so that you may get a feeling for the deductive nature of geometry and for the abstract viewpoint that characterizes modern geometry. Like the ancient Egyptians, we make much of our discussion descriptive and intuitive. But we also become more deductive on occasion, like the Greeks and the modern mathematicians. The order of our presentation is somewhat the reverse of the historical development just described. We begin with an abstract view, and from it we develop some of Hilbert's assumptions about *incidence*, *separation*, and *betweenness*. Then we become more intuitive in our presentation of the geometric ideas, inserting occasional deductive proofs, where appropriate.

Exercises 11.1

Exercises 1–7 present some definitions as given by Euclid. Analyze each one in accordance with the Chapter 1 discussion (on pages 18–20) of what constitutes a good definition. (*Source*: Smith, David Eugene. *History of Mathematics*, Vol. II. New York: Dover Publications, Inc., 1958. pp. 274-278.)

1. A **point** is that which has no part.
2. A **line** is length without breadth.
3. A **straight line** is a line which lies evenly with the points on itself.
4. A **plane surface** is a surface which lies evenly with the straight lines on itself.

5. An angle ... le is a plane figure contained by one
two lines i ... ch that all the straight lines falling
and do not ... t from one point among those lying
the figure are equal to one another;

6. A figure is ... point is called the **center** of the
boundary o

11.2 Inci

tions, let us look at some con-
ns may be derived. Consider
club. This particular club has
termined by these bylaws:

members.

belong to exactly one com-

belong to the same committee.

5. Any two (distinct) committees have exactly one member in common.

From these bylaws it is possible to determine how many people belong to the club, how many committees the club has, and how many committees each club member belongs to. An analysis of the situation might proceed like this:

- By Rule 1 we know that there is at least one committee. By Rule 2 we know that this committee has exactly three members. Call these members A, B, and C, and let $[ABC]$ denote the committee to which they belong. By Rule 4 there must be at least one additional member of the club who doesn't belong to $[ABC]$; call this member D.
- Now, by Rule 3, there must be a committee with members A and D, and this committee must have one more member (by Rule 2). This additional member cannot be either B or C, because of Rule 3, so there must be a fifth member, E, giving us a committee $[ADE]$. By similar reasoning, there must be a committee with members B and D whose third member cannot be A, C, or E. (Why?) Thus, there is a sixth club member, F, and a committee $[BDF]$. In the same way, we can discover the existence of a seventh member, G, and a committee $[CDG]$. So far, we have

$$\text{Committee } 1 = [ABC]$$
$$\text{Committee } 2 = [ADE]$$
$$\text{Committee } 3 = [BDF]$$
$$\text{Committee } 4 = [CDG]$$

- There must also be committees with members A and F, A and G, B and E, B and G, C and E, and C and F, by Rule 3. (Check to see that these are the only additional committees needed.) Our needs are met by forming

$$\text{Committee } 5 = [AFG]$$
$$\text{Committee } 6 = [BEG]$$
$$\text{Committee } 7 = [CEF]$$

- If the club has seven members, arranged on committees as above, then all the bylaws are satisfied. (You might check this for yourself.) Notice also that each club member belongs to exactly three committees. It can be shown that these bylaws *require* the club to have *exactly* seven members and seven committees, but the argument would take us beyond the intended scope of this topic, so we shall not pursue it.

Now suppose that a college campus is being designed. Local zoning ordinances impose the following restrictions on the buildings and connecting sidewalks:

1. The campus has at least one sidewalk.
2. Every sidewalk connects exactly three buildings.
3. Any two (distinct) buildings are connected by exactly one sidewalk.
4. Not all buildings are connected by the same sidewalk.
5. Any two (distinct) sidewalks have exactly one building in common.

How many buildings are there, how many sidewalks are there, and on how many sidewalks is each building located? If you compare this campus problem to the club problem described before, you might notice that many facts are *irrelevant* to their solutions. For instance: Is it necessary to know the nature of the club or the names of the members? Is it necessary to know the type of college or its location? For that matter, is it really important to know whether you are dealing with the members and committees of a club or with the buildings and sidewalks of a college campus? Both problems are so similar to each other that the pattern of their solutions is identical. (Try analyzing the campus problem, and notice how the reasoning follows that of the club problem.)

The important elements of the solution(s) are the rules governing the relationships among the terms being investigated. If we replace *club*, *committee*, *member*, and *belong* in the club bylaws by *campus*, *sidewalk*, *building*, and *connect*, respectively, we get the campus zoning rules. Thus, the two situations have much in common. In order to describe the college campus we need only repeat the solution to the club problem, using the substitute terms. The solution does not depend on the interpretation, so the names given to the elements of the system really do not make any difference; the terms *plane*, *line*, *point*, and *(are) on* would serve just as well. If we make these substitutions, we get the following rules:

1. The plane has at least one line.
2. Every line is on exactly three points.
3. Any two (distinct) points of the plane are on exactly one line together.
4. Not all points of the plane are on the same line.
5. Any two (distinct) lines have exactly one point in common.

(*Note*: Rule 3 might look more familiar if written: "Two points determine a line.")

Figure 11.1 provides a diagram to help you visualize (all three interpretations of) this system. The page itself represents the plane, and the letters A, B, C, D, E, F, and G represent the points. The diagram contains seven

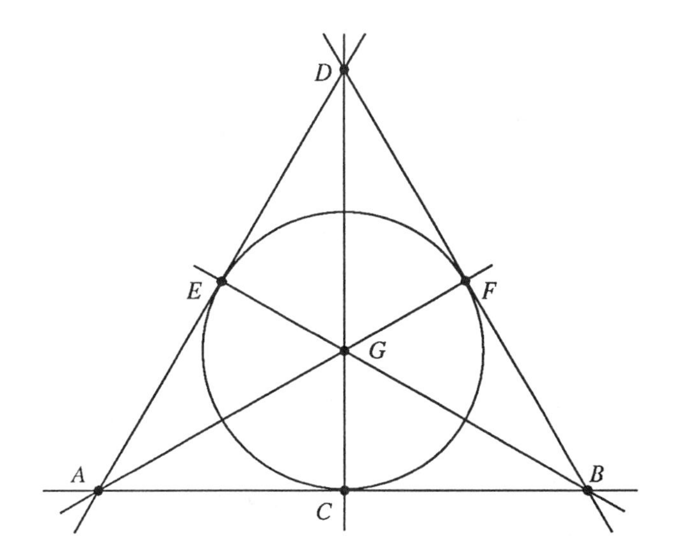

Figure 11.1 A system with seven points and seven lines.

lines, one of which is circular. (Since "line" is an *undefined term*, we can use it to refer to circles or other curves when convenient, provided the rules are satisfied.)

The (**undefined**) **terms** of this system are *point*, *line*, *plane*, and *on*. (Because we are not being completely formal, the connective word *on* appears in many disguises here, but they can all be rephrased in terms of *on*. For instance, if two lines "have a point in common," we can say that "some point is *on*" the two lines, and so forth.) The five rules are called the **postulates** of the system. Any statements that can be deduced from the postulates are called **theorems**. For instance, one theorem states that the system has exactly seven points. Another theorem asserts that the system has exactly seven lines, and a third theorem says that every point is on exactly three lines. Depending on the interpretations given to the undefined terms, Figure 11.1 could represent the committee structure of a club, the layout of a college campus, or any of a number of different situations.

Any system of postulates that contains statements about the positional relationships among *points*, *lines*, and *planes*, together with the theorems that can be deduced from them, is called an **incidence geometry**. Statements such as

"A point is on a line."

"A plane contains a line."

"A point is in a plane."

"Two lines have a point in common."

etc., are examples of **incidence statements**. The five postulates listed, together with the theorems that can be deduced from them, form an incidence geometry. All these postulates and theorems are incidence statements. The "club" and "campus" interpretations are two models of this geometry.

Being able to abstract the form of a system by using terms that have been divested of any meaning allows us to study the structure of the system itself, rather than being confined to (and confused by) any particular interpretation of it. When we find a specific situation in which some interpretation of the basic terms satisifes the rules (postulates) of the abstract system, then the deduced properties (theorems) of the abstract system are *guaranteed* to be properties of the specific situation. Thus, for example, when you find other interpretations of *point*, *line*, *plane*, and *on* that satisfy the postulates of the system just discussed, you will know immediately that each such interpretation must contain exactly seven "points" and seven "lines," and that each point must be "on" exactly three "lines." Thus, the study of a single abstract system with several—or hundreds—of interpretations is simultaneously a study of each of those interpretations!

Exercises 11.2

1. Provide two other interpretations for the undefined terms *point*, *line*, *plane*, and *on* that satisfy the postulates of the incidence geometry discussed in this section. Try to find some additional properties that your models have, and justify them.

For Exercises 2–4, form a new incidence geometry by taking the "point-line" version of the system developed in this section and replacing its fifth postulate by:

> "If **l** is a line and P is a point not on **l**, then there is exactly one line containing P that has no points in common with **l**."

2. Determine, if possible, the total number of *points* and *lines* in such a system, and the number of *lines* that are *on* any given *point*.

3. Restate this new incidence geometry using the "club" interpretation. That is, replace the undefined terms by *member*, *committee*, *club*, and *belong*. What does each postulate say? How many members are in the club? How many committees are there? How many committees is each member on?

4. Repeat Exercise 3 using the "college campus" interpretation.

5. The following four postulates describe an incidence geometry with undefined terms *point*, *line*, *plane*, and *contain*.

   (1) Every line contains at least two points.

   (2) For any two points, there is exactly one line that contains them.

   (3) There exist at least three points that are not all contained in the same line.

   (4) For any three points that are not all contained in the same line, there is exactly one plane that contains them.

   Determine the least number of *points*, *lines*, and *planes* that will satisfy the postulates. Also, draw a picture showing the relationships among these undefined terms.

6. Find at least two interpretations that satisfy the postulates of the system in Exercise 5. Rewrite each postulate, replacing the undefined terms by their new meanings in each case. What additional properties do your models have? (This last question is open-ended.)

11.3 Basic Geometric Concepts

In the preceding two sections we emphasized that words such as *point*, *line*, and *plane* can be divested of meaning and considered as undefined terms. In this section we discuss some basic geometric concepts associated with the usual interpretations of these terms as they are frequently used in the elementary and secondary schools, and we show how these interpretations reflect the postulates and theorems of geometry. Thus, a **point** will be interpreted as a location and will usually be represented by a dot. The dot is a picture of a point and, as such, it has other characteristics besides its location—for instance, color, width, length, thickness, and so forth. Except for location, none of these features is considered a property of the point itself.

The idea of a **line** is something involving infinite length in two opposite directions, as represented by a picture like Figure 11.2. (The arrowheads are there to suggest infinite length in both directions.) The concept of a line that we have in mind is "straight," a property easily imagined, but difficult to define. Just as with the picture of a point, the picture of a line has features that are not characteristic of the concept, such as color, width, and height.

Figure 11.2 A line.

Figure 11.3 suggests the idea of a **plane**. This picture is meant to convey the notion of a flat surface extending infinitely in all directions. (That is the purpose of the arrowheads.) Any flat surface—a blackboard, a floor, a sheet of paper—suggests the idea of a plane. Of course, these illustrations are somewhat inaccurate because they all possess features not common to the idea of a plane and because they are all finite in extent. As in the case of a line, we are not actually able to picture the infinite extent of a plane; its extension must be inferred.

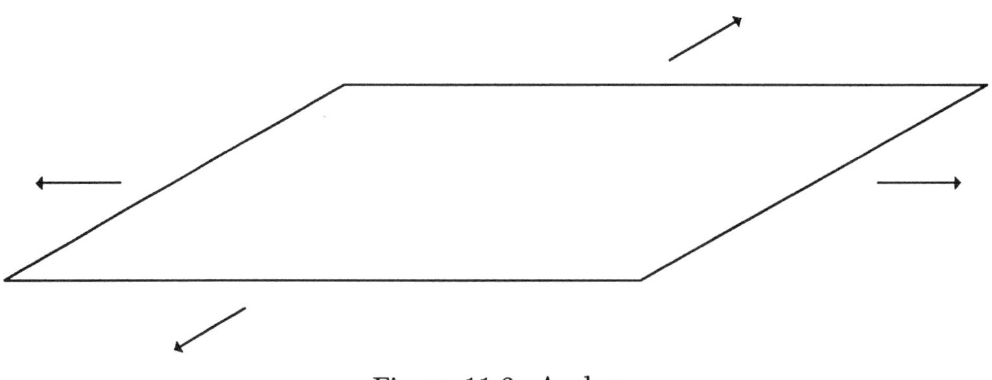

Figure 11.3 A plane.

When attempting to suggest **space** we often use a picture somewhat like Figure 11.4, intended to represent a "solid" region extending infinitely in all directions.

In general, lines, planes, and space are considered to be sets of points—lines as proper subsets of planes, and planes as proper subsets of space. We shall use uppercase italic letters A, B, C, ... to represent points, lowercase boldface letters **l**, **m**, **n**, ... to represent lines, and uppercase script letters $\mathcal{P}$, $\mathcal{Q}$, $\mathcal{R}$, ... to represent planes.

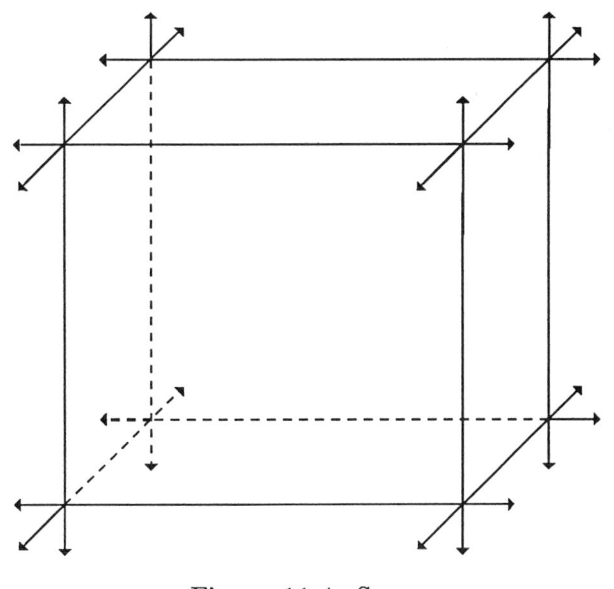

Figure 11.4 Space.

The fact that a point P is an element of a line $\mathbf{l}$ can be stated in any of the following ways:

- $P \in \mathbf{l}$
- P lies on line $\mathbf{l}$.
- P is contained in line $\mathbf{l}$, or $\mathbf{l}$ contains P.
- Line $\mathbf{l}$ passes through point P.
- Point P is incident with line $\mathbf{l}$, or $\mathbf{l}$ is incident with P.

Analogously, the fact that a point R is an element of plane $\mathcal{P}$ can be stated in any of the following ways:

- $R \in \mathcal{P}$
- R lies on plane $\mathcal{P}$.
- R is contained in plane $\mathcal{P}$, or $\mathcal{P}$ contains R.
- Plane $\mathcal{P}$ passes through point R.
- Point R is incident with plane $\mathcal{P}$, or $\mathcal{P}$ is incident with R.

Throughout the rest of this chapter we shall list postulates and theorems of this (usual) incidence geometry, using the prefixes "P-" and "T-" to distinguish one type of statement from the other. This classification distinguishes assumed statements from proved statements *in our discussion.* It does *not* imply that none of the postulates we list can be proved from the others. It is common practice in teaching introductory geometry to treat as postulates some statements (theorems) that could be proved, but whose proofs add little to the understanding of the concepts being discussed. We follow this practice here.

We begin with four postulates about the relationships between points and lines. Other postulates will be added later.

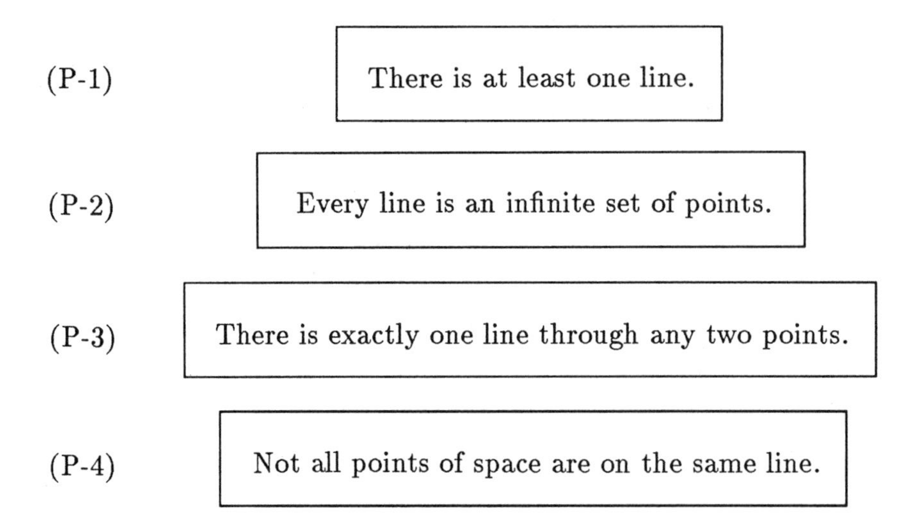

(P-1) There is at least one line.

(P-2) Every line is an infinite set of points.

(P-3) There is exactly one line through any two points.

(P-4) Not all points of space are on the same line.

Because of Postulate P-3 we can specify a line by naming any two points on it. If P and Q are points on line **l**, then we can also represent that line by the symbol $\overleftrightarrow{PQ}$, which is read "line PQ." The order in which the two points are listed can be disregarded in this notation; the line extends through them infinitely far in both directions.

Example 1 $\overleftrightarrow{QP}$, $\overleftrightarrow{PR}$, $\overleftrightarrow{RP}$, and $\overleftrightarrow{RQ}$ all represent the line shown in Figure 11.5. □

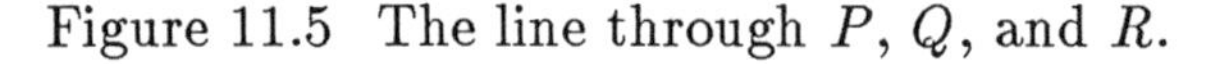

Figure 11.5 The line through P, Q, and R.

Example 2 If you are told that there exists a line $\overleftrightarrow{AB}$, then you know that points A and B are on this line. If you are told that $\overleftrightarrow{AB} = \overleftrightarrow{AC}$, then you know that a third point C is also on this line. □

In searching for additional postulates about points and lines, it is tempting to state one about the number of points that two distinct lines can have in common. Intuitively, we believe it is at most one, as illustrated by Figure 11.6(a). This is true, but we do not have to state it as a postulate because it can be deduced from the ones we already have. If there were two distinct lines containing two distinct points in common, say P and Q, then P and Q would be points belonging to two distinct lines, say **l** and **t**, as in Figure 11.6(b). But that violates Postulate P-3, which states that there is exactly one line through any two distinct points, so it cannot happen in this system. Thus, we have proved our first theorem:

(T-1) Two distinct lines have at most one point in common.

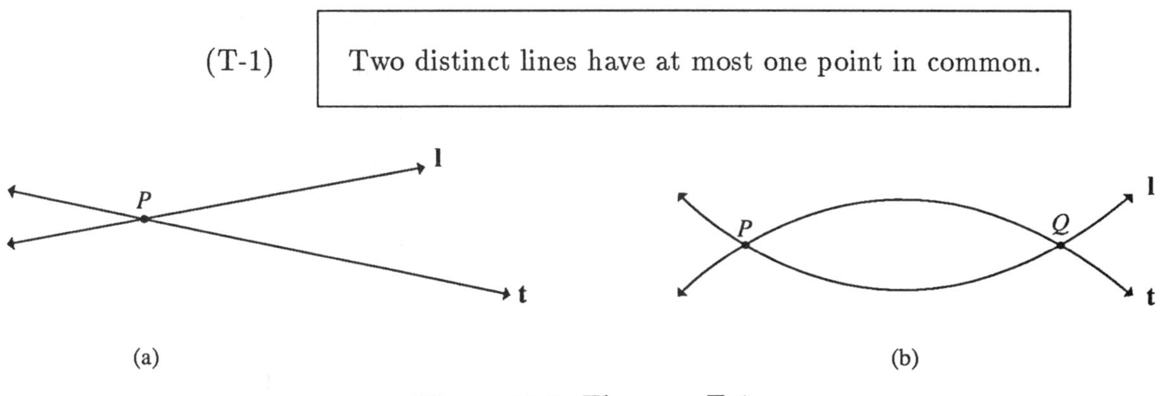

Figure 11.6 Theorem T-1.

Since lines are sets of points, we borrow set terminology and notation to describe this relationship.

Definition Two distinct lines **intersect** if they have one point in common.

We can describe the situation in Figure 11.6(a) by saying "lines **l** and **t** intersect at P," or "the intersection of **l** and **t** is P." In symbols, we write

$$\mathbf{l} \cap \mathbf{t} = \{P\}$$

Many other theorems can be deduced from our postulates, some more interesting than others. Since we are not trying to develop all of geometry in these few pages, but rather give you a taste of how the subject is developed, we confine our attention to a few basic questions that can be answered with the machinery at hand.

For instance, Postulate P-2 says that there are infinitely many points on a line, so a natural question is: How many lines go through a given point? To answer this, suppose that A represents a point of our space. Then, by Postulate P-1, we know that there is at least one line; call it **l**. Now, our postulates do not specify whether or not $\mathbf{l}$ contains A, so the situation may be depicted in either of two ways, as shown by Figures 11.7(a) and 11.7(b).

If A is on line $\mathbf{l}$, as in Figure 11.7(a), then we know (by P-4) that there is some other point—say B—that is not on $\mathbf{l}$. By P-2, there is some point besides A on line $\mathbf{l}$; call it C. Because B and C are distinct, there is a line $\overleftrightarrow{BC}$. Now, $\overleftrightarrow{BC}$ contains an infinite number of points (by P-2) and all the points of $\overleftrightarrow{BC}$ are on distinct lines containing A (by P-3), as shown in Figure 11.8.

The case of Figure 11.7(b) is similar and is left as an exercise. Thus, we have our second theorem:

(T-2)	There are infinitely many lines through each point in space.

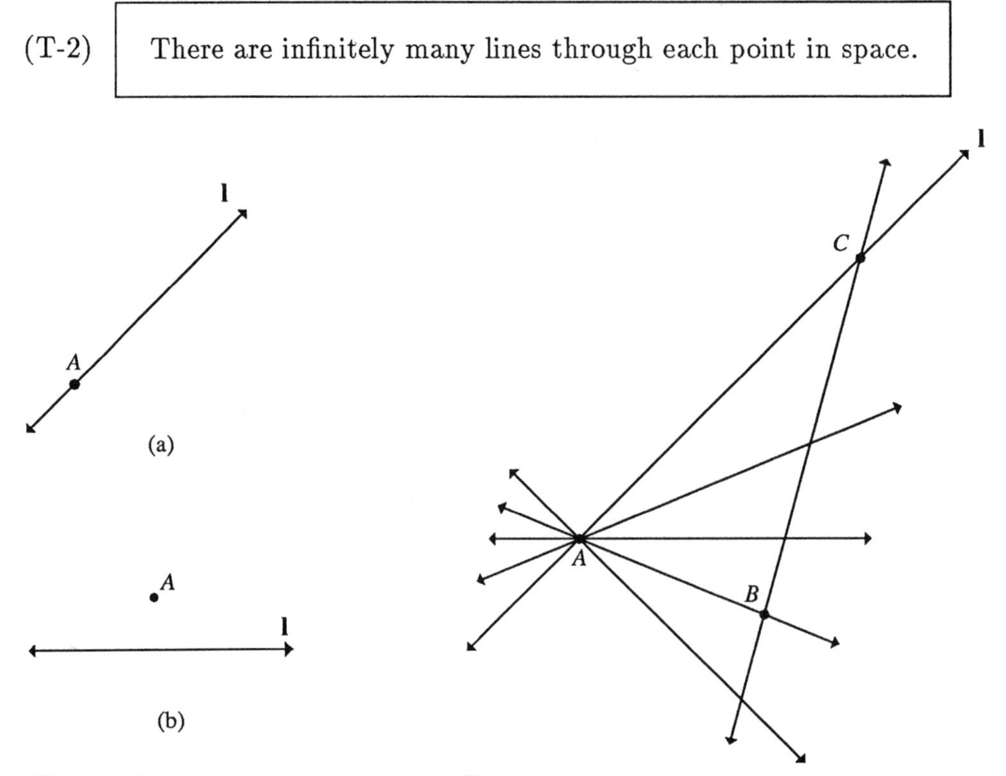

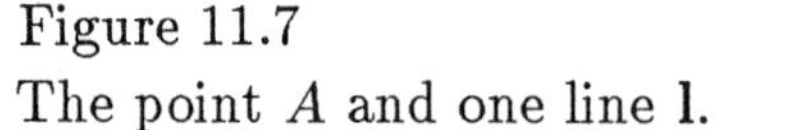

Figure 11.7
The point A and one line $\mathbf{l}$.

Figure 11.8
Infinitely many lines through A.

We have already observed, informally, that a line is a proper subset of a plane, which can be thought of as a flat surface. The next definition and postulate make this more precise.

Definition A line $\mathbf{l}$ is **in** a plane $\mathcal{P}$ if all points of $\mathbf{l}$ are also points of $\mathcal{P}$; that is, if $\mathbf{l}$ is a subset of $\mathcal{P}$.

(P-5)

> If two points of a line are in a plane, then the line is in the plane.

By the contrapositive of Postulate P-5, if a line is *not* in a plane, then the line has at most one point in common with the plane.

Example 3 In Figure 11.9, line $\mathbf{l}$ and plane $\mathcal{P}$ have only point A in common, (that is, $\mathbf{l} \cap \mathcal{P} = \{\mathcal{A}\}$), so $\mathbf{l}$ is not in the plane ($\mathbf{l} \not\subseteq \mathcal{P}$). □

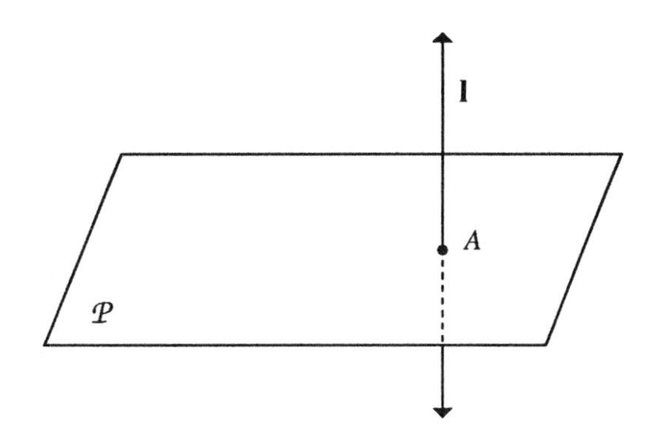

Figure 11.9 A line and a plane with one point in common.

Example 4 In Figure 11.10, $\mathbf{l}$ and $\mathcal{P}$ have no points in common ($\mathbf{l} \cap \mathcal{P} = \emptyset$), so $\mathbf{l}$ is not in the plane. □

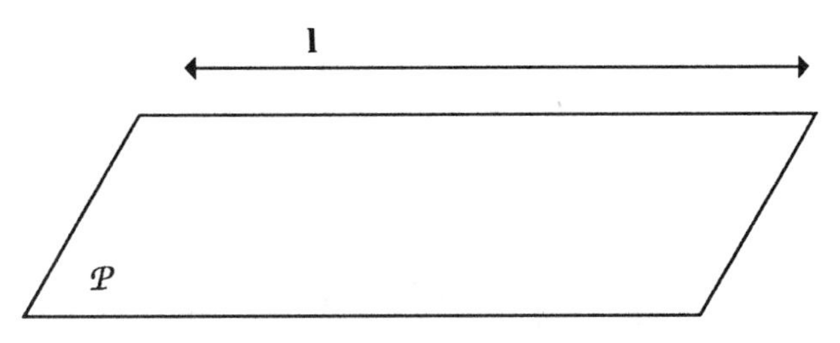

Figure 11.10 A line and a plane with no points in common.

If two distinct lines $\mathbf{l}$ and $\mathbf{t}$ do not intersect, then they do not have any points in common; that is, $\mathbf{l} \cap \mathbf{t} = \emptyset$. In this case, $\mathbf{l}$ and $\mathbf{t}$ may or may not be in the same plane. This prompts us to define a new term for each possibility:

Definition Two (distinct) lines are **parallel** if they are in the same plane and do not intersect. In symbols, we write "$\mathbf{l}$ is parallel to $\mathbf{t}$" as $\mathbf{l} \parallel \mathbf{t}$.

Definition Two lines are **skew** if they are not in the same plane and do not intersect.

Example 5 Figure 11.11 shows **l** parallel to **t** (that is, **l** || **t**). They are in the same plane and have no points in common. □

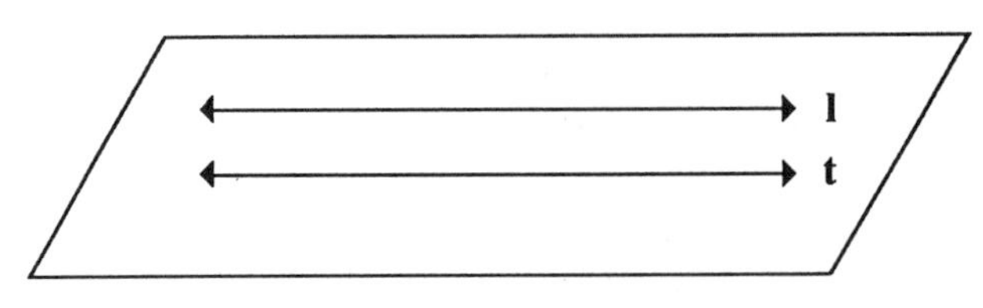

Figure 11.11 Parallel lines.

Example 6 Figure 11.12 represents **l** and **t** as skew lines in different planes and with no common points. To visualize this, think of **l** as going from north to south in the plane of the ceiling, while **t** goes from east to west in the plane of the floor. □

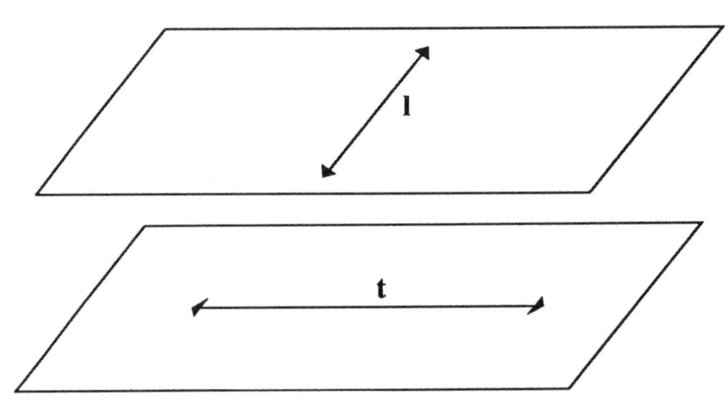

Figure 11.12 Skew lines **l** and **t**.

These definitions allow us to state conveniently three more postulates about the relationship between lines and planes.

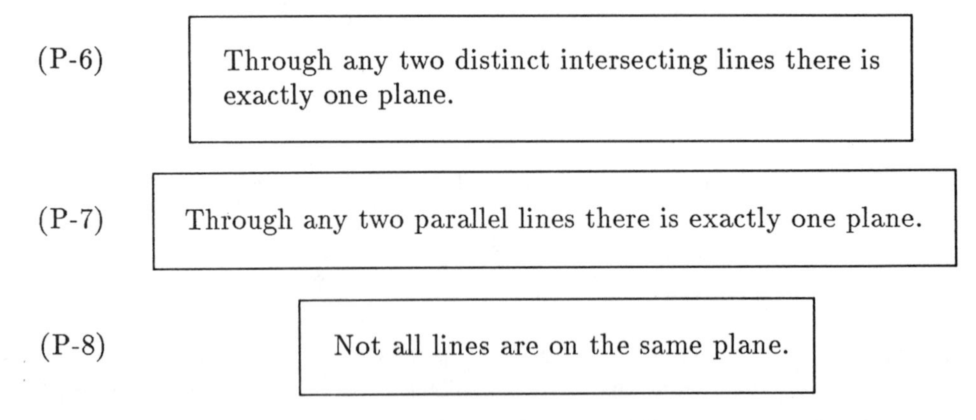

(P-6) Through any two distinct intersecting lines there is exactly one plane.

(P-7) Through any two parallel lines there is exactly one plane.

(P-8) Not all lines are on the same plane.

Postulates P-6 and P-7 establish minimum conditions for determining a plane (uniquely). These postulates, along with one more definition, lead to two theorems that provide two other ways to determine a plane. These theorems express a fundamental relationship among points, lines, and planes.

Definition Three or more points in space are said to be **collinear** if they are on the same line. If they are not on the same line, they are said to be **noncollinear**.

(T-3) Through any three distinct noncollinear points there is exactly one plane.

Notation Because of this theorem, an alternate way to name a plane is to specify three distinct noncollinear points that are contained in it.

(T-4) Through a line and a point not on the line there is exactly one plane.

To prove Theorem T-3, we begin by establishing some notation. Suppose that A, B, and C are three noncollinear points. Then $\overleftrightarrow{AB}$ and $\overleftrightarrow{CB}$ are two distinct lines that intersect at B, as shown in Figure 11.13. By Postulate P-6, there is exactly one plane containing $\overleftrightarrow{AB}$ and $\overleftrightarrow{CB}$. But then all points of these lines are in that plane, by the definition of a line in a plane, so A, B, and C are in the plane. Thus, we have at least one plane containing the three points.

To show that there is only one such plane, suppose there were a second plane containing A, B, and C. In that case, the lines $\overleftrightarrow{AB}$ and $\overleftrightarrow{CB}$ would

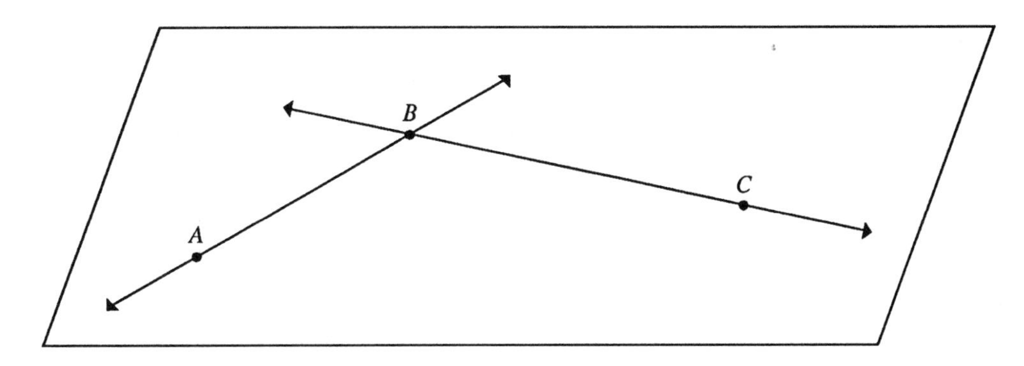

Figure 11.13 Three noncollinear points determine a plane.

also be in this plane (by P-5), in violation of P-6. Thus, such a second plane cannot exist, and the proof of Theorem T-3 is complete. The proof of T-4 is similar to that of T-3; it is left as an exercise.

A few more terms round out the basic vocabulary for points, lines, and planes in space.

Definition Four or more points in space are said to be **coplanar** if they are all in the same plane. If they are not all in the same plane, they are said to be **noncoplanar**. Two or more lines in space are said to be **coplanar** if they are all in the same plane, and **noncoplanar** if they are not.

Examples 7 and 8 refer to Figure 11.14, where points P, Q, R, S, and T are in plane PQR, points M, Q, R, and U are in plane MQR, and planes MQR and PQR are distinct.

Example 7 P, Q, R, S, and T all are in the same plane, so this plane can be represented by any three of these points that are noncollinear, in any order:

$$PQR, \quad RPQ, \quad QST, PTR, \quad \text{etc.}$$ □

Example 8 P, Q, R, S, and T are coplanar, as are M, R, Q, and U. Points M, R, Q, and P are noncoplanar. □

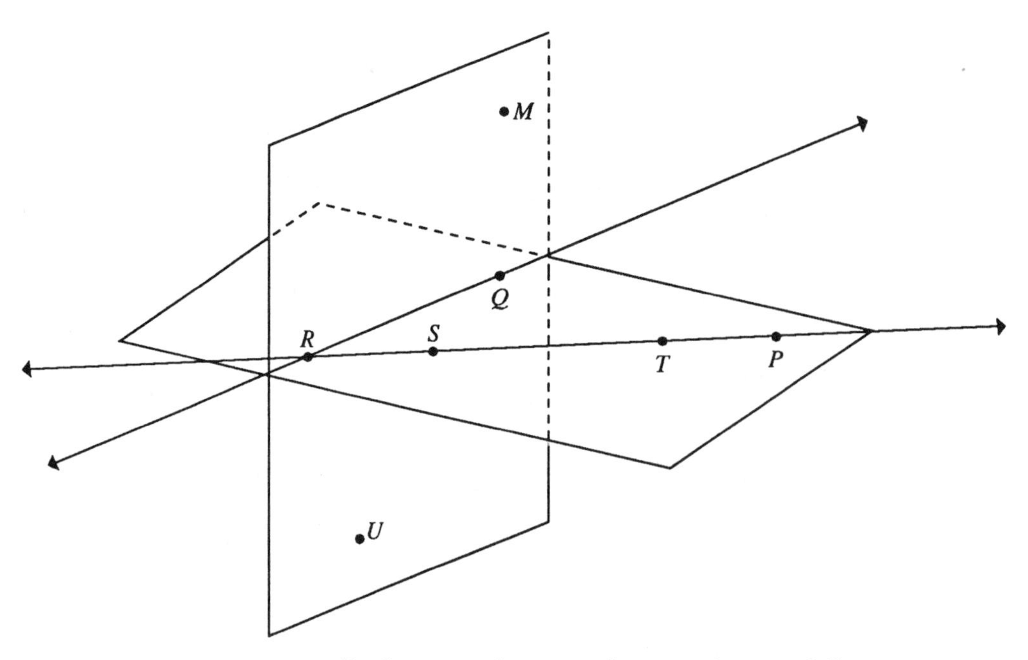

Figure 11.14 Coplanar and noncoplanar points and lines.

Definition If a line has no points in common with a plane, then **the line is parallel to the plane.** Two planes that have no points in common are called **parallel planes.** As with parallel lines, we use the symbol "||" to represent "is parallel to."

Examples 9–12 refer to the boxlike shape in Figure 11.15.

Example 9 $\overleftrightarrow{AB} \parallel \overleftrightarrow{FG}$, $\overleftrightarrow{DC} \parallel \overleftrightarrow{EH}$, $\overleftrightarrow{AD} \parallel \overleftrightarrow{EF}$, and $\overleftrightarrow{BC} \parallel \overleftrightarrow{GH}$ □

Example 10 $\overleftrightarrow{AB} \parallel FGH$, $\overleftrightarrow{EH} \parallel ABC$, $\overleftrightarrow{AF} \parallel CDE$, and $\overleftrightarrow{CH} \parallel ABG$ □

Example 11 $ABC \parallel EFG$, $ADE \parallel BCG$, and $ABG \parallel CDE$ □

Example 12 Although $\overleftrightarrow{AB}$ and $\overleftrightarrow{GH}$ are in parallel planes (ABC and FGH, respectively), they are not parallel lines; they are skew. □

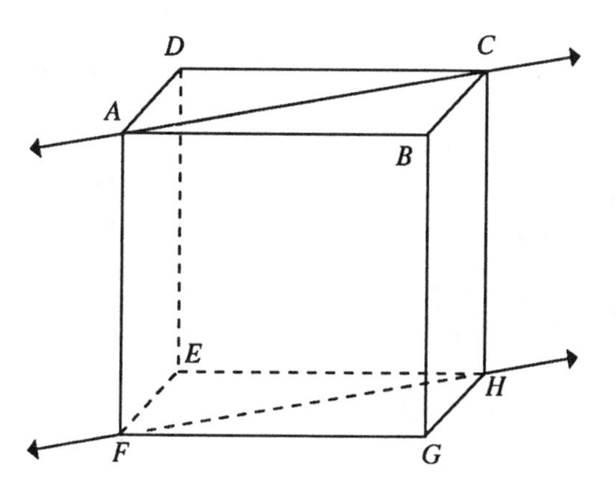

Figure 11.15 Parallel planes and lines.

Exercises 11.3

1. Name all the distinct lines in Figure 11.16. For each line, specify a point not on it.

For Exercises 2–6, draw a diagram showing a line **t** and points P, Q, and R satisfying the given conditions.

2. $\mathbf{t} = \overleftrightarrow{PQ}$ and $R \notin \mathbf{t}$

3. $\mathbf{t} = \overleftrightarrow{PQ}$ and $R \in \mathbf{t}$

4. $\mathbf{t} \cap \overleftrightarrow{PQ} = \{R\}$

5. $\{P, Q\} \subset \mathbf{t}$ and $R \notin \mathbf{t}$

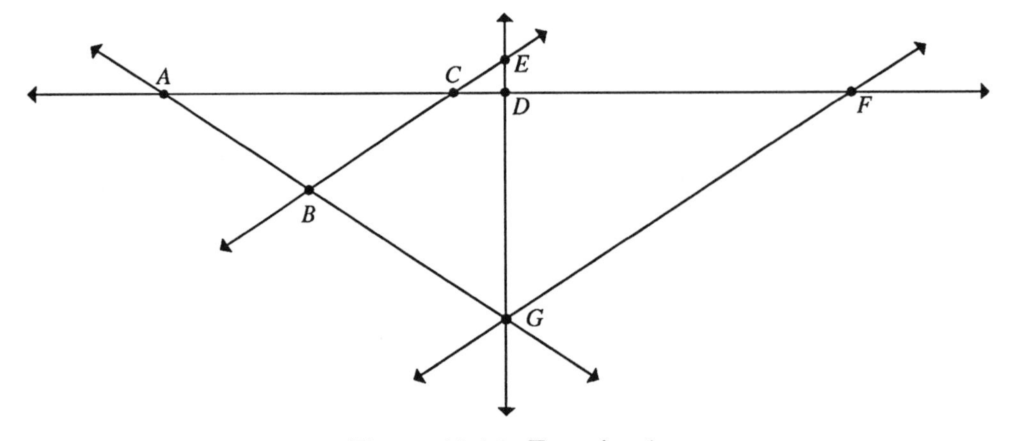

Figure 11.16 Exercise 1.

6. $\mathbf{t} \cap \overleftrightarrow{PQ} = \emptyset$

7. Complete the proof of Theorem T-2. [See Figure 11.7(b).]

8. Often it is useful to suggest to students that the properties stated in Postulates P-7 and P-8 and Theorems T-3 and T-4 have simple applications in the physical world. For example, using a tripod to hold a camera in a fixed position is an application of T-3. List two applications of each of these properties.

The front and top of the boxlike shape in Figure 11.17 are in planes $\mathcal{P}$ and $\mathcal{S}$, respectively. Exercises 9–18 refer to this figure.

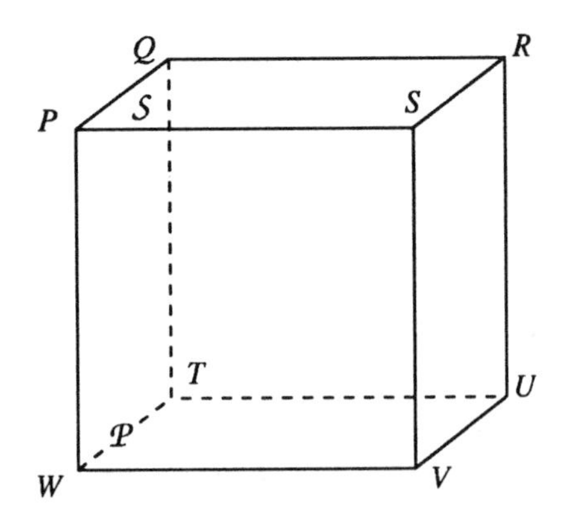

Figure 11.17 Exercises 9–18.

9. Name four points contained in plane $\mathcal{P}$.

10. Name six distinct lines in plane $\mathcal{S}$.

11. Determine $\mathcal{P} \cap \mathcal{S}$.

12. Name plane $\mathcal{S}$ in two different ways.

13. Name all other distinct planes determined by the eight labeled points.

14. Name two lines that do not intersect but are in the same plane. What kind of lines are they?

15. Name two lines that do not intersect and are not in the same plane. What kind of lines are they?

16. Name a line and a point that determine $\mathcal{P}$.

17. Name two intersecting lines that determine $\mathcal{S}$.

18. Name two parallel lines that determine $\mathcal{P}$.

For Exercises 19 and 20, let **l**, **m**, and **n** be three distinct lines in space, arranged so that **l** intersects both **m** and **n**.

19. Must **m** and **n** lie in the same plane?

20. If there is a fourth line **k** that intersects all three of the lines **l**, **m**, and **n**, must **m** and **n** lie in the same plane?

21. Prove Theorem T-4.

For Exercises 22–31, let $\mathbf{l}$, $\mathbf{m}$, and $\mathbf{n}$ be distinct lines and let $\mathcal{P}$, $\mathcal{R}$, and $\mathcal{S}$ be distinct planes in space. Justify your answer to each question by drawing a diagram. (*Hint*: Think of the sides of a box or the walls of a room.)

22. If $\mathbf{l}$ and $\mathbf{m}$ are in the same plane and $\mathbf{m}$ and $\mathbf{n}$ are in the same plane, must $\mathbf{l}$ and $\mathbf{n}$ be in the same plane? (That is, is

    being in the same plane

    a transitive relation?)

23. Can $\mathbf{l}$ and $\mathbf{m}$ be coplanar, $\mathbf{m}$ and $\mathbf{n}$ be coplanar, and $\mathbf{l}$ and $\mathbf{n}$ be coplanar, but no plane exist containing all three lines?

24. If $\mathbf{l}$, $\mathbf{m}$, and $\mathbf{n}$ all intersect at the same point, must they be coplanar?

25. If $\mathbf{l} \parallel \mathbf{m}$ and $\mathbf{m} \parallel \mathbf{n}$, does it follow that $\mathbf{l} \parallel \mathbf{n}$? (That is, is the relation $\parallel$ transitive?)

26. If $\mathbf{l} \parallel \mathbf{m}$, $\mathbf{m} \parallel \mathbf{n}$, and $\mathbf{l} \parallel \mathbf{n}$, must $\mathbf{l}$, $\mathbf{m}$, and $\mathbf{n}$ be coplanar?

27. If $\mathbf{l} \cap \mathbf{m} \neq \emptyset$ and $\mathbf{m} \cap \mathbf{n} \neq \emptyset$, must $\mathbf{l} \cap \mathbf{n} \neq \emptyset$?

28. If $\mathbf{l} \parallel \mathcal{P}$ and $\mathbf{l} \parallel \mathcal{R}$, must it follow that $\mathcal{P} \parallel \mathcal{R}$?

29. If $\mathcal{P} \cap \mathcal{R} = \mathbf{l}$, $\mathcal{P} \cap \mathcal{S} = \mathbf{n}$, and $\mathbf{l} \parallel \mathbf{n}$, does it follow that $\mathcal{R} \parallel \mathcal{S}$?

30. If $\mathcal{P} \cap \mathcal{R} = \mathbf{l}$, $\mathcal{R} \cap \mathcal{S} = \mathbf{m}$, and $\mathcal{P} \cap \mathcal{S} = \mathbf{n}$, can $\mathbf{l}$, $\mathbf{m}$, and $\mathbf{n}$ be parallel?

31. If $\mathbf{l} \parallel \mathbf{m}$ and $\mathbf{m} \parallel \mathbf{n}$, can $\mathbf{l}$ and $\mathbf{n}$ be skew?

For Exercises 32–43, indicate whether the given statement is *true* or *false*. Base your answers on the postulates, definitions, and theorems, and on your intuition.

32. Two skew lines determine a plane.

33. If $\mathbf{l}$ is a line and $P \notin \mathbf{l}$, then there exists a line $\mathbf{m}$ such that $P \in \mathbf{m}$ and $\mathbf{l} \parallel \mathbf{m}$.

34. If $\mathbf{l}$ is a line and $P \notin \mathbf{l}$, then there is at most one line containing P that is parallel to $\mathbf{l}$.

35. If $\mathbf{l}$ is a line and $P \notin \mathbf{l}$, then there exists a line $\mathbf{m}$ such that $P \in \mathbf{m}$ and $\mathbf{l} \cap \mathbf{m} \neq \emptyset$.

36. If $\mathbf{l}$ is a line and $P \notin \mathbf{l}$, then there is at most one line containing P that intersects $\mathbf{l}$.

37. If $\mathbf{l}$ is a line and $P \notin \mathbf{l}$, then there exists a line $\mathbf{m}$ such that $P \in \mathbf{m}$, and $\mathbf{m}$ and $\mathbf{l}$ are skew.

38. If $\mathbf{l}$ is a line and $P \notin \mathbf{l}$, then there is at most one line $\mathbf{m}$ containing P such that $\mathbf{l}$ and $\mathbf{m}$ are skew.

39. If $\mathcal{P}$ is a plane and $P \notin \mathcal{P}$, then there is a plane $\mathcal{S}$ with $P \in \mathcal{S}$ such that $\mathcal{P} \parallel \mathcal{S}$.

40. If $\mathcal{P}$ is a plane and $P \notin \mathcal{P}$, then there is at most one plane containing P and parallel to $\mathcal{P}$.

41. Any two distinct lines intersect in at most one point.

42. If P and Q are any two points in a plane $\mathcal{P}$, then $\overleftrightarrow{PQ}$ is a proper subset of $\mathcal{P}$.

43. If $\mathbf{l}$ is a line and $\mathcal{P}$ is a plane with $\mathbf{l} \not\subseteq \mathcal{P}$, then $\mathbf{l} \cap \mathcal{P}$ contains at most one point.

11.4 Separation

Early in this century, David Hilbert observed that Euclid had made some assumptions about incidence, separation, and betweenness that were not specifically addressed in his postulates. In this section we consider the concepts of *separation* and *betweenness*. Our discussion is somewhat selective, covering mainly ideas needed for the development of topics later in this chapter and the next. As you study this material, notice that the added formality reinforces, rather than contradicts, your intuitive understanding of separation and betweenness. Thus, this section illustrates how the logical structure of geometry justifies and extends the informal ideas presented in elementary-school mathematics.

You have already seen that lines, planes, and space are three basic kinds of sets of points. Without giving a formal definition of "dimension," we shall refer to these three kinds of sets as 1-, 2-, and 3-dimensional figures, respectively. This distinction of dimensions is important in considering the *separation* concept. The separation postulates in all three cases are exactly analogous and could, in fact, be treated as a single postulate, but we shall forego the rigor and elegance of that approach in order to provide an intuitive understanding of the idea. We begin with the simplest case, describing a relationship between a line and a point on it.

(P-9) Each point of a line separates the line into two disjoint subsets—two **half-lines**. The point, which is called the **boundary** of each half-line, is not an element of either one.

Definitions A **ray** is the union of a half-line and its boundary, which is called the **endpoint** of the ray. The two distinct rays of a line that are determined by a common endpoint are called **opposite rays**.

Notation A ray with endpoint P containing a (different) point Q is denoted by $\overrightarrow{PQ}$, and is read "ray PQ." The first letter *always* designates the endpoint, and the second letter identifies any other point of the ray.

Figure 11.18 shows points P, Q, R, S, and T on a line $\mathbf{l}$. Point P separates $\mathbf{l}$ into two half-lines, by (P-9). Q and R are in one half-line; S and T are in the other. P is the boundary of the half-lines, and is not contained in either one. We can say that two points are on the **same side** of a point P if they are in the same half-line determined by P, and they are on **opposite sides** of P if they are in different half-lines determined by P. Thus, Q and R are on the same side of P, but Q and S are on opposite sides of P.

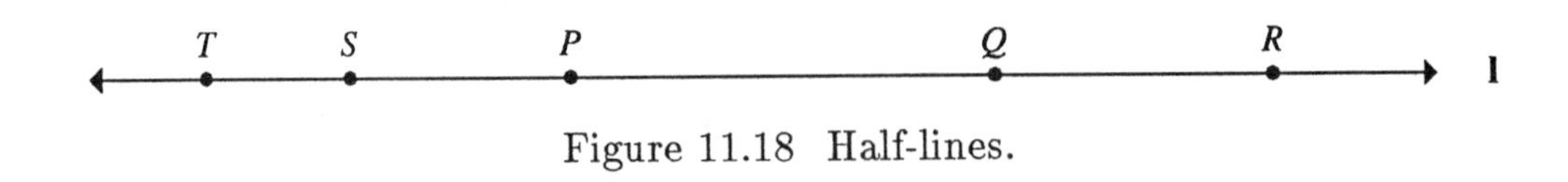

Figure 11.18 Half-lines.

Example 1 In Figure 11.18, Q also separates $\mathbf{l}$ into two half-lines, with Q as boundary. Likewise, R, S, and T also separate $\mathbf{l}$ into half-lines. □

Example 2 The two rays determined by P in Figure 11.18 can be denoted by $\overrightarrow{PQ}$ and $\overrightarrow{PS}$; they are opposite rays. $\overrightarrow{PQ}$ and $\overrightarrow{PR}$ denote the same ray, as do $\overrightarrow{PS}$ and $\overrightarrow{PT}$. □

Most of us have a good intuitive idea of what it means to say that one point is "between" two others. The next definition and two postulates make this idea more precise.

Definition A point P is **between** points Q and R if
(1) P, Q, and R are collinear, and
(2) Q and R are on opposite sides of P.

Notation We use either of the symbolic forms Q-P-R or R-P-Q to indicate that P is between Q and R; thus, R-P-Q is read as "P is between R and Q." By definition, R-P-Q if and only if Q-P-R.

We can gain some insight into *betweenness* as it applies to points on a line by considering some analogous ideas related to order and the number line in the rational-number system (described in Section 9.5). Recall that a rational number b is "between" rational numbers a and c if either $a < b < c$ or $c < b < a$.

Now, suppose that a, b, and c are distinct rationals. By the Trichotomy Law, either $a < b$ or $b < a$. If $a < b$, then either

$$c < a \text{ and hence } c < a < b$$

or

$$a < c \text{ and hence either } a < c < b \text{ or } a < b < c.$$

Likewise, if $b < a$, we have $c < b < a$, or $b < c < a$, or $b < a < c$. That is, if a, b, and c are distinct rational numbers, then exactly one of these three statements must be true:

a is between b and c; b is between a and c; c is between a and b

This suggests an analogous property for points on a line, which we state as a postulate.

(P-10) If three points are distinct and collinear, then exactly one of the points is between the other two.

That is, if P, Q, and R are any three distinct noncollinear points, exactly one of the following three statements must be true:

$$P\text{-}Q\text{-}R, \text{ or } Q\text{-}P\text{-}R, \text{ or } Q\text{-}R\text{-}P$$

We also know that, given any two distinct rational numbers, we can always find another rational number between them. Moreover, we can also find rational numbers bigger or smaller than any given rational. Thus, if a and b are rational numbers such that $a < b$, then there exist rational numbers c, d, and e such that

$$a < c < b, \quad d < a < b, \text{ and } a < b < e$$

This suggests our second betweenness postulate for points on a line.

(P-11)

> If A and B are two distinct points, then there exist points C, D, and E such that
> $$A\text{-}C\text{-}B,\ D\text{-}A\text{-}B, \text{ and } A\text{-}B\text{-}E$$

Informally, Postulate P-11 says that there are points between and on either side of every pair of distinct points. It does not imply that point A is to the left of point B; the notational suggestion of order is just an accidental consequence of the fact that we write English from left to right. Also, notice that P-11 is a postulate. It is not a fact that we can prove; we assume that this is how points are arranged on a line.

Examples 3–8 refer to Figure 11.19.

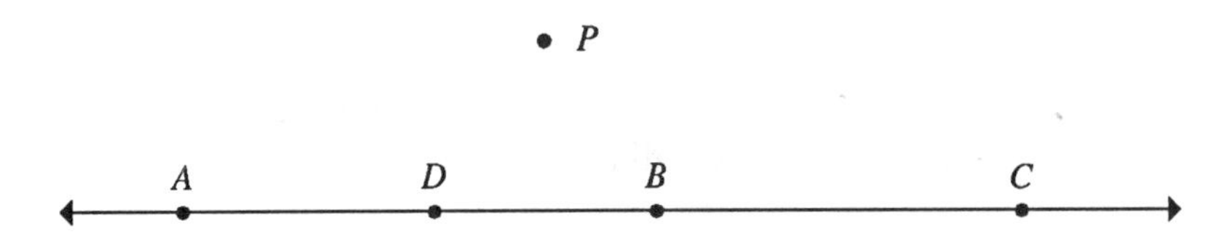

Figure 11.19 Examples of betweenness.

Example 3 Point D is between A and B; it is also between A and C. In symbols, A-D-B (or, equivalently, B-D-A) and A-D-C (or C-D-A). □

Example 4 Point P is not between A and B because P, A, and B are noncollinear. □

Example 5 By Postulate P-10, since D is between A and B, then A cannot be between D and B, nor can B be between A and D. □

Example 6 By the first part of Postulate P-11, we know that there must be a point between A and D, even though none is labeled in the figure. There must also be points between D and B and between B and C. □

Example 7 By Postulate P-11, we also know there is a point between P and A. Such a point must be on line $\overleftrightarrow{PA}$, which must exist (why?) but is not shown in the diagram. □

Example 8 The second part of Postulate P-11 implies that C is between B and some point of the ray opposite to $\overrightarrow{CB}$. Likewise, A is between D and some point of the ray opposite to $\overrightarrow{AD}$. This means that a line has no "last point on either end"; more formally, we say that a line is **unbounded.** □

Example 9 Postulate P-10 rules out the possibility that a "line" might actually look like a circle or some other kind of loop. As you can see from Figure 11.20, if P, Q, and R are three points on a circle or loop, then *all three* of the statements P-Q-R, Q-R-P, and R-P-Q are true. That is, *each* of the points is between the other two. □

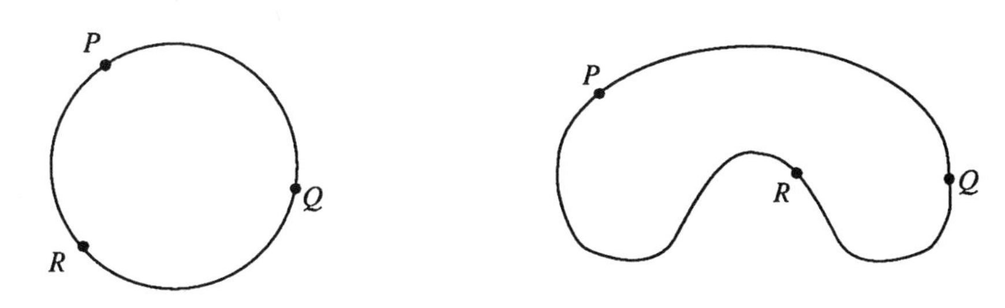

Figure 11.20 Betweenness on circles and loops.

These betweenness postulates establish how points on a line are situated relative to each other. They are essential ingredients in the process of making intuitive geometric ideas come together into a logically consistent system. The concepts in P-10 and P-11 are very simple—almost too simple to seem worth mentioning—but they are basic to formal geometry and *they cannot be proved.* However, once we assume them, many other natural geometric ideas become easy to describe rigorously. The following definitions exemplify this; they describe the concept of a *line segment*, a fundamental building-block in geometry, solely in terms of elementary set language and betweenness.

Definitions The union of two points and all points between them is called a **line segment.** If A and B are the two points, then the segment is denoted $\overline{AB}$ (read "line segment AB" or simply "segment AB"). The points A and B are called the **endpoints** of the segment, and the set of points between them is called the **interior** of the segment.

In set notation, we can write the line segment $\overline{AB}$ as

$$\underbrace{\{A, B\}}_{\text{endpoints}} \cup \underbrace{\{X \mid A\text{-}X\text{-}B\}}_{\text{interior}}$$

Example 10 In Figure 11.19, $\overline{AC}$ is a line segment; its endpoints are A and C, and points D and B are in its interior. This segment can also be written as $\overline{CA}$. Some other (distinct) segments on line $\overleftrightarrow{AC}$ are $\overline{AD}$, $\overline{AB}$, $\overline{DB}$, $\overline{DC}$, and $\overline{BC}$. □

As a consequence of Postulate P-11, the points of a line are **dense**. This means that, for any two distinct points P and Q (no matter how close they are to each other), the interior of $\overline{PQ}$ must always contain points. If you compare this usage with the idea of the denseness of the rational numbers as described in Section 9.5, you will see that the key fact there was that we could always find a rational number *between* two others. In $\mathbf{Q}$, this was done by arithmetic; we found an "average." The same kind of fact is required here. However, in incidence geometry we don't have any arithmetic operations to work with, so we need a postulate about betweenness (P-11) to insure that lines do not have "gaps" in them.

Example 11 Figure 11.21 shows how the ideas of *line* and *line segment* are presented at the sixth-grade level. □

So far, we have limited our discussion of separation to the case of a point on a line. Now we turn to the analogous descriptions of separation in the plane and in 3-dimensional space.

(P-12) Each line of a plane separates the plane into two disjoint subsets, called **half-planes**. The line, which is called the **boundary** of each half-plane, is not a subset of either one.

(P-13) Each plane of space separates the space into two disjoint subsets, called **half-spaces**. The plane, which is called the **boundary** of each half-space, is not a subset of either one.

PROBLEM-SOLVING COMMENT

Postulates P-9, P-12, and P-13 are **analogous** statements. Which terms in P-12 and P-13 are the analogues of *point* and *line* in P-9? Examples 12 and 13 are analogous in the same way. *Before* you read Example 13, try to construct it from Example 12 by analogy. ◇

B. A *plane* is a flat surface that extends without end in all directions.

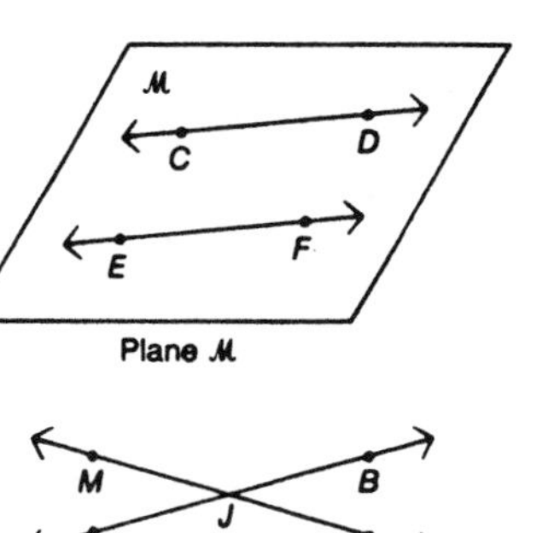

Lines in a plane that never meet are ***parallel lines***. $\overleftrightarrow{CD}$ and $\overleftrightarrow{EF}$ are parallel.

Intersecting lines meet at a point. $\overleftrightarrow{MN}$ and $\overleftrightarrow{AB}$ intersect at point J.

Try Use the diagram at the right.

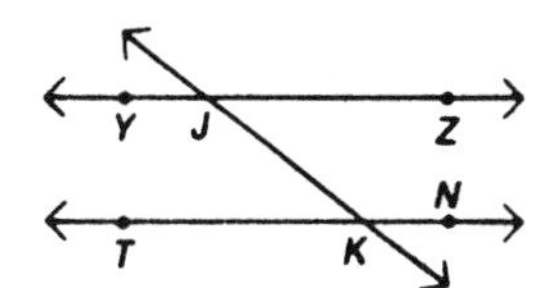

a. Name 3 points on the same line.

b. Name 3 segments on $\overleftrightarrow{YZ}$.

c. Name a pair of intersecting lines.

Practice Use the diagram at the right.

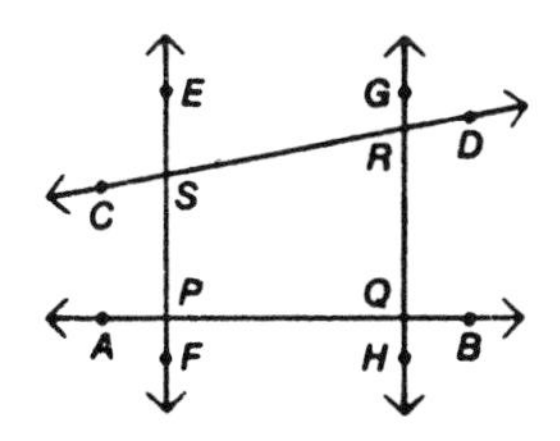

1. Which line appears to be parallel to $\overleftrightarrow{EF}$?

2. Which lines intersect at point P?

3. Name a point that is not on $\overleftrightarrow{EF}$.

4. Name 3 points on $\overleftrightarrow{CD}$.

5. Name 3 rays with endpoint P.

6. Which lines does $\overleftrightarrow{GH}$ intersect?

7. Give another name for $\overline{GH}$.

8. What is the endpoint of $\overrightarrow{CS}$?

9. Name 3 segments on $\overleftrightarrow{CD}$.

10. Name 2 rays on $\overleftrightarrow{CD}$.

11. Name 2 segments with endpoint A.

Apply Use the diagram at the right of a portion of the painting by Klee.

12. Name 3 segments on $\overline{KQ}$.

13. Name the intersection of $\overline{KQ}$ and $\overline{MN}$.

14. Give another name for $\overrightarrow{JK}$.

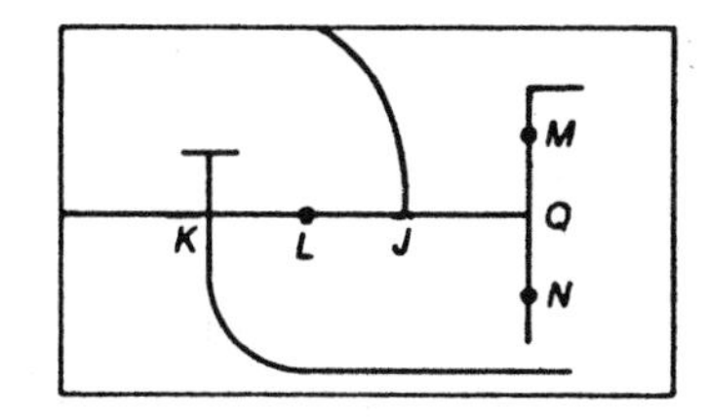

Figure 11.21 INVITATION TO MATHEMATICS, Grade 6, p. 271

Example 12 In Figure 11.22, line **l** separates plane $\mathcal{P}$ into two half-planes. Points P, Q, and R (in plane $\mathcal{P}$) are not on line **l**. We say that points P and Q are on opposite sides of **l** (or **in opposite half-planes**) because the segment $\overline{PQ}$ contains a point, S, of **l**. We say that Q and R are **on the same side** of **l** (or **in the same half-plane**) because $\overline{QR}$ does not contain a point of **l**.

We can refer to the half-plane of $\mathcal{P}$ containing the point Q as the "Q-side" of **l**. In this case, the Q-side and the R-side of **l** are the same half-plane, but the Q-side and the P-side of **l** are opposite half-planes. □

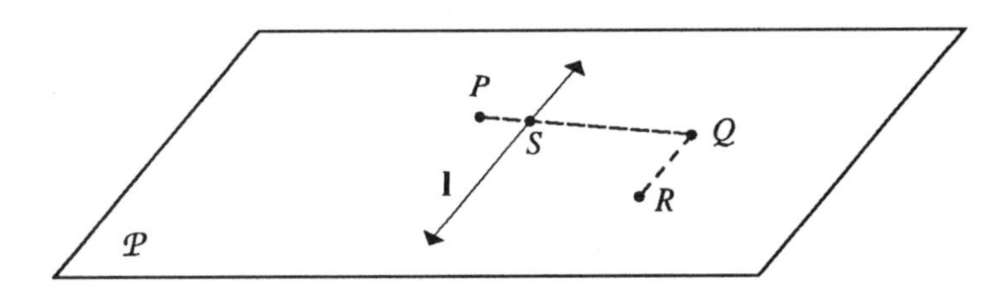

Figure 11.22 Opposite sides of a line.

Example 13 In Figure 11.23, plane $\mathcal{P}$ separates 3-dimensional space into two half-spaces. Points P, Q, and R are not on plane $\mathcal{P}$. We say that points P and Q are **on opposite sides** of $\mathcal{P}$ (or **in different half-spaces**) because the segment $\overline{PQ}$ contains a point, S, of $\mathcal{P}$. We say that Q and R are **on the same side** of $\mathcal{P}$ (or **in the same half-space**) because $\overline{QR}$ does not contain a point of $\mathcal{P}$. We can refer to the half-space containing the point Q as the "Q-side" of $\mathcal{P}$. In this case, the Q-side and the R-side of $\mathcal{P}$ are the same half-space, but the Q-side and the P-side of $\mathcal{P}$ are opposite half-spaces. □

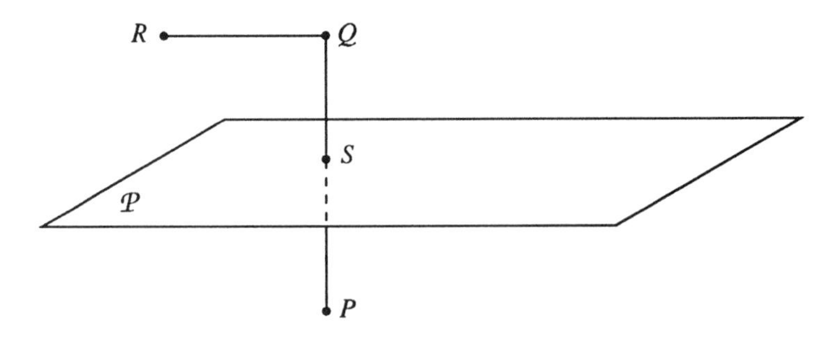

Figure 11.23 Opposite sides of a plane.

Exercises 11.4

For Exercises 1–10, complete the given statement by filling in the blank(s).

1. Every point of a line is the endpoint of exactly ______ ray(s) of that line.

2. The boundary of a half-line is a ______; the boundary of a half-plane is a ______; the boundary of a half-space is a ______.

3. The union of a half-line and its boundary point is called a ______.

4. The union of two opposite rays is a ______, and their intersection is a ______.

5. Let R, S, and T be distinct collinear points. If R is not between S and T and if S is not between R and T, then ________.

6. If F, G, and H are distinct points and none of them is between the other two, then you know that ________.

7. If points A, B, C, and line $\mathbf{l}$ are coplanar, if A and B are on opposite sides of $\mathbf{l}$, and if B and C are on opposite sides of $\mathbf{l}$, then A and C are _______.

8. If points A, B, C, and line $\mathbf{l}$ are coplanar, if A and B are on the same side of $\mathbf{l}$, and if B and C are on the same side of $\mathbf{l}$, then A and C are _______.

9. If points A, B, C, and line $\mathbf{l}$ are coplanar, if A and B are on opposite sides of $\mathbf{l}$, and if B and C are on the same side of $\mathbf{l}$, then A and C are _______.

10. If points A, B, C, and line $\mathbf{l}$ are coplanar, and if no two of these points are in the same half-plane, then at least one of the points must be _______.

11. Answer *all* the questions on the sixth-grade text page shown in Figure 11.21.

Exercises 12–14 refer to the diagram associated with Questions 1–11 of the sixth-grade text page shown in Figure 11.21.

12. What is the union of $\overline{SR}$ and $\overline{RD}$?

13. What is the union of $\overrightarrow{SR}$ and $\overrightarrow{RD}$?

14. What is the union of $\overline{SR}$ and $\overrightarrow{RD}$?

Exercises 15–19 refer to Figure 11.24.

15. Name four points on the A-side of $\mathbf{l}$.

16. Name four points on the C-side of $\mathbf{t}$.

17. Name two points in the intersection of the A-side of $\mathbf{l}$ and the C-side of $\mathbf{t}$.

18. Name a point of the plane that is not on either half-plane determined by $\mathbf{l}$.

19. Name a point of the plane that is on opposite sides of both $\mathbf{t}$ and $\mathbf{l}$ from A.

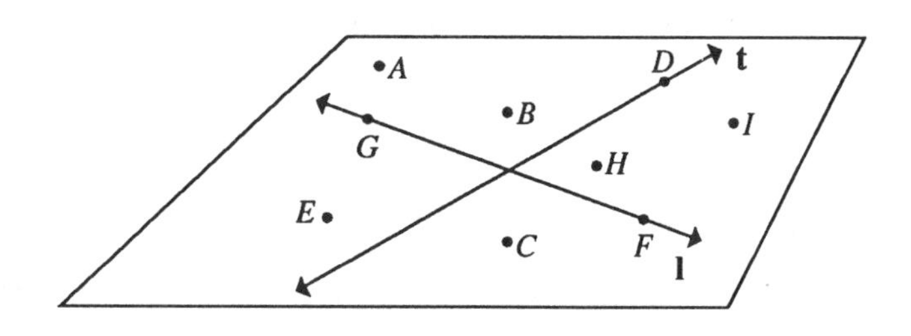

Figure 11.24 Exercises 15–19.

Exercises 20–24 refer to Figure 11.25.

20. Name two points on the A-side of $\mathcal{P}$.

21. Name two points on the F-side of $\mathcal{P}$.

22. Name two points that are in neither half-space determined by $\mathcal{P}$.

23. What is $\overline{FB} \cap \mathcal{P}$?

24. What is $\overline{AB} \cap \mathcal{P}$?

25. Given three noncollinear points, how many rays have one of these points as an endpoint and contain one of the other two points?

26. Given n points, no three of which are collinear, how many rays have one of these points as an endpoint and also contain one of the other $n - 1$ points? (*Hint*: Try some problem-solving strategies. Can you solve a simpler problem?...construct examples?...look for patterns?...)

27. The concepts of ray and segment are natural outgrowths of Postulate P-9. Sketch and/or describe what you believe to be the analogous concepts arising from Postulates P-12 and P-13.

28. In Postulates P-9, P-10, and P-11, replace the word "line" by "circle." Which of these three new statements are true?

29. Replace the words "line" and "plane" in Postulate P-12 by "great circle" and "sphere," respectively. Is this new statement true? (*Note*: On a sphere of radius r, a *great circle* is a circle of radius r, the largest possible circle that can be drawn on the sphere.)

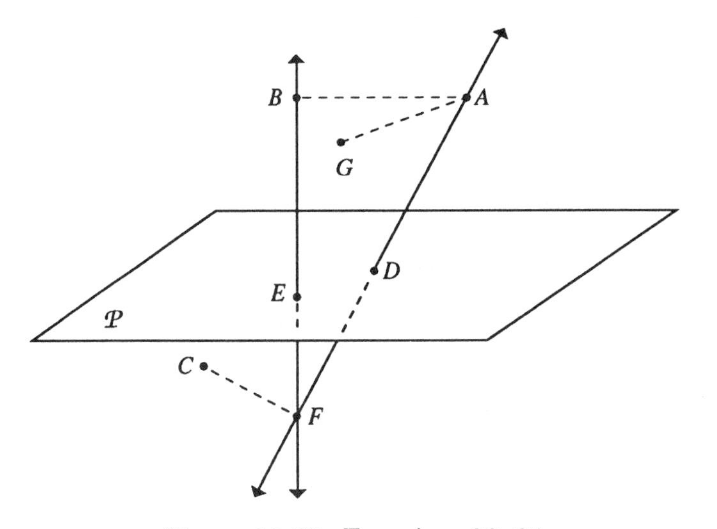

Figure 11.25 Exercises 20–24.

11.5 Angles, Polygons, and Polyhedra

The concepts of incidence, separation, and betweenness enable us to describe some of the common geometric shapes as subsets of planes or of 3-dimensional space. Angles are among the most basic of these shapes; they are the building blocks for polygons and polyhedra.

Definition An **angle** is the union of two rays with a common endpoint. The common endpoint is called the **vertex** and the rays are called the **sides** of the angle.

An angle is usually symbolized by three letters that correspond to three points on the angle, the middle letter naming the vertex and the other two (in conjunction with the middle one) specifying the rays. Sometimes, if it is clear from the context, only the letter for the vertex is used. In either case, the letter or letters are preceded by the angle sign, $\angle$. For example, the angle in Figure 11.26 can be written as $\angle BAC$ (*read*: "angle BAC") or as $\angle A$ (*read*: "angle A"). $\angle CAB$ also represents this angle. Point A is the vertex; rays $\overrightarrow{AB}$ and $\overrightarrow{AC}$ are the sides.

The definition of angle includes the possibility that the rays may be on the same line. This may happen in two ways that look the same in a diagram, but are quite different geometrically; both are shown by Figure 11.27. $\angle PQR$ is an angle whose sides are opposite rays; such an angle is called a **straight angle.** $\angle QPT$ is an angle whose sides ($\overrightarrow{PQ}$ and $\overrightarrow{PT}$) are coincident; such an angle is called a **zero angle**.

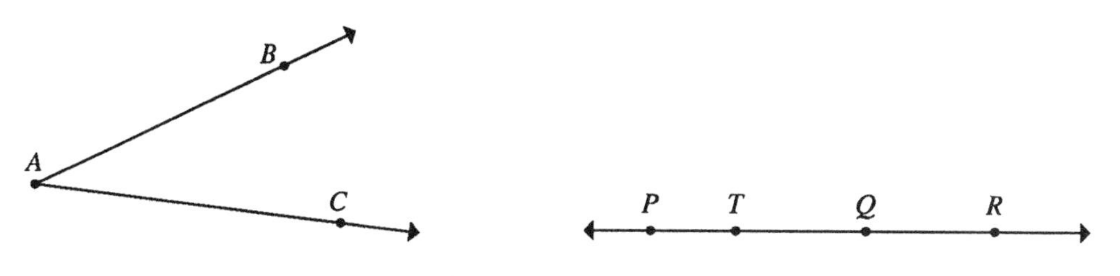

Figure 11.26 Angle BAC.

Figure 11.27 Straight and zero angles.

In our definition of *angle* there are no restrictions on the two rays other than that they have a common endpoint, so the definition includes straight angles and zero angles. In some texts, straight angles and/or zero angles are excluded by the definition of an angle. Some books require the rays to be "non-opposite," which excludes straight angles but not zero angles; others require the rays to be noncollinear, which excludes both of these special types. The choice of whether or not to include these types under the general definition of *angle* is just a matter of personal preference; there are good reasons for each approach. One reason for including them is that they are important geometric objects encountered in the formal study of figures. The main reason for excluding them is that many subsequent definitions and properties are not valid when applied to such angles; the next definition is a good example of this.

A straight angle is, of course, simply a line. Postulate P-12 tells us that this figure separates any plane containing it into two disjoint subsets. It is not hard to see that any other angle except the zero angle also separates its plane into two disjoint subsets. In the case of the straight angle, the two planar regions are both half-planes, but for any other angle they are not; in all other cases, one of the regions is "inside" the angle and the other is "outside." Formally:

Definition If $\angle ABC$ is an angle with noncollinear sides, then the **interior** of $\angle ABC$ is the intersection of the C-side of $\overleftrightarrow{BA}$ with the A-side of $\overleftrightarrow{BC}$. The **exterior** of $\angle ABC$ is the set of all points in the plane of the angle that are neither in the interior nor on the angle.

Example 1 In Figure 11.28, the C-side of line $\overleftrightarrow{BA}$ is shaded horizontally; the A-side of $\overleftrightarrow{BC}$ is shaded vertically. The interior of $\angle ABC$ is the intersection of these two half-planes and is shown by the cross-hatch shading. Point X is in the interior, Y is in the exterior, and Z is on the angle. □

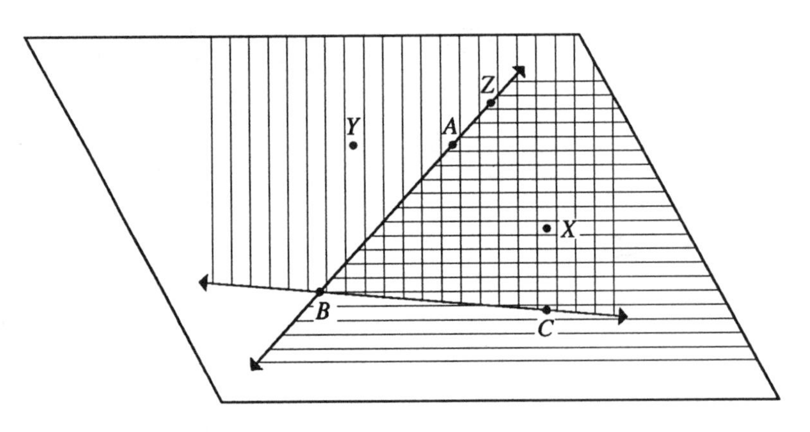

Figure 11.28 The interior and the exterior of an angle.

There is another important class of subsets of a plane that separate the plane into three disjoint parts. These subsets are called **simple closed curves**. It is not an easy task to give a precise definition of such figures, and we shall not even try to do so here, settling instead for an intuitive description of the concept.

The prototype of the simple closed curve is a circle. (In this chapter, we discuss circles informally, presuming that you are familiar with them from prior experience. A more precise discussion of circles appears in Chapter 12.) Consider the circle in Figure 11.29, with four of its points labelled W, X, Y, and Z. To form other simple closed curves from this figure, we may distort the circle in certain ways, but not in others. We are allowed to stretch, shrink, or bend the curve, provided that we do not change the

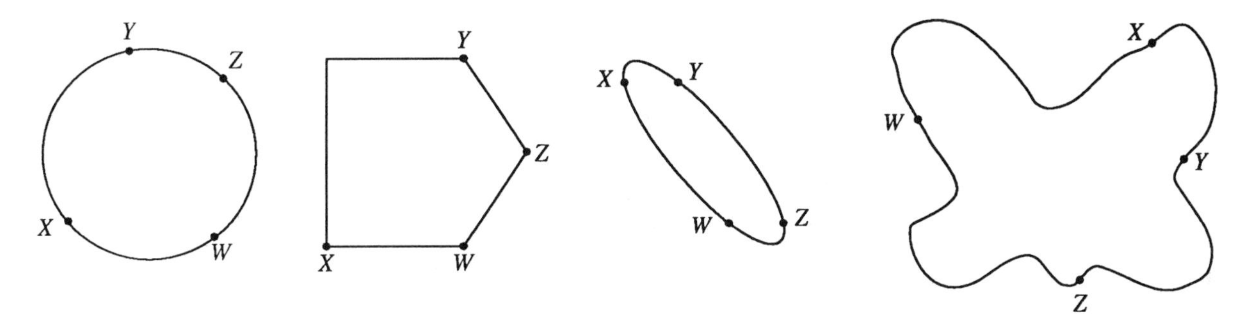

Figure 11.29: Simple closed curves.

order of the points W, X, Y, and Z or the separation properties of the circle (that is, its separation of the plane into two disjoint subsets). Conversely, a planar figure is a simple closed curve if it can be stretched, shrunk, or bent into a circle. We are not permitted to "break" or "tear" the curve, nor can we distort it in such a way that distinct points are made to coincide.

Example 2 All the curves in Figure 11.29 are simple closed curves. Notice that, beginning at any point, you can trace the entire curve, ending where you began, without crossing it. □

Example 3 None of the curves in Figure 11.30 are simple closed curves. The first curve is not closed because there is a "break" in it. In all the rest, two points coincide. All but the fourth curve violate the separation property. In general, a tracing of a simple closed curve (starting from anywhere on it and ending at the same point) must never cross itself. □

Definitions A **polygon** is a simple closed curve consisting of n points P_1, P_2, P_3, ..., P_n (the **vertices**) and of n line segments $\overline{P_1P_2}$, $\overline{P_2P_3}$, $\overline{P_3P_4}$, ..., $\overline{P_nP_1}$ (the **sides**), where $n \geq 3$ and no three consecutive points are collinear. Any

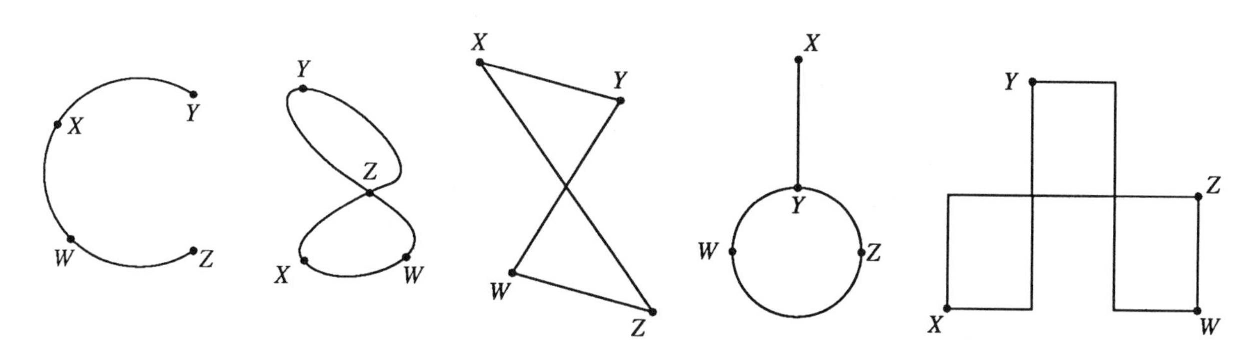

Figure 11.30 None of these are simple closed curves.

segment that has vertices as endpoints but is not a side is called a **diagonal** of the polygon. A **triangle** is a polygon with three sides.

Notation A polygon is usually denoted by listing its vertices consecutively (starting at any vertex). The 3-vertex notation for a triangle is distinguished from the label for an angle by preceding it with the symbol $\triangle$. Thus, a triangle whose vertices are A, B, and C is written $\triangle ABC$.

Example 4 All the shapes in Figure 11.31 are polygons. The first one is a triangle, and may be represented as $\triangle ABC$ or $\triangle BAC$ or $\triangle BCA$, etc. The second polygon in Figure 11.31 can be written as $EFGHIJ$; two of its diagonals are $\overline{GE}$ and $\overline{FI}$. □

Every polygon has as many sides as it has vertices. Each vertex of a polygon together with the pair of sides that share that vertex form a subset of an angle, usually called an **angle of the polygon**. A polygon also has as many angles as it has vertices (or sides).

Just as there is a special name for 3-sided polygons, polygons with more than 3 sides are often named by words that refer to the number of sides or angles they contain. Some of the common names are given in the following list. With the exception of the first one, which is Latin, the names are Greek derivatives. In particular, the suffix *-gon* comes from the Greek words for *angle* and *knee*. The prefix *poly-* means *many*, so a "poly-gon" is literally a figure of many angles.

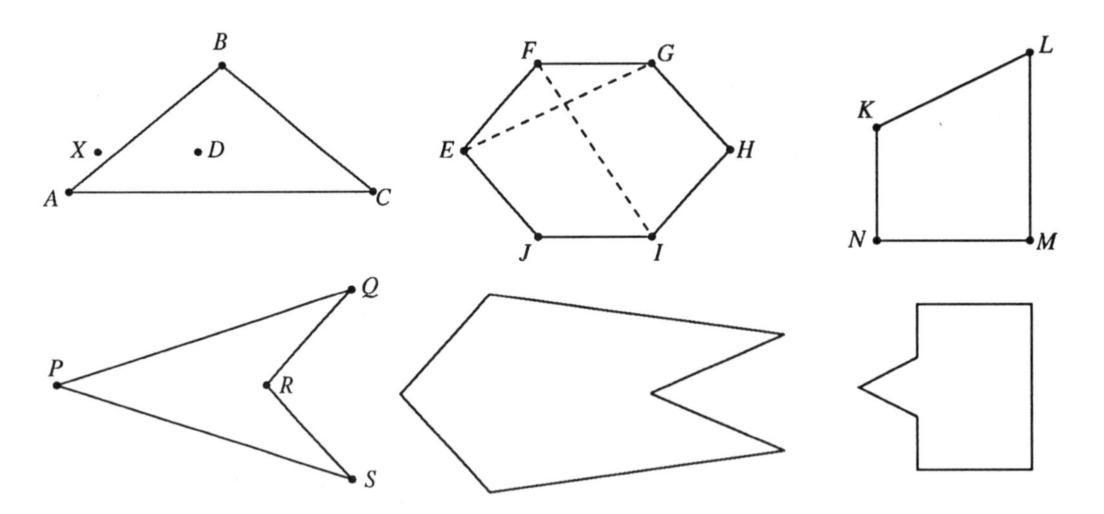

Figure 11.31 Polygons.

Number of sides	Name
4	quadrilateral
5	pentagon
6	hexagon
7	heptagon
8	octagon
9	nonagon
10	decagon
n	n-gon

Parallelograms, rectangles, squares, rhombuses, and trapezoids are important special kinds of quadrilaterals. We can only define two of them formally at this time. The others will be defined in Chapter 12, after we have introduced the concept of a *measure*.

Definitions Two sides of a quadrilateral are **adjacent** if they share a common vertex; the sides are **opposite** if they are not adjacent.

Example 5 In the quadrilateral $KLMN$ of Figure 11.31, $\overline{KL}$ and $\overline{LM}$ are adjacent sides, as are $\overline{KL}$ and $\overline{KN}$. Sides $\overline{KN}$ and $\overline{LM}$ are opposite, as are $\overline{KL}$ and $\overline{MN}$. □

Definitions Two sides of a polygon are **parallel** if the lines that contain them are parallel lines. A **trapezoid** is a quadrilateral with exactly one pair of opposite sides parallel; a **parallelogram** is a quadrilateral with both pairs of opposite sides parallel.

Example 6 Quadrilateral $ABCD$ in Figure 11.32 is a trapezoid with parallel opposite sides $\overline{AB}$ and $\overline{CD}$. Quadrilateral $EFGH$ is a parallelogram; $\overline{EF}$ and $\overline{GH}$ are parallel opposite sides, as are $\overline{EH}$ and $\overline{FG}$. □

Some polygons possess an interesting property known as *convexity*. Intuitively, a figure is *convex* if any two points inside it can be joined by a

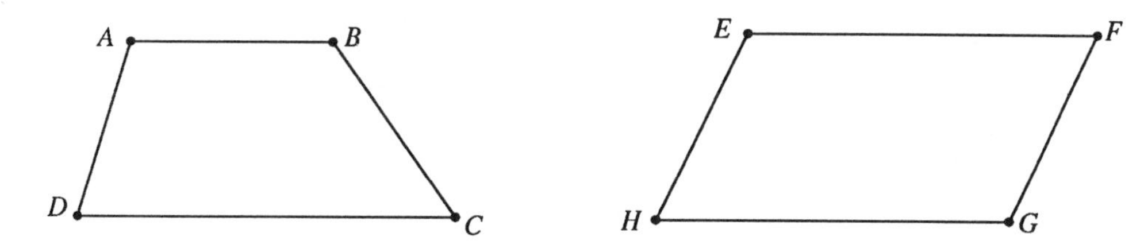

Figure 11.32 A trapezoid and a parallelogram.

straight line segment that lies completely inside it. Every triangle is convex, but a quadrilateral may or may not be convex, as Figure 11.31 shows. (See Example 7.) The formal definition of convexity avoids the still-undefined notion of the "inside" of a figure.

Definition A polygon is **convex** if each side determines a half-plane that includes all the other sides of the polygon. Polygons that are not convex are said to be **concave.**

Example 7 The first three polygons in Figure 11.31 are convex; the other three are concave. In quadrilateral $PQRS$, sides $\overline{PQ}$ and $\overline{PS}$ determine half-planes that contain all the other sides, but $\overline{QR}$ and $\overline{RS}$ do not. □

Example 8 Figure 11.33 shows the types of polygons studied in the sixth grade. (Which of these polygons is concave?) □

As is the case with a line or an angle, a simple closed curve in a plane separates the plane into two disjoint sets, the interior and the exterior, with the curve forming the boundary. Formal, general definitions of the interior and the exterior of a simple closed curve are beyond the scope of this course, but we can give an easy characterization of these regions for convex polygons. In this case, the *interior* of the polygon is the intersection of the interiors of all the angles of the polygon; the *exterior* is the set of all points in the plane that are neither in the interior nor on the polygon itself. The union of a polygon and its interior is called a **polygonal region**. The union of a circle with its interior is called a **circular region** or a **disk**.

Example 9 In Figure 11.31, points A, B, and C are on the triangle, point D is in the interior, and point X is in the exterior. Points A, B, C, and D are in the triangular region. □

Up to now we have only considered figures that are subsets of a plane. We next examine some figures in 3-dimensional space. The 3-dimensional analogue of the polygon is the *polyhedron*, a figure of "many surfaces." It is a special type of **simple closed surface**. The prototype of a simple closed surface is the sphere. (You might think of a sphere as the covering of a tennis ball or as a ping-pong ball, a round shell with nothing inside it.) Any distortion that does not break or tear the sphere—such as bending, shrinking, stretching, or twisting—produces a simple closed surface. Hollow cones, pyramids, cylinders, and cubes are examples of simple closed surfaces. These shapes are shown in Figures 11.34 and 11.35.

Polygons

A. A *polygon* is a figure whose sides are all line segments.

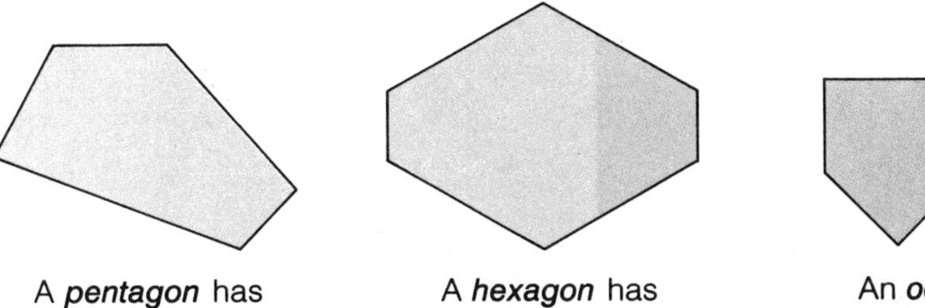

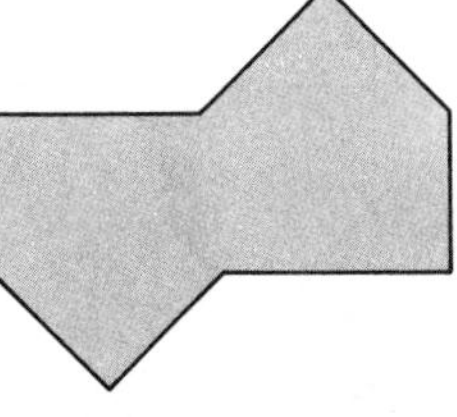

A ***pentagon*** has 5 sides.

A ***hexagon*** has 6 sides.

An ***octagon*** has 8 sides.

B. A triangle is a polygon with 3 sides.

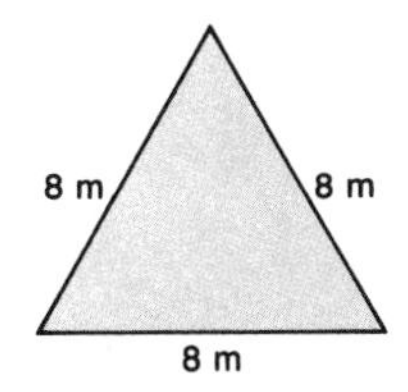

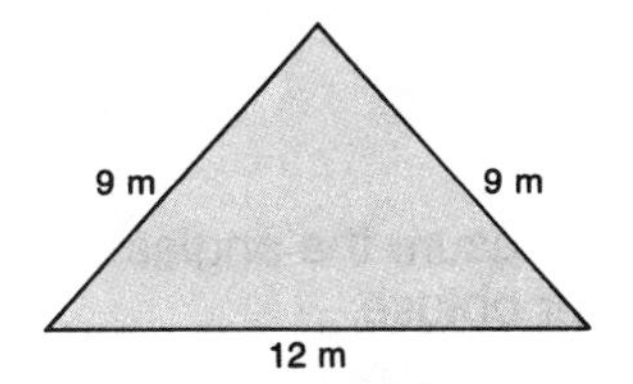

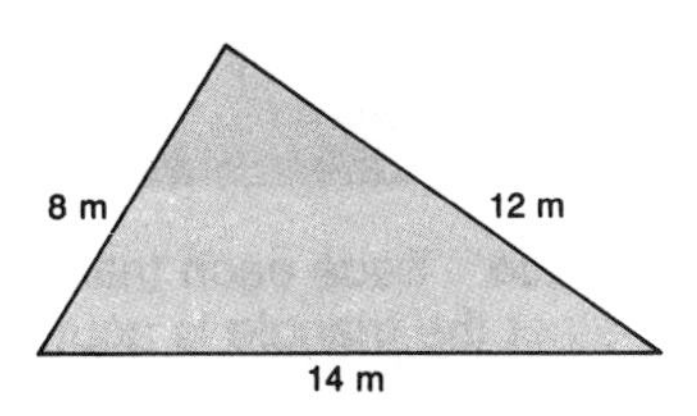

An ***equilateral triangle*** has 3 congruent sides.

An ***isosceles triangle*** has at least 2 congruent sides.

A ***scalene triangle*** has no congruent sides.

C. A polygon with 4 sides is a ***quadrilateral***. Some quadrilaterals have special names.

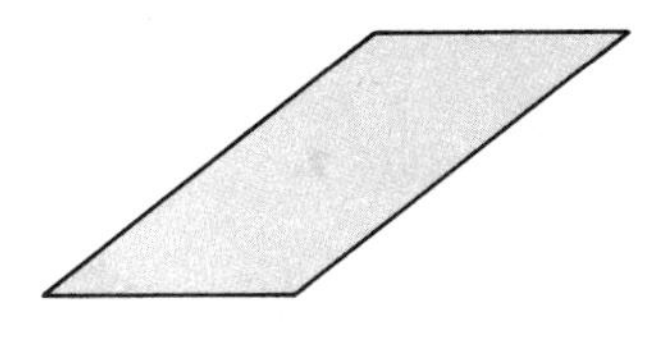

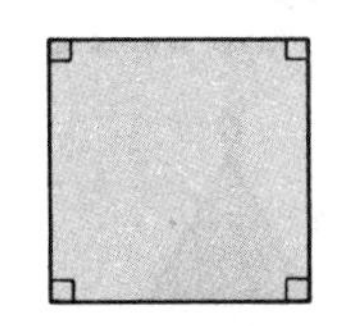

A ***parallelogram*** is a quadrilateral with opposite sides parallel.

A ***rectangle*** is a parallelogram with 4 right angles.

A ***square*** is a rectangle with 4 congruent sides.

Discuss Is a square a parallelogram?

Figure 11.33 INVITATION TO MATHEMATICS, Grade 6, p. 278

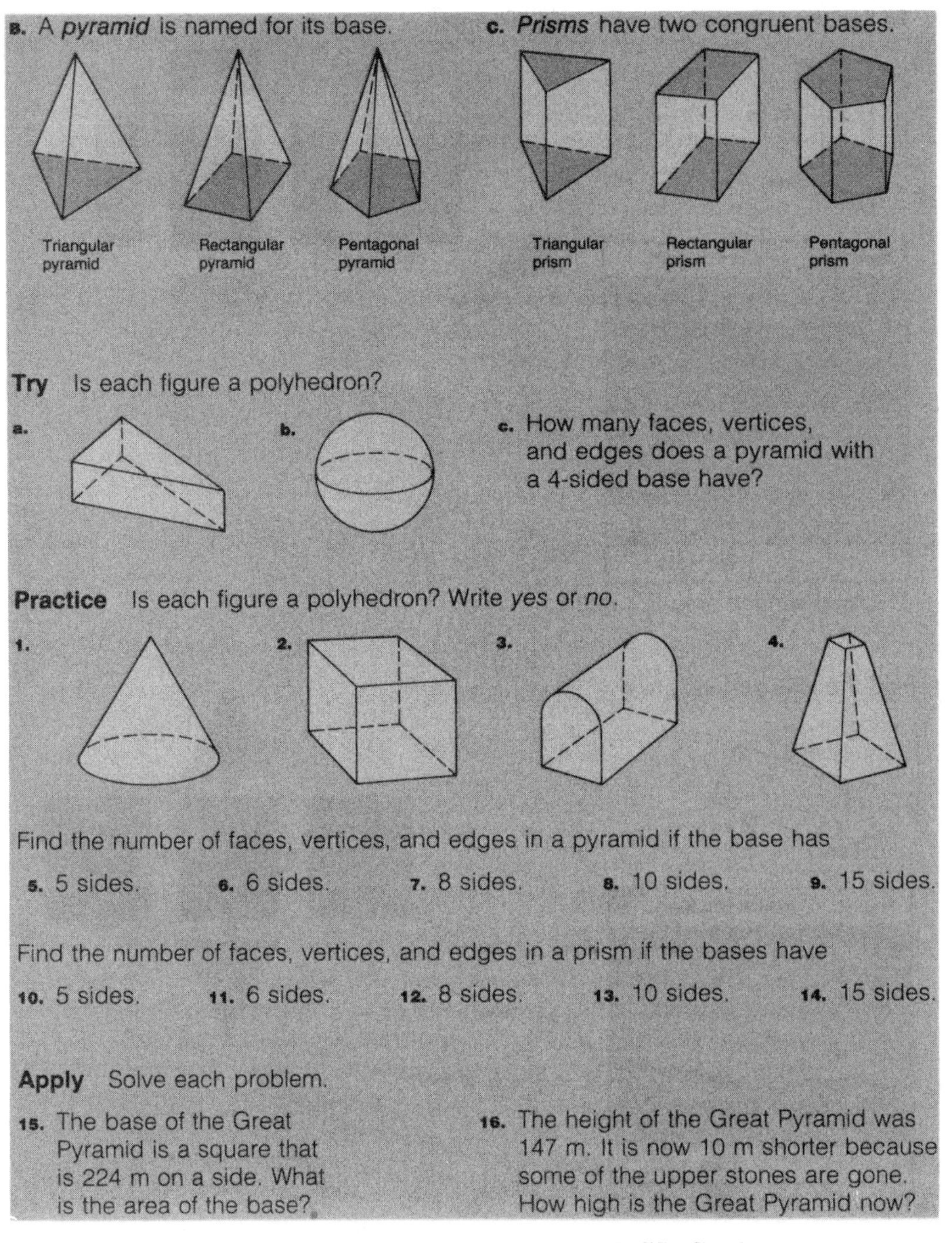

B. A *pyramid* is named for its base.

c. *Prisms* have two congruent bases.

Try Is each figure a polyhedron?

a.

b.

c. How many faces, vertices, and edges does a pyramid with a 4-sided base have?

Practice Is each figure a polyhedron? Write *yes* or *no*.

1. **2.** **3.** **4.**

Find the number of faces, vertices, and edges in a pyramid if the base has

5. 5 sides. **6.** 6 sides. **7.** 8 sides. **8.** 10 sides. **9.** 15 sides.

Find the number of faces, vertices, and edges in a prism if the bases have

10. 5 sides. **11.** 6 sides. **12.** 8 sides. **13.** 10 sides. **14.** 15 sides.

Apply Solve each problem.

15. The base of the Great Pyramid is a square that is 224 m on a side. What is the area of the base?

16. The height of the Great Pyramid was 147 m. It is now 10 m shorter because some of the upper stones are gone. How high is the Great Pyramid now?

Figure 11.34 INVITATION TO MATHEMATICS, Grade 6, p. 295

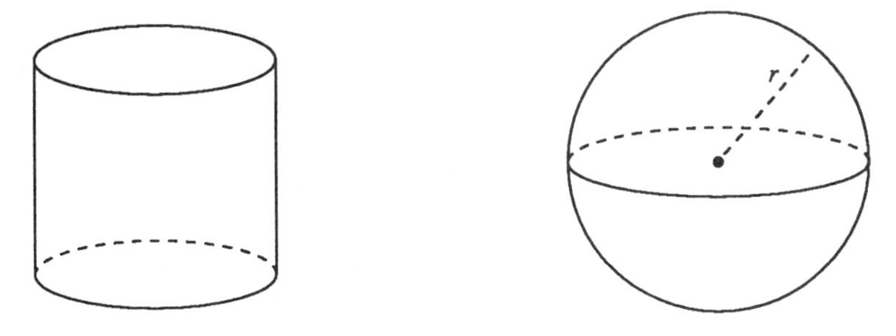

Figure 11.35 A cylinder and a sphere.

PROBLEM-SOLVING COMMENT

As you study the rest of this section, observe the ways in which many of the 3-dimensional concepts are *analogous* to corresponding ideas in the plane. For instance, the sphere is analogous to the circle, not only in the way these figures are usually defined, but also because they are the most typical simple closed surfaces of their dimensions. For each of the figures in 3-dimensional space that we define, try to find (or formulate a definition of) an analogous figure in the plane. ◇

Definitions

A **polyhedron** is a simple closed surface that is the union of a finite number of polygonal regions. The polygonal regions that have sides in common and are noncoplanar are called **faces** of the polyhedron, the common sides are called **edges**, and the common points of two or more edges are called **vertices**.

Note: The traditional plural form of "polyhedron" is "polyhedra," reflecting the Greek origin of the word. It is grammatically like the words "phenomenon" [an observable thing] and "phenomena" [observable things], also from ancient Greek. Recent books sometimes use "polyhedrons," acceptable in English, but not faithful to the etymology.

A simple closed surface along with all the points of space that lie "inside" it is called a **solid** figure. Thus, a solid figure in three-dimensional space is analogous to a region in a plane. Often the names for the polyhedra are also used for the corresponding solids, leaving the distinction to the context. For instance, a **cube** refers both to a polyhedron composed of six mutually perpendicular squares and to the solid figure enclosed by these six squares.

Here are the definitions of some other polyhedral forms commonly discussed in elementary-school mathematics. A **pyramid** is a polyhedron with one face (the **base**) a polygon and the other faces (the **lateral faces**) triangles with a common vertex. (See Part B of Figure 11.34.) A **prism** is a polyhedron with two parallel faces of the same size and shape (the **bases**), whose other faces (the **lateral faces**) are parallelograms formed by joining corresponding vertices of the bases. (See Part C of Figure 11.34.)

As Figure 11.34 shows, pyramids and prisms are named according to the shapes of their polygonal bases. The famous pyramids of Egypt are square pyramids. In elementary physics, the glass prism through which light is refracted is usually a triangular prism; the shape of a shoebox is a rectangular prism. Since a polygon is a simple closed curve, it can be deformed into a circle without breaking or tearing it. If you deform the polygonal base of a pyramid into a circle in this way and "smooth out" the edges between the lateral faces, the resulting figure is a **cone**. (See Practice Figure 1 of Figure 11.34.) Similarly, if you deform both bases of a prism so that they are circles of the same size and shape and also "smooth out" the edges between its lateral faces, the resulting 3-dimensional figure is a **(right circular) cylinder**. (See Figure 11.35.) The 3-dimensional analogue of a circle is a **sphere**, the set of all points in space that are a fixed distance (the radius) from a particular point (the center). (See Figure 11.35.)

Polyhedra are usually classified according to the number of faces they have. It is not possible for a polyhedron to have fewer than four faces, so the first type consists of the four-sided figures. Each of these figures is called a **tetrahedron**, again using Greek word forms. (Literally, *tetra-* means *four*, and *-hedron* means *side* or *base*.) The first pyramid in Part B of Figure 11.34 is a tetrahedron; the others are not. Similarly, a **pentahedron** is a five-sided polyhedron and a **hexahedron** is a six-sided polyhedron. A cube is an example of a hexahedron. You will be asked in an exercise to identify the pentahedra and hexahedra in Figure 11.34.

Exercises 11.5

Exercises 1–9 refer to Figure 11.36 on page 428.

1. Name four angles that are not straight angles.
2. Name a pair of angles that share a common side and a common vertex but have disjoint interiors. (Such angles are called **adjacent**.)
3. Name a pair of nonadjacent angles that are formed by two intersecting lines. (Such angles are called **vertical**.)
4. Name two straight angles.
5. Name two zero angles.
6. Name a point in the interior of $\angle XPS$.
7. Name all the labeled points in the exterior of $\angle XPS$.
8. Name the sides of $\angle XPS$.
9. Name two angles that are neither adjacent nor vertical.

HISTORICAL NOTE: THE PLATONIC BODIES

From the viewpoint of symmetry, the "nicest" polygons are those in which all sides and all angles are congruent; they are called **regular**. For example, a regular triangle is one that is equilateral; a regular quadrilateral is a square. There are regular polygons with any number of sides (although some are harder to construct than others). Similarly, in 3-dimensional space, a polyhedron is *regular* if all its faces are congruent regular polygons. Thus, for example, a cube is a regular polygon; all its faces are squares of the same size.

It is a remarkable fact of geometry, proved as the last proposition of Euclid's *Elements*, that there are *not* regular polyhedra with any number of faces; in fact, there are exactly five different types:

tetrahedron — 4 faces (triangles)
hexahedron — 6 faces (squares)
octahedron — 8 faces (triangles)
dodecahedron — 12 faces (pentagons)
icosahedron — 20 faces (triangles)

The regularity of these figures implies that each one can be *inscribed* in a sphere; that is, it can be placed inside a spherical shell in such a way that *all* its vertices "touch" (are points of) the sphere. This fact was of great significance to the Pythagoreans, who regarded the sphere as the most perfect 3-dimensional figure. It appears that Pythagoras may have learned of the tetrahedron, the cube, and the octahedron from the Egyptians. In studying these figures, the Pythagoreans discovered the (regular) icosahedron. These four figures were used to represent the four "elements" of the physical world:

fire — tetrahedron
earth — hexahedron (cube)
air — octahedron
water — icosahedron

Later the Pythagoreans discovered the regular dodecahedron, which they then used to represent the universe. Plato (in *Timaeus*) used this representation of the universe and its elements as the foundation for an elaborate theory of matter in which everything is composed of right triangles.[1] Because of this theory, these five polyhedra are known as "the Platonic solids" or "the Platonic bodies."

Plato's triangular theory of matter has not stood the test of time, but the Platonic bodies are still to be found in the earth's elements. The crystalline structures of lead ore and rock salt are cubic; fluorite forms octahedral crystals; garnet forms dodecahedral crystals; iron pyrite crystals come in all three of these forms—and the basic crystalline form of the silicates (which form about 95% of the rocks in the Earth's crust) is the smallest of the regular triangular solids, the tetrahedron. ◇

[1]Carl B. Boyer, *A History of Mathematics*, New York: John Wiley & Sons, 1968, p. 97.

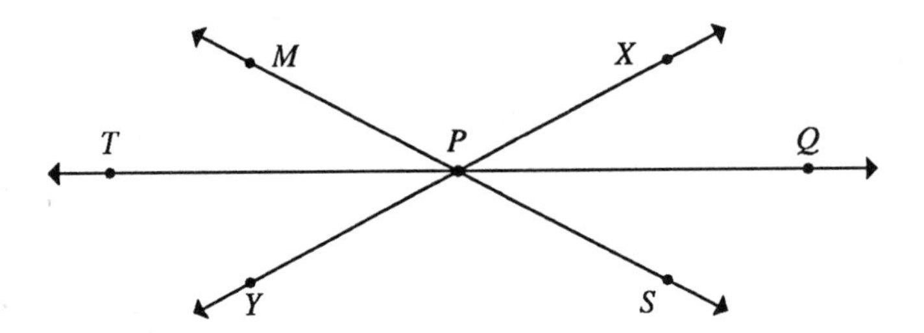

Figure 11.36 Exercises 1–9.

10. (a) Draw a diagram showing three noncollinear points P, Q, and S, and also showing $\angle PQS$.

(b) In your diagram for part (a), draw $\overline{PS}$; then choose any point X such that P-X-S and draw $\overrightarrow{QX}$. Let Y be any point of $\overrightarrow{QX}$. Is Y in the interior of $\angle PQS$?

11. Which parts of Figure 11.37 represent simple closed curves? For those that do not, explain why not.

12. Classify the curves in Figure 11.38 as *polygons* or *nonpolygons*. Classify the polygons as *convex* or *concave*. For each polygon, indicate the number of vertices and the number of sides. Name the type of polygon it is, if possible.

13. Draw convex polygons having 4, 5, 6, 7, and 8 sides; then draw all their diagonals.

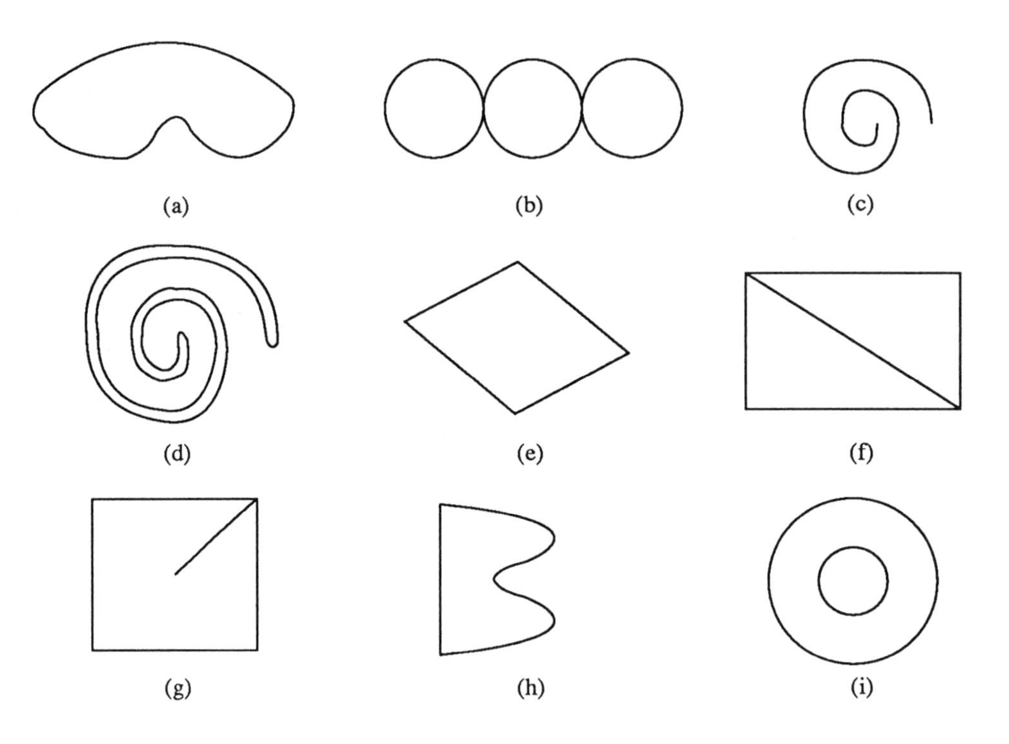

Figure 11.37 Exercise 11.

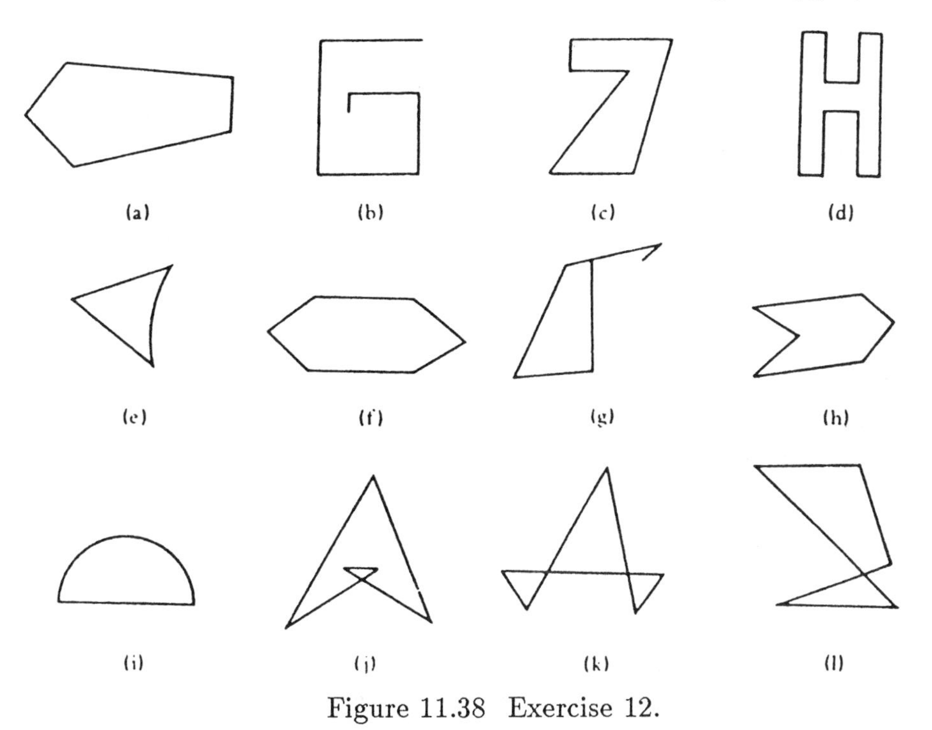

Figure 11.38 Exercise 12.

14. Draw concave polygons having 4, 5, 6, 7, and 8 sides; then draw all their diagonals.

15. Judging from your drawings for Exercises 13 and 14, what appears to be true for diagonals of convex polygons, but false for diagonals of concave polygons?

Exercises 16–20 refer to the sixth-grade text page shown in Figure 11.34.

16. Answer *all* the questions on the page.

17. Which of the diagrams represent tetrahedra? pentahedra? hexahedra? Do any of the polyhedra appear to be regular? If so, which ones?

18. Cover up the names of the polyhedra at the top of the page; then classify by base type the pyramids and the prisms.

19. For each of the six polyhedra at the top of the page, count the number of vertices (V), the number of faces (F), and the number of edges (E), and complete the following table:

Polyhedron	F	V	E
Triangular pyramid	4	4	6
Rectangular pyramid			
Pentagonal pyramid			
Triangular prism			
Rectangular prism			
Pentagonal prism	7	10	15

20. Leonhard Euler (pronounced "oiler") (1707–1783) discovered a relationship among the numbers of vertices, faces, and edges that is true for any polyhedron. He expressed this relationship as an equation, now called **Euler's Formula**. Can you discover this relationship? Use the table you constructed in Exercise 19; then test your conjecture with the remaining polyhedra in Figure 11.34, and also with your answers to Practice Exercises 5–14 of that figure.

Review Exercises for Chapter 11

For Exercises 1–10, indicate whether the given statement is *true* or *false*.

1. The earliest study of geometry is credited to the Greeks.
2. In modern geometry, no attempt is made to define the terms *point*, *line*, and *plane*.
3. There is very little difference between a postulate and a theorem.
4. If one point of a line **l** is in a plane $\mathcal{P}$, then all points of **l** are in $\mathcal{P}$.
5. If two lines do not intersect, then they must be parallel.
6. A line segment is composed of two distinct points and all the points between these two points.
7. "On the same side of a line" is an equivalence relation on the set of all points in a plane.
8. The union of any two rays forms an angle.
9. All prisms have triangular bases.
10. There are only five different types of regular polyhedra.

For Exercises 11–15, sketch and label a diagram to show a line **t** and points P, Q, and R that satisfy the given condition(s).

11. $\mathbf{t} \neq \overleftrightarrow{PQ}$, $P \in \mathbf{t}$, $R \in \mathbf{t}$
12. $\mathbf{t} \cap \overline{PQ}$, $R \in \mathbf{t}$, $R \notin \overline{PQ}$
13. $\mathbf{t} \cap \overrightarrow{PQ} = \emptyset$, $R \in \overline{PQ}$
14. $\mathbf{t} \cap \overline{PQ} = \emptyset$, $\mathbf{t} \cap \overleftrightarrow{PQ} = \{R\}$
15. $\mathbf{t} \cap \angle PRQ = \{P, Q\}$

In Exercises 16–20, complete the given statement by filling in the blank(s).

16. Every point of a line separates the line into two disjoint subsets, called two _______.
17. A point P is between points Q and R if P, Q, and R are _______, and Q and R are _______.
18. If three distinct points are collinear, then _______.
19. Every _______ of a plane separates the plane into two disjoint subsets, called two half-planes.
20. An angle is the _______ of two _______ with a common _______.

For Exercises 21–25, draw a sketch of each figure.

21. A convex quadrilateral.
22. A concave quadrilateral.
23. A convex hexagon.
24. A concave octagon.
25. A convex pentagon with two parallel sides.

For Exercises 26–31, determine the numbers of faces, edges, and vertices in each polyhedron.

26. A hexahedron.
27. A triangular prism.
28. A hexagonal prism.
29. A square pyramid.
30. An octagonal pyramid.
31. An octahedron. (Think of two square pyramids glued together at their bases.)

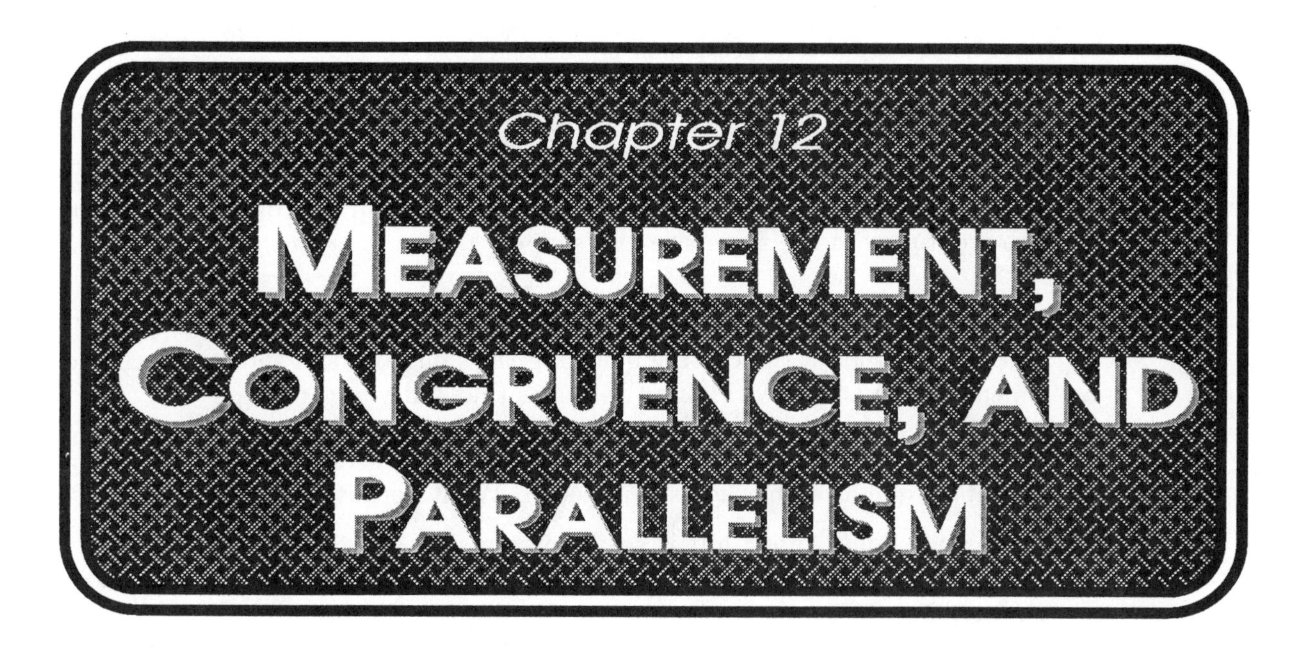

12.1 Linear Measure

Note: As we extend our discussion of geometry to the ideas of measurement, congruence, and parallelism, our presentation varies from an informal, intuitive development of ideas to a somewhat more rigorous one, depending on the topic at hand. In selected situations, we return to the postulate/theorem format, reflecting an approach commonly found in textbooks at the junior and senior high-school levels. In all cases, the level of formality has been chosen to help you understand the kinds of questions and concepts that underlie the topic being discussed.

In studying the geometric ideas in Chapter 11, you might have noticed that one prominent high-school geometry concept was missing—congruence.

"I was wondering about that," says Terry (who is still with us). "Sometimes it seemed like all we did in high-school geometry was prove that triangles were congruent to each other—you know; SAS and ASA and all that stuff. It seemed like a pretty important idea at the time. How come we haven't seen anything about it yet?"

Question 1: Do you remember what *congruence* means?

Terry: Two triangles are congruent if one can be put on top of the other so that all their points and lines match up.

Question 2: How can you put one triangle on top of another? Aren't triangles sets of points in a plane, and aren't points specific locations?

Terry: You don't really pick up the triangles, of course. You measure them to see if they're the same size and shape.

Question 3: How do you measure the size and shape of a triangle?

Terry: You use a ruler to see how long the sides are, and something else to measure the angles with. It's the angles that tell you the shape.

Question 4: I thought the only tools Euclidean geometry allowed were a compass and an unmarked straightedge. What permits you to use a ruler?

Terry: I don't know, but it seems pretty silly not to. In real life, if you want to see how long something is, you use a ruler or a tape measure, or something.

Question 5: Where do rulers come from?

Terry: The store.

Question 6: Thanks a lot. Let me try to get at my question another way: If you and I each have a ruler and we both measure the same side of the same triangle, will we always get the same number?

Terry: Sure. Unless I'm measuring by inches and you're measuring by centimeters, or something like that.

Question 7: That's a good point, but not what I had in mind. Let's leave the question of inches versus centimeters until later, and agree for now that we are both measuring in inches. Are you sure we'll always get the same answer?

Terry: This is starting to sound like a trick question. Why wouldn't we get the same answer?

Question 8: Here's my point: If you and I buy different rulers at different stores at different times, how do we know that your inch marks and mine will match up? Have you ever bothered to check your ruler against anyone else's to see if they match?

Terry: Oh. I've never checked, but I know they all match. Like roses, an inch is an inch is an inch.

Question 9: Cute. I didn't think your generation knew about Gertrude Stein. Why are you so sure that all rulers made by all companies will match?

Terry: They have to. Isn't there some government agency, like the Bureau of Standards or something, that guarantees that sort of thing?

Question 10: So you're saying that any company that wants to make a ruler has to match its inch length with some standard inch length. Isn't that *congruence*?

Terry: I *knew* there was a catch! We're back where we started!

Question 11: Almost, but not quite. From what you've told me, it seems that congruence of triangles somehow depends on the congruence of line segments representing some standard unit of measure, like an inch or a centimeter. Is that right?

Terry: I think so. But I don't remember talking about any of this in high school. We did congruence by proving theorems, not by measuring.

Question 12: But *you* suggested measuring lengths to compare the sides of triangles. What about other figures?

Terry: I guess it works the same way, if you can measure.

Question 13: Even in three-dimensional space?

Terry: Sure. It's even better to be able to measure in that case. I don't know how to do compass-and-straightedge constructions anywhere except in a plane.

Question 14: What about comparing angles?

Terry: Well, I know you don't measure angles with a ruler, but there's a half-moon-shaped thing that works like a ruler for measuring angles.

Question 15: Yes; it's called a *protractor*. If we can measure lengths and angles, can we determine whether or not any two geometric figures are congruent?

Terry: I think so. Geometric figures don't have weight or anything like that. Size and shape are all we need to compare, and lengths and angles determine size and shape.

Rather than continue this dialogue, let us summarize and clarify some of the valuable ideas it contains. Terry remembered correctly that the congruence of geometric figures can be determined without measuring. However, the "real-world" process of comparing objects, whether they be planar (like tracts of land) or 3-dimensional (like boxes or buildings), is based on measurement. But measurement itself depends on congruence, at least with respect to the basic units of measure. Thus, one important connection between geometric theory and its physical applications is based on the relationship between congruence and measurement. The present chapter is devoted to exploring this relationship.

Measurement is the process of quantifying properties of an object by comparison with some standard unit. In geometry, the objects are sets of points, and the properties we seek to quantify are length, area, volume, and angle size. Some geometric figures have only one of these properties. For instance, a line segment only has length. A polyhedron, on the other hand, has all four properties; we can consider the length of an edge, the area of a face, the volume of its interior, or the size of the angle between two of its edges. Polyhedral objects in the physical world, such as cardboard boxes or metal blocks, have at least two other measurable properties—mass and temperature. As Terry suggests, the simplest place to start is with the lengths of line segments; that is, with **linear measure**.

In Chapters 8, 9, and 10, we made extensive use of the number line to illustrate various numerical concepts. In particular, we described the absolute value of a number as the "distance" or "number of units" between that number and zero on the number line. We extended this idea by defining the "distance" between any two rational numbers p and q as $|p - q|$, the absolute value of their difference. Now, since the numbers p and q correspond to points on the number line, say P and Q, it seems quite natural to define the distance between the *points* P and Q, or simply the **length** of segment $\overline{PQ}$, to be the distance between the *numbers* p and q; that is, $|p - q|$. This is a reasonable way to define the length of any line segment, *provided* that every point of a line can be made to correspond to some rational number, which represents the distance of that point from the zero point.

Unfortunately, that is not the case! Even when all the rational numbers are properly positioned on a number line, relative to some unit length, there will be points of the line that have no rational numbers associated with them (a fact that will be demonstrated later in this section). The ability to measure the length of any line segment, then, requires that the rational numbers be extended to a larger set of numbers (the *real numbers*) that contains the rationals. The construction of this extension is the main subject of Chapter 13; in this chapter we simply *assume* that it can be done. For now, we formulate what we need into a postulate, which we call the "Ruler Postulate." Before stating it formally, let us clear up a few details.

To measure length on a straight line, we must first choose a **unit length**; that is, we must decide what length will be assigned the number 1. This choice can be made in any way we please. The measurement theory will be the same, regardless of our choice of unit, so long as we are consistent throughout the process. We could use a foot, an inch, a meter, a mile, or we could even make up a unit, such as ————. Different units are convenient for different purposes—inches or centimeters for cabinet makers; miles or kilometers for drivers; light years for astronomers; angstroms or microns for physicists; and so on. Unit lengths are even used to represent time intervals—seconds, minutes, hours, etc. To emphasize that our theoretical work is independent of the particular choice of unit length, we shall often just use the word *unit* to denote the chosen basis for measurement.

Example 1

0 P

The distance of P from 0 on the line shown here can be described (approximately) as:

2, if the unit length is an inch;
5, if the unit length is a centimeter;
$\frac{1}{6}$, if the unit length is a foot;
3, if the unit length is ________. □

Example 1 illustrates that the numerical label given to a point on a line depends on both the choice of the unit length and the location of 0. Thus, it is necessary to distinguish between the points themselves and the numbers that correspond to them, which are called the **coordinates** of the points. The point with coordinate 0 is called the **origin**. Now we state formally the assumption that the choice of a point of origin and a unit length determines a coordinate (a numerical label) for every point on a line.

(P-14) **Ruler Postulate:** There exists a set of numbers that can be put in 1-1 correspondence with the points on a line in such a way that:

(a) the two points with coordinates 0 and 1 can be chosen arbitrarily; and

(b) the distance between any two points equals the absolute value of the difference of their coordinates.

Notation Following common practice, we use AB to represent the distance between two points A and B. In other words, the number AB is the length of the line segment $\overline{AB}$. (Be careful of this notational distinction! It is used in many elementary and secondary textbooks, but is a potential source of confusion for students.)

The Ruler Postulate does not prescribe a method for setting up the 1-1 correspondence; it just says that one can be constructed. But how? A common way is to begin by choosing a specific object and calling the length of that object a "unit length." For example, an ancient unit of measure called a *cubit* was defined to be the distance from one's elbow to the tip of one's middle finger. Of course, this varies from person to person, so it is not very exact. One way around that problem historically was to choose a king or other prominent person as the basis for the standard. As more precision became necessary, these human objects were replaced, first by carefully crafted physical objects (such as metal bars kept in a controlled environment), then by observably constant physical phenomena (such as the wavelength of a radioactive isotope).

Once an origin and a unit length have been chosen, the line is "coordinatized" by choosing a positive direction and marking off the integer points in successive unit lengths. If the line is horizontal, we usually regard *right* as the positive direction; if the line is vertical, we usually think of *up* as positive. Subdividing these unit segments into halves, thirds, quarters, tenths, etc. identifies other points of the line with rational coordinates. This subdivision process can be done in various ways; the Greeks had a geometric-construction method capable of locating the point for any rational number. Labeling the remaining points numerically poses a somewhat more delicate problem (which normally does not arise in the elementary-school classroom); we discuss that problem in Chapter 13.

In the early elementary grades, students are asked to pick some unit object (a pencil, a piece of chalk, a finger, etc.) and measure various things by marking off copies of this unit object end to end along the thing being measured. Later, such activities can be formalized with the aid of **Cuisinaire Rods**. These are wooden rods of different colors; each color corresponds to a different length, which is a multiple of the length of the shortest rod, the white one. If the white rod is chosen as the unit length, then the length of any other object can be measured with this white rod, producing integer measurements. The length of the red rod is twice the length of the white one, so if the red rod is chosen as the unit object, then we have a measurement system with integers and halves of integers. The orange rod is ten times as long as the white rod, so if the orange rod is chosen as the unit object, then we have a measurement system that includes tenths of integers. Other colors and other lengths allow for a wide range of manipulative illustrations about measurement. (You undoubtedly will see more about Cuisinaire Rods in your "methods" course; they are useful for explaining many numerical and geometric ideas.)

Notice that the activities just described are based on the notion of congruence of line segments. In a strictly formal treatment, the phrase *congruent line segments* is treated as an undefined term because it is difficult to define "copies" of segments without the notion of congruence, and just as difficult to define congruent segments without referring to measurement. We begin, instead, with measurement on a line. We assume that a chosen unit of measure can be applied uniformly to all lines everywhere, and then define two segments to be **congruent** if they have equal length. In symbols,

$$\overline{AB} \cong \overline{CD} \quad \text{if} \quad AB = CD$$

Note: The distinction between congruence of segments and equality of lengths is like the one between equivalence of sets and equality of cardinal numbers. Moreover, just as equivalent sets are *equal* only when they contain exactly the same elements, so also are congruent segments *equal* only when they contain exactly the same points.

Examples 2–5 refer to Figure 12.1, which shows a line on which some points with integer coordinates have been identified.

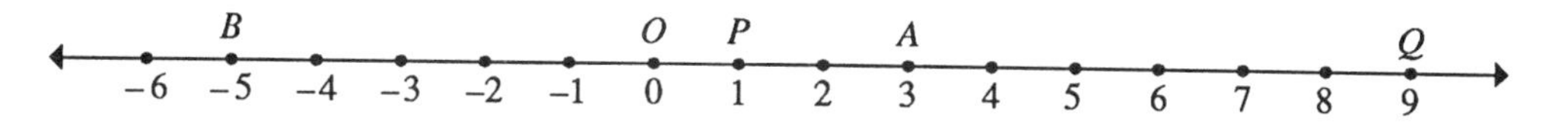

Figure 12.1 Coordinates of some points on a line.

Example 2 O is the origin and the coordinate of P is 1, so $OP = 1$. Thus, $\overline{OP}$ is a unit segment. □

Example 3 Since the coordinates of A and B are 3 and -5, respectively, $AB = |3 - (-5)| = 8$. Also, $BA = |-5 - 3| = 8$, so $\overline{AB} \cong \overline{BA}$ because their lengths are equal. (The definition of absolute value insures that the length of every line segment is positive.) These two segments contain exactly the same points, so we can also say that $\overline{AB} = \overline{BA}$. □

Example 4 Since PQ also equals 8, $PQ = AB$, so $\overline{PQ} \cong \overline{AB}$. However, $\overline{PQ}$ contains points that are not in $\overline{AB}$ (Q is one such point), so $\overline{PQ} \neq \overline{AB}$. □

Example 5 For points X and Y not specified in Figure 12.1, if you are told that $\overline{XY} \cong \overline{AQ}$, then you know that $XY = AQ = 6$. You don't know the coordinates of X and Y, but you do know that they are 6 units apart. □

Two natural and necessary assumptions underlie the process of placing copies of a unit length end to end to determine length. One is that, beginning at any point on any line, it is always possible to construct a segment congruent to a given segment. The other is that when two segments are placed end to end to cover a third segment, the length of the third segment equals the sum of the lengths of the other two. We state these more carefully as postulates.

(P-15) **Segment-Construction Postulate:** Given a segment $\overline{CD}$ and a ray $\overrightarrow{AB}$, there is a unique point E on $\overrightarrow{AB}$ such that $\overline{AE} \cong \overline{CD}$. [See Figure 12.2(a).]

(P-16) **Segment-Addition Postulate:** If B is between A and C, then $AB + BC = AC$. [See Figure 12.2(b).]

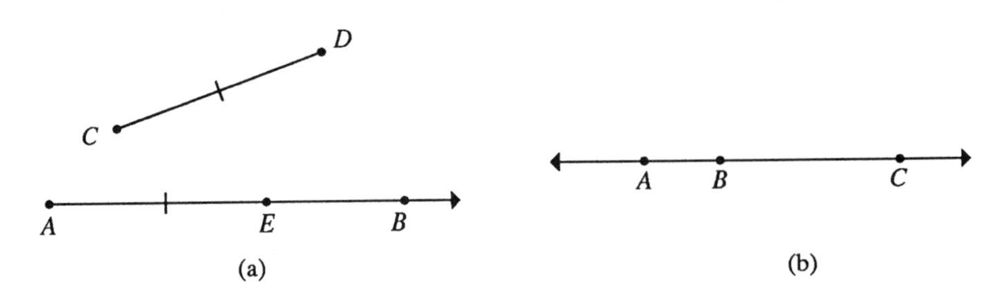

Figure 12.2 The Segment-Construction and Segment-Addition Postulates.

Example 6 In Figure 12.3, suppose that $\overline{AB} \cong \overline{CD}$ and that $CD = 3$ inches. Then $AB = 3$ inches, so $AX + XB = 3$ inches, by P-16. □

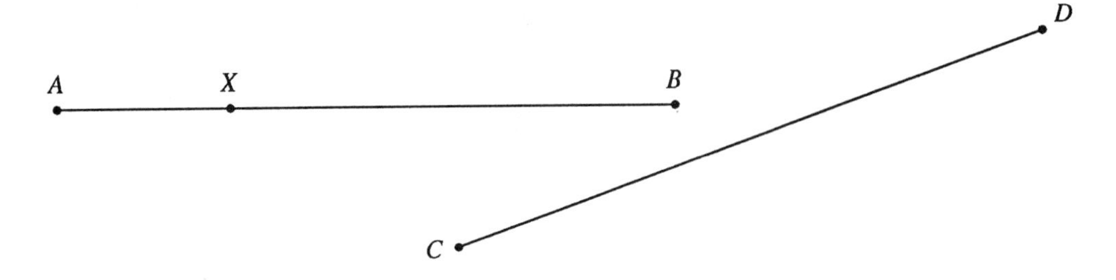

Figure 12.3 Using the Segment-Addition Postulate.

Definition If C is a set of real numbers, then a **graph** of C is the set of all points of a number line whose coordinates are in C.

Example 7 This example refers to Figure 12.1.

$\{B, O, P, A\}$ is the graph of $\{-5, 0, 1, 3\}$.
The graph of $\{x \mid 0 \leq x \leq 3\}$ is the segment $\overline{OA}$.

The graph of $\{x \mid x \geq 3\}$ is the ray $\overrightarrow{AQ}$. The graph of $\{x \mid x < 1\}$ is the half-line that is the B-side of the point P. □

Exercises 12.1

All of these exercises refer to Figure 12.4.

For Exercises 1–8, determine each length.

1. CD **2.** AC **3.** BD **4.** BC

5. CB **6.** AD **7.** AB **8.** BA

9. Find a pair of congruent line segments that are not equal.

10. Find a pair of congruent line segments that are equal.

For Exercises 11–14, identify the graph of the given set. (*Hint*: Look at Example 7.)

11. $\{x \mid -3 \leq x \leq 2\}$ **12.** $\{x \mid x \geq -1\}$

13. $\{x \mid x < 5\}$ **14.** $\{x \mid -1 < x < 5\}$

For Exercises 15–20, identify the set of coordinates for the given graph.

15. $\overline{BD}$ **16.** $\overrightarrow{AC}$ **17.** $\overrightarrow{CA}$

18. The D-side of C.

19. All of $\overline{CD}$ except for its endpoints.

20. All of $\overline{AB}$ except for A.

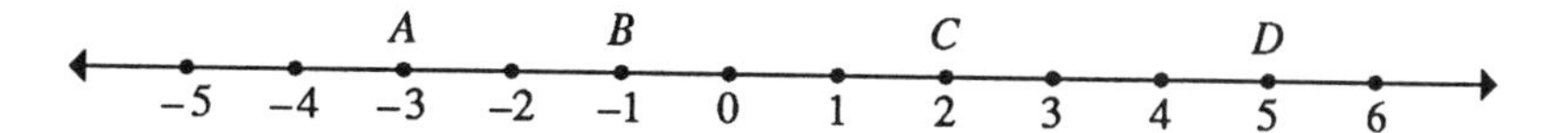

Figure 12.4 Exercises for Section 12.1.

12.2 Polygons and Circles

In Section 12.1, we saw that linear measure allows us to describe congruence of line segments in terms of the length of these segments. Now we use congruence of segments to describe some common kinds of polygons.

Definitions A polygon is **equilateral** if all of its sides are congruent (to each other). A **rhombus** is an equilateral quadrilateral.

A triangle is **isosceles** if at least two of its sides are congruent; it is **scalene** if no two of its sides are congruent.

The **perimeter** of a polygon is the sum of the lengths of its sides.

Example 1 An equilateral hexagon that is 5 units on a side has a perimeter of $6 \cdot 5 = 30$ units. In general, the perimeter p of an equilateral n-gon with each side s units long is given by $p = ns$. □

Example 2 If the lengths of the sides of a triangle are 12, 14, and 20 units, then the perimeter is $12 + 14 + 20 = 46$ units. □

The concept of linear measure also enables us to give a formal definition of a *circle* and some important related terms.

Definition A **circle** is the set of all points in a plane that are a given distance from a fixed point in that plane. The fixed point is called the **center**, and the given distance is the **radius**.

There is some ambiguity in common usage of the term *radius*. Sometimes a radius is considered to be a line segment with one endpoint at the center of the circle and the other endpoint on the circle, thus making the radius a line segment, rather than a number (the length of such a segment). A similar situation exists with the term *diameter*. Sometimes the diameter of a circle is defined to be the number that is twice the radius (and hence is a measurement), and sometimes it is defined as a line segment. In the latter sense, we say that a **chord** of a circle is a line segment with both its endpoints on the circle, and a **diameter** is a chord that contains the center of the circle. These "double meanings" rarely cause a problem because the context usually clarifies the intended meaning.

A circle is a simple closed curve. As such, it separates its plane into two disjoint sets, with the circle as the boundary. If we remove the points of a circle from the plane, then the two remaining sets are distinguished by calling the one containing the center the **interior** of the circle and calling the other one the **exterior**.

Example 3 In Figure 12.5, points A, B, C, D, E, and F are on the circle, and O is the center. O, H, and J are interior points; U and V are exterior points. Segments $\overline{OB}$, $\overline{OD}$, and $\overline{OE}$ are radii of the circle; $\overline{AC}$ is a chord; and, if D, O, and E are collinear (as they appear to be), then $\overline{DE}$ is the diameter. □

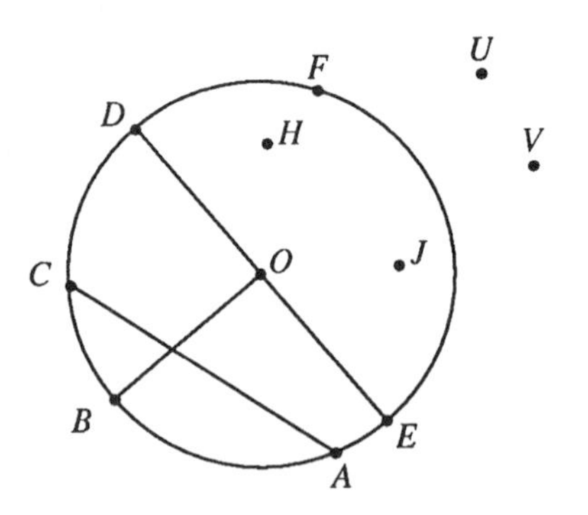

Figure 12.5 A circle and some related line segments.

Example 4 If the measure of $\overline{OB}$ is 3 cm, then we can say that the radius of the circle is 3 cm, and its diameter is 6 cm. □

Example 5 Often a circle is identified by its center. The circle in Figure 12.5 may be called circle O. Also, a particular circle may be specified by naming any three points on it. Thus, circle O is also called circle ACD, circle DBE, etc. □

Definitions A **central angle** of a circle centered at O is any angle with O as vertex. If A and B are points on the circle, with A, O, and B noncollinear, then the **minor arc** AB is the set containing A, B, and all circle points in the interior of $\angle AOB$, whereas the **major arc** AB is the set containing A, B, and all circle points in the exterior of $\angle AOB$. If A, O, and B are collinear, the two arcs are called **semicircles**.

Notation Let C and D be points on a circle centered at O. If C is in the interior of $\angle AOB$ and D is in its exterior, then the minor and major arcs can be written as $\overset{\frown}{ACB}$ and $\overset{\frown}{ADB}$, respectively.

Example 6 In Figure 12.5, $\angle DOB$ and $\angle BOE$ are central angles, $\overset{\frown}{DCB}$ is a minor arc, and $\overset{\frown}{DAB}$ is a major arc. $\overset{\frown}{DBE}$ and $\overset{\frown}{DFE}$ are the two semicircles determined by diameter $\overline{DE}$. □

Although the arcs of a circle are curved, we can apply linear measure to them. Think of bending a line segment (or a piece of string) along an arc so that the endpoints of the arc and the segment coincide, or flattening the arc into a line segment, or rolling the circle along a line segment beginning and ending at the endpoints of the arc. The measure of the arc is the measure of the corresponding line segment. This intuitive idea of arc length can be used to describe the measure of the entire circle, which we can think of as the distance a circle covers in one complete revolution. This distance is called the **circumference** of the circle.

A basic property of the circle is the relation between its circumference and its diameter. A simple experiment demonstrates this relationship: Take some circular objects (jar lids, wheels, plates, etc.) and measure their circumferences by wrapping a string around the edge of each, then measuring the length of string used. Next, measure the diameter of each object, and divide each circumference by its diameter. Careful measurement results in a ratio just a little more than 3 each time. If the measurements were exact, the ratio would come out *exactly the same* for every circle. The proof of this property is beyond the scope of this book, but we label it as a theorem to signify that it is a *proven* characteristic of all circles.

(T-5) | The ratio of the circumference of a circle to its diameter is a constant.

We use the Greek letter **pi** (π) to represent this fixed number. Thus, if C is the circumference and d is the diameter, then $\frac{C}{d} = \pi$; that is,

$$C = \pi \cdot d$$

If r is the radius of the circle, it follows that

$$C = \pi \cdot 2r = 2\pi r$$

The number π is commonly approximated as 3.14 or as $\frac{22}{7}$, but these values are not exactly equal to π. In fact, π is an *irrational* number; that is, it cannot be expressed as the ratio of two integers. It does have a decimal form—an infinite sequence of digits that begins

$$3.14159265358979823846264 3\ldots$$

(See Chapter 13 for a detailed explanation of the distinction between rational and irrational numbers.)

Example 7 To determine the circumference of a circle with a diameter of 4.70 units, we substitute $d = 4.70$ into the formula $C = \pi \cdot d$, getting $C = \pi \cdot (4.70)$, which is usually written as 4.70π. This is approximately $4.70 \times 3.14 = 14.758 \approx$ 14.8 units, to 3 significant digits. □

Example 8 To determine the radius of a circle whose circumference is 1.5 units, we write $1.5 = 2\pi r$ and solve this equation for r. Thus,

$$r = \frac{1.5}{2\pi} \text{ units}$$

which is approximately $\frac{1.5}{6.28}$ or .24 units, to two significant digits. If we want a more precise answer, then we would need a more precise measurement of the radius, perhaps to the nearest thousandth unit, and we would also have to use an approximation for π that is correct to 4 significant digits or more. If the circumference is 1.500 units, then

$$r = \frac{1.500}{2\pi} \approx \frac{1.500}{2 \times 3.14159} \approx .2387 \text{ units}$$ □

Now we can describe a process for finding some more points on the number line that do not correspond to rational numbers. Take a circle with radius $\frac{1}{2}$ unit; its circumference is $2 \cdot \pi \cdot \frac{1}{2} = \pi$ units. If you rest that circle on the origin and roll it (to the right) along the number line one full revolution, it will be resting on a point exactly π units from the origin. Thus, this is a point that does not have a rational coordinate. If you continue rolling the circle along, the full revolutions would end on the points with coordinates 2π, 3π, 4π, ..., which are not rational numbers. Similarly, by rolling it to the left from the origin, you obtain the points with coordinates $-\pi$, -2π, -3π, etc. (See Figure 12.6.)

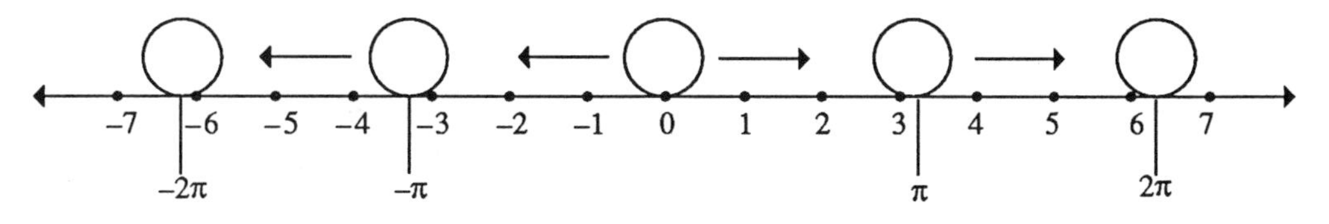

Figure 12.6 Rolling a circle of diameter 1 along the number line.

Exercises 12.2

In Exercises 1–4, the numbers r, s, and t are the lengths of the sides of a triangle. Find the perimeter; then classify the triangle as *equilateral, isosceles,* or *scalene.*

1. $r = 15, \quad s = 13, \quad t = 22$
2. $r = 15, \quad s = 15, \quad t = 15$
3. $r = 15, \quad s = 15, \quad t = 22$
4. $r = 15, \quad s = 22, \quad t = 15$

For Exercises 5–8, compute the perimeter of the given polygon.

5. An equilateral pentagon with each side 7 units long.
6. An equilateral octagon with each side 10 units long.
7. A rhombus with each side 90 feet long. (What very familiar real-world figure fits this description?)
8. An equilateral hexagon with each side 6 units long.

Exercises 9–14 contain statements about circles as you might find them in an elementary-geometry lesson. For each use of *diameter* or *radius,* state whether the term represents a line segment or a measurement.

9. The circumference is π times the diameter, or 2π times the radius.
10. Draw a circle with a radius of 3 centimeters; then draw in the radius.
11. A diameter of a circle is a chord that contains the center of the circle.
12. If a radius is drawn to a tangent at the point of tangency, then the radius is perpendicular to the tangent.
13. The area of a circle is π times the square of the radius.
14. The diameter of a circle is twice its radius.

In Exercises 15–22, the letters C, r, and d represent the circumference, the radius, and the diameter of a circle, respectively. Approximate your answers to 3 significant digits.

15. If $r = 3.00$, find C.
16. If $r = 2.56$, find C.
17. If $d = 4.32$, find C.
18. If $d = 1.20$, find C.
19. If $C = 2.15$, find r.
20. If $C = 432.1$, find r.

21. If $C = 67.3$, find d.

22. If $r = 30.7$, find d.

Exercises 23–28 refer to Figure 12.7(a), which shows a circle with center Q. Indicate whether the given arc is a major arc, a minor arc, or a semicircle.

23. $\widehat{PST}$ **24.** $\widehat{SPR}$ **25.** $\widehat{PTR}$

26. $\widehat{STR}$ **27.** $\widehat{SRP}$ **28.** $\widehat{RSP}$

Exercises 29–32 refer to Figure 12.7(b), which shows a circle with center O.

29. Name all the diameters shown.

30. Name all the radii shown.

31. Name all the chords shown.

32. Name all the central angles shown.

33. (a) Express $\frac{22}{7}$ as a decimal with at least 8 significant digits.

(b) Using 3.1415926536 as an "almost-true value" for π, to how many significant digits of accuracy is $\frac{22}{7} \approx \pi$?

(c) If you use $\frac{22}{7}$ as an approximation of π, how accurate can your answers be? How will this accuracy compare with using 3.14 as an approximation? With using 3.14159?

34. (a) Continue with the activity described in conjunction with Figure 12.6 to show how to identify points with coordinates of

i. -3π and 3π

ii. $\ldots, \frac{-5\pi}{2}, \frac{-3\pi}{2}, \frac{-\pi}{2}, \frac{\pi}{2}, \frac{3\pi}{2}, \frac{5\pi}{2}, \ldots$

(b) Sketch a number line and approximately locate the points having the coordinates listed in part (a).

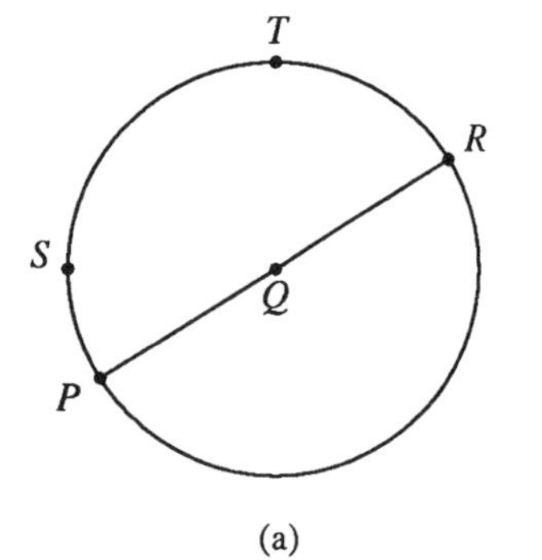

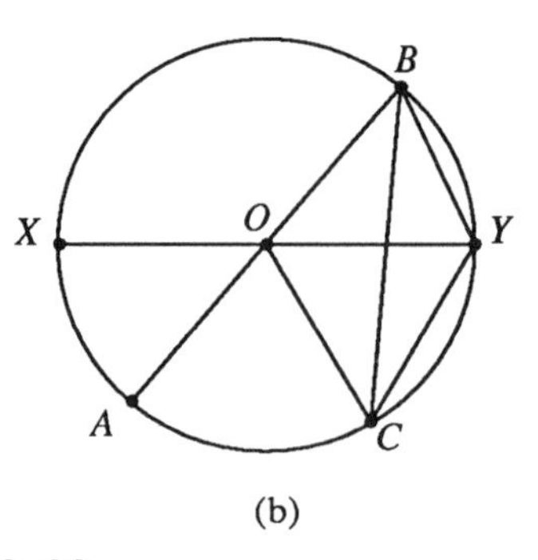

Figure 12.7 Exercises 23–32.

12.3 Angular Measure

Our approach to measuring angles follows much the same pattern as the approach to linear measure. Recall that we began the process of measuring line segments by considering a typical instrument for measuring length, the ruler, and abstracting its properties. That led to a formal basis for linear measure, a statement we called the Ruler Postulate. In this section we begin with a common instrument for measuring angles, the protractor. We shall abstract its properties to formulate a "Protractor Postulate," thereby providing a formal basis for angular measure.

PROBLEM-SOLVING COMMENT

Watch for both similarities and differences in the developments of linear and angular measure; the processes are *analogous* in some ways, but very different in others. One immediate difference is that linear measure had to be developed "from scratch"; we had no measurement concept on which to build. Now that we have *solved the simpler problem* of linear measure, we shall be able to adapt and modify that solution to help us derive angular measure. ◇

A protractor is a flat object in the shape of a half-moon, outlined by a semicircle and the diameter that connects the endpoints. (See Figure 12.8.) Its straight edge (the diameter) has a mark at the middle to indicate the center of the semicircle. The protractor's utility as a measuring device comes from the fact that each point of its semicircular edge is paired with a unique real number from 0 to some largest positive number. (If the unit of measure is chosen to be degrees, that largest number is 180.) An angle is measured by placing the protractor so that the center of its semicircle is at the vertex of the angle and the two sides of the angle intersect the semicircle at two points. Denoting the coordinates of those two points by p and q, the measure of the angle is $|p-q|$. (A slightly easier, but less general, procedure taught in elementary school requires that one side of the angle intersect at the 0 mark of the protractor, thereby making the coordinate of the other intersection point the measure of the angle. This is convenient, but not essential.)

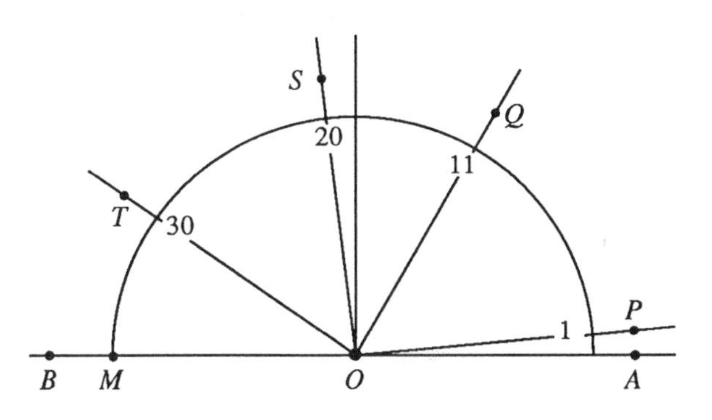

Figure 12.8 A protractor, used in Examples 1–6.

The following postulate, P-17, does for angular measure what the Ruler Postulate did for linear measure, formalizing much of what we have just described.

(P-17) **Protractor Postulate:** Let M be a positive real number and let $\overleftrightarrow{AB}$ be a line in a plane, with O a point of $\overleftrightarrow{AB}$ between A and B. The set of rays consisting of $\overrightarrow{OA}$, $\overrightarrow{OB}$, and all rays with endpoint O on a given side of $\overleftrightarrow{AB}$ can be put in 1-1 correspondence with the real numbers between 0 and M in such a way that

(a) $\overrightarrow{OA}$ is paired with 0 and $\overrightarrow{OB}$ is paired with M;

(b) if $\overrightarrow{OP}$ is paired with p and $\overrightarrow{OQ}$ is paired with q, then the measure of $\angle POQ$ equals $|p - q|$.

Notation The measure of $\angle POQ$ will be written as $m\angle POQ$.

Examples 1–3 refer to Figure 12.8.

Example 1 Since $\overrightarrow{OQ}$ is paired with 11 and $\overrightarrow{OS}$ is paired with 20,

$$m\angle QOS = |11 - 20| = 9$$

Likewise, $m\angle POQ = |1 - 11| = 10$. □

Example 2 $\angle AOP$ is a unit angle because $m\angle AOP = 1$. It takes 9 adjacent congruent copies of $\angle AOP$ to "cover" $\angle QOS$. □

Example 3 $m\angle AOA = 0$ and $m\angle AOB = M$. In general, the measure of every zero angle is 0, and the measure of every straight angle is M. □

For convenience, let us assume from now on that the maximum number, M, is at least 1. Then we can have a "unit" angle, an angle with measure 1. Just as linear measure can be described in terms of placing "copies" of the unit segment end to end, so, too, we can describe angular measure in terms of "copies" of the unit angle placed next to each other. As with segments, this view requires the concept of congruence. Once again, we ease the formal axiomatic approach by simply defining two angles to be **congruent** if their measures are equal. In symbols:

$$\angle ABC \cong \angle DEF \quad \text{if} \quad m\angle ABC = m\angle DEF$$

As in the case of linear measure, two postulates are needed to allow us to measure an angle by making adjacent "copies" of a unit:

(P-18) **Angle-Construction Postulate:** Given $\angle ABC$ and one of the half-planes determined by $\overleftrightarrow{XY}$, there is a unique ray $\overrightarrow{XZ}$ such that $\angle YXZ \cong \angle ABC$. (See Figure 12.9.)

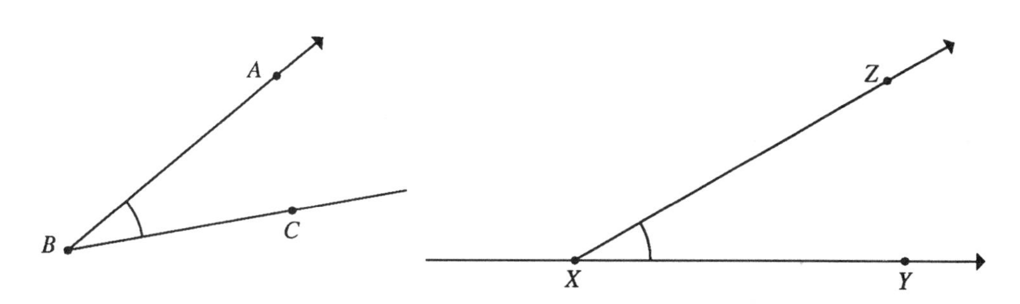

Figure 12.9 The Angle-Construction Postulate.

(P-19) **Angle-Addition Postulate:** If $\angle ABC$ is adjacent to $\angle CBD$ and if $m\angle ABC + m\angle CBD \leq M$, then

$$m\angle ABC + m\angle CBD = m\angle ABD$$

(See Figure 12.10.)

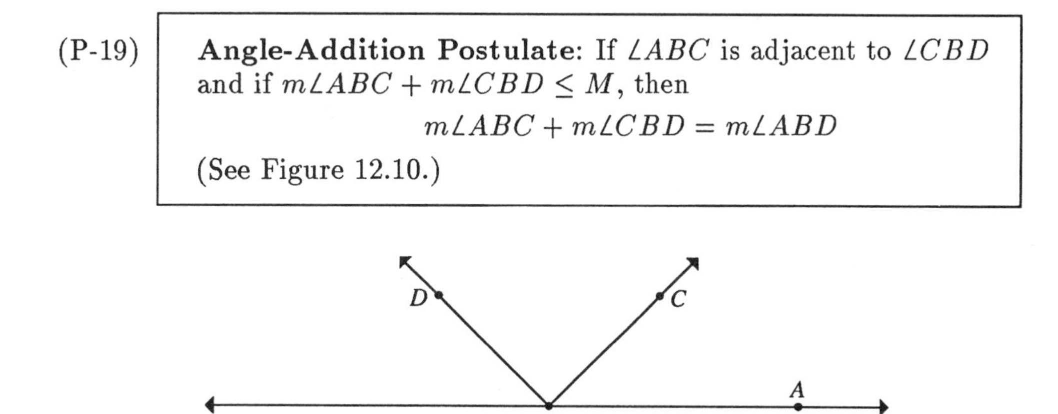

Figure 12.10 The Angle-Addition Postulate.

Examples 4–6 refer to Figure 12.8.

Example 4 $\angle TOS \cong \angle POQ$ because $m\angle TOS = 10 = m\angle POQ$. However, since $\angle TOS$ contains points that are not on $\angle POQ$ (in fact, only O is on both angles), $\angle TOS$ does not *equal* $\angle POQ$. (*Remember*: Angles, like line segments, are sets of points.) □

Example 5 $m\angle SOQ = |20 - 11| = m\angle QOS$, so $\angle SOQ \cong \angle QOS$. Moreover, these two angles contain exactly the same points, so we can say that $\angle SOQ = \angle QOS$. □

Example 6 Since $\angle QOS$ and $\angle SOT$ are adjacent, $m\angle QOT = 9 + 10 = 19$ by the Angle-Addition Postulate. □

The main difference between linear and angular measurement is that there is no longest line segment, but there is a largest angle—the straight angle, whose measure is M. Thus, any standard unit for angular measure is some portion of the maximum value, M; in other words, the choice of M determines the unit of angular measure.

By far the most common value for M in elementary mathematics is 180. In this case the unit of measure is called a **degree**, and the measure of a unit angle is one degree (1°). Thus, a straight angle has measure 180°. This means that it takes 180 adjacent copies of a 1°-angle to "cover" a straight angle. In Figure 12.11, $\angle ONE$ is a unit angle of 1° and $\angle UNE$ is a straight angle.

Angular measure allows us to define a number of geometric objects that arise in elementary-school mathematics. In keeping with common practice, we define these objects in terms of degrees. However, keep in mind that each one has an analogous definition for each choice of the maximum angular measure M.

Definitions A **right angle** is any angle with measure 90°. An angle with measure greater than 0° but less than 90° is called an **acute angle**; an angle with measure greater than 90° but less than 180° is called an **obtuse angle**.

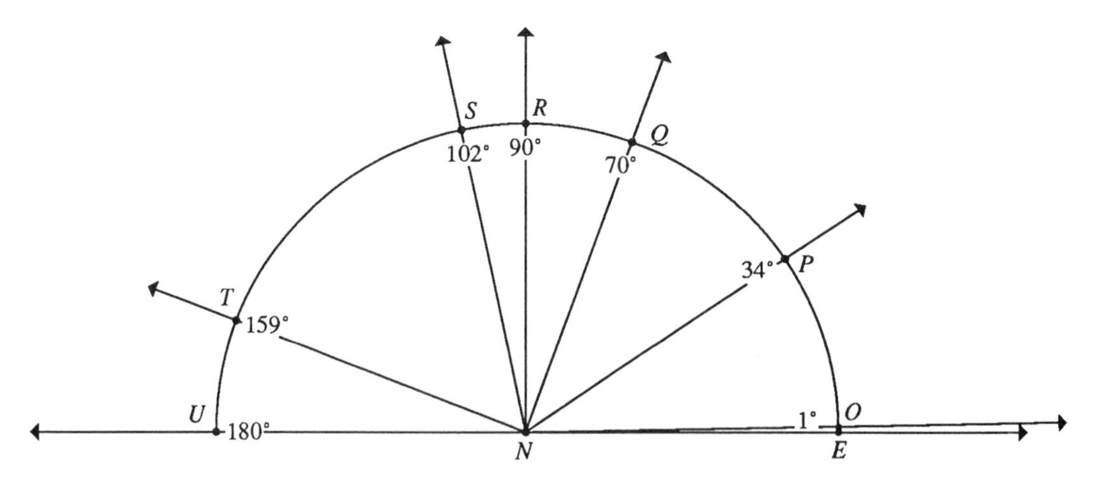

Figure 12.11 Angles measured in degrees.

Two lines that intersect to form right angles are said to be **perpendicular**. If lines **l** and **m** are perpendicular, we write $\mathbf{l} \perp \mathbf{m}$.

If the sum of the measures of two angles is 180°, then the angles are said to be **supplementary**, and each angle is called the **supplement** of the other. If the sum of the measures of two angles is 90°, then the angles are said to be **complementary**, and each angle is called the **complement** of the other.

Example 7 This example refers to Figure 12.11.

(a) $\angle ENR$ and $\angle UNR$ are right angles. Since all right angles have the same measure (90°), it follows that all right angles are congruent, so $\angle ENR \cong \angle UNR$. Also, $\overleftrightarrow{EU} \perp \overleftrightarrow{NR}$.

(b) $\angle PNE$, $\angle ENQ$, $\angle UNT$, $\angle UNS$, $\angle TNS$, and $\angle PNQ$ are acute angles.

(c) $\angle ENS$, $\angle ENT$, $\angle UNQ$, and $\angle UNP$ are obtuse angles.

(d) By the Angle-Addition Postulate,

$$m\angle ENQ + m\angle QNU = 180°$$

so $\angle ENQ$ and $\angle QNU$ are supplementary. Similarly,

$$m\angle ENQ + m\angle QNR = 90°$$

so $\angle ENQ$ and $\angle QNR$ are complementary. □

Definitions An **acute triangle** is any triangle with three acute angles. A **right triangle** is any triangle with one right angle. An **obtuse triangle** is any triangle with one obtuse angle.

Example 8 In Figure 12.12, $\triangle ABC$ is acute, $\triangle XYZ$ is a right triangle, and $\triangle DEF$ is an obtuse triangle. □

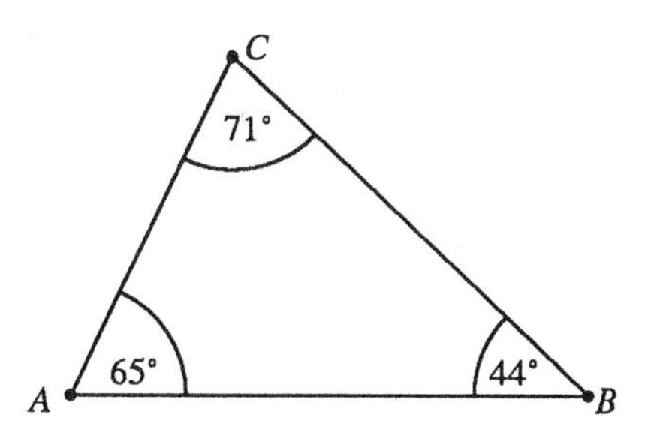

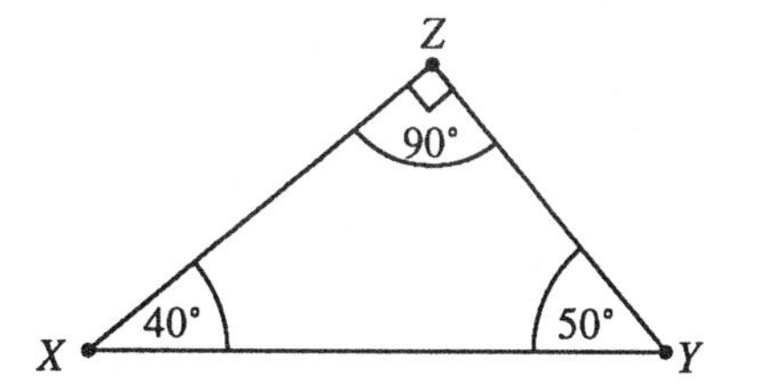

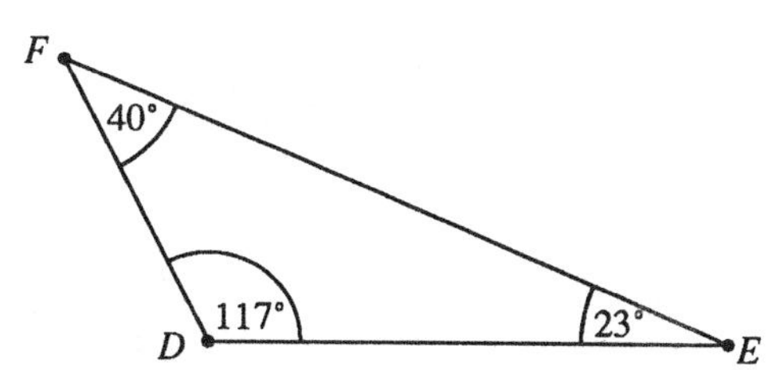

Figure 12.12 Three types of triangles.

Definitions A **rectangle** is any quadrilateral with four right angles; a **square** is a rectangle that is also a rhombus.

Example 9 In Figure 12.13, $ABCD$ and $EFGH$ are rectangles, $EFGH$ and $IJKL$ are rhombuses, and $EFGH$ is a square. □

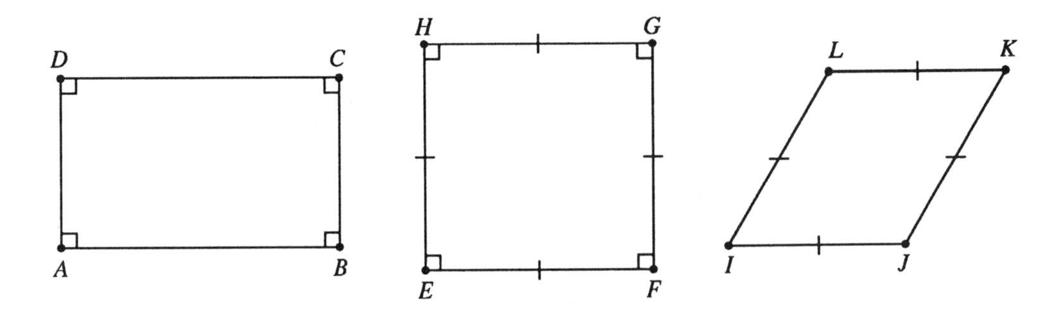

Figure 12.13 Three types of quadrilaterals.

Definitions An **equiangular polygon** is any polygon with all of its angles congruent to each other; a **regular polygon** is any polygon that is both equiangular and equilateral.

Example 10 In Figure 12.13, $ABCD$ is an equiangular quadrilateral that is not equilateral; $EFGH$ is both equiangular and equilateral; and $IJKL$ is equilateral, but not equiangular. In Figure 12.14 you see a regular triangle, a regular pentagon, and a regular hexagon. □

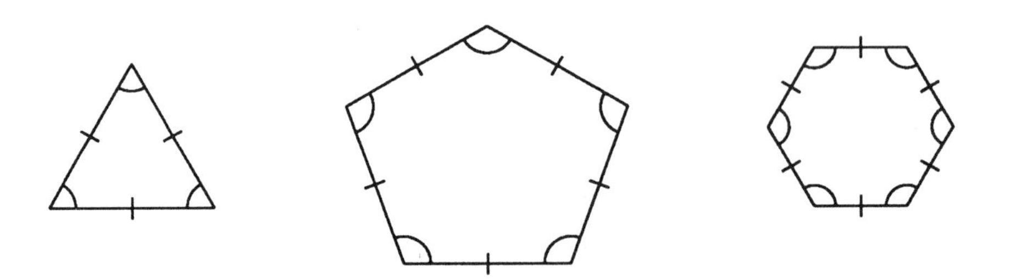

Figure 12.14 Regular polygons.

Despite its familiarity, 180 is neither the only value nor always the most convenient value of M to be used in measuring angles. Another common value is π, the length of a semicircle of radius 1. When π is used for M, the unit of measure is called a **radian**. Thus, the measure of a unit angle is 1 radian (1 rad), and the measure of a straight angle is π rad. Replacing 180° by π in the foregoing definitions yields the angle and figure types as they are defined with radian measure. For instance, a *right angle* has a measure of $\frac{\pi}{2}$ rad. This measurement scale is important in many applications of geometry, for reasons that will be explained shortly.

Let us return now to the relationship between linear and angular measure. These two approaches to measurement are linked through the application of linear measure to the semicircle of radius measure r, whose circumference then measures πr. When a protractor is used to measure an angle, that angle is a central angle of the semicircle forming the protractor. Now, the angular measure of a central angle is directly proportional to the linear measure of the intercepted arc. Since the measure of a straight angle is M and the measure of the intercepted arc is πr, we have the proportion

$$\frac{M}{\pi r} = \frac{a}{s}$$

where a is the measure of a given angle and s is the measure of its intercepted arc. Cross multiplying and solving for a, we get

$$a = \frac{Ms}{\pi r}$$

which is a relationship between the angular measure a and the linear measures s and r. In degree measure, this yields the formula

$$a = \frac{180s}{\pi r} \tag{12.1}$$

The formula in radian measure is even more interesting:

$$a = \frac{\pi s}{\pi r} = \frac{s}{r} \tag{12.2}$$

That is:

> If a central angle of a circle of radius r determines a minor arc of length s, then the radian measure of the angle is $\frac{s}{r}$, the ratio of the arc length divided by the radius.

Thus, in a **unit circle**—a circle of radius 1—we have $a = \frac{s}{1} = s$; that is, the measure of the angle equals the (linear) measure of the intercepted arc. Because of this very simple relationship between angular and linear measure, the unit circle is fundamental to the development of trigonometry and calculus.

Formulas 12.1 and 12.2 can also be used to relate the degree measure of an angle with its radian measure. If we let D and R represent the degree and radian measures of a given angle, then 12.1 and 12.2 yield

$$D = \frac{180s}{\pi r} \quad \text{and} \quad R = \frac{s}{r}$$

so the ratio of the radian measure to the degree measure is given by

$$\frac{R}{D} = \frac{s/r}{180s/\pi r} = \frac{s}{r} \cdot \frac{\pi r}{180s} = \frac{\pi}{180}$$

Thus:

> The ratio of the radian measure of an angle to its degree measure is always $\frac{\pi}{180}$.

Example 11 In Figure 12.15, $\angle AOB$ is an angle of 1 radian because the length of minor arc AB equals the radius of the circle. $\angle AOC$ has a measure of 1.5 radians; the length of arc $\widehat{ABC}$ is 1.5 times the radius. □

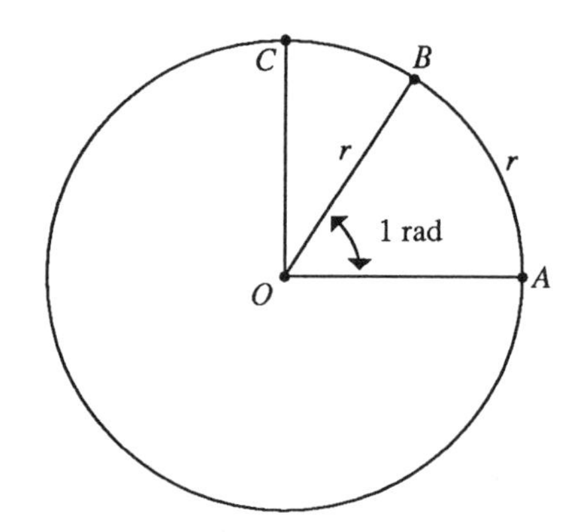

Figure 12.15 Radian measure in a circle of radius r.

Example 12 If a central angle in a circle of radius 4.8 units determines a minor arc of 11.04 units, then the measure of the angle is

$$\frac{11.04}{4.8} = 2.3 \text{ rad}$$

□

Example 13 To find the radian measure R of a 30° angle, we solve $\frac{R}{30} = \frac{\pi}{180}$ for R, getting $R = \frac{\pi}{6}$ rad. □

We end this section with a brief presentation of one of the most famous statements of mathematics, the Pythagorean Theorem. This theorem states a fundamental relationship among the lengths of the sides of a right triangle, thereby connecting linear and angular measure. Particular cases of it were known and used by the ancient Egyptians, but it was first stated in general terms and proved by Pythagoras, around 500 B.C.

In its original form, the theorem related the areas of three squares with sides that form the sides of a right triangle. (See Figure 12.16.) It stated that the area of the square on the hypotenuse (the side opposite the right angle) equals the sum of the areas of the squares on the other two sides (the legs). At that time, the sense of the term *equals* was explicitly geometric; it meant that the areas enclosed in the two smaller squares could be cut up and reassembled to form the large square. (This meaning forms the basis for an upper elementary-grade activity about the Pythagorean Theorem: students are asked to trace a diagram like Figure 12.16 on graph paper and "approximately verify" the theorem by counting the "grid" squares inside the three major squares.)

With the introduction of modern algebraic notation and terminology (almost 2000 years later), the sides of the figures were replaced by their numerical lengths, the term *square* came to mean multiplying a number by itself, and "squares *of* the sides" replaced "squares *on* the sides." Today the theorem is stated as follows:

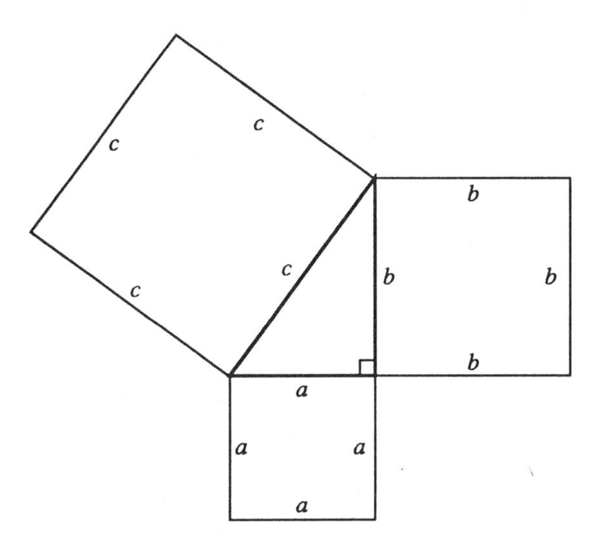

Figure 12.16 The Pythagorean Theorem.

The Pythagorean Theorem: In any right triangle, the square of the hypotenuse is equal to the sum of the squares of the other two sides. Thus, if c is the length of the hypotenuse, and if a and b are the lengths of the other two sides, then

$$c^2 = a^2 + b^2$$

Example 14 If the legs of a right triangle are 3 and 4 units long, then the square of the hypotenuse is

$$3^2 + 4^2 = 9 + 16 = 25$$

so the hypotenuse is $\sqrt{25} = 5$ units long. □

Example 15 If the hypotenuse of one triangle is 25.1 units long and one leg is 4.20 units long, then the length of the other leg is found as follows:

$$25.1^2 = 4.20^2 + b^2$$

$$630.01 = 17.64 + b^2$$

Subtracting, we get

$$b^2 = 630.01 - 17.64 = 612.37$$

Thus, $b = \sqrt{612.37} \approx 24.7$ units, to 3 significant digits. □

The proof of this theorem probably has commanded more attention than any other in all of mathematics. There is a book[1] containing 366 different proofs of it, one even credited to a President of the United States (James A. Garfield)! Since a proof is not essential to our discussion, we do not present one here, but turn instead to an interesting application of the theorem itself.

If we apply the Pythagorean Theorem to an isosceles right triangle with each leg 1 unit long, then the square of the hypotenuse is $1^2 + 1^2 = 2$, so its length is $\sqrt{2}$. Figure 12.17 shows how to construct a series of right triangles with hypotenuse lengths $\sqrt{2}$, $\sqrt{3}$, $\sqrt{4}$, etc. If each of these line segments were placed on the positive side of a number line with one end at the origin, then the other ends would be on a point whose coordinate was the indicated square root. Figure 12.18 shows the points corresponding to the square roots of the numbers 2 through 6. Except for those of the perfect squares 4, 9, 16, ..., these square roots are not rational numbers (as you will see in Chapter 13). Thus, we have another example of an infinite "family" of number-line points that do not have rational coordinates.

---

[1]Elisha Scott Loomis, *The Pythagorean Proposition*, Washington, D.C.: National Council of Teachers of Mathematics, 1968.

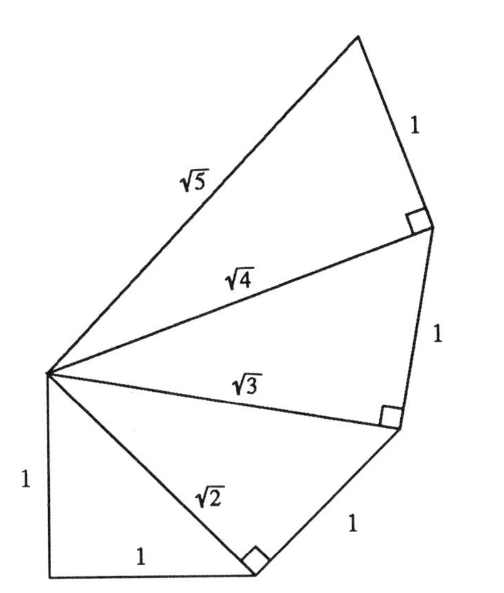

Figure 12.17
An application of the Pythagorean Theorem.

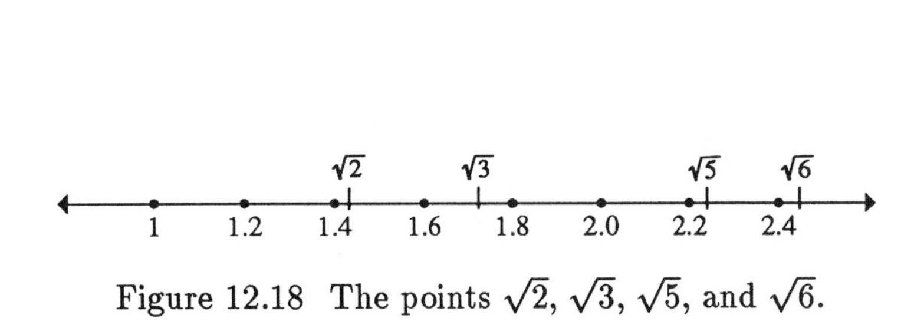

Figure 12.18 The points $\sqrt{2}$, $\sqrt{3}$, $\sqrt{5}$, and $\sqrt{6}$.

Exercises 12.3

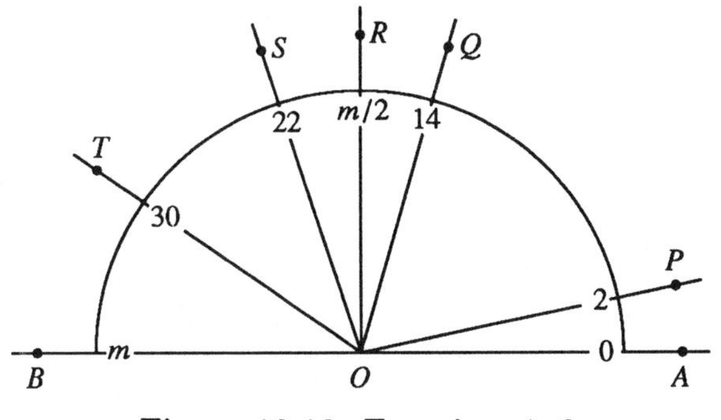

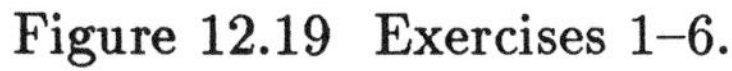

Figure 12.19 Exercises 1–6.

Exercises 1–6 refer to Figure 12.19.

1. Determine each of the following measures:

   $m\angle AOB$, $m\angle POQ$, $m\angle POS$,
   $m\angle QOS$, $m\angle AOT$

2. Find a pair of congruent angles that are not equal.
3. Find a pair of equal angles.
4. Find a pair of supplementary angles.
5. Find a pair of complementary angles.
6. Identify a right angle, an acute angle, and an obtuse angle.
7. Classify the angles in Figure 12.20 (on page 456) as acute, right, or obtuse, and estimate their measures in degrees and in radians.
8. Use radian measure to describe each of the following terms:

   right angle, acute angle, obtuse angle,
   supplementary angles, complementary angles

For Exercises 9–12, convert to radian measure

9. 45° **10.** 60° **11.** 150° **12.** 72°

For Exercises 13–16, convert to degree measure.

13. π rad **14.** $\frac{5\pi}{6}$ rad
15. $\frac{3\pi}{5}$ rad **16.** 2 rad

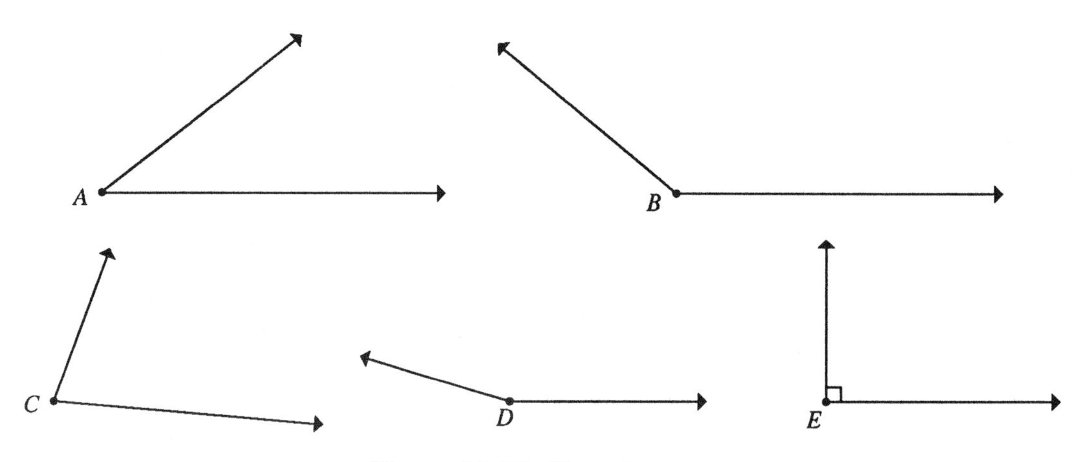

Figure 12.20 Exercise 7.

For Exercises 17–22, find the supplement of the angle, stating your answer using the given measurement unit.

17. 25° **18.** 179° **19.** 127°

20. $\frac{\pi}{3}$ rad **21.** 3 rad **22.** $\frac{\pi}{2}$ rad

For Exercises 23–28, find the complement of the angle, stating your answer using the given measurement unit.

23. 22° **24.** 78° **25.** $\frac{\pi}{4}$ rad

26. $\frac{\pi}{6}$ rad **27.** .5 rad **28.** $\frac{\pi}{2}$ rad

Exercises 29–42 all refer to a central angle of a circle of radius r units. The measure of the minor arc determined by that angle is s units, and the measure of the angle is a radians.

For Exercises 29–32, find the measure of the central angle in radians.

29. $r = 2, \quad s = 5$ **30.** $r = 1.5, \quad s = .5$

31. $r = 1, \quad s = 1$ **32.** $r = \pi, \quad s = \frac{\pi}{2}$

For Exercises 33–37, find the radius r.

33. $a = 1.5$ rad, $s = 3$

34. $a = .75$ rad, $s = 2.3$

35. $a = 2.8$ rad, $s = 5.7$

36. $a = \frac{\pi}{2}$ rad, $s = \pi$

37. $a = \frac{\pi}{6}$ rad, $s = 3$

For Exercises 38–42, find the arc length s.

38. $a = 1.5$ rad, $r = 3$

39. $a = .75$ rad, r = 2.3

40. $a = 2.8$ rad, $r = 5.7$

41. $a = \frac{\pi}{2}$ rad, $r = \pi$

42. $a = \frac{\pi}{6}$ rad, $r = 3$

For Exercises 43–45, h is the length of the hypotenuse of a right triangle, and k and m are the lengths of its legs. Use the Pythagorean Theorem to find the missing measure. Approximate your answer to the appropriate number of significant digits.

43. $k = 3.2, \quad m = 4.1$

44. $h = 7.5, \quad k = 2.56$

45. $h = 32.78, \quad m = 21.472$

Any set of three whole numbers satisfying the Pythagorean Theorem is called a **Pythagorean triple**. For Exercises 46–57, determine whether the given numbers form a Pythagorean triple. Use your answers to conjecture how you might form infinitely many Pythagorean triples if you know one of them.

46. 3, 4, 5 **47.** 4, 5, 6

48. 6, 8, 10 **49.** 9, 12, 15

50. 4, 6, 8 **51.** 5, 12, 13

52. 10, 24, 26 **53.** 15, 36, 39

54. 10, 13, 17 **55.** 7, 24, 25

56. 70, 240, 250 **57.** 8, 10, 12

12.4 The Metric System

Historically, there have been many different standard units of measure, varying from culture to culture and even within cultures. Some of the earliest standards were identified with parts of the human body, such as the **span** (the distance from the tip of the thumb to the tip of the little finger with the fingers spread out), the **palm** (the breadth of the four fingers held close together), and the **digit** (the breadth of the first finger or the middle finger). The problem with "standard" measuring units like these is obvious; dimensions of the human body vary from person to person. Thus, it was natural to select a king or other prominent person on whom the standard units would be based. This probably was the source of the term *ruler* as applied to the instrument for measuring length.

In England, King Henry I (1068-1135) declared that a **yard** was to be the distance from the tip of his nose to the tip of his thumb with his arm stretched out. That was the basis of the **English system of measurement**, a system still commonly used in the United States as well as in England. Users of this system often struggle with the task of converting from one measuring unit to another, using

1 yard = 36 inches
1 mile = 5280 feet
1 gallon = 231 cubic inches
1 pound = 16 ounces
1 cubic foot of water weighs 62.4 pounds

and so forth.

Until late in the 18th century many different standard measuring units were used throughout the world. However, as commerce among countries increased, the need for a single universally accepted system of measurement became evident, and in 1791 a committee of the French Academy recommended a new system, called the **metric system**:

- The standard unit of linear measure is the **meter**, which was defined to be one ten-millionth of the Earth's meridian quadrant at sea level (the distance from the North Pole to the Equator).

- Larger or smaller units of length are obtained by multiplying or dividing the meter by powers of ten. The Greek prefixes **deka-**, **hecto-**, and **kilo-** are used to indicate the multiples of 10, 100, and 1000, respectively; the Latin prefixes **deci-**, **centi-**, and **milli-** indicate subdivisions of .1, .01, and .001, respectively. Thus, a **centimeter** is one-tenth of a meter, whereas a **kilometer** is one thousand meters, and so forth. (Table 12.3 on page 464 provides a more complete list of terms.)

- Unit conversion within the metric system is done by multiplying or dividing by an appropriate power of ten, and this can be done in our numeration system just by moving the decimal point.

Example 1

$$\begin{aligned} 25.2 \text{ centimeters} &= 25.2 \times .01 \text{ meters} \\ &= .252 \text{ meters} \\ &= .0252 \text{ dekameters} \\ &= 252 \text{ millimeters} \end{aligned}$$ □

Example 2

$$\begin{aligned} 1.34 \text{ kilometers} &= 1.34 \times 1000 \text{ meters} \\ &= 13.4 \text{ hectometers} \\ &= 1340 \text{ meters} \\ &= 134{,}000 \text{ centimeters} \end{aligned}$$ □

In order to measure the area of regions in a plane, the standard unit of area must be a figure that has area and is shaped "nicely" enough so that copies of it fit together without leaving gaps or spaces. One simple shape (though not the only one) that satisfies these requirements is the square. The square is particularly easy to describe in terms of linear measure, so it is a natural choice for a standard unit of area measure.

In the metric system the basic unit of area is a square with each side one meter long, called (not surprisingly) a **square meter**.

Other suitable units of area measure are the squares with side length equal to standard metric units—square millimeters, square centimeters, square kilometers, etc. Because we know the relationships among the standard multiples and subdivisions of the meter, conversion from one metric area unit to another is easy.

Example 3 One square millimeter is the area of a square with each side one millimeter long; one square centimeter is the area of a square with each side one centimeter long. Since one centimeter equals 10 millimeters and the area of a square is found by squaring the length of one side, one square centimeter is equivalent to 100 square millimeters. □

Example 4 One square kilometer is the area of a square with each side 1000 meters long, so one square kilometer must equal 1000^2 meters; that is,

$$1 \text{ square kilometer} = 1{,}000{,}000 \text{ square meters}$$ □

Example 5 Children are introduced to the concept of area at a very early age. Figure 12.21 shows how this is done in kindergarten. □

How large?

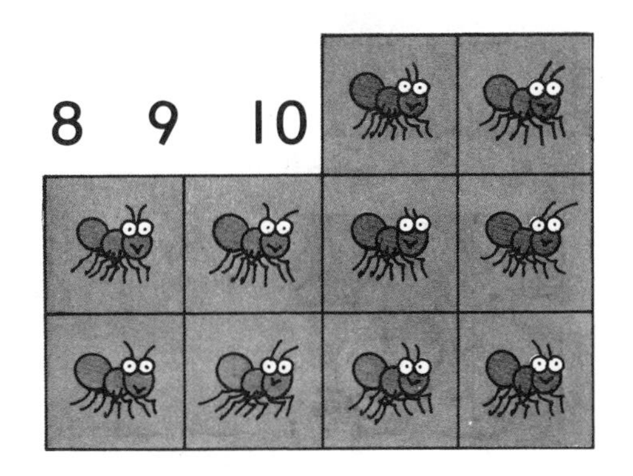

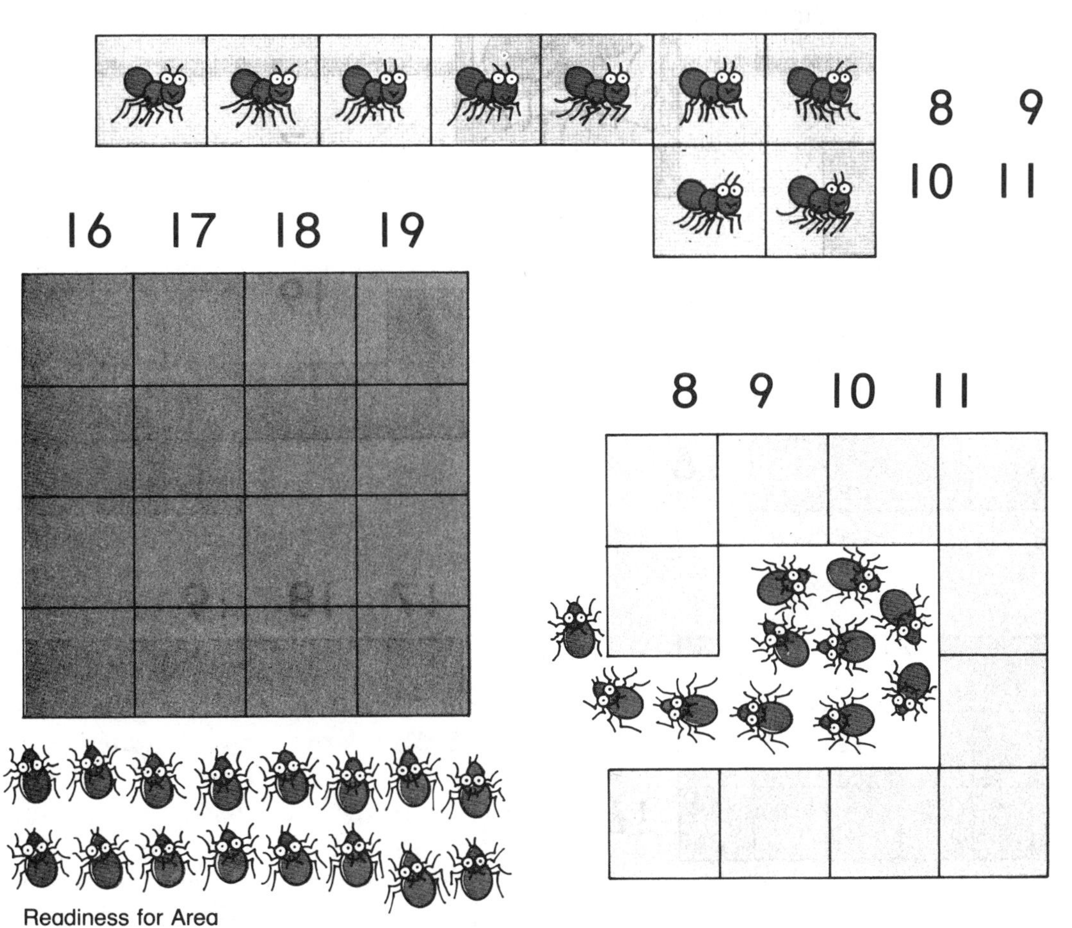

Figure 12.21 INVITATION TO MATHEMATICS, Kindergarten, p. 137

The analogous treatment of volume is based on the **cube**, a hexahedron with (six) square faces.

> The standard metric unit of volume is the **cubic meter**, a cube with edges 1 meter long.

Cubes with edges equal to any of the standard multiples or subdivisions of meters are also useful units of volume. Two of these are of particular interest:

> The **cubic centimeter**, the volume of a cube with edges one centimeter long, is a common measure for small volumes, especially of medicines and chemicals; it is commonly abbreviated as **cc**.
>
> The **cubic decimeter**, the volume of a cube with edges one decimeter (or 10 centimeters) long.

A cubic decimeter is 10^3 centimeters, or 1000 cubic centimeters. This unit of volume is also called a **liter**. Thus, .001 liter equals 1 cubic centimeter, and, because the prefix *milli-* means *one thousandth*, we have

$$.001 \text{ liter} = 1 \text{ milliliter} = 1 \text{ cc}$$

This relationship allows us to convert easily back and forth between cubic measure and liters.

Example 6

$$\begin{aligned} 23 \text{ milliliters} &= 23 \text{ cubic centimeters} \\ &= 23{,}000 \text{ cubic millimeters} \end{aligned}$$

□

Example 7

$$\begin{aligned} 4578 \text{ cubic meters} &= 4578 \times 100^3 \text{ centimeters} \\ &= 4578 \times 10^6 \text{ cc} \\ &= 4578 \times 10^6 \text{ milliliters} \\ &= 4578 \times 10^3 \text{ liters} \\ &= 4578 \text{ kiloliters} \end{aligned}$$

□

The standard temperature unit in the metric system is the **Celsius degree.** 0° Celsius is defined to be the freezing point of water at sea level, and 100° Celsius is defined as the boiling point of water at sea level. Thus, 1° Celsius is $\frac{1}{100}$ of the difference in temperature between the boiling point and the freezing point of water at sea level.

Before discussing metric units of *mass*, let us distinguish between weight and mass. **Weight** is the measure of the gravitational force exerted on a body. **Mass** is the measure of the tendency of a body to oppose changes in velocity—that is, of a body's resistance to changes in motion. For example, as astronauts move from the earth to an orbit around the earth, their weight decreases from what it was on earth to 0 when in orbit. If they proceed to the moon, their weight will increase again as they approach the moon, but on the moon their weight will be far less than when they were on earth because the moon's gravitational pull is less than that of the earth. However, throughout their journey the astronauts' *mass* is unaffected by the changing effects of the gravitational fields. For practical purposes, we can say that mass remains constant.[2] As long as two objects of equal mass remain in the same gravitational field, they will also have equal weight; if they are subjected to different gravitational forces, then they will have different weights, but their masses will remain the same.

> In the metric system, the standard unit of mass is the **gram**, defined to be the mass of one cubic centimeter of water in its densest state, at about 4° Celsius.

(This is roughly equivalent to the mass of two thumb tacks.) As before, standard multiples and subdivisions of the gram are indicated by the same Latin and Greek prefixes for the various powers of ten. This definition is convenient for estimating volumes or masses of liquids, but the accuracy of such estimates is affected by temperature (because water expands as the temperature varies from $4°C$) and by the relative densities of water and of the liquid being measured.

Example 8 30 cubic centimeters of cool water has a mass of about 30 grams. □

Example 9

$$1 \text{ kilogram} = 1000 \text{ grams} \approx \text{the mass of 1 liter of cool water}$$ □

Example 10 The mass of a cup of hot coffee, which has a volume of about 240 milliliters, is a little less than 240 grams. □

The metric system has been a marked improvement over all previous measurement systems because its standard units are related to measurable properties of the Earth and one of its most readily available resources, water. Furthermore, one can convert easily from one subunit of measure to another, merely by moving a decimal point. It has gradually gained acceptance by nearly all major nations; the United States is the last holdout.

Recently, there have been some changes in the metric system as advances in technology have required the development of more accurate standard measurement units. What has evolved is the **International System of Units**, abbreviated (in many different languages) as **SI**. The system is basically metric, but many of the standard units are based on scientific formulas and natural constants. For instance, the SI meter is defined to be 1,650,763.73

[2]Technically, this is not quite true. Relativity theory says that the mass of an object changes relative to its velocity; however, that change only becomes significant as the speed of the object approaches the speed of light.

HISTORICAL NOTE: THE METRIC MEASURE OF ANGLES

An interesting bit of history surrounds the selection of the standard metric unit of angular measure. In 1791, when the French Academy established the *meter* as one ten-millionth of the Earth's quarter meridian at sea level, an accurate measurement of that distance was needed to determine precisely the length of 1 meter. Two French surveyors were commissioned for the task. Now, up to this time the most common unit of angular measure had been the *degree*. However, in the spirit of the French Academy's desire to use units based on powers of 10, the surveyors formulated a new unit of angular measure, the **grade** (or **grad**)—$\frac{1}{100}$th of a right angle. Thus, the measure of a straight angle is 200 grades, and all the angle and figure types described in Section 12.3 can be defined in terms of grades simply by replacing "180°" with "200 grad." Like the other metric units, the grade was subdivided into hundredths (centigrades) and thousandths (milligrades).

The surveyors proceeded to use the grade as the unit of angular measure in their work to determine the precise length of 1 meter. However, when the French Academy accepted the expedition's findings for the precise length of a meter, it passed over the angular unit that helped to produce it, choosing instead the *radian* as the standard metric unit of angular measure! The unique relationship between angular measure a in radians and linear arc measure s ($a = \frac{s}{r}$, shown in Section 12.3) was judged to far outweigh any advantage gained by having a unit angle that is $\frac{1}{100}$th of a right angle.

Since it was used in the surveying that produced an accurate measurement of the meter, the *grade* has been referred to as the "oldest metric unit," but it is not officially a metric unit at all! Nevertheless, it was used extensively in the 19th century as a common unit of angular measure. Today it is rarely used, except by French surveyors. ◇

wavelengths, in a vacuum, of the radiation corresponding to the transition between two electronic energy levels of the krypton-86 atom.[3] The kilogram, rather than the gram, is the standard SI unit of mass; it is defined to be the mass of a prototype object stored in Sevres, France. This is the only remaining standard unit defined by a model. Probably it will eventually be redefined in terms of atomic mass.

The SI system also changes the standard temperature unit from 1° Celsius to 1 **Kelvin** (*not* " 1 degree Kelvin"). The Kelvin scale is like the Celsius scale in that one Celsius degree change in temperature equals one Kelvin change; however, **absolute zero** (the complete absence of heat, which occurs at −273.15 degrees Celsius) corresponds to 0 Kelvin. Thus, the freezing and boiling points of water (0° and 100° Celsius) are 273.15 and 373.15 Kelvin, respectively. Despite the official shift to Kelvin, the Celsius scale probably will still be used in weather forecasts, medical reports, and so on.

[3]For this and other SI unit definitions, see *Dictionary of Physics and Mathematics*, Daniel N. Lapedes, ed., New York: McGraw-Hill, 1978.

Converting between Celsius and Kelvin is easy—the Kelvin temperature is always exactly 273.15 more than the corresponding reading in degrees Celsius. The relation between Celsius and Fahrenheit is based on the difference between the freezing and boiling points of water, which is 100 Celsius degrees, but 180 Fahrenheit degrees (212°−32° F). Thus, the ratio of Fahrenheit degrees to Celsius degrees is $\frac{180}{100}$, or $\frac{9}{5}$. This, along with the fact that 0° Celsius is 32° Fahrenheit, yields a conversion formula:

$$d \text{ degrees Celsius} = \left(\frac{9}{5} \cdot d + 32\right) \text{ degrees Fahrenheit}$$

However, it is far better to learn these temperature scales by association with familiar situations than by algebraic manipulation. Table 12.1 shows some common temperatures in Celsius, Fahrenheit, and Kelvin.

	Celsius	*Fahrenheit*	*Kelvin*
Absolute zero	−273.15°	−459.67°	0
Cold winter day	−17.8°	0°	255.37
Freezing point of water	0°	32°	273.15
Room temperature	20°	68°	293.15
Warm summer day	30°	86°	303.15
Body temperature	37°	98.6°	310.15
Hot cup of coffee	50°	122°	323.15
Boiling point of water	100°	212°	373.15

Table 12.1 Some common temperatures.

Strictly speaking, the SI system of measurement is different from the metric system. However, because there are so many similarities and because the differences between the sizes of the corresponding standard units are so slight (from a nonscientist's viewpoint), it is still frequently called *the metric system.* The SI system is rapidly gaining worldwide acceptance. It has been adopted by Great Britain, and the United States has set a timetable for conversion to it. Most American school systems now include the SI system as part of the K-12 curriculum. Tables 12.2 and 12.3 summarize the main SI units of measure, the multiple and subdivision prefixes, and the standard symbols.

Note: The liter ("l") is not an official SI unit, but it is often used as an alternative volume measure.

Example 11 45.6 kg = 45.6 kilograms = 45,600 grams = 45,600 g □

quantity	*unit name*	*unit symbol*
length	meter	m
mass	kilogram	kg
temperature	Kelvin	K
area	square meter	m^2
volume	cubic meter	m^3
angle	radian	rad

Table 12.2 SI measurement units.

prefix:	tera-	giga-	mega-	kilo-	hecto-	deka-
symbol:	T	G	M	k	h	da
power:	10^{12}	10^9	10^6	10^3	10^2	10^1
prefix:	deci-	centi-	milli-	micro-	nano-	pico-
symbol:	d	c	m	μ	n	p
power:	10^{-1}	10^{-2}	10^{-3}	10^{-6}	10^{-9}	10^{-12}

Table 12.3 Prefixes for units in SI.

Example 12

$$2.378 \text{ megameters} = 2.378 \text{ Mm} = 2.378 \times 10^6 \text{ m}$$ □

Example 13

$$57 \text{ micrograms} = 57\ \mu\text{g} = 57 \times 10^{-6} \text{ g} = 57 \times 10^{-9} \text{ kg}$$ □

Example 14 When a unit is raised to a power, the exponent applies to the whole unit, including the prefix. For instance, cm^3 represents $(\text{cm})^3$—that is, "centimeters cubed," more commonly called "cubic centimeters." Thus,

$$34.5 \text{ cm}^3 = 34.5 \times (10^{-2}\text{ m})^3 = 34.5 \times 10^{-6}\text{m}^3$$

Similarly, 7 square kilometers is

$$7 \text{ km}^2 = 7 \times (10^3\text{ m})^2 = 7 \times 10^6\text{m}^2$$ □

Example 15 Sometimes, adults think that the metric system is difficult to learn. That's largely because they are unable to associate metric units with the sizes of familiar objects. Figure 12.22 shows how first graders are taught to associate the size of a centimeter with common things. □

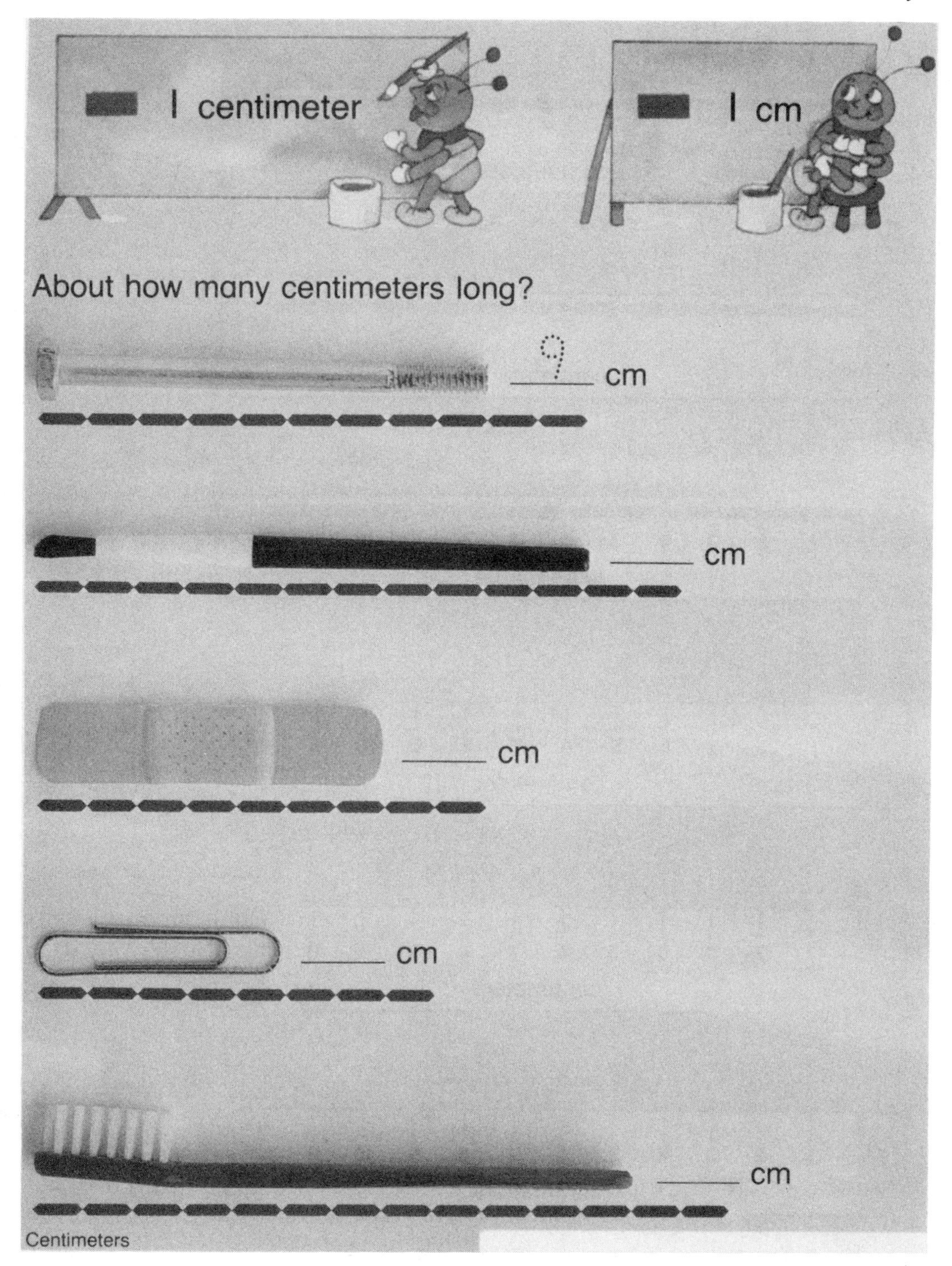

Figure 12.22 INVITATION TO MATHEMATICS, Grade 1, p. 107

PROBLEM-SOLVING COMMENT

The metric system provides a good example of the use of *appropriate notation.* It uses standard prefix symbols to indicate the power of 10 being used for each unit. However, until you get used to it, the very efficiency of the notational system can be its most confusing aspect. Thus, as you begin a problem involving metric units, be sure to *check the definitions* of the units and the symbols for them to guarantee that you fully understand the question asked and the information available. Table 12.3 provides a reference list for this purpose. ◇

Exercises 12.4

The measurement 3.145 m can be read "3 m, 1 dm, 4 cm, and 5 mm." For Exercises 1–6, interpret the given measurement in a similar way.

1. 5.276 m **2.** 2.003 m **3.** .364 m

4. 8.19 km **5.** 3.768 g **6.** 1.408 g

For Exercises 7–10, combine the given measurements into a single metric measurement that represents their sum.

7. 7 m, 3 dm, and 5 cm

8. 2 g, 3 dg, 5 cg, and 7 mg

9. 4m and 2 mm

10. 8 kg, 8 g, and 8 mg

For Exercises 11–42, complete each statement.

11. 6.25 km = ______ hm

12. 32.4 hm = ______ dam

13. 412 dam = ______ m

14. .63 m = ______ dm

15. 7.54 dm = ______ cm

16. 1207 cm = ______ mm

17. 325.1 m = ______ hm

18. 2.34 m = ______ km

19. 214.6 hm = ______ dm

20. 347.9 cm = ______ m

21. 23.5 g = ______ mg

22. .127 kg = ______ g

23. 12.7 mg = ______ g

24. 68,590 kg = ______ dag

25. 34.67 l = ______ ml

26. 350.1 cl = ______ l

27. .6572 kl = ______ ml

28. 56.7 hl = ______ l

29. 32.4 ml = ______ kl

30. .0078 dal = ______ ml

31. 234 cm^2 = ______ mm^2

32. 1.28 m^2 = ______ mm^2

33. 57.23 m^2 = ______ km^2

34. 5.67 hm^2 = ______ mm^2

35. 2589.5 cm^2 = ______ hm^2

36. 1 m^3 = ______ cm^3

37. .057 km^3 = ______ m^3

38. 1.795 dm^3 = ______ dam^3

39. 3.67 cm^3 = ______ ml

40. 24.7 l = ______ cm^3

41. 26.98 kl = ______ m^3

42. 65.21 mm^3 = ______ ml

For Exercises 43–48, estimate the mass of the given volume of cool water.

43. 3 cm^3 **44.** 57 mm^3 **45.** 258 m^3

46. 54.7 l **47.** 46 ml **48.** 5.2 kl

For Exercises 49–54, estimate the volume of the given mass of cool water in cubic measure and in liters.

49. 2 kg **50.** 45.6 g **51.** 78.5 mg
52. .34 hg **53.** 21.7 dg **54.** 44.4 cg

For Exercises 55–60, *estimate* the temperatures of the given items in degrees Celsius, in degrees Fahrenheit, and in Kelvin. Check the relationship among your three estimates by computation.

55. A lukewarm cup of coffee.

56. Someone with a high fever.

57. A hot day in Death Valley.

58. Antarctica during a storm.

59. A glass of iced tea.

60. A comfortable summer day.

12.5 Congruence

Previously, we described congruence intuitively by saying that two figures are congruent if they have the same size and shape. We then tied this together with the concept of measure by saying that two segments or two angles are congruent if they have the same measure. These references to "sameness" should lead you to anticipate the next theorem.

(T-6) Congruence of segments and of angles is an equivalence relation.

Since congruence of both segments and angles is defined by equality of measures, Theorem T-6 is a direct consequence of the fact that *equality* is an equivalence relation on numbers. (You might find it useful to check for yourself that congruence is reflexive, symmetric, and transitive.)

The Angle-Addition Postulate (P-19) allows us to establish three more theorems about congruence of angles. The first two of them are illustrated by Figure 12.23.

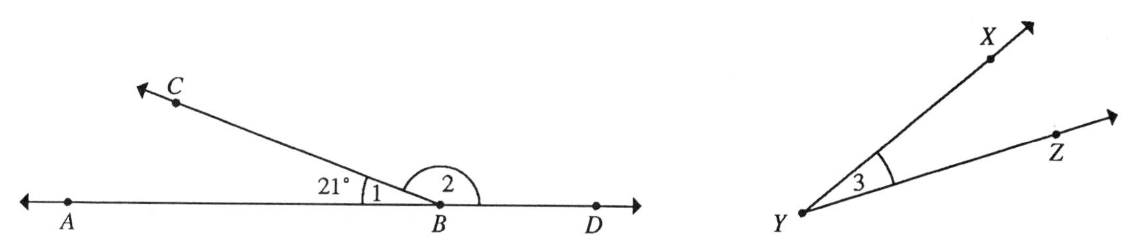

Figure 12.23 Theorems T-7 and T-8.

Notation It is often convenient to represent an angle in a diagram by using a single numeral written in the interior of the angle. In Figure 12.23, for example, $\angle ABC$ is called $\angle 1$, $\angle CBD$ is $\angle 2$, and $\angle XYZ$ is $\angle 3$.

Referring to Figure 12.23, suppose that $\angle 2$ is a supplement of $\angle 3$, that $m\angle 1 = 21°$, and that $\angle ABD$ is a straight angle. Then, by the Angle-Addition Postulate,

$$21° + m\angle 2 = 180°$$

Since $\angle 3$ and $\angle 2$ are supplementary,

$$m\angle 2 + m\angle 3 = 180°$$

implying that

$$21° + m\angle 2 = m\angle 2 + m\angle 3$$

By subtraction, $21° = m\angle 3$, so $\angle 1 \cong \angle 3$.

This example illustrates two simple, but important, theorems; their proofs are left as exercises:

(T-7) Two adjacent angles whose exterior sides form a straight angle are supplementary. (Such angles are called a **linear pair**.)

(T-8) Two angles that are supplements of congruent angles are themselves congruent.

These two theorems can be used to prove an important theorem about **vertical angles**—nonadjacent angles formed by two intersecting lines.

(T-9) Vertical angles are congruent.

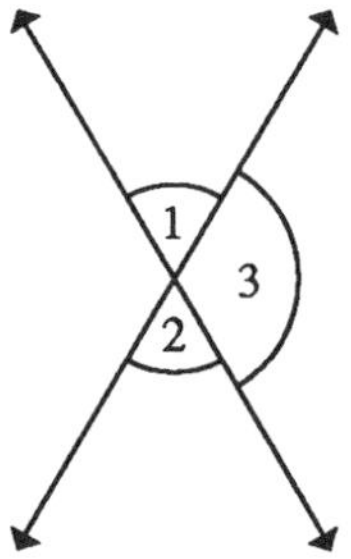

Using Figure 12.24 to help visualize the proof, suppose that $\angle 1$ and $\angle 2$ are vertical angles. Then by Theorem T-7, $\angle 1$ and $\angle 3$ are supplementary, as are $\angle 2$ and $\angle 3$. Since $\angle 3$ is congruent to itself, $\angle 1 \cong \angle 2$, by Theorem T-8.

Theorem T-9 tells us that when two lines intersect, the nonadjacent angles are always congruent. Sometimes the adjacent angles may also be congruent, but only in a very special circumstance.

Figure 12.24 Vertical angles.

(T-10) Two lines are perpendicular if and only if they intersect to form congruent adjacent angles.

(In many treatments of geometry, T-10 is used as the definition of perpendicularity.) This theorem is a biconditional, so there are two conditional statements to be proved:

(T-10a) If two lines are perpendicular, then they intersect to form congruent adjacent angles.

(T-10b) If two lines intersect to form congruent adjacent angles, then they are perpendicular.

To prove T-10(a), observe that perpendicular lines intersect at right angles, all of which are 90°. Therefore, these angles are congruent. The proof of T-10(b) follows immediately from Theorem T-7; it is left as an exercise.

In order to extend the congruence idea to triangles, we need some additional terminology. For each angle of a triangle, its **opposite side** is the side that (except for its endpoints) is in the interior of the angle. It is customary to let an uppercase letter represent a vertex of the triangle (and hence of an angle), and to let the corresponding lowercase letter represent *the measure of* its opposite side. Thus, in Figure 12.25, the side opposite $\angle A$ is $\overline{BC}$; its length is a.

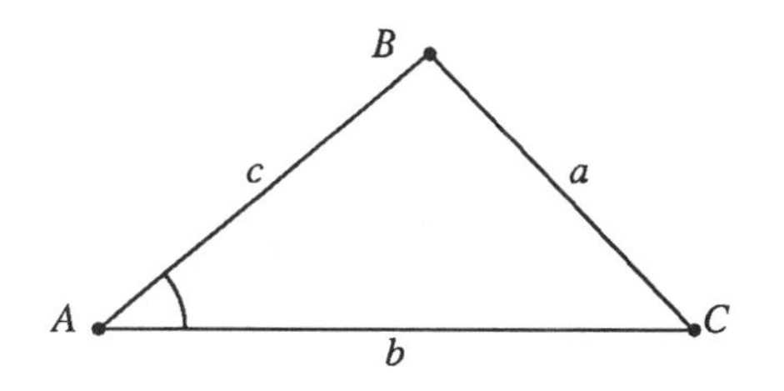

Figure 12.25 Angles and their opposite sides.

Intuitively, we regard two congruent triangles as triangles of the same "size" and "shape." More formally:

Definition Given a 1-1 correspondence between the sets of vertices of two triangles, two angles with corresponding vertices are called **corresponding angles**, and the sides opposite corresponding angles are called **corresponding sides**. Two triangles are **congruent** if there exists a 1-1 correspondence between their sets of vertices under which the corresponding angles and the corresponding sides are congruent.

Notation We write $\triangle ABC \cong \triangle XYZ$, where the order of the letters specifies the correspondence between the sets of vertices; in this case, $A \leftrightarrow X$, $B \leftrightarrow Y$, $C \leftrightarrow Z$. Thus, $\angle A \cong \angle X$, $\overline{AB} \cong \overline{XY}$, $a = x$, and so forth.

Example 1 Suppose $\triangle BDE \cong \triangle PRQ$, $m\angle B = \frac{\pi}{3}$ rad, $b = 3$ cm, and $r = 7$ cm. Then you also know that $m\angle P = \frac{\pi}{3}$ rad, $p = 3$ cm, and $d = 7$ cm. □

Example 2 "Hash marks" are often used in diagrams to identify congruent sides of triangles. Congruent angles are commonly marked by drawing the same number of arc lines. Figure 12.26 illustrates these conventions. According to the markings, $\triangle FCG \cong \triangle KME$. □

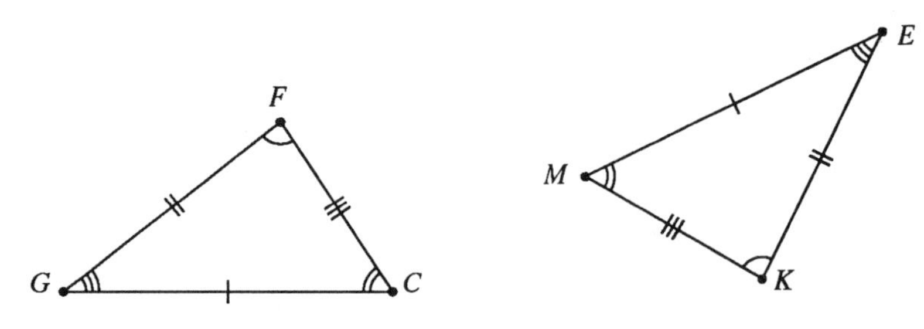

Figure 12.26 Congruent sides and angles.

Every pair of sides of a triangle is a subset of one of the angles, called the **included angle** between the two sides. Similarly, every pair of angles of a triangle has a side in common, called the **included side** between the two angles. Two fundamental congruence statements, perhaps familiar from high school, use this terminology. In many treatments of geometry, the first is given as a postulate and the second as a theorem. We follow that practice here, also omitting the proof of Theorem T-11 for the sake of brevity. Some exercises at the end of this section are designed to make these statements credible. Other ways to prove triangles congruent are also addressed in the exercises.

(P-20) **Side-Angle-Side (SAS)**: If two sides and the included angle of one triangle are congruent, respectively, to two sides and the included angle of another, then the two triangles are congruent.

(T-11) **Angle-Side-Angle (ASA):** If two angles and the included side of one triangle are congruent, respectively, to two angles and the included side of another, then the two triangles are congruent.

Example 3 In Figure 12.27, $\angle B$ is included between sides $\overline{AB}$ and $\overline{BC}$, and $\angle T$ is included between $\overline{RT}$ and $\overline{ST}$. From the congruence of the parts marked in the diagram we can conclude that $\triangle ABC \cong \triangle STR$, by SAS. □

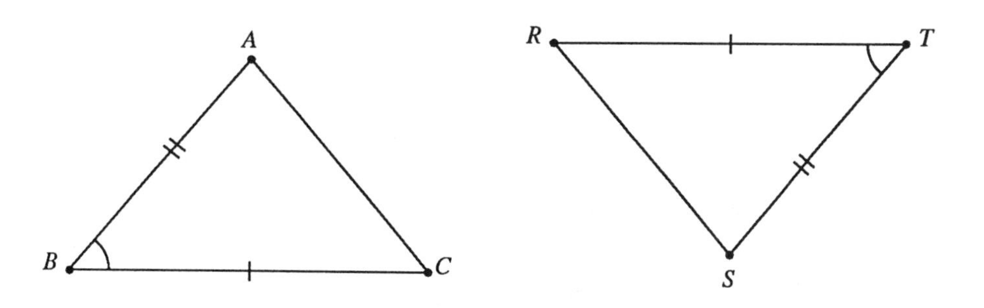

Figure 12.27 Side-Angle-Side.

Example 4 In Figure 12.28, $\overline{DP}$ is included between $\angle D$ and $\angle P$, and $\overline{XY}$ is included between $\angle X$ and $\angle Y$. The congruences marked in the diagram show that $\triangle RPD \cong \triangle QXY$, by ASA. □

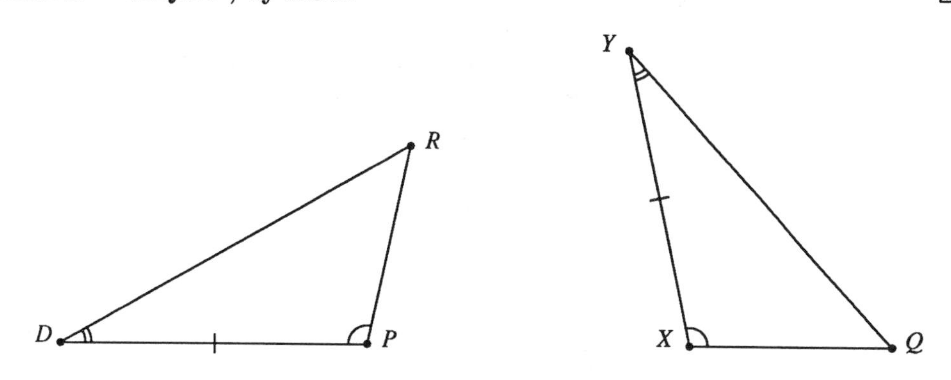

Figure 12.28 Angle-Side-Angle.

In an isosceles triangle (a triangle with two sides congruent), the two congruent sides are called its **legs**, and the remaining side is called its **base**. The angles opposite the legs are called the **base angles**, the angle opposite the base is called the **vertex angle**, and the vertex of that angle is called the

vertex of the triangle. (An equilateral triangle is a special kind of isosceles triangle, one in which any side can be considered as the base and any pair of angles as base angles.) An important two-part theorem relates the legs and base angles of isosceles triangles.

(T-12a) | If a triangle is isosceles, then the base angles are congruent.

(T-12b) | If two angles of a triangle are congruent, then it is isosceles.

This theorem is often (loosely) stated as a biconditional, to make it easier to remember:

(T-12) | A triangle is isosceles if and only if its base angles are congruent.

We prove T-12(a) using an argument that is believed to be the one that Euclid used when he originally proved the theorem. It involves showing that $\triangle ABC$ is congruent to itself by means of a correspondence *other* than the trivial one ($A \leftrightarrow A$, $B \leftrightarrow B$, $C \leftrightarrow C$). As you examine this proof, refer to Figure 12.29, which shows two copies of $\triangle ABC$ arranged so that the matching of the proof is visually obvious. The proof of T-12(b) is an adaptation of this argument; it left as an exercise.

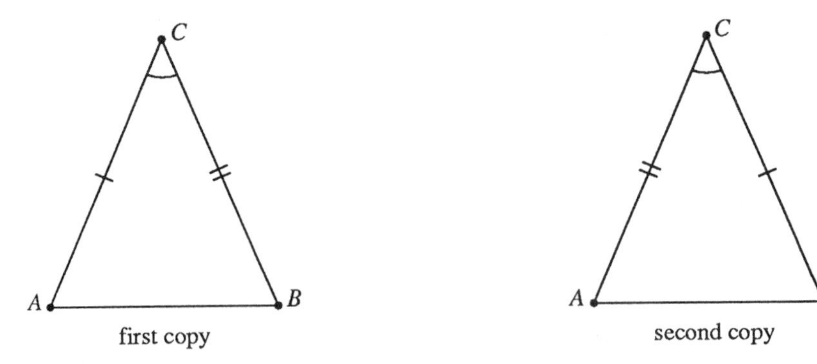

Figure 12.29 Theorem T12(a).

Proof: Assume that $\triangle ABC$ is isosceles, with $\overline{AC}$ and $\overline{BC}$ its legs. By the definition of *isosceles*, $\overline{AC} \cong \overline{BC}$. (We show this in Figure 12.29 by putting a single hash mark on $\overline{AC}$ in the first copy of the triangle and on $\overline{BC}$ in the second copy.) By the symmetric property, $\overline{BC} \cong \overline{AC}$. (We put a double hash mark on $\overline{BC}$ in the first copy and on $\overline{AC}$ in the second copy.) Now, $\angle C \cong \angle C$ (by what property of $\cong$?) Thus, by SAS (Postulate P-20), we have $\triangle ABC \cong \triangle BAC$. Therefore, $\angle A \cong \angle B$ (why?), as required.

One other important property of isosceles triangles should be mentioned, but before we can do that we need to define two more terms.

Definitions

A point M is a **midpoint** of $\overline{AB}$ if A-M-B and $\overline{AM} \cong \overline{MB}$. A **perpendicular bisector** of a line segment is a line perpendicular to the given line segment that goes through its midpoint.

(T-13) The line that joins the vertex of an isosceles triangle with the midpoint of its base is the perpendicular bisector of the base.

Proof: Theorem T-13 is proved by using congruent triangles. In Figure 12.30, $\triangle ABC$ is isosceles, with $\overline{AC} \cong \overline{BC}$. D is the midpoint of the base $\overline{AB}$, so $\overline{AD} \cong \overline{BD}$. By Theorem T-12, $\angle A = \angle B$. Thus, $\triangle ADC \cong \triangle BDC$, by SAS. In fact, $\angle ADC$ corresponds to $\angle BDC$, so $\angle ADC \cong \angle BDC$. Since these angles are adjacent and congruent, $\overleftrightarrow{CD} \perp \overleftrightarrow{AB}$ by Theorem T-10. Therefore, $\overleftrightarrow{CD}$ is the perpendicular bisector of $\overline{AB}$.

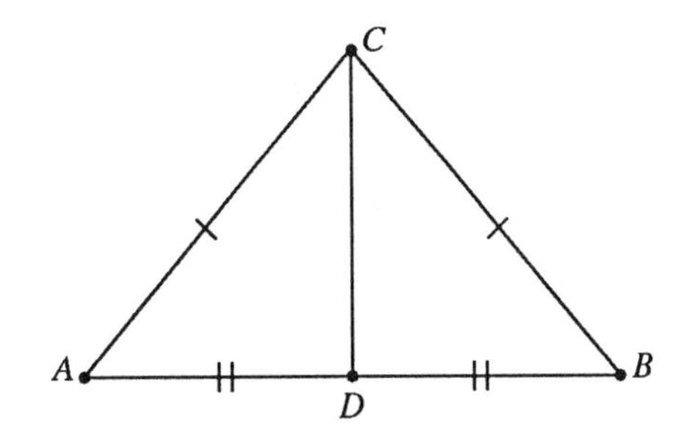

Figure 12.30 Theorem T-13.

An important question about perpendicularity is: How many lines can be drawn through a given point perpendicular to a given line? If the point is on the line, the Protractor Postulate tells us there is exactly one. (Think of using a protractor to draw a right angle with its vertex at the given point and one side along the given line.) Thus:

(T-14)	Through a given point on a given line, there is exactly one line perpendicular to the given line.

If the point is not on the line, the question is not so easily resolved. Although it can be proved rigorously, we shall use an intuitive argument to show that there is at least one perpendicular line. Suppose that C is a point not on a given line $\overleftrightarrow{PQ}$. (See Figure 12.31.) Then, by T-14, there is a line $\overleftrightarrow{XY}$ determined by the center of the semicircle that forms a protractor aligned with $\overleftrightarrow{PQ}$ and the point on the semicircle that corresponds to a right angle (90°). Envision sliding the edge of the protractor along $\overleftrightarrow{PQ}$, causing $\overleftrightarrow{XY}$ to slide along until it passes through C. At that instant, $\overleftrightarrow{XY}$ is a perpendicular to $\overleftrightarrow{PQ}$ through C, so there is at least one such line.

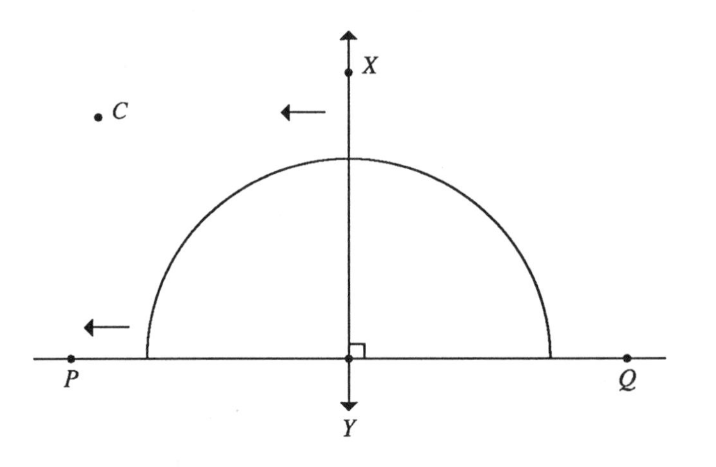

Figure 12.31 Theorem T-14.

There remains the question of whether there is more than one such perpendicular line. For instance, what if you were to approach C by sliding the protractor from the other direction? Would you arrive at the same perpendicular line or a different one? Perhaps the situation might look like Figure 12.32. If that were the case, then $\triangle ABC$ would contain two congruent right angles, so it would be isosceles, by Theorem T-12(b). Then, by T-13, if we connected C with the midpoint M of $\overline{AB}$, we would have a third perpendicular through C. We could then locate additional perpendiculars through the midpoints of $\overline{AM}$ and $\overline{MB}$, etc., implying that there are infinitely many perpendiculars to $\overleftrightarrow{PQ}$ through C. While this seems a bit strange, it is not *obviously* false. For the moment, then, we shall have to be content with this statement:

> Through a given point not on a given line, there is *at least* one line perpendicular to the given line.

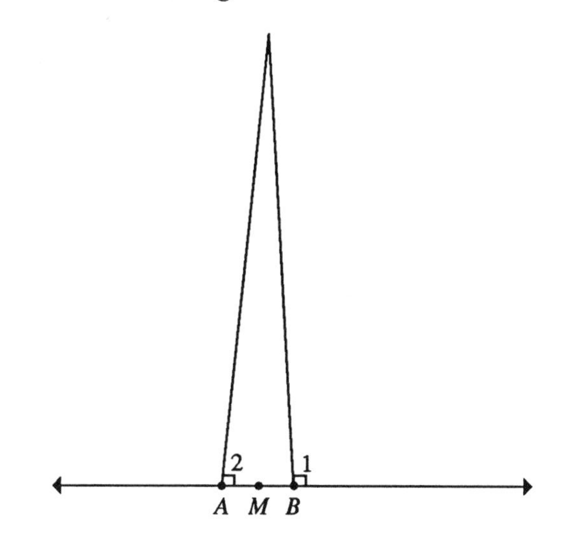

Figure 12.32 Perpendiculars from a point to a line.

To prove that there is no more than one perpendicular through a point not on a given line, we must discuss the notion of *exterior angles.*

Definition An **exterior angle** of a triangle is an angle that forms a linear pair with some angle of the triangle. Each of the other two angles of the triangle are called **remote interior angles** with respect to this exterior angle.

Example 5 In Figure 12.33, $\angle ABD$ is an exterior angle of $\triangle ABC$. It is adjacent and supplementary to $\angle ABC$. Relative to $\angle ABD$, angles A and C are remote interior angles. □

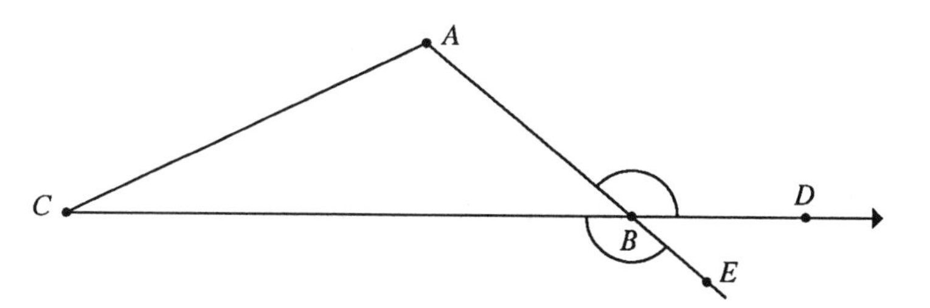

Figure 12.33 Exterior angles.

Example 6 Referring again to Figure 12.33, $\angle CBE$ is also an exterior angle of the triangle; it is congruent to $\angle ABD$ because they are vertical angles. At each vertex of the triangle there is one pair of congruent exterior angles. $\angle DBE$ is not an exterior angle because it does not form a linear pair with any angle of $\triangle ABC$. □

(T-15) **Exterior-Angle Theorem:** The measure of an exterior angle of a triangle is greater than the measure of either remote interior angle.

In Figure 12.34, $\angle CBD$ is an exterior angle of $\triangle ABC$; we want to show that its measure is greater than both $m\angle A$ and $m\angle C$. To prove $m\angle CBD > m\angle C$, let M be the midpoint of $\overline{BC}$, and let F be a point on $\overline{AM}$ such that A-M-F and $\overline{AM} \cong \overline{MF}$. Then

$$\triangle AMC \cong \triangle FMB$$

by SAS, implying that

$$\angle FBC \cong \angle C$$

Since F is in the interior of $\angle CBD$, we have

$$m\angle CBF + m\angle FBD = m\angle CBD$$

by angle addition. Thus, since $\angle CBF \cong \angle C$, we have

$$m\angle CBD > m\angle CBF = m\angle C$$

as required. The argument for $\angle A$ is similar and is left as an exercise.

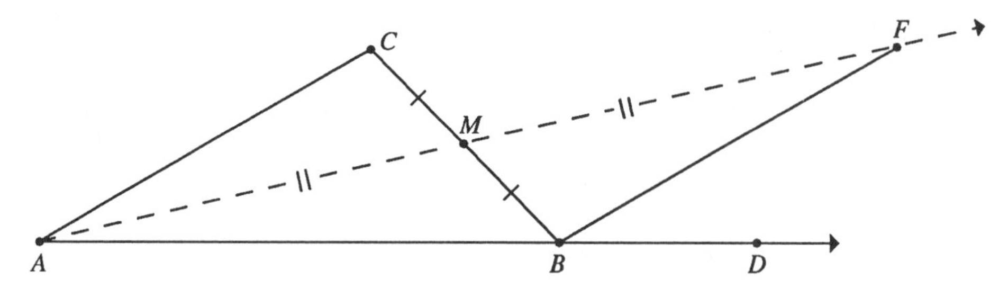

Figure 12.34 The Exterior-Angle Theorem.

The proof just given depends on the subtle, but important, assumption that point F is in the interior of $\angle CBD$. The diagram is drawn that way, of course, but the proof that the point F as constructed *always* lies in the interior of the angle is not easy. It depends on some betweenness and separation assumptions, whose implications will be discussed in more detail in Section 12.7.

Now we can prove the uniqueness of a perpendicular to a given line from a point not on that line. Suppose (as in Figure 12.32) that there were two perpendicular lines from point C to $\overleftrightarrow{AB}$. Then $\angle 1$ and $\angle 2$ are both right angles and therefore are congruent. But $\angle 1$ is an exterior angle of $\triangle ABC$, implying that $m\angle 1 > m\angle 2$, by Theorem T-15. This is a contradiction. Thus, the supposition must be false, and the theorem is proved. Therefore:

(T-16) Through a given point not on a given line, there is *exactly* one perpendicular to the line.

Exercises 12.5

1. Suppose that two angles are adjacent, that their exterior sides are collinear, and that the measure of one of the angles is 32°. What is the measure of the other? Why?

2. Suppose that two adjacent angles have collinear exterior sides and are congruent. What can you conclude about the angles? Why?

3. Suppose two angles are adjacent, their exterior sides are perpendicular, and the measure of one angle is 32°. What is the measure of the other? Why?

4. Draw a diagram that satisfies the conditions of Theorem T-7. Explain why the angles are supplementary.

For Exercises 5–8, assume that $\triangle PQR \cong \triangle VWU$.

5. Which side of $\triangle VWU$ corresponds to $\overline{PR}$? to $\overline{PQ}$? to $\overline{QR}$?

6. Which angle of $\triangle PQR$ corresponds to $\angle U$? to $\angle V$? to $\angle W$?

7. In $\triangle PQR$, which angle is included between $\overline{PR}$ and $\overline{QP}$?

8. In $\triangle VWU$, which side is included between $\angle U$ and $\angle V$?

9. This exercise provides the argument that proves Theorem T-8; all parts are related.

   (a) Suppose that $\angle 1 \cong \angle 2$ and $m\angle 1 = x°$. What is $m\angle 2$? Why?

   (b) Suppose that $\angle 3$ is the supplement of $\angle 1$. What is the measure of $\angle 3$? Why?

   (c) Suppose that $\angle 4$ is the supplement of $\angle 2$. What is the measure of $\angle 4$? Why?

   (d) Why are $\angle 3$ and $\angle 4$ congruent?

10. Using a protractor and a ruler:

    (a) Draw a triangle so that a 25°-angle is included between a side 3 cm long and a side 5 cm long.

    (b) Draw a second triangle with the same properties as the one in part (a). Compare the two triangles; measure the corresponding sides and angles. What do you observe? Which theorem does this support?

    (c) Draw a third triangle containing a side 3 cm long, a side 5 cm long, and a 25°-angle *not* included between these two sides. Is this triangle congruent to either of the other two? Why or why not?

11. Using a protractor and a ruler:

    (a) Draw a triangle containing a 22°-angle, a 55°-angle, and an included side 4 cm long.

    (b) Draw a second triangle with the same properties as the one in part (a). Compare the two triangles; measure the corresponding sides and angles. What do you observe? Which theorem does this support?

(c) Draw a third triangle with a 55°-angle, a 103°-angle, and an included side 4 cm long. Measure the third angle. Is this triangle congruent to either of the other two? Why or why not?

12. (a) Draw a triangle with sides of lengths 3 cm, 4 cm, and 6 cm.

(b) Draw a second triangle with the same side lengths as in part (a).

(c) Can you draw a triangle that has the same side lengths as in part (a), but is not congruent to either of the triangles you drew in parts (a) and (b)? Does this suggest a theorem? If so, state what you think it should be.

13. (a) Draw a triangle with a side 4 cm long, a side 3 cm long, and a 30°-angle that is not between these two sides.

(b) Draw a second triangle with the same three measures as the ones given in part (a), but with its third side a different length than the corresponding side of the first triangle.

(c) Do you think there should be a theorem labeled SSA? Why or why not?

14. Should there be a theorem labeled AAA? Why or why not?

15. Prove that if two angles of a triangle are congruent, then their opposite sides are congruent (that is, the triangle is isosceles).

16. Prove that a triangle is equilateral if and only if it is equiangular.

17. Is every equilateral quadrilateral equiangular? Why or why not?

18. Complete the proof of the Exterior-Angle Theorem by showing that $m\angle CBD > m\angle A$. (*Hint*: In Figure 12.34, draw exterior angle ABE, and take the midpoint of $\overline{AB}$.)

19. Use Theorem T-7 to prove Theorem T-10(b).

12.6 Parallelism

Parallelism has been one of the most controversial topics in all of mathematics. That may be hard to believe, since the common-sense idea of lines running side by side and never meeting (like the rails of a railroad track) seems simple and natural. Nevertheless, as soon as Euclid introduced the concept formally into his geometry, it became the focus of a logical dispute that has had consequences even to this day. We shall discuss this dispute in the next section; in this one we develop some of the basic properties of parallel lines in a plane. We begin, as Euclid did, by examining the situation of two lines crossed by a third one.

Definition A **transversal** is a line that intersects two other lines at two distinct points.

Line **t** in Figure 12.35 is a transversal that intersects $\mathbf{l}_1$ and $\mathbf{l}_2$ at points A and B, respectively. Of the eight angles formed (four at each intersection point), angles 1, 2, 3, and 4 are called **interior angles**, and angles 5, 6, 7, and 8 are **exterior angles**.

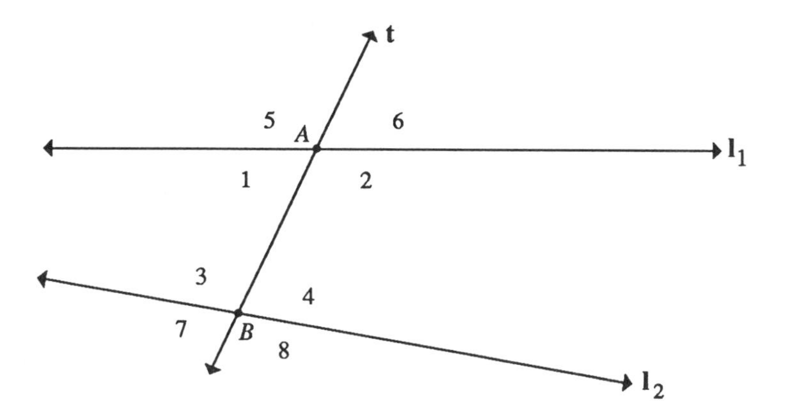

Figure 12.35 Two lines crossed by a transversal.

Definitions With respect to a transversal that intersects two other lines: Two nonadjacent, interior angles whose interiors are on opposite sides of the transversal are called **alternate interior angles**. Two nonadjacent angles, one interior and one exterior, but both on the same side of a transversal, are called **corresponding angles**.

Example 1 In Figure 12.35, angles 1 and 4 are alternate interior angles, as are angles 2 and 3. They are the only two such pairs in the diagram. □

Example 2 In Figure 12.35, angles 1 and 7 are corresponding angles, as are angles 3 and 5, 2 and 8, and 4 and 6. Angles 6 and 8 are not corresponding because both are exterior angles; angles 1 and 8 are not corresponding because their interiors are not on the same side of the transversal; angles 1 and 5 are not corresponding because they are adjacent. Angles 1 and 3 are called, naturally enough, *interior angles on the same side of the transversal.* □

The next theorem establishes a fundamental connection between alternate interior angles and parallelism, often seen in high-school geometry.

(T-17) If two lines are cut by a transversal so that a pair of alternate interior angles are congruent, then the lines are parallel.

To prove this, let **t** be a transversal that intersects lines **l** and **m** at points A and B, as shown in Figure 12.36, and let $\angle 1$ and $\angle 2$ be a pair of alternate interior angles. We must show that, if $\angle 1 \cong \angle 2$, then $\mathbf{l} \parallel \mathbf{m}$. We use an indirect argument: Suppose that the lines are not parallel. Then they must intersect on one side of the transversal. Assume they intersect

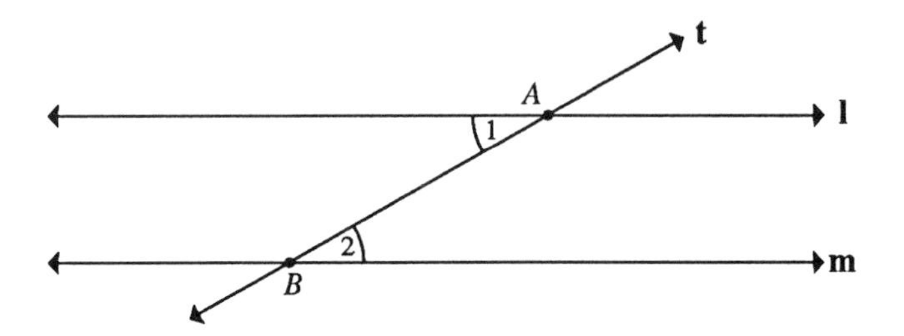

Figure 12.36: Parallel lines cut by a transversal.

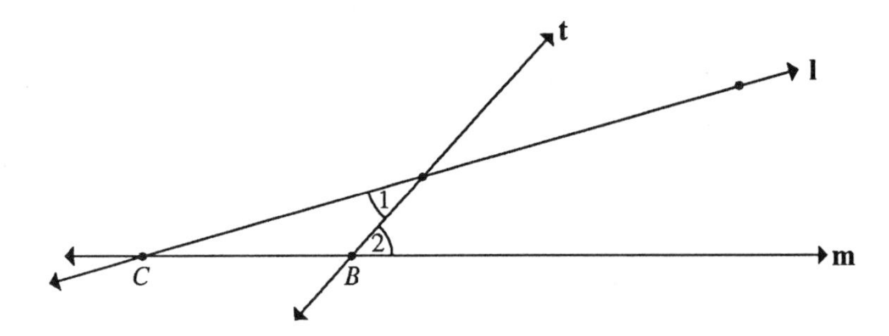

Figure 12.37: Intersecting lines cut by a transversal.

as in Figure 12.37, forming $\triangle ABC$. Then $\angle 2$ is an exterior angle of that triangle that is congruent to $\angle 1$ (a remote interior angle), contradicting the Exterior-Angle Theorem. Thus, the original supposition must be false, so the theorem is true. That is, **l** and **m** are parallel.

Two related results follow easily from this theorem; their proofs are left as exercises.

(T-18) If two lines are cut by a transversal so that two corresponding angles are congruent, then the lines are parallel.

(T-19) If two lines are cut by a transversal so that two interior angles on the same side of the transversal are supplementary, then the lines are parallel.

We can use Theorem T-17 to construct a line parallel to a given line through a point not on that line. Let **l** be a line and P a point not on it, as shown in Figure 12.38. Draw a line **t** through P intersecting **l** at some point

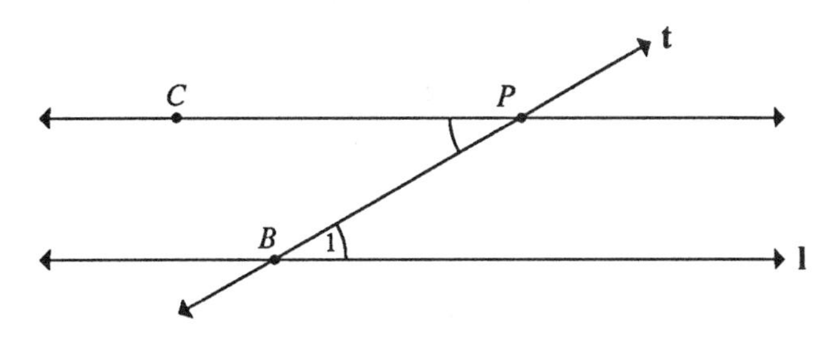

Figure 12.38 Theorem T-20.

B, forming $\angle 1$. Then at P construct $\angle CPB$ (for some point C) so that $\angle 1$ and $\angle CPB$ are alternate interior angles and $\angle 1 \cong \angle CPB$. (Can you justify these construction steps?) Then $\overline{PC}$ must be parallel to l, by T-17. This provides our next theorem.

(T-20) Through a given point not on a given line there is at least one line parallel to the given line.

Theorem T-20 leaves open the uniqueness question; it does not say that there is *only one* such parallel line. For now, we shall adopt a postulate stating that such a line is unique. The question of whether or not this can be proved as a theorem will be handled in the next section.

(P-21) **Playfair's Postulate:** Through a given point not on a given line there is at most one line parallel to the given line.

Postulate P-21 is named after the British mathematician John Playfair (1748–1819), who first stated it in this form. The converses of T-17, T-18, and T-19 are direct consequences of Playfair's Postulate:

(T-21) If two parallel lines are cut by a transversal, then any two alternate interior angles are congruent.

(T-22) If two parallel lines are cut by a transversal, then any two corresponding angles are congruent.

(T-23) If two parallel lines are cut by a transversal, then any two interior angles on the same side of the transversal are supplementary.

To prove Theorem T-21, consider a transversal **t** cutting parallel lines **l** and **m** at points A and B, as in Figure 12.39. Let $\angle 1$ and $\angle 2$ be alternate interior angles. If these two angles are not congruent, use the Angle-Construction Postulate to construct a line **k** through A to form $\angle 3$ in such a way that $\angle 2$ and $\angle 3$ are congruent and form alternate interior angles. $\mathbf{k} \parallel \mathbf{m}$, by Theorem T-17. But we also have $\mathbf{l} \parallel \mathbf{m}$ (by hypothesis), so there are two lines through A parallel to **m**, contradicting Playfair's Postulate.

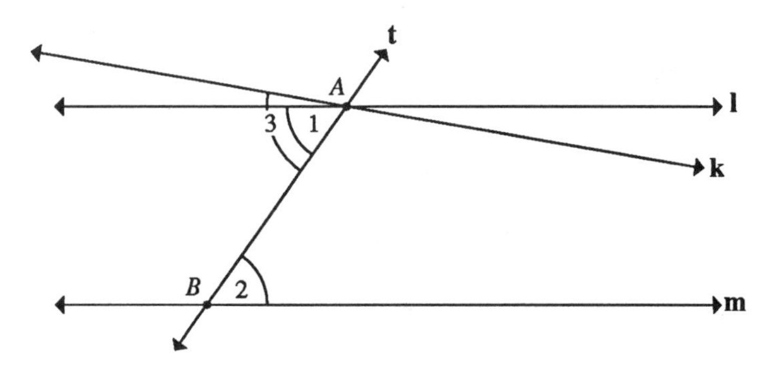

Figure 12.39 Theorems T-21, T-22, and T-23.

Theorem T-21 can be used to prove T-22 and T-23; these proofs are left as exercises. We end this section by proving a fundamental property of triangles:

(T-24) The sum of the measures of the angles of a triangle is 180°.

Proof: Consider a triangle $\triangle ABC$, as shown in Figure 12.40. By Theorem T-20, we can construct a line through C parallel to $\overline{AB}$. Since $\angle 1$ and $\angle A$ are alternate interior angles, as are $\angle 3$ and $\angle B$, we have

$$\angle 1 \cong \angle A \quad \text{and} \quad \angle 3 \cong \angle B$$

by Theorem T-21. Since $\angle DCE$ is a straight angle,

$$m\angle 1 + m\angle 2 + m\angle 3 = 180°$$

Thus, by substitution,

$$m\angle A + m\angle 2 + m\angle B = 180°$$

as required.

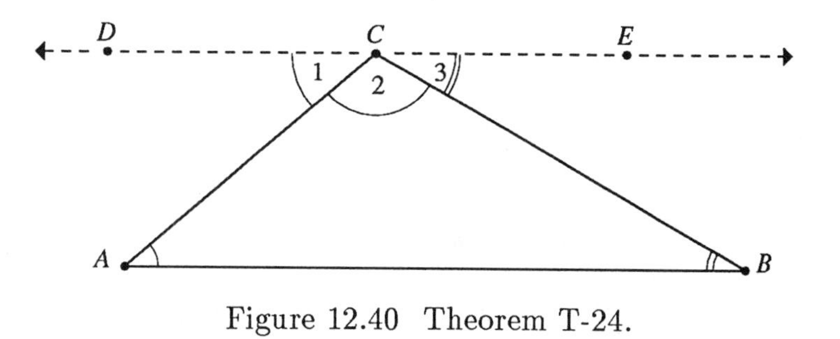

Figure 12.40 Theorem T-24.

Example 3 Figure 12.42 on page 484 shows how fifth graders are introduced to Theorem T-24. Compare this concrete activity with the abstract proof just given and note the similarities. □

Exercises 12.6

In Exercises 1–6, each statement about the diagram in Figure 12.41 can be justified by one of Theorems T-17 through T-23. In each case, indicate which theorem supports the statement.

1. If $\angle 2 \cong \angle 6$, then $\mathbf{l} \parallel \mathbf{m}$.
2. If $\mathbf{l} \parallel \mathbf{m}$, then $\angle 8 \cong \angle 6$.
3. If $\mathbf{l} \parallel \mathbf{m}$, then $\angle 2$ and $\angle 3$ are supplementary.
4. If $\angle 2 \cong \angle 4$, then $\mathbf{l} \parallel \mathbf{m}$.
5. If $\angle 6$ and $\angle 7$ are supplementary, $\mathbf{l} \parallel \mathbf{m}$.
6. If $\mathbf{l} \parallel \mathbf{m}$, then $\angle 3 \cong \angle 7$.
7. Prove Theorem T-18.
8. Prove Theorem T-19.
9. Prove Theorem T-22.
10. Prove Theorem T-23.
11. State and prove a theorem about alternate exterior angles formed by two lines and a transversal.
12. State and prove a theorem about nonadjacent exterior angles on the same side of a transversal.

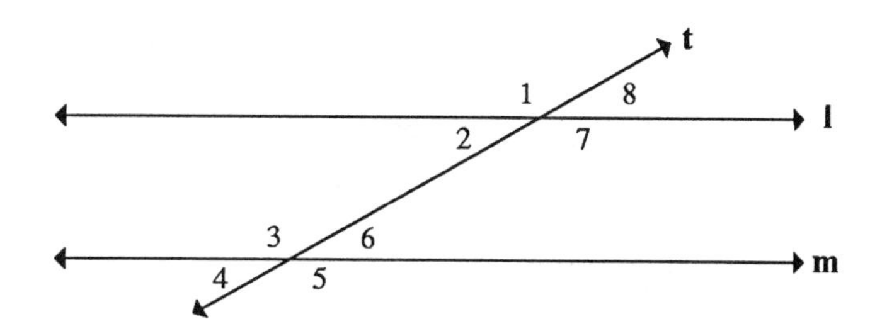

Figure 12.41 Exercises 1–6.

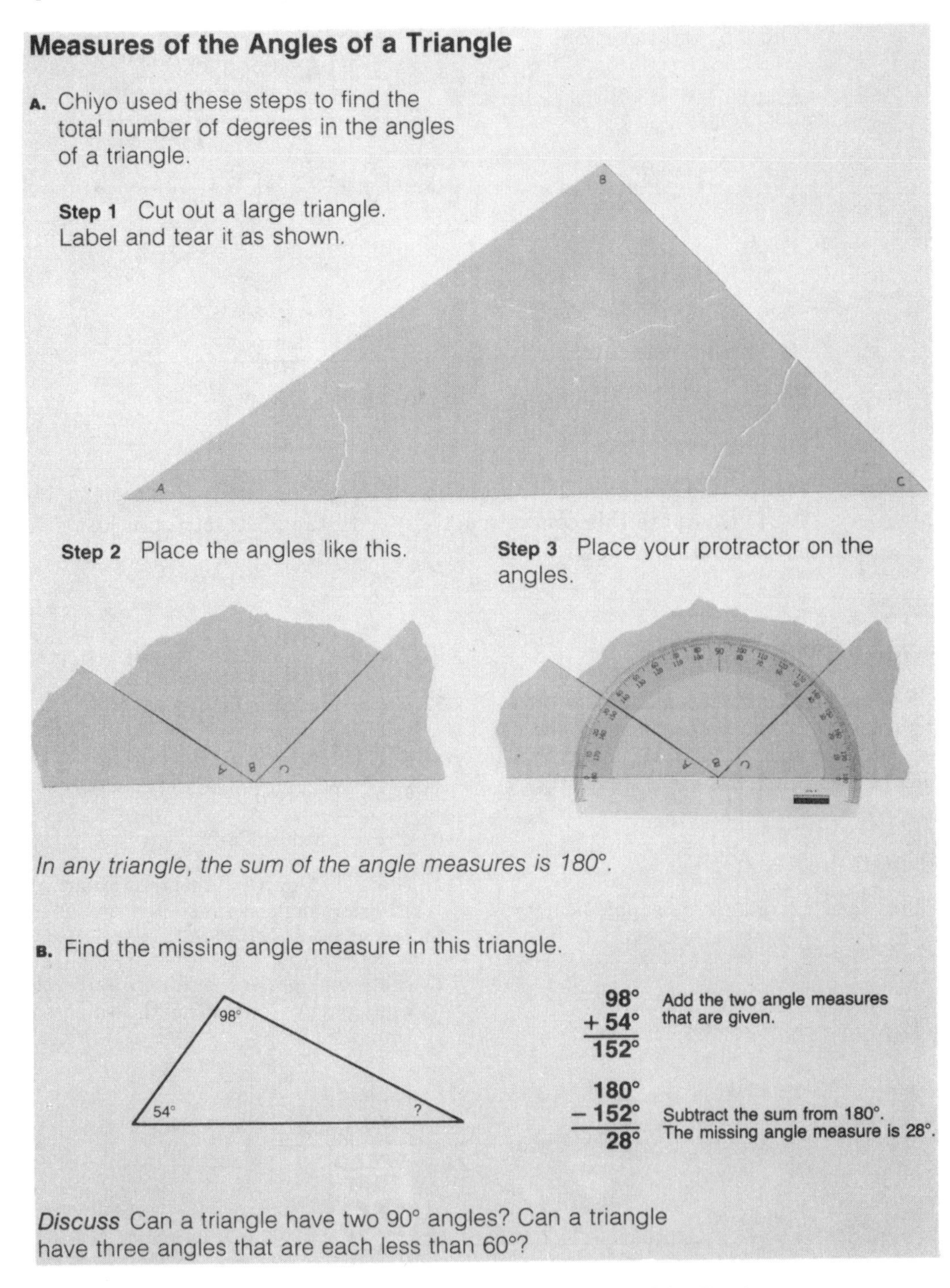

Measures of the Angles of a Triangle

A. Chiyo used these steps to find the total number of degrees in the angles of a triangle.

Step 1 Cut out a large triangle. Label and tear it as shown.

Step 2 Place the angles like this.

Step 3 Place your protractor on the angles.

In any triangle, the sum of the angle measures is 180°.

B. Find the missing angle measure in this triangle.

$$\begin{array}{r} 98^\circ \\ +\ 54^\circ \\ \hline 152^\circ \end{array}$$ Add the two angle measures that are given.

$$\begin{array}{r} 180^\circ \\ -\ 152^\circ \\ \hline 28^\circ \end{array}$$ Subtract the sum from 180°. The missing angle measure is 28°.

Discuss Can a triangle have two 90° angles? Can a triangle have three angles that are each less than 60°?

Figure 12.42 INVITATION TO MATHEMATICS, Grade 5, p. 318

13. Complete the following statement so that it is the contrapositive of T-23 on page 482:

    "If two lines are cut by a transversal so that ..."

14. A common activity used to help students discover the sum of the angles of a triangle is to have them measure each angle and add. The protractor commonly used in elementary schools measures angles to the nearest degree.

    (a) Using this protractor, what is the interval estimate for the sum of the angles of any triangle?

    (b) Listed below are some individual results recorded by students who were measuring different triangles. Which of these experiments may have been done correctly? Why? (Assume all measurements were done with the type of protractor described above.)

    Ann: $\angle A = 46°$, $\angle B = 31°$, $\angle C = 104°$
    Ben: $\angle A = 126°$, $\angle B = 17°$, $\angle C = 37°$
    Cal: $\angle A = 58°$, $\angle B = 58°$, $\angle C = 62°$
    Dot: $\angle A = 77°$, $\angle B = 19°$, $\angle C = 83°$
    Eve: $\angle A = 6°$, $\angle B = 153°$, $\angle C = 22°$

    (c) For each experiment in part (b) that may have been done correctly, suggest a possible "true value" for each angle so that the sum is exactly 180°.

    (d) On the basis of the experiment described above, what valid conclusion could students draw about the sum of the angles of a triangle?

12.7 The Great Euclidean Controversy

Early in Chapter 11 we pointed out that Euclid deduced all of the great body of plane geometry known in his time from five postulates. These five postulates, paraphrased a bit, are:

1. Two points determine a line segment.
2. Any line segment can be extended continuously to form a line.
3. Given any point and any line segment, there exists a circle with that point as center and the length of that line segment as radius.
4. All right angles are congruent.
5. If two straight lines in a plane are cut by a transversal making the sum of the measures of two interior angles on the same side of the transversal less than 180°, then the two straight lines will meet on that side of the transversal.

The sharpest criticisms of Euclid's work centered on the fifth postulate. Unlike the other four, this statement seemed to be much too complicated an idea to be a postulate—that is, to be an "obvious" assertion of fact. Other statements as complex as this one were proved as theorems. Why, then, did Euclid not prove it? Thus began a 2100-year history of attempts by mathematicians to prove Euclid's fifth postulate. As one mathematician after another attacked this problem and failed, it came to be recognized as a very difficult problem, one that would bring prominence to the person who solved it. Some mathematicians literally devoted their entire lives to it, trying one strategy after another without success.

It is instructive to discuss some of the statements that were recognized as equivalent to the fifth postulate. Recall that two statements are equivalent if each one implies the other. Exercise 8 of Section 12.6 indicates that the contrapositive of Theorem T-23 is, in fact, Euclid's fifth postulate; thus, T-23 and the fifth postulate are equivalent statements. In an exercise you will be asked to show that T-21, T-22, and T-23 are all equivalent. Thus, the fifth postulate can be proved if any one of those three theorems can be proved.

"So, what's the problem?" you ask. "We proved all of those theorems in the last section." That's true, but we did it *using Playfair's Postulate*, which is *also* equivalent to the fifth postulate! In fact, because of this equivalence, Euclid's fifth postulate often is called the "Parallel Postulate" and is written in Playfair's form. Another equivalent form of the fifth postulate is Theorem T-24. Thus, it is apparent that, postulate or theorem, this statement plays a major role in plane geometry.

Some mathematicians tried to prove the Parallel Postulate or one of its equivalent statements by contradiction. To prove Playfair's Postulate by contradiction, one must deny that there is at most one line parallel to a given line through a given point not on that line. This is done by assuming that there are at least two such parallels through the point, and then attempting to show that this assumption leads to a contradiction. Although such attempts based on the "at least two parallels" assumption led to many strange theorems that departed radically from the intuitive geometric view of what our world is "really" like, no contradiction was found—except, of course, for contradictions of those statements that are actually equivalent to the parallel postulate itself). For example, one consequence is that the sum of the angle measures of a triangle is less than 180°. This contradicts T-24 (the sum equals 180°), but T-24 is equivalent to the parallel postulate.

Early in the 19th century, Janos Bolyai (1802–1860) in Hungary, Karl Friedrich Gauss (1777–1855) in Germany, and Nikolai Lobachevsky (1793–1856) in Russia, each working independently of the others, began to entertain the radical idea that perhaps there were no contradictions to be found. That is, in spite of the strangeness of the theorems they were proving, perhaps they were, in fact, creating a new geometry, different from Euclid's but nevertheless, a consistent body of knowledge! Lobachevsky is acknowledged as the first actually to publish these findings (in 1829), but many historians believe that Gauss was the first to derive the results. The new geometry is called **non-Euclidean** because it contradicts one of Euclid's postulates; this specific form is called **Lobachevskian** (or **hyperbolic**) **geometry**.

In 1854 Bernhard Riemann (1826–1866) showed there is a second type of non-Euclidean geometry, now called **Riemannian** (or **elliptic**) **geometry**, in which there are no parallels at all. At first you may think that this is not possible because we had proved the existence of parallels in T-20. However, if you check the proof of Theorem T-20, you will find that T-17 was used in it, and the Exterior-Angle Theorem (T-15) was used to prove T-17. As you may recall, our discussion of the Exterior-Angle Theorem (in Section 12.5) ended with a caution that its proof depends on the fairly subtle issue of whether or not a point (F) was in the interior of the exterior angle, as it appeared to be from the diagram. Riemann showed that this proof depends on certain assumptions about betweenness and separation, and, if these assumptions were denied, then the Exterior-Angle Theorem would be false and parallel lines would not exist. In this geometry, it can be shown that the sum of the angle measures of a triangle is *greater than* 180°.

Thus, the 2100-year-old controversy finally was resolved. Euclid had been vindicated. He was, after all, quite correct in stating his fifth postulate, for indeed it was not possible to prove it from the others. In so doing he had successfully derived a geometry that describes our world in a way consistent with our perceptions and intuition. However, it now has been shown that there actually are three different (mutually contradictory) geometries—Euclidean, Lobachevskian, and Riemannian. Each one has useful applications in the "real world." The many applications of Euclidean geometry are obvious, ranging through architecture, astronomy, engineering, surveying, and so on. Not so obvious, but also important, are the applications of the other two geometries. Albert Einstein developed the general theory of relativity using the assumption that we live in a non-Euclidean universe. Non-Euclidean geometries are being applied in the study of atomic physics and of space travel; it seems that very small and very large environments do not conform well to Euclid's view of our medium-sized world.

In Chapter 11 we observed that different applications of a geometry can be obtained by different interpretations of the undefined terms. It was the discovery of the non-Euclidean geometries that led to this abstract view of geometry and of mathematical systems in general. Initially, mathematicians were reluctant to accept these new and strange ideas largely because they were contrary to the traditional interpretations of the terms *point*, *line*, and *plane*. However, when different interpretations of the basic terms were used, the other geometries became plausible and useful in physical science. A major result of these discoveries was the realization that postulates in a system need not be viewed as "evident truths"; they are just assumptions, not required to conform to one's intuition about "reality." Thus, it was the parallel-postulate controversy and its resolution that led to the current practice of studying an abstract postulate system stripped of interpretation, and applying a variety of interpretations to that system only after its theorems have been developed.

The following summary of properties provides a basis for comparison and contrast among Euclidean geometry and the two major non-Euclidean systems.

Properties common to all three geometries:

- T-1 through T-10
- T-12(a)
- T-13
- T-14 (unique perpendicular from a point on a line)
- T-16a (existence of perpendicular from point not on a line)

Properties of Euclidean and Lobachevskian geometries only:

- T-11 (ASA)
- T-12(b)
- T-15 (Exterior-Angle Theorem)
- T-16(b) (unique perpendicular from a point not on a line)
- T-17, T-18, T-19 (transversal theorems about angle properties implying parallelism)
- T-20 (existence of parallel lines)

Properties of Euclidean geometry only:

- Parallel Postulate (uniqueness of parallels)
- T-21, T-22, T-23 (transversal theorems about parallelism implying angle properties)
- T-24 (Angle sum in a triangle equals 180°.)
- Pythagorean Theorem

Properties of Lobachevskian geometry only:

- Two or more parallels (in fact, infinitely many) through a point not on a line
- Angle sum in a triangle is less than 180°.

Properties of Riemannian geometry only:

- No parallel lines.
- Two or more perpendiculars (in fact, infinitely many) through a point to a line.
- Angle sum in a triangle is greater than 180°.

Exercises 12.7

For Exercises 1–17, place E, L, or R beside each statement to indicate if it is true in Euclidean, Lobachevskian, or Riemannian geometry, respectively. Use as many of these letters as apply.

1. Vertical angles are congruent.
2. The base angles of an isosceles triangle are congruent.
3. If two lines are parallel, then alternate interior angles are congruent.
4. Supplements of congruent angles are congruent.
5. Through a point not on a line at least one perpendicular to the line may be drawn.
6. Through a point not on a line at least one parallel to the line may be drawn.
7. An exterior angle of a triangle has greater measure than either of its remote interior angles.
8. Two lines cut by a transversal so that alternate interior angles are congruent must be parallel.
9. Two lines cut by a transversal so that corresponding angles are not congruent cannot be parallel.
10. If two intersecting lines are cut by a transversal, then the interior angles on the same side of the transversal are not supplementary.
11. Through a point not on a line there is at most one parallel to that line.
12. Through a point not on a line there are at least two parallels to that line.
13. Through a point not on a line there is at most one perpendicular to that line.
14. Through a point not on a line there are at least two perpendiculars to that line.
15. The sum of the measures of the angles of a triangle is not 180°.
16. The sum of the measures of the angles of a triangle is at least 180°.
17. The sum of the measures of the angles of a triangle is at most 180°.

Exercises 18 and 19 together show that Theorems T-21 and T-22 are equivalent.

18. Use Theorem T-21 to prove T-22.
19. Use Theorem T-22 to prove T-21.
20. Prove that Theorems T-22 and T-23 are equivalent. (*Hint*: Prove that each implies the other.)
21. Assuming the results of Exercises 18–20, is there an easy way to conclude that Theorems T-21 and T-23 are equivalent? Explain.

Review Exercises for Chapter 12

For Exercises 1–10, indicate whether the given statement is *true* or *false*.

1. There is a 1-1 correspondence between the points on a line and the set of rational numbers.
2. If two line segments are congruent, then they are equal.
3. A chord of a circle is any line segment that intersects the circle in two distinct points.
4. Pi (π) is the ratio of the circumference of a circle to its diameter.
5. Because π is an irrational number, it does not correspond to a point on the number line.
6. Any triangle that contains an acute angle is an acute triangle.
7. The *degree* is the standard metric unit of angular measure.
8. A centimeter is $\frac{1}{100}$th of a meter.
9. Finally, after centuries of hard work and unsuccessful attempts, mathematicians have proved Euclid's fifth postulate.
10. The proof that the sum of the angles of a triangle equals 180° can only be proved by assuming Euclid's fifth postulate or a statement equivalent to it.

Exercises 11–17 refer to the number line shown in Figure 12.43.

11. Name a unit segment.
12. Name two congruent segments.
13. What is the measure of $\overline{BF}$?
14. Name a segment whose measure is 13.
15. Name two points that are 6 units from C.
16. What is the graph of $\{x \mid -5 \leq x \leq 5\}$?
17. What is the graph of $\{x \mid x \leq -5\}$?

For Exercises 18–29, complete each statement by filling in the blank(s).

18. An equilateral quadrilateral is called a _____.
19. If a triangle has two or more congruent sides, it is _____.
20. If the measure of an angle is greater than 90° and less than 180°, the angle is _____.
21. The ratio of the radian measure of an angle to its degree measure is always _____.
22. In the metric system, the standard unit of linear measure is the _____.
23. In the metric system, the prefix for 1000 is _____, whereas the prefix for $\frac{1}{1000}$th is _____.
24. In the metric system, the prefix for 10 is _____, whereas the prefix for $\frac{1}{10}$th is _____.
25. Two nonadjacent angles formed by two intersecting lines are called _____.
26. A line that is perpendicular to a line segment and goes through the midpoint of the segment is called a _____.
27. The measure of an exterior angle of a triangle is always _____ the measure of either remote interior angle.
28. Any geometry that contradicts one of Euclid's postulates is called _____.
29. Water freezes at _____C and boils at _____C.

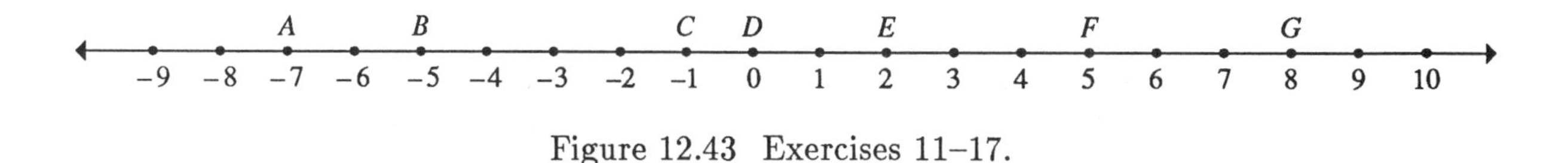

Figure 12.43 Exercises 11–17.

30. Which basic metric unit would most likely be used by a butcher to measure meat?

31. Which basic metric unit would most likely be used by a seamstress to measure cloth?

32. Which basic metric unit would most likely be used by a gas station to measure fuel?

For Exercises 33–38, answer the question by moving the decimal point an appropriate number of places to the right or left.

33. 5.25 kg = ? g

34. 235 mm = ? dm

35. 47.2 hl = ? cl

36. 42 cm = ? m

37. 8.95 km = ? mm

38. 342,587 g = ? kg

In Exercises 39–42, the letters C, r, and d represent the circumference, the radius, and the diameter of a circle, respectively. Approximate your answers to 3 significant digits.

39. If $r = 3.10$, find C.

40. If $r = 42.3$, find C.

41. If $C = 54.1$, find r.

42. If $C = 1.57$, find d.

Exercises 43–47 refer to Figure 12.11 on page 448.

43. What is the measure of $\angle RNQ$?

44. What is the measure of $\angle TNP$?

45. Name two obtuse angles.

46. Name two acute angles.

47. Name two supplementary angles.

48. What is the radian measure of an angle of $210°$?

49. What is the degree measure of an angle of $\frac{2\pi}{3}$ radians?

In Exercises 50 and 51, h is the length of the hypotenuse of a right triangle, and k and m are the lengths of the two legs. Use the Pythagorean Theorem to find the missing measure. Approximate your answer to the appropriate number of significant digits.

50. $k = 3.4$, $m = 2.56$

51. $h = 52.4$, $k = 21.4$

For Exercises 52–56, sketch figures that satisfy the given conditions.

52. Two noncongruent triangles such that the three angles of one are congruent to corresponding angles of the other.

53. Two noncongruent triangles such that two sides and an angle of one are congruent to corresponding parts of the other.

54. An equilateral quadrilateral that is not equiangular.

55. A pair of lines and a transversal such that two alternate interior angles are not congruent.

56. A triangle $\triangle XYZ$ such that $\triangle XYZ \cong \triangle YXZ$, but $\triangle XYZ \ncong \triangle ZXY$. (Recall that the order in which the vertices are listed indicates their correspondence.)

13.1 The Problem

By the end of Chapter 9 we had developed a number system that allows us to perform all four elementary arithmetic operations—addition, subtraction, multiplication, and division. This system is called the *rational*-number system because it is based on *ratios* of two numbers, which we commonly call *fractions*. Ratios of natural numbers have been studied and used in mathematics since ancient Greek times.

Our story begins with a familiar statement from geometry, known to the early Greeks and credited to Pythagoras:

> **The Pythagorean Theorem**: If a and b are the lengths of the two legs of a right triangle and c is the length of its hypotenuse, then
> $$a^2 + b^2 = c^2$$

Ironically, one of the most significant ideas to come from the Pythagoreans was the verification of a fact that destroyed their own philosophy! They proved that *some lengths cannot be expressed as ratios of natural numbers*,

Historical Note: The Pythagoreans

The Pythagoreans were a secret society whose founder, Pythagoras, apparently lived from about 575 to 500 B.C. Reliable information about Pythagoras and his followers is obscured by legends and the lack of written records from that time, but it appears that Pythagoras was born on the Greek island of Samos. He left there as a young man to travel and study in Egypt and Babylon, and possibly in India, then returned many years later to found a "school" at Crotona, a town of Magna Graecia on the southeast coast of the Italian peninsula.

The followers of Pythagoras consisted of several hundred young aristocrats who formed themselves into a secret society built around a strict disciplinary code and the belief that the pursuit of knowledge was the guiding principle of a moral life. They were strict vegetarians because they believed that, at death, human souls sometimes inhabited the bodies of animals. Like many other secret societies, the Pythagoreans had a variety of rituals and strange customs, including the refusal to eat lentils or drink wine, or to pick up anything that had fallen, or to stir a fire with a piece of iron. Their special sign was the *pentagram*, the five-pointed star.[1]

The Pythagoreans based their entire philosophy of reality on the assumption that the world can be totally explained by natural numbers and their ratios. For example, they asserted such things as "five is the cause of color, six of cold, seven of health, and eight of love."[2] Their motto was said to be "All is number," and Pythagoras is credited with coining the word *mathematics* to mean "that which is learned."[3] As they pursued mystical properties of numbers, the Pythagoreans made fundamental contributions to number theory, music theory, astronomy, and geometry. They were among the first scholars to insist on deductive proof as the way to insure the truth of a statement. The most famous theorem of geometry (about right triangles) still bears their name. ◇

thereby discovering the existence of "irrational quantities." Specifically, one of the Pythagoreans proved that the diagonal of a square one unit on a side cannot be expressed as a ratio of natural numbers. The proof, outlined here, is interesting for its logical form as well as for its content; it is a clear, brief example of "proof by contradiction," a classical proof technique developed by Greek logicians and used by mathematicians even today.

Suppose that a square one unit on a side has a diagonal whose length d can be expressed as a rational number. (See Figure 13.1.) Then d can be written in lowest terms, say as $\frac{x}{y}$. (Recall that *lowest terms* means that the

---

[1]David M. Burton, *The History of Mathematics*, Boston: Allyn and Bacon, Inc., 1985, p. 99.

[2]David Eugene Smith, *History of Mathematics*, Vol. I, New York: Dover Publications, Inc., 1958, p. 74.

[3]Carl B. Boyer, *A History of Mathematics*, New York: John Wiley & Sons, 1968, pp. 53–54.

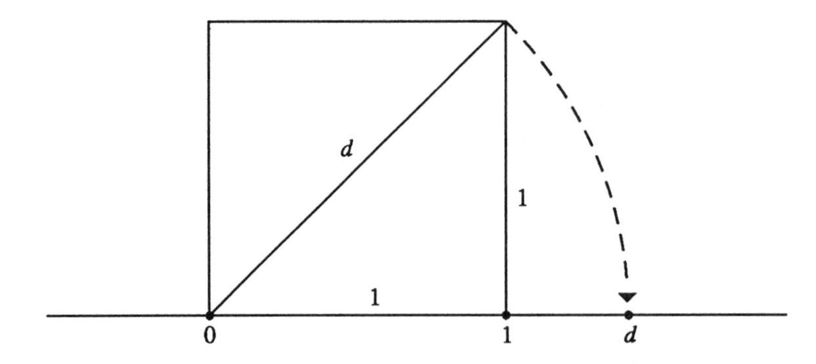

Figure 13.1 The diagonal of a unit square.

numerator and denominator are relatively prime, and that every rational number can be expressed in lowest terms.) By the Pythagorean Theorem,

$$\left(\frac{x}{y}\right)^2 = 1^2 + 1^2$$

so

$$\frac{x^2}{y^2} = 2$$

which implies

$$x^2 = 2y^2$$

This means that x^2 is even, so x must be even. (See Exercise 1.) Write x as $2s$, where s is some natural number. Then, substituting $2s$ for x in the preceding equation, we have

$$(2s)^2 = 2y^2$$

or

$$4s^2 = 2y^2$$

so

$$2s^2 = y^2$$

This implies that y^2 is even, which, in turn, means that y must be even. The fact that x and y must both be even contradicts the hypothesis that $\frac{x}{y}$ is in lowest terms (because 2 is a common factor). The only logical escape from this absurdity is to recognize the original hypothesis as false. That is, the diagonal d *cannot* be expressed as the ratio of two natural numbers!

The unfortunate Pythagorean who discovered this numerical heresy was drowned in a shipwreck. His colleagues tried to suppress the offensive idea he had found, but eventually word leaked out and people began shopping for a new philosophy of nature.

In modern terminology, the positive number whose square is 2 is called "the square root of 2" and is denoted by the symbol $\sqrt{2}$. Today we describe the fact that there is no rational number whose square is 2 by saying that $\sqrt{2}$ is "irrational." This language implies, however, that $\sqrt{2}$ is a number *of some sort*, a fact that was by no means obvious to the Greeks. The only numbers they ever dealt with were (positive) rational numbers; the thing we call $\sqrt{2}$ was for them a geometric object rather than a numerical one. It presented the Greeks with a numerical problem that might be phrased something like this:

> The diagonal of a unit square represents a specific segment of a line. If one end of that segment is used to represent zero and all the rationals are placed at their appropriate points on the line (each rational representing the point that is that distance from zero), then *no rational number will be exactly at the other end of the original segment.* But that other endpoint must represent *some* distance from zero, so it must have some numerical meaning.

The difficulty the Greeks faced rests on the implicit assumption that there are no "gaps" in a continuously produced line. That assumption was never explicitly stated by the Greeks because its acceptance was never questioned. When the irrationality of $\sqrt{2}$ was proved, the Pythagoreans *automatically* inferred the collapse of their philosophy, even though that proof only showed that their claim of the "universality" of whole-number ratios was incompatible with the belief that there are no gaps in a continuous line. To resolve the incompatibility, one might as easily have discarded one assumption as the other because there was no proof of either one. However, the thought that there might be holes or gaps in a continuous line was uncomfortable for the Greeks (as it is for most people). Thus, the makers of mathematics chose the path of greater comfort; they discarded the Pythagorean assumption and set out to expand their suddenly incomplete system of describing magnitudes.

The problem we face is this: If we choose to agree (as we do) that there are no gaps in a line, then to each point on a line there should correspond some unique numerical label that indicates its distance from some fixed starting point, 0, with respect to some chosen unit of measure. We have seen that the diagonal of a unit square determines a point whose distance from 0 is not a rational number, and it is not hard to show that there are other such points. (See Exercises 5–9.) Therefore, we must expand our system of numbers to include labels for these "extra" points in some convenient, consistent way. Ideally, we would like our labeling system to allow us to find the length of any line segment, to decide which of two segments is longer, and to add, subtract, multiply, or divide these labels in much the same way as we handle the rational numbers. Sections 13.2 – 13.4 are devoted to constructing the elements of such a system.

PROBLEM-SOLVING COMMENT

Notice that the instructions and hints to the following exercises contain both explicit and implicit references to the problem-solving tactics we have studied. For instance, Exercise 1 depends on *checking the definition* of an even number in part (a), *arguing by analogy* in part (b), and *generalizing these solutions* in part (c). Other exercises in this set also use these tactics, as well as suggesting that you *draw diagrams* and *reason backwards* in an indirect argument form. ◇

Note: For an abbreviated treatment of the real-number system, it is possible to go directly to the beginning of Section 13.5, skipping the construction of the infinite decimals as described in Sections 13.2 – 13.4. In so doing, you would simply have to accept the fact that every point on the number line can be labeled by an infinite decimal in a "reasonable" way. While this approach is not recommended unless you are thoroughly familiar with the geometric interpretation of decimals, we realize that time constraints or other considerations may make this option necessary. Therefore, the next three sections have been labeled as "optional."

Exercises 13.1

1. Prove that whenever a whole number of the form n^2 is even, then n itself must be even. (*Hint*: When we say that a whole number is even, we mean that it has 2 as a factor. How are the prime factorizations of n and n^2 related?)

2. Prove that whenever a whole number of the form n^2 is divisible by 3, then n itself must be divisible by 3. (*Hint*: Notice the analogy with Exercise 1; try to adapt the argument you used there.)

3. Generalize the arguments you used in Exercises 1 and 2 to prove that whenever a whole number of the form n^2 is divisible by a prime p, then n itself must be divisible by p.

4. Draw a line and label two points as 0 and $\sqrt{2}$. Then mark the points that should be labeled as $5\sqrt{2}$ and $6\sqrt{2}$.

5. Prove that $5\sqrt{2}$ cannot be a rational number. (*Hint*: Think about the fact that the product of two rational numbers must be rational.)

6. Prove that $6\sqrt{2}$ cannot be a rational number. (*Hint*: Argue by analogy with Exercise 5.)

7. Generalize the arguments you used in Exercises 5 and 6 to prove that $n\sqrt{2}$ cannot be a rational number for any natural number n. (*Note*: This exercise shows that there must be infinitely many points on the number line that do not have rational-number labels.)

8. Suppose you have a right triangle whose shorter leg is 1 cm long and whose hypotenuse is 2 cm long. Prove that the length of the other leg of this triangle cannot be expressed (in centimeters) as a rational number. (*Hint*: Begin by drawing a sketch of the situation. Then look back at the proof

of the irrationality of $\sqrt{2}$ given in this section and mimic it as closely as you can. Start by applying the Pythagorean Theorem.)

9. Suppose you have a right triangle whose legs are 1 cm and 2 cm long, respectively. Prove that the length of the hypotenuse cannot be expressed (in centimeters) as a rational number.

10. The proof that $\sqrt{2}$ is not a rational number can be simplified by using the Fundamental Theorem of Arithmetic. The equation

$$x^2 = 2y^2$$

is in conflict with that theorem, and hence immediately provides the desired contradiction. Explain how this equation conflicts with the Fundamental Theorem.

13.2 Approximation by Binary Fractions [optional]

(Some material in this section is required for Sections 13.3 and 13.4.)

Here is our situation:

- A number line (as seen in Chapter 9) represents a 1-1 correspondence between the set of rational numbers (our most comprehensive number system so far) and a subset of the points on a line. Each number represents the direction and distance of its corresponding point from some point of origin, called 0.
- Some points of the line do not have rational-number labels; that is, the distances between some points and 0 are not expressible as rational numbers. We have called these points *irrational.* One such point is denoted by $\sqrt{2}$; the existence of infinitely many others was established in the exercises of Section 13.1 (and in Chapter 12).

Although we cannot state exactly the distance between any one of these irrational points and 0 using rational numbers, we can use rationals to approximate that distance *as closely as we please.* In fact, we do not even need all the rationals to do this; we can approximate distances to any desired degree of accuracy using any one of many different subsets of the rational numbers. For instance, we can approximate all distances as closely as we please using only fractions whose denominators are powers of 2; such fractions are called **binary fractions**. One or two detailed examples should be enough to show you how this process works.

PROBLEM-SOLVING COMMENT

Distances on the number line are usually approximated by using decimal fractions (fractions whose denominators are powers of 10). Decimal approximation is the topic of Section 13.3; it is the key to our development of the real-number system. Approximation by binary fractions is a *simpler problem* that is *analogous* to the problem of approximating distances by decimal fractions. (It is much simpler to divide something in half than into tenths.) Moreover, the very fact that real numbers are *not* usually approximated by binary fractions should make it easier for you to see the main idea of the process without getting distracted or misled by your familiarity with decimal notation. ◇

Let us approximate the length $\frac{2}{3}$ using only fractions with denominators that are powers of 2. Suppose we have a metal rod that is $\frac{2}{3}$ of a meter long and we are trying to measure it with a meter stick marked off only in halves, quarters, eighths, etc. The powers of 2 are

$$1, 2, 4, 8, 16, 32, 64, 128, 256, 512, \ldots$$

It is easy to see (in Figure 13.2) that the rod is not as long as the meter stick, but is more than half as long. Let us use $\frac{1}{2}$ as a crude first approximation of its length. We might as easily use 1 as a first approximation, but it will be easier in the long run if we agree *always* to approximate by a number *less than* the actual length. (The reason for this agreement will emerge shortly.)

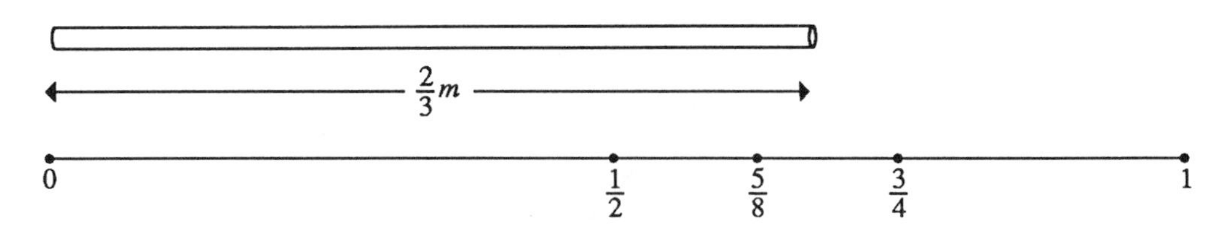

Figure 13.2 A metal rod two-thirds of a meter long.

A closer look reveals that the rod is not as long as $\frac{3}{4}$ of the meter stick, so the approximation in fourths is $\frac{2}{4}$. The next approximation is $\frac{5}{8}$. To be sure that our eyes are not deceiving us, we can check that $\frac{5}{8}$ is indeed less than $\frac{2}{3}$ by comparing the values of the two fractions expressed with a common denominator:[4]

$$\frac{5}{8} = \frac{15}{24} \quad < \quad \frac{2}{3} = \frac{16}{24}$$

This approximation cannot be improved by using 16ths because

$$\frac{11}{16} = \frac{33}{48} \quad > \quad \frac{32}{48} = \frac{2}{3}$$

[4]Notice that the numerators of the common-denominator forms of these fractions can be found by "cross multiplying," reflecting the fact (established in Section 9.5) that, for positive fractions,

$$\frac{a}{b} < \frac{c}{d} \quad \text{if and only if} \quad ad < bc$$

The denominators are shown here to remind you of why this comparison works.

but we can get closer using 32nds:

$$\frac{21}{32} = \frac{63}{96} \quad < \quad \frac{64}{96} = \frac{2}{3}$$

(You might find a calculator helpful as you work through these routine arithmetic checks.) Continuing in this way, we get successively finer approximations of $\frac{2}{3}$ by binary fractions, and each approximation is greater than or equal to the one before it. Moreover, the approximations get larger *by adding some binary fraction to a previous binary approximation*:

$$\begin{array}{llcl}
\text{1st:} & \frac{1}{2} & = & \frac{1}{2} \\
\text{2nd:} & \frac{2}{4} & = & \frac{1}{2} + \frac{0}{4} \\
\text{3rd:} & \frac{5}{8} & = & \frac{1}{2} + \frac{0}{4} + \frac{1}{8} \\
\text{4th:} & \frac{10}{16} & = & \frac{1}{2} + \frac{0}{4} + \frac{1}{8} + \frac{0}{16} \\
\text{5th:} & \frac{21}{32} & = & \frac{1}{2} + \frac{0}{4} + \frac{1}{8} + \frac{0}{16} + \frac{1}{32} \\
\text{6th:} & \frac{42}{64} & = & \frac{1}{2} + \frac{0}{4} + \frac{1}{8} + \frac{0}{16} + \frac{1}{32} + \frac{0}{64} \\
\text{7th:} & \frac{85}{128} & = & \frac{1}{2} + \frac{0}{4} + \frac{1}{8} + \frac{0}{16} + \frac{1}{32} + \frac{0}{64} + \frac{1}{128} \\
 & \vdots & &
\end{array}$$

Notice that the 7th approximation is within $\frac{1}{128}$th of the actual value $\frac{2}{3}$, and $128 = 2^7$. The 8th approximation would be within $\frac{1}{256}$th of the actual value, and $256 = 2^8$; the 9th approximation would be within $\frac{1}{512}$th of the actual value, and $512 = 2^9$; etc. Because the denominators of these approximating fractions are growing, the maximum possible error between the approximations and the actual value must be shrinking. This means that we can get as close to $\frac{2}{3}$ as we please by using binary fractions.

There is a pattern in this list of successive approximations. Each approximation is obtained from the previous one by adding either 0 or 1 of the next smaller fraction unit, as shown by the numerators used in the listed sums. Thus, once we know what the denominators of the successive approximations are, we can represent those sums of fractions by just listing their numerators. That is, we can efficiently represent the seven successive approximations listed previously by lists of 0s and 1s, as follows:

$$\begin{array}{rl}
\frac{1}{2}: & 1 \\
\frac{2}{4}: & 1, 0 \\
\frac{5}{8}: & 1, 0, 1 \\
\frac{10}{16}: & 1, 0, 1, 0 \\
\frac{21}{32}: & 1, 0, 1, 0, 1 \\
\frac{42}{64}: & 1, 0, 1, 0, 1, 0 \\
\frac{85}{128}: & 1, 0, 1, 0, 1, 0, 1
\end{array}$$

A reasonable guess (and at this point only a guess) at how the successively closer binary approximations of $\frac{2}{3}$ proceed from here would lead you to an unending string of alternating 1s and 0s; that is, we might reasonably say that an abbreviated way of describing the successive approximations of $\frac{2}{3}$ by binary fractions is the infinite string of digits

$$1, 0, 1, 0, 1, 0, 1, 0, \ldots$$

Example 1 (You might find a calculator useful as you work through this example.) Let us approximate $\frac{5}{6}$ by binary fractions until a pattern of numerators emerges. Both $\frac{1}{2}$ and $\frac{3}{4}$ are less than $\frac{5}{6}$, but $\frac{7}{8}$ $(= \frac{21}{24})$ is greater than $\frac{5}{6}$ $(= \frac{20}{24})$, so the third approximation is

$$\frac{6}{8} = \frac{1}{2} + \frac{1}{4} + \frac{0}{8}$$

The next four approximations result from

$$\begin{aligned} \frac{13}{16} = \frac{39}{48} \quad &< \quad \frac{40}{48} = \frac{5}{6} \\ \frac{27}{32} = \frac{81}{96} \quad &> \quad \frac{80}{96} = \frac{5}{6} \\ \frac{53}{64} = \frac{159}{192} \quad &< \quad \frac{160}{192} = \frac{5}{6} \\ \frac{107}{128} = \frac{321}{384} \quad &> \quad \frac{320}{384} = \frac{5}{6} \end{aligned}$$

Thus, the seventh approximation is

$$\frac{106}{128} = \frac{1}{2} + \frac{1}{4} + \frac{0}{8} + \frac{1}{16} + \frac{0}{32} + \frac{1}{64} + \frac{0}{128}$$

so the sequence of numerators begins

$$1, 1, 0, 1, 0, 1, 0$$

Were it not for the doubled 1s at the beginning, the alternating 1s and 0s would be quite convincing by now. As it is, we still might be tempted to settle for this pattern as the right one, especially if we try a few more terms and it continues to work (as it will). Either way, however, we are just guessing; for all we *really* know, there might not even be a pattern. □

By the method just illustrated, any rational number between 0 and 1 can be approximated to any desired degree of accuracy by a binary fraction. Moreover, all the other rationals can then be approximated by adding appropriate integers. For instance, since $\frac{2}{3}$ can be approximated by $\frac{85}{128}$, the numbers

$$\frac{17}{3} = 5\tfrac{2}{3}, \quad \frac{44}{3} = 14\tfrac{2}{3}, \quad \text{and} \quad \frac{-61}{3} = -21 + \tfrac{2}{3}$$

can be approximated by

$$5 + \frac{85}{128}, \quad 14 + \frac{85}{128}, \quad \text{and} \quad -21 + \frac{85}{128}$$

respectively. This relationship between the set of binary fractions and the set of all rational numbers exemplifies an important general property of some sets of points on the number line.

Definition Let A and B be sets of points on the number line such that A is a subset of B. If there is always some point of A within any given distance (no matter how small) of any point in B, then we say that A is **dense in** B.

The foregoing discussion illustrates that

(13.1) The set of binary fractions is dense in the set of rational numbers.

More importantly, the set of points with binary-fraction labels is dense in the set of *all* points on the number line, including those without rational-number labels. That is, the position of any point on the line can be approximated to within any desired degree of accuracy by a fraction with denominator 2. Moreover, in some cases we can find the approximations by arithmetic techniques similar to the ones shown above. For example, let us look at successively closer approximations of the point called $\sqrt{2}$.

Once we locate the point $\sqrt{2}$ geometrically and see that its distance from 0 corresponds to the length of the diagonal of a unit square, it is clear that this distance is more than 1 but less than 2. (See Figure 13.3.) To find the nearest half-unit point less than $\sqrt{2}$, we construct the midpoint $\frac{3}{2}$ of the line segment [1, 2] and observe that $\sqrt{2}$ lies to the left of $\frac{3}{2}$. Similarly, the nearest smaller quarter-unit point, $\frac{5}{4}$, may be found by observing that $\sqrt{2}$ lies to the right of the middle of the segment $[1, \frac{3}{2}]$.

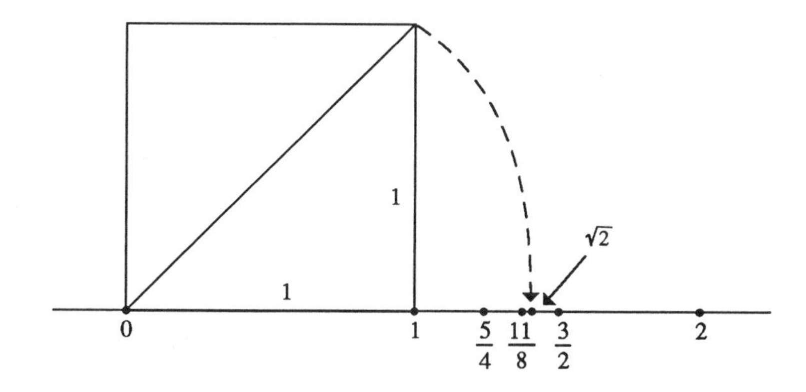

Figure 13.3
Approximating the position of the square root of 2.

For more exact approximations the picture is an unreliable guide, so a numerical approach must be used to decide whether a number lies to the left or right of $\sqrt{2}$ on the number line. In this case, the approach is fairly easy. As we have seen, $\sqrt{2}$ is the name for something which when multiplied by itself yields 2. Thus, "$\sqrt{2}$" is just a name that tells us how any numerical label for that point *ought to behave* with respect to multiplication by itself. A high-school algebra fact provides just what we need:

(13.2) For any two numbers x and y greater than 0, $x < y$ if and only if $x^2 < y^2$

This tells us that a number x is greater than or less than $\sqrt{2}$ if and only if x^2 is greater than or less than 2, respectively. Now it is easy to get very close approximations to $\sqrt{2}$, especially if a calculator is available. The next approximation—by eighths—is determined by squaring the midpoint of the segment $[\frac{5}{4}, \frac{3}{2}]$, which we have already shown to contain $\sqrt{2}$:

$$\left(\frac{11}{8}\right)^2 = \frac{121}{64} < 2, \quad \text{so} \quad \frac{11}{8} < \sqrt{2}$$

Repeating the process, we can "trap" $\sqrt{2}$ within smaller and smaller intervals:

$$\left(\frac{23}{16}\right)^2 = \frac{529}{256} > 2, \quad \text{so} \quad \frac{11}{8} < \sqrt{2} < \frac{23}{16}$$

$$\left(\frac{45}{32}\right)^2 = \frac{2025}{1024} < 2, \quad \text{so} \quad \frac{45}{32} < \sqrt{2} < \frac{23}{16}$$

$$\left(\frac{91}{64}\right)^2 = \frac{8281}{4096} > 2, \quad \text{so} \quad \frac{45}{32} < \sqrt{2} < \frac{91}{64}$$

$$\left(\frac{181}{128}\right)^2 = \frac{32,761}{16,384} < 2, \quad \text{so} \quad \frac{181}{128} < \sqrt{2} < \frac{91}{64}$$

$$\left(\frac{363}{256}\right)^2 = \frac{131,769}{65,536} > 2, \quad \text{so} \quad \frac{181}{128} < \sqrt{2} < \frac{363}{256}$$

We can summarize these approximations of $\sqrt{2}$ in the same form as we used earlier, so that the numerator sequence is easy to see:

$$1 + \frac{0}{2} + \frac{1}{4} + \frac{1}{8} + \frac{0}{16} + \frac{1}{32} + \frac{0}{64} + \frac{1}{128} + \frac{0}{256} \cdots$$

Thus, the sequence of numerators begins

$$1., 0, 1, 1, 0, 1, 0, 1, 0, \ldots$$

(The first number in this string is followed by a dot to indicate that it represents the integer part of the approximation.)

It is tempting to guess that, after a slight irregularity in the first few places, the numerator sequence settles down to a regular alternating pattern of 1s and 0s. However, one more computation tells us that this is not the case:

$$\left(\frac{725}{512}\right)^2 = \frac{525,625}{262,144} > 2$$

so the next numerator is also 0 A natural question arises at this point: Is there ever a repeating pattern in this numerator sequence? How can we be *sure* of an answer to this question? That's worth thinking about for awhile. (Later sections of this chapter will shed some light on the problem.)

PROBLEM-SOLVING COMMENT

Notice the repeated use of two tactics here—*constructing examples* and *finding patterns*. Notice also that neither of these techniques provides any logical *certainty* about how things work in general. What we get from them is some guidance about how to formulate general statements, which must then be proved before they become reliable. ◇

Convenient numerical techniques for finding binary-fraction approximations, like the one just shown for $\sqrt{2}$, do not always apply to every point on the number line. Sometimes the approximation process is quite intricate or sophisticated. Investigation of other processes, although interesting in itself, is not relevant to elementary-school mathematics, so we shall resist the temptation to go further in that direction. The foregoing examples and discussion should make it plausible that *the location of every point can be approximated (somehow) as closely as we please by a binary fraction*; that is:

(13.3) | The set of all binary-fraction points is a dense subset of the number line. |

Exercises 13.2

For Exercises 1–5, use the methods of this section to approximate the given rational number by binary fractions to within $\frac{1}{32}$ of its actual value. Then represent the successive approximations as a list of 0s and 1s.

1. $\frac{1}{3}$ **2.** $\frac{4}{7}$ **3.** $\frac{3}{5}$

4. $\frac{3}{4}$ **5.** $\frac{1}{6}$

For Exercises 6–10, use a calculator to help you approximate the given rational number by binary fractions to within $\frac{1}{512}$ of its actual value. Represent the successive approximations as a list of 0s and 1s in each case. Can you find a pattern for any of them? Can you justify your guess that the pattern will continue? (See Example 1.)

6. $\frac{1}{3}$ **7.** $\frac{4}{5}$ **8.** $\frac{35}{128}$

9. $\frac{8}{9}$ **10.** $\frac{3}{10}$

11. Approximate $\sqrt{3}$ by binary fractions to within $\frac{1}{16}$ of a unit. If you have a calculator, approximate it to within $\frac{1}{512}$ of a unit. Represent your result as a numerator list.

12. Approximate $\sqrt{7}$ by binary fractions to within $\frac{1}{16}$ of a unit. If you have a calculator, approximate it to within $\frac{1}{512}$ of a unit. Represent your result as a numerator list. (*Note*: The first term in your list might not be either 0 or 1.)

13. Approximate $\sqrt[3]{2}$ by binary fractions to within $\frac{1}{8}$ of a unit. If you have a calculator, approximate it to within $\frac{1}{128}$ of a unit. Represent your result as a numerator list.

14. Approximate $\sqrt[3]{11}$ by binary fractions to within $\frac{1}{16}$ of a unit. If you have a calculator, approximate it to within $\frac{1}{512}$ of a unit. Represent your result as a numerator list. (*Note*: The first term in your list might not be either 0 or 1.)

13.3 Approximation by Decimals [optional]

(Some material in this section is required for Section 13.4.)

The process for approximating quantities by fractions with denominators that are powers of 2, described in Section 13.2, can easily be adapted to approximate quantities by fractions whose denominators are powers of any fixed number. We chose to begin with 2 because the computations are relatively simple in that case. However, the theory is the same if we use powers of 3, of 7, of 12, of 68, or of any number; the only differences are ones of convenience. Whatever number we choose, we have to subdivide intervals into that many equal parts. This is always possible geometrically, but the actual construction gets awkward as the number gets large, and taking successive powers of a large number rapidly complicates the arithmetic. Thus, it is advantageous to keep the number small, and from that viewpoint 2 is the best choice.

However, from another viewpoint, our numeration system makes the number 10 an especially convenient choice. Because successive powers of 10 are easily represented (by just annexing 0s), much of the computation becomes quite easy. Fractions whose denominators are powers of 10 are called **decimal fractions** (from the Latin word for 10, *decem*). To see how approximation by decimal fractions works, let us approximate the length $\sqrt{2}$ by units + tenths + hundredths + ...

Note: The rational-number labels of (some) points on the number line often are being viewed in two ways throughout this chapter—as indicators of locations on the line and as objects that obey the laws of rational-number arithmetic. As we move back and forth between these two views, we often rely on the context of the discussion to indicate which view you should be taking. This approach permits a clear, simple explanation of the main ideas of the chapter.[5]

The point labeled $\sqrt{2}$ lies between the points 1 and 2, so the largest unit value less than $\sqrt{2}$ is 1. Next, we divide the interval [1, 2] into ten equal pieces and see that the point $\sqrt{2}$ lies in the fifth piece. The accuracy of our geometric observation may need to be checked, as before, by squaring the values of the interval's endpoints:

$$\left(\frac{14}{10}\right)^2 = \frac{196}{100} \quad < \quad 2 \quad < \quad \frac{225}{100} = \left(\frac{15}{10}\right)^2$$

Now we divide the interval $[\frac{14}{10}, \frac{15}{10}]$ into tenths and repeat the procedure, as illustrated by Figure 13.4. The computations that justify the accuracy of the interval endpoints are also shown. (If you have a calculator you might check them just to see that, despite their formidable appearance, they are easy to do. The calculator does the squaring for you, and deciding whether the resulting fraction is greater or less than 2 is then a simple matter of observation.)

$$\left(\frac{141}{100}\right)^2 = \frac{19{,}881}{10{,}000} \quad < \quad 2 \quad < \quad \frac{20{,}164}{10{,}000} = \left(\frac{142}{100}\right)^2$$

$$\left(\frac{1414}{1000}\right)^2 = \frac{1{,}999{,}396}{1{,}000{,}000} \quad < \quad 2 \quad < \quad \frac{2{,}002{,}225}{1{,}000{,}000} = \left(\frac{1415}{1000}\right)^2$$

$$\left(\frac{14{,}142}{10{,}000}\right)^2 = \frac{199{,}996{,}164}{100{,}000{,}000} \quad < \quad 2 \quad < \quad \frac{200{,}024{,}449}{100{,}000{,}000} = \left(\frac{14{,}143}{10{,}000}\right)^2$$

The only significant formal difference in method occurs when we try to write the "shorthand" sequence of numerators. A sequence of 1s and 0s doesn't work in this situation because it is not enough just to say whether or not to add to the previous approximation *one* tenth, *one* hundredth, etc. The construction sometimes allows us to add *more than one* interval of a particular size, so this time we must have a sequence of symbols that tells us how many pieces to add at each stage.

---

[5]If you are curious about the pitfalls of maintaining explicit distinctions between objects and their names, you might consult Chapter VIII of Lewis Carroll's *Alice Through The Looking Glass*. The relevant passage is in the conversation between Alice and the White Knight, beginning with "The name of the song is called '*Haddocks' Eyes*'."

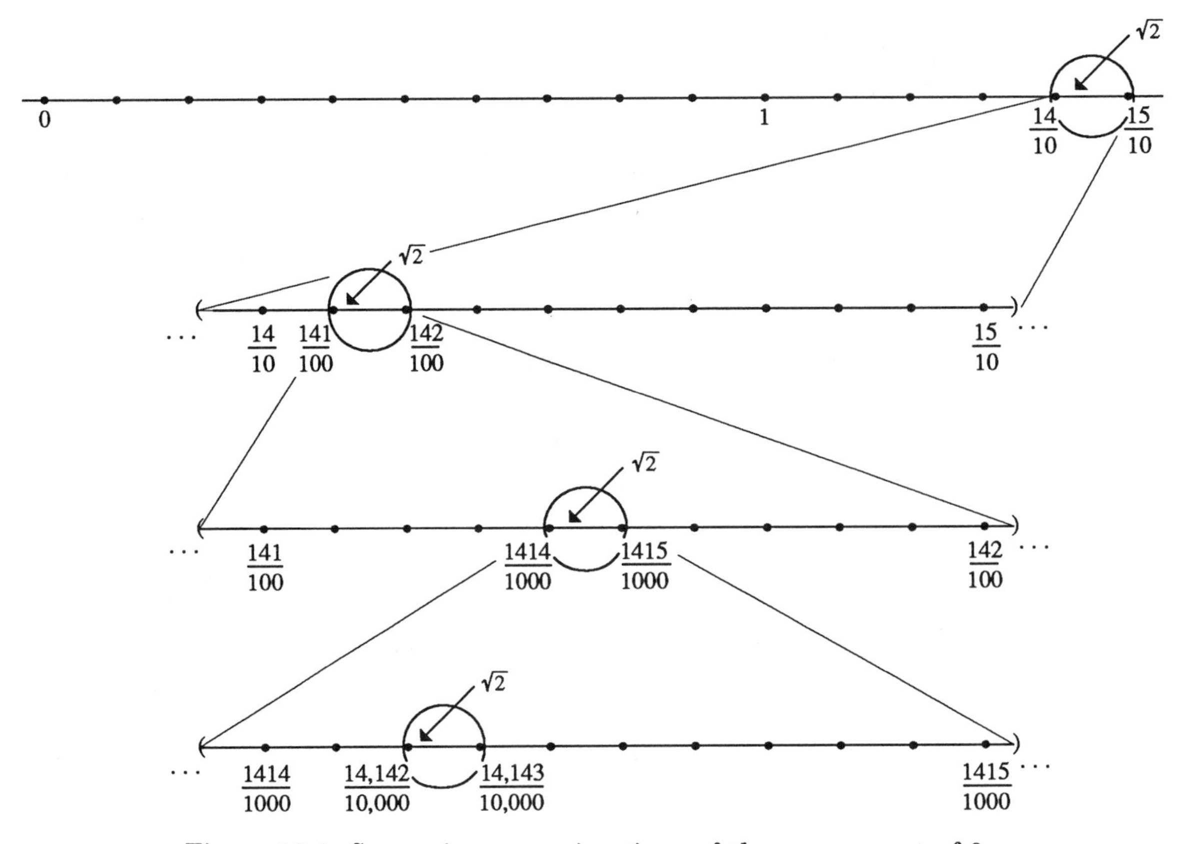

Figure 13.4 Successive approximations of the square root of 2.

Now, we cannot possibly add more than nine pieces at any one step, so a sequence of single digits is enough to give us all the information we need. Thus, the list of successive approximations of $\sqrt{2}$ by decimal fractions is

$$
\begin{array}{rcl}
\text{1st:} & \dfrac{14}{10} & = & 1 + \dfrac{4}{10} \\
\text{2nd:} & \dfrac{141}{100} & = & 1 + \dfrac{4}{10} + \dfrac{1}{100} \\
\text{3rd:} & \dfrac{1414}{1000} & = & 1 + \dfrac{4}{10} + \dfrac{1}{100} + \dfrac{4}{1000} \\
\text{4th:} & \dfrac{14,142}{10,000} & = & 1 + \dfrac{4}{10} + \dfrac{1}{100} + \dfrac{4}{1000} + \dfrac{2}{10,000} \\
& & \vdots &
\end{array}
$$

We can summarize this information by the numerator list

$$1., 4, 1, 4, 2, \ldots$$

This sequence of digits, usually written without commas, is nothing more (or less!) than the usual decimal representation of $\sqrt{2}$: 1.4142... In general:

> A **(lower) decimal approximation** of a quantity is an approximation by a sequence of fractions whose denominators are successively larger powers of 10, and that sequence is abbreviated by simply listing the numerators of the added fractional amount at each step.

(The word "lower" indicates that all the fractions used in these approximations are less than or equal to the quantity being approached. We shall drop that word from here on because this is the only type of decimal approximation we discuss.)

Example 1 To find the decimal approximation of $\frac{2}{3}$, we begin by observing that $\frac{2}{3}$ is between 0 and 1. The tenths approximation can be found using the common denominator 30; we observe that $\frac{20}{30}$ is between $\frac{6}{10}$ $(= \frac{18}{30})$ and $\frac{7}{10}$ $(= \frac{21}{30})$. The hundredths approximation can be found by using the common denominator 300:

$$\frac{66}{100} = \frac{198}{300} < \frac{200}{300} < \frac{201}{300} = \frac{67}{100}$$

Further approximations can be found similarly:

$$\frac{666}{1000} = \frac{1998}{3000} < \frac{2000}{3000} < \frac{2001}{3000} = \frac{667}{1000}$$

$$\frac{6666}{10,000} = \frac{19,998}{30,000} < \frac{20,000}{30,000} < \frac{20,001}{30,000} = \frac{6667}{10,000}$$

$$\frac{66,666}{100,000} = \frac{199,998}{300,000} < \frac{200,000}{300,000} < \frac{200,001}{300,000} = \frac{66,667}{100,000}$$

and so on. Thus, the decimal fraction $\frac{66,666}{100,000}$ is an approximation of $\frac{2}{3}$ with an error of less than $\frac{1}{100,000}$. Since

$$\frac{66,666}{100,000} = \frac{6}{10} + \frac{6}{100} + \frac{6}{1000} + \frac{6}{10,000} + \frac{6}{100,000}$$

we can abbreviate this approximation by the numerator list

$$0., 6, 6, 6, 6, 6$$

which is further abbreviated as 0.66666. The pattern developed so far suggests that closer decimal approximations can be found by adjoining more 6s, but no convincing justification for that has been provided yet. An efficient method for testing such guesses will be discussed in Section 13.5. □

Example 2 To see why 2.236 is a decimal approximation of $\sqrt{5}$ with an error of less than $\frac{1}{1000}$, we observe first that $\sqrt{5}$ lies between 2 and 3 because 5 lies between 4

($= 2^2$) and 9 ($= 3^2$). Then a little arithmetic (preferably with the help of a calculator) yields the following:

$$\left(\frac{22}{10}\right)^2 = \frac{484}{100} \quad < \quad 5 \quad < \quad \frac{529}{100} = \left(\frac{23}{10}\right)^2$$

$$\left(\frac{223}{100}\right)^2 = \frac{49,729}{10,000} \quad < \quad 5 \quad < \quad \frac{50,176}{10,000} = \left(\frac{224}{100}\right)^2$$

$$\left(\frac{2236}{1000}\right)^2 = \frac{4,999,696}{1,000,000} \quad < \quad 5 \quad < \quad \frac{5,004,169}{1,000,000} = \left(\frac{2237}{1000}\right)^2 \qquad \square$$

The examples above illustrate how any point on the number line can be approximated by decimal fractions, to within any desired degree of accuracy, by dividing the appropriate unit interval into tenths, then into hundredths, and so forth. In other words:

(13.4) | The set of all points represented by decimal fractions is a dense subset of the number line.

This means that for all practical measurement purposes, *quantities can be represented by decimal approximations with as little error as a particular situation requires.* It also follows from the definition of a *dense subset* that any subset of the number line that contains a dense subset of the line must itself be a dense subset of the line. In particular, then:

> The set of all points with rational-number coordinates (labels) is a dense subset of the number line.

Exercises 13.3

1. Find the sixth digit in the decimal approximation of $\sqrt{2}$. Interpret this information by specifying the interval that contains the point $\sqrt{2}$ and whose endpoints are successive hundred thousandths.

For Exercises 2–7, use the method of this section to approximate the given point by a fraction whose denominator is 1000; then express that fraction as a decimal.

2. $\sqrt{3}$
3. $\sqrt{7}$
4. $\frac{1}{6}$
5. $\frac{22}{7}$
6. $\frac{3}{4}$
7. $\sqrt[3]{2}$

8. Let A, B, and C be subsets of the number

line such that $A \subseteq B \subseteq C$. Justify the claim that, if A is dense in C, then B must also be dense in C.

9. (a) Use the statement in Exercise 8 to justify the claim that all fractions whose denominators are *multiples* of 10 are labels for a dense subset of the number line.

(b) Do all fractions whose denominators are multiples of 5 label a dense subset of the number line? Why or why not?

13.4 What Is a Real Number? [optional]

The decimal-approximation idea, described in Section 13.3, may be summarized in this way: Every point to the right of zero on the number line can be approached as closely as we please by an unending sequence of numbers

$$s_0, s_1, s_2, \ldots s_n, \ldots$$

such that

1. s_0 is a positive integer or zero, and
2. for each positive integer n, the nth term is greater than the term preceding it by a fraction with denominator 10^n and numerator one of the integers 0 through 9. (A term may actually equal the one before it, using the numerator 0 for the added fraction.) That is, for each $n > 0$,

$$s_n = s_{n-1} + \frac{a}{10^n}$$

where a is a single digit.

We call a sequence of rational numbers with these properties a **special sequence.** (This is not standard mathematical terminology for the idea, but it should be easy to remember and will suit our purpose well enough.)

Example 1 0, .6, .66, .666, .6666, ...is a special sequence. For each n, the nth term is greater than the term preceding it by $\frac{6}{10^n}$; that is,

$$s_n = s_{n-1} + \frac{6}{10^n}$$ □

Example 2 $1, \frac{1}{10}, \frac{1}{100}, \frac{1}{1000}, \frac{1}{10{,}000}, \ldots$ is not a special sequence because each term is not greater than or equal to the one preceding it. A special sequence never decreases. □

Example 3 $0, \frac{1}{2}, \frac{3}{4}, \frac{5}{6}, \frac{7}{8}, \ldots$ is not a special sequence:

$$s_2 = \frac{1}{2} + \frac{1}{4} = s_1 + \frac{25}{100}$$

and, although the denominator of the added fraction is of the correct kind (10^2), the numerator is not a single digit. □

The decimal approximation of $\sqrt{2}$ in the previous section is a good illustration of how a special sequence arises from closer and closer rational approximations of a location on the number line. In general, let p represent any point to the right of zero. To get the first term of the special sequence that "approaches" p, observe which unit interval contains p and choose its left endpoint number as s_0. Divide this unit interval into ten equal parts, observe which of these parts contains p, and choose its left endpoint number as the next term of the sequence, s_1. Then divide that interval of length $\frac{1}{10}$ into ten parts, observe which part contains p, and choose its left endpoint number as s_2. Each further subdivision yields a new term of the sequence, and it is theoretically possible to continue subdividing as many times as we like, so there is no "last" term in the succession of numbers we can get.

It is not hard to see from the way we constructed special sequences that different points must have different approximating sequences:

> If p and q are different points on the number line, then there is some distance between them; call this distance d. Clearly, there is some exponent n such that $\frac{1}{10^n}$ is less than d. Now, if we let n represent the smallest such exponent, then the terms
>
> $$s_0, s_1, \ldots s_{n-1}$$
>
> will be the same in the sequences for both p and q, but the term s_n will necessarily be different.

For example, look at the two points pictured in Figure 13.5. The upper line shows them both in the same tenth of a unit interval, but the magnified picture on the lower line shows that there is a distance between them that contains the interval $\left[\frac{143}{100}, \frac{144}{100}\right]$. Thus, the special sequences for these two points would begin

$$p: \quad 1, \frac{14}{10}, \frac{142}{100}, \ldots = 1, 1.4, 1.42, \ldots$$

$$q: \quad 1, \frac{14}{10}, \frac{144}{100}, \ldots = 1, 1.4, 1.44, \ldots$$

This means that, regardless of what digits appear later, any decimal approximations of these two points that are within hundredths of being accurate will necessarily differ in the second decimal place.

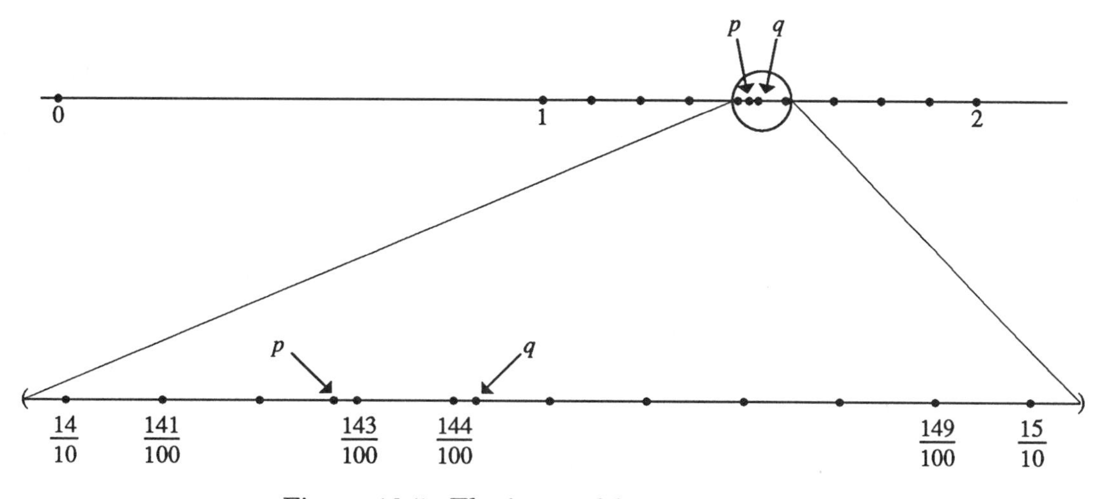

Figure 13.5 The interval between p and q.

Example 4 The number π (pi), used so often in high-school geometry, is usually approximated by $\frac{22}{7}$ or 3.14. These two rational numbers, though close together, actually label different points on the number line. Moreover, both are different from π itself, which is not a rational number. The special sequence that approaches $\frac{22}{7}$ can be found by division; the one for 3.14 is obvious:

$$\frac{22}{7}: \quad 3, 3.1, 3.14, 3.142, 3.1428, \ldots$$
$$3.14: \quad 3, 3.1, 3.14, 3.140, 3.1400, \ldots$$

The special sequence that approaches π (which is not easily computed, but can be looked up) begins

$$\pi: \quad 3, 3.1, 3.14, 3.141, 3.1415, \ldots$$

All three of these points are within the same hundredth interval—$\left[\frac{314}{100}, \frac{315}{100}\right]$; they differ by thousandths of a unit. Thus, the special sequences for these three points are different from s_3 on. □

The foregoing discussion explains an important fact about the relationship between points on the number line and special sequences:

(13.5) If two points on the number line are different, then the special sequences that approximate them are different.

If two special sequences differ in the nth place, their terms from there on represent points in different intervals of length $\frac{1}{10^n}$, so they cannot approach the same point—except in one very special situation. It may happen that the first difference between the two sequences puts them in adjoining intervals, and from there on one sequence represents the common boundary point of those two intervals, whereas the other sequence "approaches" that same point from the left. This case is easily recognized from the decimal expressions of the two sequences. The only way this can happen is if the first difference between the decimal expressions occurs as successive digits (e.g., 5 and 6, or 2 and 3), and from there on the approximations with the larger of these two digits have all 0s from that digit on, while the corresponding approximations in the other sequence have all 9s. For example, the infinite decimals 2.35000... and 2.34999... are both "shorthand" representations of special sequences that approach the number $\frac{235}{100}$:

$$2.35000\ldots: \quad 2, 2.3, 2.35, 2.350, 2.3500, 2.35000, \ldots$$
$$2.34999\ldots: \quad 2, 2.3, 2.34, 2.349, 2.3499, 2.34999, \ldots$$

The first sequence reaches the point $\frac{235}{100}$ by its third term and stays there from then on, whereas the other sequence approaches the point from the left, staying within the interval [2.34, 2.35] from the third term on.

We can summarize this discussion conveniently if we recall (from Section 13.3) that every special sequence can be expressed in "shorthand" form as an infinite decimal. In this form, the special sequences that lead to ambiguity are the ones whose infinite decimals are either all 0s or all 9s from some point on. If we agree to discard all of one of these types of sequences—say the ones with repeated 0s—from our collection of special sequences, then,

(13.6) (Except as just noted) there is a 1-1 correspondence between the set of special sequences of positive rational numbers and the set of all points on the number line to the right of zero.

We can find an approximating sequence for each point to the left of zero simply by "folding the number line over" at 0. More precisely, given any point q to the left of 0, find the point p that is the same distance from 0 on the right. There is a special sequence

$$s_0, s_1, s_2, \ldots s_n, \ldots$$

that approximates p; so the sequence

$$-s_0, -s_1, -s_2, \ldots - s_n, \ldots$$

will approximate q (from the right). (Notice that this sequence of negative numbers is *not* a *special* sequence. Why not?)

Now, each special sequence can be abbreviated as an infinite decimal, and points to the left of 0 can be approximated by negatives of special sequences, so Property 13.6 can be expanded to the yield the fundamental fact of this section: If we agree to ignore infinite decimals that are all 0s from some digit on, except for 0.000 ..., then:

(13.7) Every point on the number line can be represented unambiguously by an infinite decimal.

In other words,

there is a 1-1 correspondence between the set of all points on a line and the set of all infinite decimals (as adjusted).

Thus, the problem described in Section 13.1 can be solved by using the set of infinite decimals as the numerical labels for the points on the number line. The quantities represented by these infinite decimals are called "real" numbers (though they are no more or less real than any other numbers we have seen). For our purposes, then, the set of **real numbers** is the set of all infinite decimals. We shall denote this set by **R**.

Note: Despite the risk of occasional ambiguity, it is sometimes useful to have both the repeated-0 and the repeated-9 types of infinite decimals at our disposal. We shall continue to discuss both types, with the understanding that a decimal of either kind can be converted to the other form, as needed.

Exercises 13.4

For Exercises 1–10, answer these two questions for the given sequence $s_0, s_1, s_2, \ldots s_n \ldots$ of rational numbers.

(a) $s_3 = s_2 + \_\_\,?$

(b) Is this a special sequence? Why or why not?

1. $2, \frac{21}{10}, \frac{212}{100}, \frac{2121}{1000}, \ldots$
2. $0, \frac{1}{4}, \frac{2}{5}, \frac{3}{6}, \ldots, \frac{n}{n+3}, \ldots$
3. $7, 7, 7, 7, \ldots$
4. $1, \frac{11}{10}, \frac{122}{100}, \frac{1333}{1000}, \frac{14{,}444}{10{,}000}, \ldots$
5. $3, 3+\frac{1}{10}, 3+\frac{2}{10}, \ldots, 3+\frac{n}{10}, \ldots$
6. $5, 5.5, 5.55, 5.555, 5.5555, \ldots$
7. $9, 9.1, 9.2, 9.2, 9.2, \ldots, 9.2, \ldots$
8. $4, 4.2, 4.02, 4.002, 4.0002, 4.00002, \ldots$
9. $4, 4.2, 4.22, 4.222, 4.2222, 4.22222, \ldots$
10. $4, 4.2, 4.20, 4.200, 4.2000, 4.20000, \ldots$
11. Verify geometrically that the sequence
$$0, .9, .99, .999, .9999, \ldots$$
approaches 1. (Draw a picture.)
12. Verify geometrically that the sequence
$$0, -.9, -.99, -.999, -.9999, \ldots$$
approaches -1. Is this a special sequence?

In Exercises 13–16, construct the special sequences that approach each of the two given number-line points until you find the first place where they differ. (A calculator might help.)

13. $\frac{1}{2}$ and $\frac{2}{3}$

14. $\frac{10}{7}$ and $\frac{5}{4}$

15. $\frac{3}{4}$ and $\frac{743}{990}$

16. $\frac{433}{250}$ and $\sqrt{3}$

For Exercises 17–20, answer these three questions for the given pair of infinite decimals.

(a) Write each infinite decimal as a special sequence.

(b) Find the first term s_n where the two sequences differ. What is n in this case?

(c) With respect to the integer n found in part (b), find, if possible, an interval that lies to the right of all points labeled by terms of the first sequence after s_n and to the left of all points labeled by terms of the second sequence after s_n. Justify your answer.

17. 2.5333... and 6.5333...

18. 0.147222... and 0.195111...

19. 1.84999... and 1.85000...

20. 4.2000... and 4.3999...

For Exercises 21–24, replace the given infinite decimal with one containing no 0s but representing the same point on the number line.

21. 2.000...

22. 1.5000...

23. 6.459000...

24. .176000...

13.5 Rational and Irrational Numbers

We have shown (in Sections 13.2 – 13.4) that all the points on the number line can be labeled using the set **R** of all real numbers, which are represented by infinite decimals. However, some of those points already have *rational* labels; that is, some (but not all) real numbers are also rational numbers. In this section we characterize the subset of R that represents the rationals.

Every rational number can easily be converted to a decimal by dividing its numerator by its denominator. If we agree that any finite decimal can be made infinite simply by attaching an infinite "tail" of zeros, then every rational number can be written as an infinite decimal. For instance,

$$\frac{1}{3} = 0.33333\ldots \qquad\qquad \frac{3}{4} = 0.75000\ldots$$

and so forth. In other words, the set **Q** of rational numbers can be considered a subset of **R**.

However, as we have already seen, not every infinite decimal represents a rational number. To distinguish those decimals that are rational numbers from those that are not, consider first an example of the division process.

Example 1 To convert $\frac{47}{22}$ to a decimal, we divide 47 by 22, regarding each computation of a quotient digit and a remainder as a "step" in the division process.

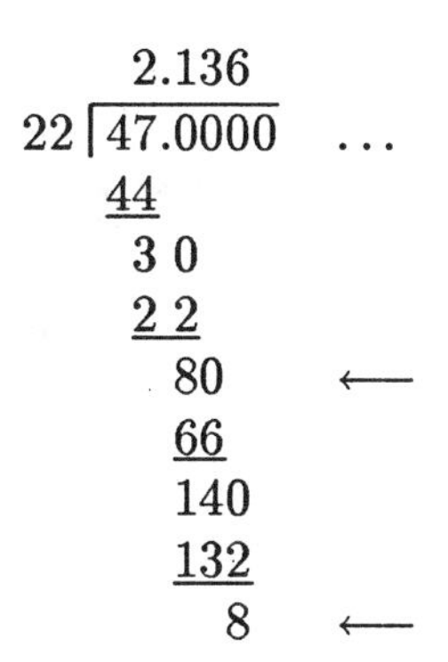

Since the remainder at the fourth step, 8, is the same as the one at the second step, the third and fourth digits of the quotient will repeat again and again from there on, without interruption. Thus,

$$\frac{47}{22} = 2.1363636\ldots$$ □

Must such a repetition of remainders happen with any division example, or it this just a carefully chosen special case? The answer is *Yes* to both parts of that question. A repetition must always occur *eventually* (as we shall show), but the fact that it happens so conveniently soon in this example is the result of careful choice.

Any case in which the denominator is bigger than the numerator can be simplified by first taking care of the integer part of the quotient. For instance, we could have begun Example 1 by writing

$$\frac{47}{22} = 2 + \frac{3}{22}$$

then converting $\frac{3}{22}$ to the decimal .1363636... afterwards. In this way we can safely confine our attention to fractions in which the denominator is smaller than the numerator.

In general, when one integer is divided by another one using long division, the remainder at any step must be less than the divisor. Thus, if we are converting the rational number $\frac{p}{q}$ to a decimal, the largest possible remainder is $q - 1$; that is,

(13.8) | The number of different nonzero remainders in the division problem $p \div q$ can be no larger than $q - 1$.

This means that, once all the nonzero digits of p have been used, after at most q more steps in the division process, the remainder must repeat an earlier remainder, so the digits in the quotient that occur in the steps between those two equal remainders will repeat over and over from there on without interruption. Sometimes this repetition occurs early, as in Example 1. Sometimes all the possible nonzero remainders are used; for instance,

$$\frac{4}{7} = 0.571428\ 571428\ 571428\ldots$$

Thus, we have established an important fact:

(13.9)

Every rational number can be expressed as an infinite decimal in which, from some specific digit on, a finite sequence of digits repeats again and again in the same order, without interruption.

Such decimals are called **repeating decimals**. To denote the repeating sequence of digits without having to write them several times over, we just put a bar over the first occurrence of the sequence. For instance, the results of the two examples above are written

$$\frac{47}{22} = 2.1\overline{36} \qquad \text{and} \qquad \frac{4}{7} = 0.\overline{571428}$$

Two natural questions arise at this point:

1. Does *every* repeating decimal represent a rational number?
2. Are there really any (or many) such things as infinite nonrepeating decimals?

It is surprisingly easy to show that the answer to the first question is *Yes.* To do this we use a simple method for converting any repeating decimal to fraction form. The next two examples illustrate how that method is used to convert repeating decimals to fractions.

Example 2 Suppose $d = 0.\overline{35}$. Since there are two digits in the repeating sequence, we multiply d by 100 and then subtract d:

$$\begin{array}{rcl} 100d &=& 35.353535\ldots \\ -\quad d &=& 0.353535\ldots \\ \hline 99d &=& 35.000000\ldots \\ d &=& \dfrac{35}{99} \end{array}$$

□

Example 3 Suppose $d = 2.8\overline{473}$. Because there are three digits in the repeating sequence, we multiply d by 1000 and then subtract d:

$$\begin{array}{rcl} 1000d & = & 2847.3473473473\ldots \\ -\quad d & = & 2.8473473473\ldots \\ \hline 999d & = & 2844.5000000000\ldots \\ d & = & \dfrac{2844.5}{999} \\ & = & \dfrac{28,445}{9990} \\ & = & \dfrac{5689}{1998} \end{array}$$

□

The general method for converting infinite decimals to fractions may be described like this:

> Suppose d is a repeating decimal with n digits in its repeating sequence. If we multiply d by 10^n and then subtract d from $10^n \cdot d$, we will get a *finite* decimal that equals $(10^n - 1) \cdot d$. Then d is the fraction obtained by dividing that finite decimal by $10^n - 1$. To obtain an integer numerator, multiply both numerator and denominator by the same power of ten, if necessary.

Perhaps this description reads more like a magic incantation than a logical procedure. It was stated in general terms to show that there is a procedure that works all the time. Look back at the preceding examples, then reread the general method; you should find it much clearer.

Example 4 Figure 13.6 shows a sixth-grade text page in which repeating decimals are introduced. How do the authors try to convince the students that the decimals repeat?

□

Since every rational number can be expressed as a repeating decimal, and vice versa, we have just characterized all those real numbers (in decimal form) that are rational—they are just the repeating decimals. The other real numbers are called **irrational numbers**; they are the infinite nonrepeating decimals. But are there any such numbers? Certainly—here are some examples that should suggest how to construct many different irrational numbers.

Repeating Decimals

A. North America makes up about $\frac{1}{6}$ of the earth's land area. Write the fraction as a decimal.

Find 1 ÷ 6.

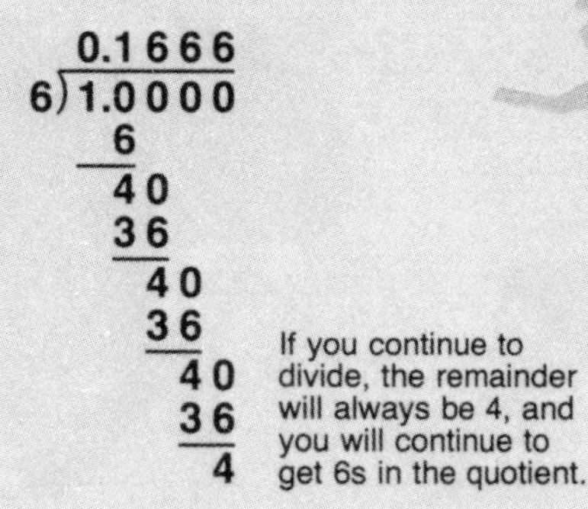

$\frac{1}{6} = 0.1666\ldots$

This decimal is a *repeating* decimal. To write the decimal, put a bar over the digit that repeats.

$\frac{1}{6} = 0.1\overline{6}$

B. Write $\frac{3}{11}$ as a repeating decimal.

Find 3 ÷ 11.

```
    0.2727
11)3.0000
   22
    80
    77
     30
     22
      80
      77
       3
```

If you continue, the remainder will always be 8 or 3, and you will continue to get 27s in the quotient.

To write the decimal, put a bar over the digits that repeat.

$\frac{3}{11} = 0.\overline{27}$

Try

a. Write the first 6 digits of $0.\overline{12}$.

b. Write the first 6 digits of $0.1\overline{2}$.

c. Write 0.8777 . . . using a bar over the digit that repeats.

Write each fraction as a repeating decimal.

d. $\frac{1}{3}$ **e.** $\frac{5}{12}$

Figure 13.6 INVITATION TO MATHEMATICS, Grade 6, p. 210

Example 5 Consider 0.101001000100001..., where the ellipsis ("...") implies that the continuing pattern requires one more 0 after each successive 1. Because there is no finite sequence of digits that repeats again and again in the same order without interruption, this is not a repeating decimal. □

Example 6 For another example of an infinite nonrepeating decimal, consider a sequence of successive multiples of some integer greater than 1, such as

$$3, 6, 9, 12, 15, 18, 21, 24, \ldots$$

and form a decimal using these digits in order:

$$0.3691215182124\ldots$$

(Can you convince yourself that the digit sequence in such cases is nonrepeating?) □

Sometimes irrational numbers arise in other contexts, and when they do, it often is difficult to prove that they are irrational. Some familiar irrational numbers are $\sqrt{2}$ (the length of the diagonal of a unit square), π (the ratio of the circumference of a circle to its diameter), and e (the base of the natural logarithms). The proof that $\sqrt{2}$ cannot be expressed as a rational number was known to the early Greeks; a version of it appears at the beginning of this chapter. The proofs that π and e are irrational are much more difficult; they were not established until relatively modern times.

The decimal characterization of rational and irrational numbers is important and useful; we summarize it here for emphasis:

(13.10) The decimal expression for a real number is repeating if the number is rational and is (infinite) nonrepeating if the number is irrational.

Operations and Order in R

To make the set **R** of real numbers into a number system, we must define the arithmetic operations and an order relation. Because we want the reals to be an extension of the rationals, we must choose these definitions so that they agree with rational-number arithmetic when confined to the set of all repeating decimals. Now, recall that the infinite-decimal representation of a real number is a shorthand way of labeling that number by an infinite sequence of rationals—of decimal fractions, to be exact. Thus, if we want to add, subtract, multiply, or divide two real numbers, we must find a way of combining their two decimal-fraction sequences into a single "answer" sequence. The observation that provides the key is:

> Each term of the special sequence for a real number is a rational approximation of that number, and the real-number operations we want must agree with the rational-number operations we already have. Thus, the "new" operations must agree term by term on the approximating sequences with the corresponding "old" operations.

Therefore, a natural way to add, subtract, multiply, or divide two real numbers is to perform the operation term by term on the two approximating sequences and see if the result is an approximating sequence for some real number. If it is, then that number must be the answer we seek.

A rigorous development of the arithmetic structure of **R** is beyond the scope of this book; it would require first a careful discussion of "limits" of infinite sequences, a topic better left to calculus courses. Instead, let us look at one simple example of how such an approach would work for addition, and then simply observe that a development of all the arithmetic operations can be carried out from this viewpoint.

Example 7 Consider the sum

$$.333\ldots \oplus .666\ldots$$

Since these numbers are rational, we already know the sum in this case (what is it?), so the process will be easy to check. Add the approximating sequences term by term:

$$\begin{array}{rl} .333\ldots : & \begin{array}{r} .3 \\ +\ .6 \\ \hline .9 \end{array} \quad \begin{array}{r} .33 \\ +\ .66 \\ \hline .99 \end{array} \quad \begin{array}{r} .333 \\ +\ .666 \\ \hline .999 \end{array} \quad \begin{array}{r} .3333 \\ +\ .6666 \\ \hline .9999 \end{array} \quad \begin{array}{r} .33333 \\ +\ .66666 \\ \hline .99999 \end{array} \quad \ldots \\ \oplus\ .666\ldots : & \end{array}$$

Each finite-decimal sum we get is called a *partial sum* of the two real numbers, for obvious reasons. In this case, the partial sums form the sequence whose shorthand name is .999.... This result agrees with the rational number addition of these numbers because

$$.333\ldots = \frac{1}{3}, \quad .666\ldots = \frac{2}{3}, \quad .999\ldots = 1$$

and

$$\frac{1}{3} + \frac{2}{3} = 1$$

□

Although Example 7 deals with infinite decimals that represented rational numbers, the "partial-sum" approach it illustrates can be applied to adding *any* infinite decimals. Moreover, subtraction, multiplication, and division of real numbers can be handled just like addition, by computing partial differences, products, and quotients, respectively. In this way, arithmetic operations can be defined on **R** in such a way that they extend the rational-number operations, rather than conflict with them.

Finally, the order relation on **Q** is very easily extended to infinite decimals. Recall that we can compare the sizes of whole numbers written in base ten by looking at the largest (farthest left) place where the digits differ. We do the same sort of thing with decimals, finite or infinite. (Here it is essential to have exactly one infinite decimal for each real number; it is more convenient in this case to presume that the repeated-9 forms have been converted to the repeated-0 form.) Thus:

$$603.1577 < 603.2 \text{ because } 1 < 2 \text{ in the tenths place}$$

$$.5\overline{548} < .\overline{5} \text{ because } 4 < 5 \text{ in the thousandths place}$$

$$.101001000\ldots < .10110111\ldots \text{ because } 0 < 1 \text{ in the ten-thousandths place}$$

Properties of the Real Numbers

Real-number arithmetic has all the properties you have come to expect of a respectable number system:

- Addition and multiplication are both commutative and associative;
- Multiplication is distributive over addition;
- 0 is the additive identity and 1 is the multiplicative identity;
- Every real number has an additive inverse and every nonzero real number has a multiplicative inverse; therefore, **R** is closed under subtraction and under division by nonzero numbers.
- The order relation on **R** has the same properties with respect to addition and multiplication as does the relation on **I** or **Q**.
- The rational-number system is contained in the real-number system, and their respective operations agree on **Q**.

Moreover, the problem we posed at the beginning of this chapter is resolved by the structure of **R**:

> The real-number system provides a numerical label for every point on the number line in such a way that the label of each point represents the (directed) distance from the point labeled 0 to that point.

This property is the basis for all of analytic geometry, and thus it is vitally important for the development of calculus and many other areas of mathematics and physics.

Exercises 13.5

For Exercises 1–8, before doing the given division problem, indicate the maximum number of different remainders that can occur. Then use long division to find out how many different remainders actually do occur. Is the number of different remainders the same as the number of digits in the repeating sequence?

1. $3 \div 5$ **2.** $2 \div 3$ **3.** $11 \div 37$

4. $7 \div 10$ **5.** $3 \div 7$ **6.** $14 \div 6$

7. $70 \div 14$ **8.** $451 \div 999$

For Exercises 9–17, convert to repeating-decimal form. (Write the entire repeating sequence of digits.)

9. $\frac{1}{9}$ **10.** $\frac{2}{5}$ **11.** $\frac{13}{6}$

12. $\frac{15}{22}$ **13.** $\frac{5}{11}$ **14.** $\frac{2}{13}$

15. $\frac{10}{7}$ **16.** $\frac{9}{21}$ **17.** $\frac{112}{37}$

For Exercises 18–29, convert to fraction form.

18. $1.\overline{4}$ **19.** $0.0\overline{2}$ **20.** $0.\overline{02}$

21. $0.\overline{9}$ **22.** $0.\overline{09}$ **23.** $0.\overline{15}$

24. 0.15 **25.** $62.1\overline{75}$ **26.** $62.\overline{175}$

27. $0.\overline{40}$ **28.** $0.1\overline{01}$ **29.** $0.\overline{101}$

30. Write five examples of irrational numbers in infinite-decimal form.

For Exercises 31–36, add the given pair of infinite decimals by using partial sums. Wherever possible, convert the decimals to fractions and add them (as rational numbers) to check your answers.

31. $.\overline{1} \oplus .\overline{2}$ **32.** $.\overline{6} \oplus .\overline{5}$

33. $.\overline{7} \oplus .\overline{3}$ **34.** $.\overline{7} \oplus .3$

35. $.1010010001\ldots \oplus .2020020002\ldots$

36. $.1010010001\ldots \oplus .0202202220\ldots$

37. List these real numbers in size order, from smallest to largest.

$$1.41,\ .141,\ 1.414,\ 1.\overline{41},\ .1\overline{41},$$
$$.\overline{141},\ 1.4\overline{1},\ .0\overline{141},\ .\overline{0141},\ \sqrt{2}$$

38. List these real numbers in size order, from smallest to largest.

$$3.0,\ 3.3,\ 3.03,\ 3.33,\ 3.\overline{3},$$
$$3.\overline{03},\ 3.0\overline{3},\ 3.\overline{303},\ 3.333,\ \sqrt{3}$$

13.6 What Is a Complex Number?

One of the problems that led to the construction of the real-number system was the need for a number whose square is 2. We introduced the problem geometrically, by way of the Pythagorean Theorem, but it can also be stated algebraically: We wanted a solution to the equation

$$x^2 = 2$$

Earlier in this chapter we saw how such a quantity can be represented by an infinite decimal, and how the set of infinite decimals can be used to represent the square roots of other numbers, as well. In fact, the square roots of all positive real numbers are themselves real numbers. We have not given a general proof of this fact, but it is not hard to see that the method for getting an infinite-decimal representation of $\sqrt{2}$ by successive approximations might be extended to represent the square root of any positive real number. Thus, in algebraic terms, we can say that the real-number system contains a solution of any equation of the form

$$x^2 = r$$

provided that r is a nonnegative real number.

But why are we limited to square roots of *nonnegative* numbers in $\mathbf{R}$? The answer stems from a fact that we proved for integers and that has remained true in every extension:

> The product of two positive numbers or of two negative numbers is always positive.

Historical Note: Imaginary Numbers

This difficulty with the square-root process arose late in the Greek mathematical era (within the first few centuries A.D.) and continued to cause confusion for some 1500 years. In the 17th century, Descartes, Newton, and Leibniz contributed valuable ideas to the solution of the question, but they occasionally supplied more fog than light. For example, square roots of negative numbers were first called "imaginary" in Descartes' *La Geometrie*, published in 1637. These numbers are no more imaginary than the others are "real," a name also conferred by Descartes in the same sentence of the same work. Nevertheless, the names stuck, and have confused students from that day on. The problem was not fully resolved until European mathematicians in the 18th century produced a logical, workable theory of imaginary numbers, establishing them once and for all as a "real" part of mathematics. ◇

How should we treat $\sqrt{-p}$, where p is a positive real number? The square of any nonzero real number must be positive, but

$$(\sqrt{-p})^2 = -p$$

by the usual definition of square root, so $\sqrt{-p}$ cannot be a real number. Hence, if we are to treat it as a number at all, we need a more comprehensive number system.

An examination of the square roots of positive numbers provides a clue as to how this extension might be made. We know from the basic laws of exponents that for positive real numbers x and y,

$$\sqrt{x \cdot y} = \sqrt{x} \cdot \sqrt{y}$$

If we suppose that numbers in the extended system might be treated in the same way,[6] then $\sqrt{-p}$ could be factored into $\sqrt{p} \cdot \sqrt{-1}$. This seemingly trivial observation is actually a giant stride forward because $\sqrt{p}$ is *real* for any (positive) p, and thus, instead of having to deal with the behavior of infinitely many square roots of negative numbers, we need only consider $\sqrt{-1}$.

If the number system can be enlarged to contain $\sqrt{-1}$ as well as all the real numbers, then, by the closure of multiplication, it must also contain all real-number multiples of $\sqrt{-1}$; that is, it must contain all things of the form $b \cdot \sqrt{-1}$, where b is any real number. By the closure of addition, it must also contain all possible sums of multiples of $\sqrt{-1}$ and real numbers; that is, it must contain all things of the form

$$a + b\sqrt{-1}$$

where a and b are real numbers. The notation is usually simplified by using i for $\sqrt{-1}$. We might say, then, that our new number system should be the set of all things of the form $a + bi$, where a and b are real numbers and where the arithmetic operations work more or less the way they do in $\mathbf{R}$.

PROBLEM-SOLVING COMMENT

What we are doing here is a classic example of *reasoning backwards from the desired conclusion*. By examining what must happen if our problem is to have an appropriate solution, we have seen what form that solution might take. We have *not* solved the problem yet, however. The proposed extended system was described by using $\sqrt{-1}$. Hence, the existence of the system depends on $\sqrt{-1}$; but this "number" was itself derived by *assuming* the existence of an extended number system. That is, each of these things depends on the other, and nowhere have we established the existence of either! We have merely explained what we would *like* to be true. To show that such a system exists, we must construct it without using either of these assumptions. The value of this hypothetical reasoning is that it suggests a place to start. ◇

Complex Numbers

We begin the construction of the set of complex numbers with the observation that any number system containing the real numbers and some other element i must contain all things of the form $a + bi$, where $a, b \in \mathbf{R}$. Since all things of this form are determined by the choice of a and b, it makes sense to begin by considering the set $\mathbf{R} \times \mathbf{R}$ of all ordered pairs of real numbers. We want to think of the pair (a, b) as if it were $a + b\sqrt{-1}$, but we cannot

[6]It is dangerous to put too much faith in this motivating supposition. See Exercise 28 for an instance of where it causes difficulty.

assume the existence of $\sqrt{-1}$. Therefore, we *define the operations* so that the b in any ordered pair (a, b) behaves as if it were multiplied by $\sqrt{-1}$.

Definition A **complex number** is an ordered pair of real numbers. The set of all complex numbers is denoted by $\mathbf{C}$; that is,

$$\mathbf{C} = \{(a, b) \mid a, b \in \mathbf{R}\}$$

The real numbers a and b are called the **real** and **imaginary parts** of the complex number (a, b), respectively. A complex number of the form $(0, b)$ is called an **imaginary number.**

Example 1 (2, 3) is a complex number. It can be thought of as $2 + 3i$, or $2 + 3\sqrt{-1}$, or $2 + \sqrt{-9}$. Its real part is 2 and its imaginary part is 3. (2, 3) is not an imaginary number. □

Example 2 (0, 5) is a complex number. It can be thought of as $0 + 5i$, or $5\sqrt{-1}$, or $\sqrt{-25}$. Its real part is 0 and its imaginary part is 5. It is an imaginary number. □

Two complex numbers are **equal** if *and only if* they are identical ordered pairs of real numbers; that is,

$$(a, b) = (c, d) \quad \text{if and only if} \quad a = c \text{ and } b = d$$

There is no need to form equivalence classes to get the complex numbers.

Because complex numbers are ordered pairs of real numbers, operations on $\mathbf{C}$ can be defined in terms of real-number operations on the real and imaginary parts of the complex numbers. As in earlier chapters, circles will be used to distinguish new operations from old ones.

Definition The **sum** of two complex numbers (a, b) and (c, d) is the number $(a+c, b+d)$. In symbols,

$$(a, b) \oplus (c, d) = (a + c, b + d)$$

Definition The **product** of two complex numbers (a, b) and (c, d) is the number $(ac - bd,\ ad + bc)$. In symbols,

$$(a, b) \odot (c, d) = (ac - bd,\ ad + bc)$$

Example 3

$$(1, 2) \oplus (3, 4) = (1 + 3, 2 + 4) = (4, 6)$$

$$(1, 2) \odot (3, 4) = (1 \cdot 3 - 2 \cdot 4,\ 1 \cdot 4 + 2 \cdot 3) = (-5, 10)$$ □

The obvious question at this point is: Where do these operations come from? In particular, why should we choose to define multiplication like this, rather than pick some other, perhaps simpler, way to do the job? The answer comes from remembering that we want the imaginary parts of complex numbers to behave as if they were multiplied by $\sqrt{-1}$, and we want the rest of the operational arithmetic to behave as it does in the real-number system. Thus, if we think of (a,b) and (c,d) as $a+b\sqrt{-1}$ and $c+d\sqrt{-1}$, respectively, then by multiplying binomials as in elementary algebra, the product "ought" to be:

$$\begin{aligned}(a+b\sqrt{-1})\cdot(c+d\sqrt{-1}) &= ac+ad\sqrt{-1}+bc\sqrt{-1}+bd(\sqrt{-1})^2\\ &= ac+(ad+bc)\sqrt{-1}+bd(-1)\\ &= (ac-bd)+(ad+bc)\sqrt{-1}\end{aligned}$$

Example 4 Here is a particular case of this multiplication. The use of numbers instead of letters may make the process easier to follow. Compare it with the general argument we just gave. (*Note*: The *form* of the result—in terms of the original real numbers—dictates how the definition of $\odot$ should be stated.)

$$\begin{aligned}(2+3\sqrt{-1})\cdot(5+7\sqrt{-1}) &= 2\cdot5+2\cdot7\sqrt{-1}+3\cdot5\sqrt{-1}+3\cdot7(\sqrt{-1})^2\\ &= 10+(14+15)\sqrt{-1}+21\cdot(-1)\\ &= (10-21)+(14+15)\sqrt{-1}\\ &= -11+29\sqrt{-1}\end{aligned}$$

□

It is not hard to verify that both addition and multiplication are commutative and associative, and that (0, 0) and (1, 0) are the additive and multiplicative identities, respectively. Moreover, multiplication is distributive over addition. All complex numbers have additive inverses and all nonzero complex numbers have multiplicative inverses, so subtraction and division can be done just as they were in the other number systems. We shall not pursue the proofs of these facts here. Most of them are easy to check and are left as exercises.

Taking our cue from the $a+b\sqrt{-1}$ motivation, if we identify each real number r with the complex number $(r,0)$, then the real-number system is included within the complex-number system: For any two real numbers r and s,

$$r\oplus s=(r,0)\oplus(s,0)=(r+s,0+0)=(r+s,0)$$

$$r\odot s=(r,0)\odot(s,0)=(r\cdot s-0\cdot0,\ r\cdot0+0\cdot s)=(r\cdot s,0)$$

Finally, the elusive $\sqrt{-1}$ emerges as the complex number (0, 1) because

$$\begin{aligned}(0,1)^2 &= (0,1)\odot(0,1)\\ &= (0\cdot0-1\cdot1,\ 0\cdot1+1\cdot0)\\ &= (-1,0)\end{aligned}$$

The existence of this number allows us to take square roots of all negative real numbers by "factoring out" -1, as described earlier.

Order

An interesting peculiarity of the complex-number system is that there is no useful way to define a *less than* relation for all complex numbers. That is, it is not possible to define an order relation on $\mathbf{C}$ that preserves addition and multiplication by "positive" numbers, satisfies the Trichotomy Law, and agrees with $<$ on $\mathbf{R}$ when confined to the real numbers.[7] A formal proof of this fact would take us too far afield, but it is not hard to construct a simple example showing the difficulties that arise when such an order relation is attempted.

For instance, suppose that $\otimes$ were such an order relation on $\mathbf{C}$, and recall that the terms *positive* and *negative* mean *greater than 0* and *less than 0*, respectively. Now, consider the two complex numbers $i = (0, 1)$ and $0 = (0, 0)$. Since i does not equal 0, the Trichotomy Law would imply that either

$$0 \otimes i \quad \text{or} \quad i \otimes 0$$

If $0 \otimes i$, then i is positive, so $i \odot 0 \otimes i \odot i$; that is,

$$0 \otimes i \odot i$$

But

$$i \odot i = (0,1) \odot (0,1) = (-1,0) = -1$$

and we know from the order on the reals that

$$-1 < 0$$

This contradiction means "$0 \otimes i$" is false. On the other hand, if $i \otimes 0$, then $-i$ is positive, so $-i \odot i \otimes -i \odot 0$; that is,

$$-i \odot i \otimes 0$$

But

$$-i \odot i = (0,-1) \odot (0,1) = (1,0) = 1$$

and we know from the order on the reals that

$$0 < 1$$

Thus, "$i \otimes 0$" is also false, so the Trichotomy Law does not hold!

---

[7] In particular, it makes no sense to claim that a nonzero complex number must be either "positive" or "negative."

Properties of the Complex Numbers

In summary, we have constructed a new number system **C**, called *the complex numbers*, with the following features:

- The numbers themselves are ordered pairs of real numbers; that is, $\mathbf{C} = \{(a,b) \mid a,b \in \mathbf{R}\}$.
- The subset $\{(a,0) \mid a \in \mathbf{R}\}$ is a copy of the set **R** of real numbers (and hence **C** "includes" the real numbers);
- The four basic arithmetic operations can be defined in **C**, and they agree with the arithmetic of the real numbers when they are applied to (the copy of) **R**.
- The operations in **C** have the same nice properties as their counterparts in **R** (commutativity, associativity, etc.).
- This new collection also contains some element whose square is -1, and hence it contains the square roots of all negative real numbers.

Exercises 13.6

For Exercises 1–8, write the given complex number in ordered-pair form.

1. $\sqrt{-4}$ **2.** $\sqrt{-7}$ **3.** $2+\sqrt{-1}$

4. $\frac{3}{8}$ **5.** $\sqrt{3}$ **6.** $-\sqrt{2}$

7. $2+\sqrt{-2}$ **8.** $-2-\sqrt{-2}$

For Exercises 9–14, perform the computation using the definition of $\oplus$ or $\odot$ given in this section.

9. $(2,3)\oplus(1,5)$ **10.** $(-4,7)\oplus(1,-9)$

11. $(6,0)\oplus(\frac{1}{3},2)$ **12.** $(2,5)\odot(3,8)$

13. $(4,-7)\odot(0,1)$ **14.** $(3,\frac{2}{5})\odot(5,0)$

15. Use the definition of $\oplus$ to verify that
$$(5,2)\oplus(0,0)=(5,2)$$

16. Use the definition of $\odot$ to verify that
$$(5,2)\odot(1,0)=(5,2)$$

17. Prove that $(a,b)\oplus(0,0) = (a,b)$ for any $(a,b)\in C$.

18. Prove that $(a,b)\odot(1,0) = (a,b)$ for any $(a,b)\in C$.

19. Prove that $(a,b)\odot(0,0) = (0,0)$ for any $(a,b)\in C$.

For Exercises 20–24, give an example to illustrate the stated property of the complex-number operation.

20. Addition is commutative.

21. Addition is associative.

22. Multiplication is commutative.

23. Multiplication is associative.

24. Multiplication is distributive over addition.

25. (a) Find the additive inverse of $(2,3)$. Check your answer by adding it to $(2,3)$.

(b) Subtract $(2,3)$ from $(5,1)$.

26. (a) Find the reciprocal of $(2,3)$. (This may not be easy!) Check your answer by multiplying it by $(2,3)$ to get $(1,0)$.

(b) Divide $(5,1)$ by $(2,3)$.

27. If $\sqrt{-1}$ is defined to be the complex number $(0,1)$, then, by using the correspondence between the real numbers a and b and their complex counterparts $(a,0)$ and $(b,0)$, the previously vague "number" $a + b\sqrt{-1}$ becomes

$$(a,0) \oplus [(b,0) \odot (0,1)]$$

Show by computation that this expression equals (a,b), as we would expect.

28. (a) Give two numerical examples to illustrate that

$$\sqrt{x \cdot y} = \sqrt{x} \cdot \sqrt{y}$$

holds for real numbers x and y when

i. x and y are both positive;

ii. x is positive and y is negative.

(b) Comment on the following "proof" that $1 = -1$:

$$1 = \sqrt{1} = \sqrt{(-1)\cdot(-1)}$$
$$= \sqrt{-1}\cdot\sqrt{-1} = (\sqrt{-1})^2 = -1$$

13.7 The Fundamental Theorem of Algebra

Traditional algebra—the algebra studied and practiced by mathematicians until fairly modern times—is really little more than symbolized arithmetic. It is the algebra you studied in high school, and the algebra we have used throughout this book whenever we needed to make general statements about numbers. It is just the process of using letters in the place of numbers so that general questions about arithmetic processes can be asked and answered efficiently. For example, when we wanted to discuss questions such as

$$3 + ? = 5, \quad 9 + ? = 4, \quad \text{and} \quad 29 + ? = 43$$

we expressed the general form of the problem represented by these examples as

$$a + x = b$$

where the a and b were understood to be whole numbers and the x represented the answer sought in each case. Similarly, questions such as

$$2 \cdot ? = 6, \quad 7 \cdot ? = -5, \quad \text{and} \quad -3 \cdot ? = -94$$

were represented by

$$a \cdot x = b$$

for integers a and b.

Let us use this algebraic notation to review the successive expansions of the number system that we have seen:

- Because the integers are closed under subtraction, every equation of the form

$$x + a = 0$$

where $a \in \mathbf{I}$, has a solution in $\mathbf{I}$.

- Because the rationals are closed under both subtraction and division, every equation of the form

$$ax + b = 0$$

where $a, b \in \mathbf{Q}$ and $a \neq 0$, has a solution in $\mathbf{Q}$.

- The real-number expansion is a little more difficult to describe in these terms. We saw that square roots of positive real numbers are real. Thus, every equation of the form

$$x^2 = a$$

where $a \in \mathbf{R}$ and $a \geq 0$ has a solution in $\mathbf{R}$. But $\mathbf{R}$ contains much more than just square roots—it contains *every* root of every nonnegative real number and every odd root of every negative real number. That is, every equation of the form

$$x^n = a$$

where a is a nonnegative real number, has a solution in $\mathbf{R}$ for any natural number n; and if a is negative, there is a solution in $\mathbf{R}$ for any odd natural number n. $\mathbf{R}$ is also closed under addition and multiplication. This implies that many, but not all, equations built up from sums and products of powers of x will also have solutions in $\mathbf{R}$. Equations like this are called **polynomial equations**; they have the general form

$$a_n x^n + a_{n-1}x^{n-1} + \ldots + a_2 x^2 + a_1 x + a_0 = 0$$

where the subscripted a's are real numbers (called **coefficients**) and n is some natural number.

Example 1 The polynomial equation

$$2x^3 + 7x^2 + 4x + 3 = 0$$

has integer coefficients. It also has a solution in $\mathbf{I}$, namely -3, because

$$2(-3)^3 + 7(-3)^2 + 4(-3) + 3 = -54 + 63 - 12 + 3 = 0$$

□

Example 2 The polynomial equation $4x^2 - 5 = 0$ (with integer coefficients) has no solution in either $\mathbf{I}$ or $\mathbf{Q}$ because any solution requires the square root of 5. If $4x^2 - 5 = 0$, then $4x^2 = 5$, so $x^2 = \frac{5}{4}$. This implies that x must equal $\frac{\sqrt{5}}{2}$ or $\frac{-\sqrt{5}}{2}$, both of which are irrational numbers. Thus, the equation has (two) solutions in $\mathbf{R}$. □

Example 3 The polynomial equation $5x^2 + 10 = 0$ has no solution in the real-number system. If $5x^2 + 10 = 0$, then

$$x^2 = \frac{-10}{5} = -2$$

so $x = \sqrt{-2}$, which is not a real number. The complex numbers $2i$ and $-2i$ are solutions of this equation. □

Much of high-school algebra is devoted to describing precisely which polynomial equations have integer, rational, or real-number solutions, and to developing specific techniques for solving such equations in **I**, **Q**, or **R**. Those techniques, while valuable, do not concern us here; we focus, instead, on the general existence question for solutions.

At each stage in the expansion of the number systems, from the whole numbers to the real numbers, there have been polynomial questions that could be asked but not answered in that system; that is, there have been polynomial equations with coefficients in a system, but with no solution in that system. In each case, such equations have prompted the extension to a more comprehensive system. Now that we have the complex numbers, however, the situation is considerably better. In fact, *every polynomial equation that can be stated in the complex-number system can be solved in that system*! The formal assertion of that fact is called "The Fundamental Theorem of Algebra." It was proved in 1799 by Carl Friedrich Gauss, then only 22 years old, in his doctoral thesis, and it is still regarded as one of the premiere existence results in all of mathematics.

The Fundamental Theorem of Algebra: Every equation of the form

$$a_n x^n + a_{n-1} x^{n-1} + \ldots + a_2 x^2 + a_1 x + a_0 = 0$$

where n is a natural number and all the coefficients are complex numbers, has a solution in the complex-number system.

An explanation of the proof of this elegant theorem would take us beyond the scope of this book. If you are interested in pursuing it on your own, a clear discussion of the proof may be found on pages 101–103 of *What Is Mathematics?*, by Richard Courant and Herbert Robbins (Oxford University Press, 1941).

Despite its deceptively simple statement, this theorem provides the key to understanding why the development of the complex numbers is a fitting climax to our exploration of the number systems. The Fundamental Theorem of Algebra assures us that,

> *regardless of the size or intricacy of a polynomial equation, no extension of the complex-number system is needed to solve it.*

Since polynomial equations represent the entire variety of arithmetic questions that can be formed from the four basic operations and from taking roots, this means that, from the viewpoint of arithmetic, this number system is complete.

Exercises 13.7

For Exercises 1–6, verify the given statement by substitution.

1. 5 is a solution of $4x - 20 = 0$.
2. $\frac{-2}{3}$ is a solution of $3x + 2 = 0$.
3. $\frac{1}{4}$ is a solution of $8x^2 - 6x + 1 = 0$.
4. $\sqrt{3}$ is a solution of $x^3 + x^2 - 3x - 3 = 0$.
5. $-7i$ is a solution of $x^2 + 2ix + 35 = 0$.
6. $3 + 2i$ is a solution of $x^2 - 4ix - 13 = 0$.

The number systems, in order of successive extension, are

$$\mathbf{W}, \mathbf{I}, \mathbf{Q}, \mathbf{R}, \text{ and } \mathbf{C}$$

For Exercises 7–14, specify the "smallest" number system containing a solution of the given equation.

7. $x + 4 = 0$
8. $x - 4 = 0$
9. $5x + 4 = 0$
10. $5x - 4 = 0$
11. $x^2 + 4 = 0$
12. $x^2 - 4 = 0$
13. $x^3 + 4 = 0$
14. $x^3 - 4 = 0$

For Exercises 15–18, consider the polynomial-equation form

$$(\star) \qquad ax^2 + bx + c = 0$$

15. Find nonzero $a, b, c \in \mathbf{I}$ such that $(\star)$ has a solution in $\mathbf{I}$.
16. Find nonzero $a, b, c \in \mathbf{I}$ such that $(\star)$ has a solution in $\mathbf{Q}$ but not in $\mathbf{I}$.
17. Find nonzero $a, b, c \in \mathbf{I}$ such that $(\star)$ has a solution in $\mathbf{R}$ but not in $\mathbf{Q}$.
18. Find nonzero $a, b, c \in \mathbf{I}$ such that $(\star)$ has a solution in $\mathbf{C}$ but not in $\mathbf{R}$.

Review Exercises for Chapter 13

For Exercises 1–10, indicate whether the given statement is *true* or *false*.

1. The length of the diagonal of a unit square can be represented as the ratio of two integers.
2. The set of all binary-fraction points is a dense subset of the number line.
3. The number 1.999... is just a little bit smaller than 2.
4. The number 1.999 is just a little bit smaller than 2.
5. The number 1.999... is smaller than 1.999.
6. It is possible to set up a 1-1 correspondence between the set $\mathbf{R}$ of real numbers and the set of all points on a line.
7. The number of different remainders in the division problem $p \div q$ can be no larger than $q - 1$.
8. The decimal representation of every irrational number is infinite and nonrepeating.
9. The number .101001000100001... is rational because it keeps repeating 0s and 1s.
10. The square root of a negative number is undefined.

For Exercises 11–13, approximate each number by binary fractions to within $\frac{1}{32}$ of its actual value.

11. $\frac{2}{5}$ **12.** $\frac{3}{7}$ **13.** $\sqrt{5}$

For Exercises 14–16, approximate the given number by a fraction with denominator 1000 and also express that fraction as a decimal.

14. $\frac{5}{6}$ **15.** $\frac{7}{8}$ **16.** $\sqrt{11}$

For Exercises 17–22, convert the given fraction to a repeating decimal.

17. $\frac{4}{9}$ **18.** $\frac{3}{5}$ **19.** $\frac{15}{6}$

20. $\frac{17}{22}$ **21.** $\frac{22}{7}$ **22.** $\frac{113}{37}$

For Exercises 23–28, convert to fraction form.

23. $1.\overline{5}$ **24.** $0.0\overline{3}$ **25.** $0.\overline{03}$

26. $0.2\overline{35}$ **27.** $0.\overline{235}$ **28.** $5.2\overline{19}$

29. Using only the digits 2, 4, and 6, write five examples of irrational numbers in infinite-decimal form.

For Exercises 30–33, add by using partial sums. If the numbers are rational, also convert to fraction form and add, using rational-number addition.

30. $.\overline{3} \oplus 2.\overline{5}$ **31.** $.\overline{7} \oplus 7.\overline{6}$

32. $.\overline{14} \oplus .232332333\ldots$

33. $.414114111\ldots \oplus .141441444\ldots$

For Exercises 34–39, write the given complex number in ordered-pair form.

34. $\sqrt{-36}$ **35.** $-\sqrt{36}$ **36.** $\sqrt{-11}$

37. $3 + \sqrt{-5}$ **38.** $7 - \sqrt{-9}$ **39.** $7 - \sqrt{9}$

For Exercises 40–42, verify the given statement by substitution.

40. -7 is a solution for $3x + 21 = 0$.

41. $-2i$ is a solution for $x^2 + 4 = 0$.

42. $2 + 3i$ is a solution for $x^2 - 4x + 13 = 0$.

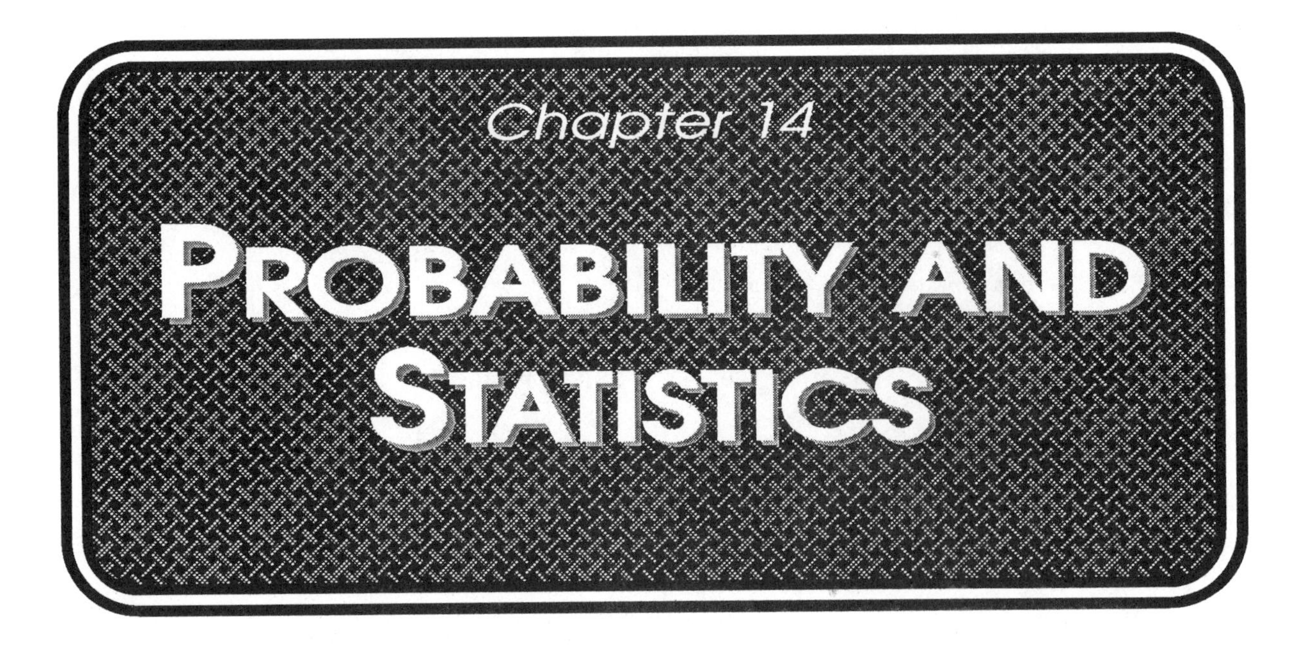

14.1 The Gamblers

Much of mathematics is concerned with deducing valid conclusions from premises that are assumed to be true. This process is logically satisfying, but any attempt to apply it to the world we live in is hampered by a major obstacle: There is no way of verifying with certainty most assumptions that deal with reality. The branch of mathematics called *probability* faces this problem squarely and tries to bridge the gap between theory and fact. It begins by considering an uncertain situation, examines all conclusions that can be drawn from the possible alternatives, and makes quantitative judgments about the reliability of those conclusions.

Probability theory and statistical probability (called from now on **probability** and **statistics**) are two distinct, but closely related, fields of "uncertain" mathematics. In fact, they are devoted to examining opposite sides of the same fundamental problem. The question asked of probability is:

> Given a known collection of objects, what can be said about the characteristics of an unknown subset of that collection?

The "problem of points," described in the Historical Note, is such a question; the set of all possible outcomes of the gambling game is known, but the particular outcome of an unfinished game is not. Statistics attempts to answer the converse question:

Historical Note: Games of Chance

The mathematical theory of probability originated in 1654 when the Chevalier de Méré, a wealthy French nobleman with a taste for gambling, proposed to the mathematician Blaise Pascal a problem involving the distribution of stakes in an unfinished game of chance. This was the famous "problem of points," in which a simple gambling game between two players is ended before either wins. The question was how to divide the prize money if the partial scores of the players are known, and its answer was based on the likelihood of each player winning the game. Pascal communicated the question to Pierre de Fermat, another prominent French mathematician, and from their correspondence a new field of mathematics emerged. Pascal and Fermat arrived at the same answer to the problem, but their methods of solution were different. They generalized the problem and its solution, then extended their investigations to other games of chance. Their work aroused interest among the European scientific community, and soon other scholars took up the challenge of analyzing gambling games.

As Pascal, Fermat, and their colleagues on the Continent were discussing games of chance, researchers in England were hard at work investigating a slightly more respectable, but equally costly, form of gambling. The many hazards of life in the sixteenth and seventeenth centuries had established an eager market for a remarkable new service called "life insurance." The groups of people who backed these first life-insurance policies must be ranked among the most daring of gamblers, for they often risked large sums of money on a single turn of fate. To help them understand this life-and-money game of chance, people began to investigate death records. The first important mortality tables were compiled by Edmund Halley in 1693 as a basis for his study of annuities. This second type of probabilistic investigation came to be known as *statistical probability* or simply *statistics*.

The first book devoted entirely to probability (both theoretical and statistical) was Jacob Bernoulli's *Ars Conjectandi*, published in 1713. In this book, Bernoulli anticipated by more than 200 years the practical importance of the subject by suggesting applications of probability to government, law, economics, and morality. As the study of statistics developed since that time, it has provided the means by which probability theory is applied not only to the fields suggested by Bernoulli, but to education, the social sciences, and many other areas as well. ◇

> Given a known sample of an unknown collection of objects, what can be said about the characteristics of the collection as a whole?

For instance, the statistical question of death rates may be phrased: "Knowing the life span of each person in a relatively small group, what can be said about the life spans of people in general?"

Example 1 Figure 14.1, taken from a seventh-grade textbook, describes an agricultural experiment that illustrates some of the ideas we have just described. Is this investigation primarily about probability or about statistics? Why? □

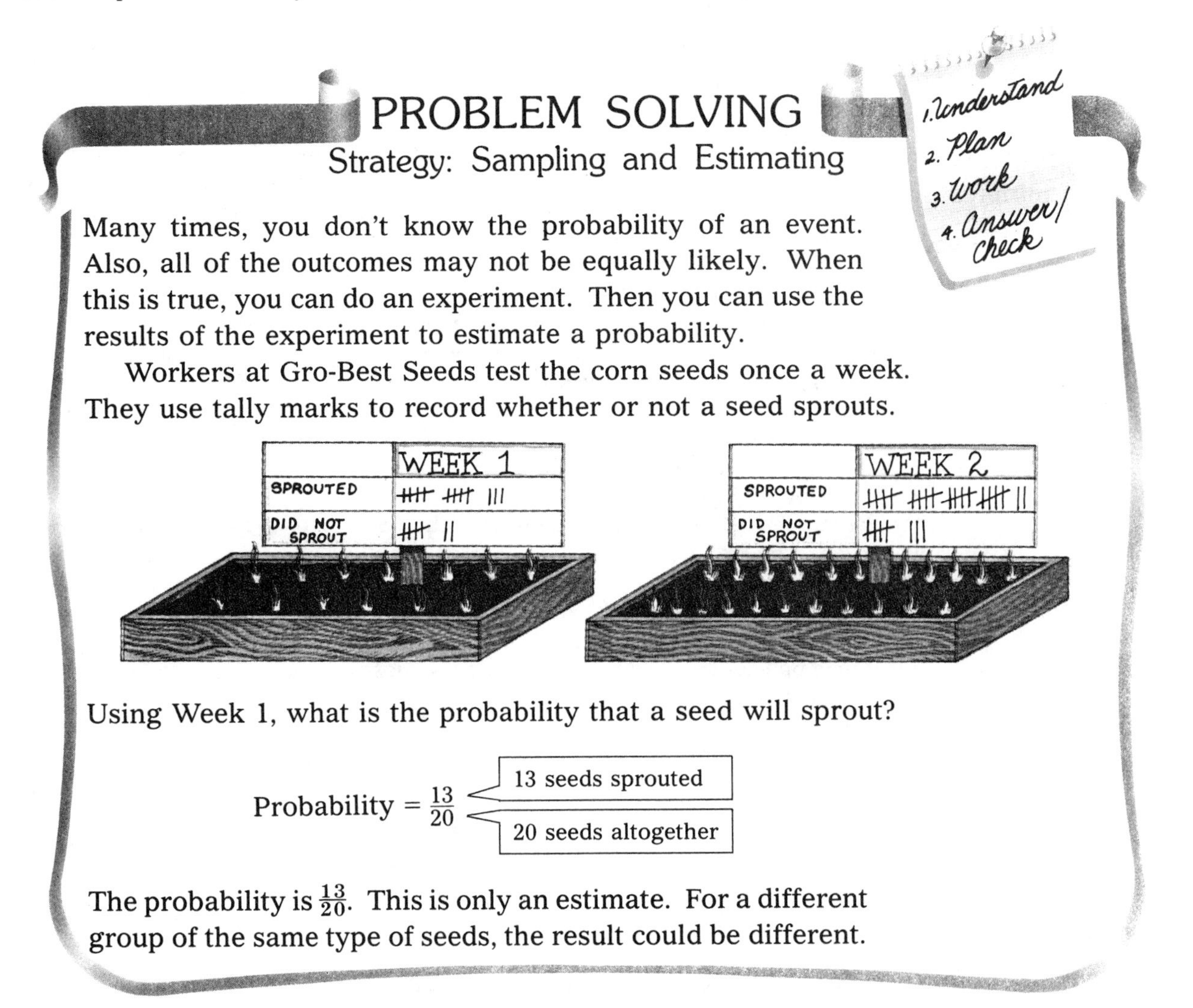

PROBLEM SOLVING

Strategy: Sampling and Estimating

Many times, you don't know the probability of an event. Also, all of the outcomes may not be equally likely. When this is true, you can do an experiment. Then you can use the results of the experiment to estimate a probability.

Workers at Gro-Best Seeds test the corn seeds once a week. They use tally marks to record whether or not a seed sprouts.

Using Week 1, what is the probability that a seed will sprout?

Probability $= \frac{13}{20}$

13 seeds sprouted

20 seeds altogether

The probability is $\frac{13}{20}$. This is only an estimate. For a different group of the same type of seeds, the result could be different.

Figure 14.1 HOUGHTON MIFFLIN MATHEMATICS, Grade 7, p. 414

The basic principles of probability are easy extensions of a common-sense approach to chance situations. A little experience with flipping a coin or playing a simple card game should provide the insight for understanding the fundamentals of the theory presented in this chapter. The study of statistics relies heavily on the principles of probability theory, but the problems of statistics are often more difficult than those of probability. Moreover, many of the more powerful results in both probability and statistics require an understanding of calculus. This chapter provides a gentle introduction to some elementary probabilistic and statistical concepts.

14.2 What Is Probability?

In general terms, the field of mathematics called **probability** is the quantitative study of "randomness." Something is **random** if it is not predictable; that is, if it cannot be determined *with certainty* from prior knowledge or experience. The possible outcomes of an unfinished gambling game and the date of death of a living person are random matters we discussed in the previous section. The results of flipping a coin or buying a lottery ticket or predicting next week's stock-market prices are also random occurrences. The uncertainty of these kinds of things is easier to cope with if we can assign a numerical value to the likelihood that a random situation will turn out one way rather than another. This allows us to measure the various alternatives, using numbers to determine which of several possible outcomes is most likely, least likely, and so forth.

One preliminary observation is in order here. Although problems involving games of chance make convenient examples for elementary probability theory and although we shall use them often to illustrate the basic ideas, they are by no means typical of the important current applications of probability. In fact, it is precisely their obvious randomness that makes these examples atypical. Most scientific applications of probability arise in situations caused or governed by physical, economic, or social principles, and as such, these situations should be predictable. (For example, when a coin is tossed, the side it lands on is determined by its weight and center of gravity, the magnitude and direction of the force applied by the thumb that tosses it, the distance it falls, the surface it lands on, and so forth.) However, the interaction of these principles is often so complex as to defy analysis, so it is convenient to think of the possible results as random occurrences. In other words, the randomness of a situation is often an *assumption* that allows a complex problem to be handled in a simple way.

Since the theory of probability is based on randomness, its applicability in particular situations depends on the validity of that assumption in each case. This and other simplifying assumptions have the effect of transforming a concrete problem of physical or behavioral science into an analogous abstract problem. The setting for the problem in its abstract form is called a **model** of the original situation. If the assumptions in this abstraction process eliminate only details that are truly irrelevant, then any solution of the model problem will also solve the original real-world problem. On the other hand, if the model is an oversimplification in some respects, its solution may have little to do with the original problem. For instance, when dice are tossed, the assumption that all of the faces are equally likely to turn up is valid *provided that* the dice have not been "loaded" (weighted off-center in some way). This is usually, but not always, a reasonable assumption to make.

The reliability of the abstraction process varies from problem to problem, and it is often complex. Investigation of this question is not essential to the development of basic probability theory, so we shall treat the relationship between concrete problems and their models as accurate and shall deal primarily with the models. Thus, for example, we shall generally assume that either side of a tossed coin is equally likely to turn up and that each face of a thrown die is as likely to turn up as any other.

Sample Spaces

We begin our formal study of probability by introducing some useful standard terminology. Any situation or problem involving uncertain results is called an **experiment** (regardless of whether or not we have control over the situation). The various possible results are called **outcomes**, and the set of all possible outcomes is called the **sample space** of the experiment. The sample space is usually denoted by S. In the language of set theory, S is the "universal set" for the experiment, and the outcomes are all the elements of that universal set. Any subset of the sample space is called an **event**.

Example 1 Tossing a coin is an experiment with two possible outcomes, *heads* or *tails*. (Our model ignores the possibility that the coin might land on its edge.) If we denote these two outcomes by h and t, then the sample space is $\{h, t\}$. There are four possible events:

$\{h\}$ – The coin lands heads up.

$\{t\}$ – The coin lands tails up.

$\{h, t\}$ – The coin lands either heads up or tails up.

$\emptyset$ – The coin lands neither heads up nor tails up. (This set is empty because we are excluding the possibility that the coin might land on edge.) □

Example 2 Rolling a single die is an experiment with six possible outcomes. Its sample space can be represented by

$$S = \{1, 2, 3, 4, 5, 6\}$$

There are 64 events in this experiment because S has 2^6 subsets. Two such events are:

$\{2, 4, 6\}$ – The die falls with an even number facing up.

$\{1, 2, 3, 4\}$ – The die falls with a number less than 5 facing up. □

The reliability of the abstraction process varies from problem to problem, and it is often complex. Investigation of this question is not essential to the development of basic probability theory, so we shall treat the relationship between concrete problems and their models as accurate and shall deal primarily with the models. Thus, for example, we shall generally assume that either side of a tossed coin is equally likely to turn up and that each face of a thrown die is as likely to turn up as any other.

Sample Spaces

We begin our formal study of probability by introducing some useful standard terminology. Any situation or problem involving uncertain results is called an **experiment** (regardless of whether or not we have control over the situation). The various possible results are called **outcomes**, and the set of all possible outcomes is called the **sample space** of the experiment. The sample space is usually denoted by S. In the language of set theory, S is the "universal set" for the experiment, and the outcomes are all the elements of that universal set. Any subset of the sample space is called an **event**.

Example 1 Tossing a coin is an experiment with two possible outcomes, *heads* or *tails*. (Our model ignores the possibility that the coin might land on its edge.) If we denote these two outcomes by h and t, then the sample space is $\{h, t\}$. There are four possible events:

$\{h\}$ – The coin lands heads up.

$\{t\}$ – The coin lands tails up.

$\{h, t\}$ – The coin lands either heads up or tails up.

$\emptyset$ – The coin lands neither heads up nor tails up. (This set is empty because we are excluding the possibility that the coin might land on edge.) □

Example 2 Rolling a single die is an experiment with six possible outcomes. Its sample space can be represented by

$$S = \{1, 2, 3, 4, 5, 6\}$$

There are 64 events in this experiment because S has 2^6 subsets. Two such events are:

$\{2, 4, 6\}$ – The die falls with an even number facing up.

$\{1, 2, 3, 4\}$ – The die falls with a number less than 5 facing up. □

14.2 What Is Probability?

In general terms, the field of mathematics called **probability** is the quantitative study of "randomness." Something is **random** if it is not predictable; that is, if it cannot be determined *with certainty* from prior knowledge or experience. The possible outcomes of an unfinished gambling game and the date of death of a living person are random matters we discussed in the previous section. The results of flipping a coin or buying a lottery ticket or predicting next week's stock-market prices are also random occurrences. The uncertainty of these kinds of things is easier to cope with if we can assign a numerical value to the likelihood that a random situation will turn out one way rather than another. This allows us to measure the various alternatives, using numbers to determine which of several possible outcomes is most likely, least likely, and so forth.

One preliminary observation is in order here. Although problems involving games of chance make convenient examples for elementary probability theory and although we shall use them often to illustrate the basic ideas, they are by no means typical of the important current applications of probability. In fact, it is precisely their obvious randomness that makes these examples atypical. Most scientific applications of probability arise in situations caused or governed by physical, economic, or social principles, and as such, these situations should be predictable. (For example, when a coin is tossed, the side it lands on is determined by its weight and center of gravity, the magnitude and direction of the force applied by the thumb that tosses it, the distance it falls, the surface it lands on, and so forth.) However, the interaction of these principles is often so complex as to defy analysis, so it is convenient to think of the possible results as random occurrences. In other words, the randomness of a situation is often an *assumption* that allows a complex problem to be handled in a simple way.

Since the theory of probability is based on randomness, its applicability in particular situations depends on the validity of that assumption in each case. This and other simplifying assumptions have the effect of transforming a concrete problem of physical or behavioral science into an analogous abstract problem. The setting for the problem in its abstract form is called a **model** of the original situation. If the assumptions in this abstraction process eliminate only details that are truly irrelevant, then any solution of the model problem will also solve the original real-world problem. On the other hand, if the model is an oversimplification in some respects, its solution may have little to do with the original problem. For instance, when dice are tossed, the assumption that all of the faces are equally likely to turn up is valid *provided that* the dice have not been "loaded" (weighted off-center in some way). This is usually, but not always, a reasonable assumption to make.

Example 3 A raffle in which a single winning ticket is drawn from a barrel containing 1000 tickets is an experiment with 1000 possible outcomes. There are 1000 elements in its sample space, and there are 2^{1000} events. □

Example 4 Tossing a coin twice is an experiment with four possible outcomes:

first *heads*, then *heads*
first *heads*, then *tails*
first *tails*, then *heads*
first *tails*, then *tails*

That is, $S = \{hh, ht, th, tt\}$. There are 16 events in this experiment. (What are they?) □

PROBLEM-SOLVING COMMENT

The importance of set-theoretic language stems from the fact that it applies to many different areas of mathematics and thereby acts as a unifying force. As we explore probability, we shall *restate the problems* in the set-theoretic language of Chapter 3. This restatement will permit us to use *appropriate notation* from that chapter, making it easy to explain new ideas with familiar terms and symbols. As a bonus, the set language here may help you see *analogies* betweeen parts of probability and other mathematical areas. ◇

The basic idea that underlies all of probability is this:

> We want to assign numbers to each event of a sample space to somehow measure the likelihood that the outcome of the experiment will lie inside, rather than outside, the event.

Don't be put off by the formal language; this is an idea that should be quite familiar from everyday life. For instance, when we say that there's a "50–50" chance of a tossed coin landing heads up, we are assigning the number 50% to the event $\{heads\}$ and 50% to the event $\{tails\}$. In Example 3, the holder of a single raffle ticket would surely say he had one chance in a thousand of winning (assuming the raffle is fair); so the number attached to the event consisting of that one ticket would be $\frac{1}{1000}$.

Equally Likely Outcomes

A raffle is fair if each ticket is as likely to be drawn as any other. A coin is fair if either side is as likely to turn up as the other when it is tossed. In general, we say that all the outcomes of a sample space are **equally likely** if each one has the same chance of happening as any other. (That's not very satisfying as a formal definition, but it will serve our purposes well enough.) Whenever possible, we shall describe the sample space of an experiment in such a way that all outcomes are equally likely because in that case the assignment of numbers to events is easy.

Definition In a sample space S of equally likely outcomes, the **probability** of an event E is the number of outcomes in E divided by the number of outcomes in S. In symbols,

$$P(E) = \frac{n(E)}{n(S)}$$

Example 5 If a fair coin is tossed once, the probability of its landing "heads up" is the number of elements in the event $\{h\}$ divided by the number of elements in the sample space, $\{h, t\}$. In symbols:

$$P(\{h\}) = \frac{n(\{h\})}{n(\{h,t\})} = \frac{1}{2}$$ □

Example 6 In Example 2, the sample space S contains six outcomes. Let us call the two events listed there E and F; that is, let

$$E = \{2,4,6\} \quad \text{and} \quad F = \{1,2,3,4\}$$

Then

$$P(E) = \frac{3}{6} = \frac{1}{2} \quad \text{and} \quad P(F) = \frac{4}{6} = \frac{2}{3}$$

In other words, if a single (fair) die is rolled, there are 3 chances in 6 of rolling an even number, and there are 4 chances in 6 of rolling a number less than 5. □

Example 7 If Ms. Chance buys 25 of the 1000 tickets sold for a single-prize raffle, then the event that she wins the raffle consists of 25 outcomes. If we call this event W, then

$$P(W) = \frac{25}{1000} = \frac{1}{40}$$

That is, she has 25 chances in 1000—or 1 chance in 40—of winning the single prize. □

Example 8 The sixth-grade text excerpt shown in Figure 14.2 illustrates how the definition of probability is treated at that level. What assumptions are being made about the coin? □

Example 9 As Example 4 shows, tossing a coin twice is an experiment with a four-element sample space. If we want the result to be one head and one tail (in either order), then the desired event, which we shall call E, is $\{ht, th\}$, and

PROBABILITY

When you toss a coin, there are two possible **outcomes**, both equally likely. There is 1 chance in 2 of the outcome being a head. The **probability** of a head is 1 out of 2 or $\frac{1}{2}$. The probability of a tail is also $\frac{1}{2}$.

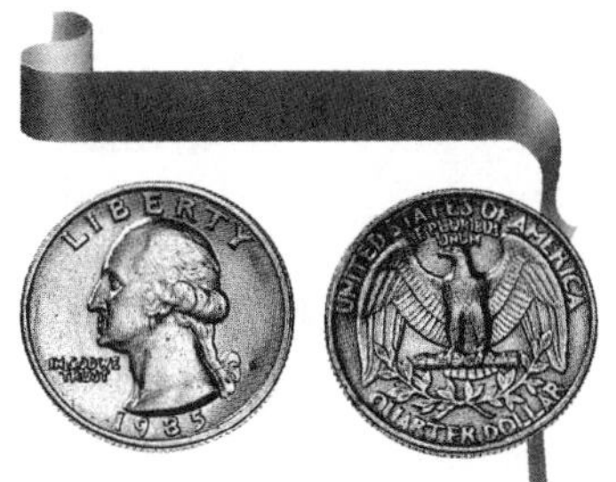

The spinner is equally likely to stop on any of the 6 equal-size sections. There are 3 chances in 6 of landing on a star. The probability of this event, P (★), is given by the following ratio.

$$P(★) = \frac{\text{number of ★ outcomes}}{\text{number of possible outcomes}}$$

$$P(★) = \frac{3}{6} = \frac{1}{2}$$

The spinner has 3 sections with a star and 2 sections with a triangle. Since there are 5 sections in all with a star or a triangle, P (★ or ▲) $= \frac{5}{6}$.

The probability of an impossible event is 0. Since there are no circles on the spinner, the probability of landing on a circle is 0. The probability of an event that must occur is 1.
Think: P (★ or ▲ or ■) = ?

Figure 14.2 HOUGHTON MIFFLIN MATHEMATICS, Grade 6, p. 380

$$P(E) = \frac{n(E)}{n(S)} = \frac{2}{4} = \frac{1}{2}$$

It is common to express a probability as a percent; in this case, then, E has a probability of 50%. That is, there is a 50% chance of getting one head and one tail (in either order) in two tosses of a fair coin. □

Example 10 There is an *incorrect*, though tempting, approach to the experiment described in Examples 4 and 9 that merits some attention. (See if you can figure out why it's wrong before we tell you.) We might describe the possible outcomes of tossing a coin twice by the three-element sample space

$$T = \{\text{both heads, one head and one tail, both tails}\}$$

In this case, the event $E = \{\text{one head and one tail}\}$ contains just one of the three outcomes in the sample space, so it appears that

$$P(E) = \frac{n(E)}{n(T)} = \frac{1}{3}$$

The probability as computed here is different from the probability of the same event in the same experiment as computed in Example 9. Why is that approach correct and this one wrong? As you might already have determined, the difficulty here is that not all the outcomes in the sample space T are equally likely. In fact, the "one head and one tail" outcome is twice as likely to happen as either of the others, so the definition of probability used here will not provide a correct measure of the likelihood of the events in this experiment. □

The Law of Large Numbers

A common misconception about the way to interpret the probability of an event deserves some comment here. Example 5 provides a simple, but typical, situation. It says that, if a fair coin is tossed once, the probability that it will come up *heads* is $\frac{1}{2}$, or 50%. Suppose you toss this coin and it does come up *heads*. Now, if you toss it a second time, is it less likely, more likely, or equally likely to come up *heads* again? The answer is that it is *equally likely*. Some people incorrectly reason that, since it came up *heads* the first time, then it is more likely to come up *tails* the second time because each outcome should occur 50% of the time. The assertion that the event {*heads*} has a 50% probability does *not* imply this. It says only that, if the coin toss is repeated a large number of times, the ratio of the number of *heads* to the total number of tosses will approach 50% *in the long run*. In other words, the more times a coin is tossed, the closer to 50% this ratio of *heads* to tosses is likely to be.

This is a special case of an important general principle of probability that connects the probability of an event with the number of repetitions, or **trials**, of an experiment:

> **The Law of Large Numbers:** As the number of repetitions of an experiment increases, the ratio of the number of occurrences of an event E to the total number of trials approaches $P(E)$.

Thus, if you toss a coin 100 times, you are not guaranteed to get precisely 50 *heads* and 50 *tails*, but the ratio of *heads* (or of *tails*) to the total number of tosses is likely to be close to 50%. Moreover, the pattern of outcomes may not even come close to alternating between *heads* and *tails.* To give you some idea of a typical *head-tail* pattern in 100 trials, we simulated this experiment on a computer; the result is shown in Figure 14.3. Notice the "strings" of *heads* and *tails*, up to 8 consecutive *tails.*

hhhttttthhthtttthtthhtthtt
hhthhhtthtthhhhhhhhttthhhh
htttttttthhhhhthhhtthhhttt
tthtthhthhthttttththhththth

Figure 14.3 100 tosses of a fair coin.

In the first 25 tosses of this experiment, there are only 10 *heads*, so the ratio of *heads* to tosses is only $\frac{10}{25}$, or 40%. In the first 50 tosses, there are 27 *heads*, so the ratio is $\frac{27}{50}$, or 54%. In the whole list of 100 tosses, there are 49 *heads* and 51 *tails*, so the ratio of *heads* to the total number of tosses in the complete run is 49%. The Law of Large Numbers tells us that, if we were to continue this experiment, say to 1000 or 10,000 tosses, it is likely that the ratio of *heads* to tosses would be even closer to 50%.

Exercises 14.2

For Exercises 1–7, the experiment consists of tossing a fair coin three times.

1. List all the equally likely outcomes for this experiment; that is, find a suitable sample space. (*Hint*: There are 8 such outcomes.)
2. List the outcomes in each of these events:
   E: The first two tosses are the same (both *heads* or both *tails*).
   F: The first and third tosses are the same.
   G: The second and third tosses are tails.
3. Find $P(E)$, $P(F)$, and $P(G)$.
4. List the elements of $E \cap F$; then find $P(E \cap F)$.
5. List the elements of $E \cup F$; then find $P(E \cup F)$.
6. List the elements of $F \cap G$; then find $P(F \cap G)$.
7. List the elements of $F \cup G$; then find $P(F \cup G)$.

For Exercises 8–12, the experiment consists of rolling a single die, then tossing a coin.

8. List all the equally likely outcomes of this experiment; that is, find a suitable sample space.
9. List the outcomes in each of these events:
   E: The number on the die is even.
   F: The number on the die is 5.
   G: The coin lands heads up.
   H: The number on the die is less than 4 and the coin lands tails up.
10. Find the probability of each event in Exercise 9.
11. List the outcomes in $E \cap G$; then find $P(E \cap G)$.
12. List the elements of $F \cup H$; then find $P(F \cup H)$.

For Exercises 13–20, the experiment consists of rolling a pair of (fair) dice.

13. List all the equally likely outcomes for this experiment; that is, find a suitable sample space.

14. Does the set $\{2, 3, 4, \ldots, 12\}$ represent a suitable sample space for this experiment? Why or why not?

15. List all the outcomes in each of these events:
E: The sum of the numbers thrown is 2.
F: The sum of the numbers thrown is 7.
G: The sum of the numbers thrown is even.
H: The sum of the numbers thrown is less than 5.

16. Find the probability of each event in Exercise 15.

17. List the elements of $E \cap F$ and those of $E \cup F$.

18. Find $P(E \cap F)$ and $P(E \cup F)$.

19. List the elements of $G \cap H$ and those of $G \cup H$.

20. Find $P(G \cap H)$ and $P(G \cup H)$.

For Exercises 21–28, the experiment consists of "rolling" three (fair) *tetrahedral* dice. (Each die has four equally likely faces, numbered 1, 2, 3, and 4.)

21. List all the equally likely outcomes for this experiment; that is, find a suitable sample space.

22. Does the set $\{3, 4, \ldots, 12\}$ represent a suitable sample space for this experiment? Why or why not?

23. List all the outcomes in each of these events:
E: The sum of the numbers thrown is 2.
F: The sum of the numbers thrown is 7.
G: The sum of the numbers thrown is even.
H: The sum of the numbers thrown is less than 5.

24. Find the probability of each event in Exercise 23.

25. List the elements of $E \cap F$ and of $E \cup F$.

26. Find $P(E \cap F)$ and $P(E \cup F)$.

27. List the elements of $G \cap H$ and of $G \cup H$.

28. Find $P(G \cap H)$ and $P(G \cup H)$.

For Exercises 29–31, suppose that a single card is drawn from a regular deck of 52 playing cards.

29. What is the probability of drawing an ace? Why?

30. What is the probability of drawing a heart? Why?

31. What is the probability of drawing the ace of hearts?

For Exercises 32–34, suppose that a jar contains seven ping-pong balls, numbered 1 through 7 Two balls are drawn from the jar, and the ball drawn first is *not* replaced before the second draw.

32. What is a suitable sample space for this experiment?

33. Find the probability of each of these events:
E: Both balls drawn have even numbers.
F: One ball has a 5 on it.
G: The sum of the numbers on the balls drawn is 6.

34. What is the probability that the sum of the numbers on the balls drawn is *not* 6? Why?

35. (a) Toss a coin 100 times and keep track of the sequence of outcomes. Compute the ratio of the number of *heads* to the total number of tosses, as was done for the experiment described in Figure 14.3 on page 543. Compare your results with the Law of Large Numbers.

(b) Toss the coin 50 more times and, adding your results to those of part (a), compute the ratio of the number of *heads* to the total number of tosses.

(c) Toss the coin 50 more times and, adding your results to those of parts (a) and (b), compute the ratio of the number of *heads* to the total number of tosses.

14.3 Some Basic Rules

Recall that we have been dealing with experiments whose sample spaces contain equally likely outcomes, and in this case the probability of an event is simply the number of outcomes in the event divided by the total number of outcomes in the sample space. Since events are just subsets of the sample space, we can make some easy observations that will be valuable for generalizing the probability concept to other kinds of experiments, in which the outcomes may not be equally likely. Some of these observations are *axioms* (postulates) of probability; that is, they are statements *assumed* to be true in every probability situation, from which the other general laws of probability are proved.

The Axioms of Probability

First of all, recall that the sample space S of an experiment plays the role of the universal set for that experiment. Now, S is a subset of itself, so we can ask what its probability is. The common-sense answer is that the actual outcome of the experiment is certain to be somewhere in the sample space, so the number assigned to S should indicate certainty. Our definition of probability says that this number must be 1 (or 100%, if you prefer), so

$$P(S) = \frac{n(S)}{n(S)} = 1$$

This becomes our first axiom:

Axiom 1: The probability of any sample space is 1.

The other extreme event is the one containing no outcomes at all—that is, the empty set. In this instance, common sense suggests that the number assigned to $\emptyset$ should indicate impossibility. By the definition of probability,

$$P(\emptyset) = \frac{n(\emptyset)}{n(S)} = \frac{0}{n(S)} = 0$$

Since any event E is a subset of S, the number of elements in E cannot exceed the number of elements in S, and the fewest elements E can have is none at all. Thus, $0 \leq n(E) \leq n(S)$, and because

$$P(E) = \frac{n(E)}{n(S)}, \quad \text{where} \quad 0 \leq \frac{n(E)}{n(S)} \leq \frac{n(S)}{n(S)} = 1$$

it follows that $P(E)$ must be a number between 0 and 1. This suggests our second axiom:

Axiom 2: For any event E, $0 \leq P(E) \leq 1$.

The third and final axiom also follows from a common-sense observation about sets: If two sets are disjoint, then the number of elements in their union can be found by counting the elements in each set separately and then adding. That is:

$$\text{If } E \cap F = \emptyset, \text{ then } n(E \cup F) = n(E) + n(F).$$

(Notice that this parallels the definition of addition of whole numbers, as stated in Chapter 4.) If these sets are events of a sample space, then saying that they are disjoint means that if the actual outcome of the experiment falls within one of these events, it cannot also fall within the other. In other words, if either of these events happens, the other cannot. Disjoint events are given a name to reflect this property; they are called **mutually exclusive** events.

In a sample space of equally likely outcomes, if we know the separate probabilities of two mutually exclusive events E and F, the probability of their union is easily found by applying the preceding counting observation:

$$P(E \cup F) = \frac{n(E \cup F)}{n(S)} = \frac{n(E)}{n(S)} + \frac{n(F)}{n(S)} = P(E) + P(F)$$

This suggests our third axiom:

Axiom 3: If E and F are mutually exclusive events, then
$$P(E \cup F) = P(E) + P(F)$$

Example 1 If we roll a single fair die (as in Example 2 of Section 14.2), it is certain that we will get 1, 2, 3, 4, 5, or 6. That is, $P(\{1,2,3,4,5,6\}) = 1$; this is the probability of the sample space. The probability that we will roll 7 is 0 because that outcome is not in the sample space, so this probability is really $P(\emptyset)$. Now, consider the events

$$E = \{2,4,6\} \text{ (rolling an even number)}$$
$$F = \{1,2,3,4\} \text{ (rolling a number less than 5)}$$

Then $P(E \cup F)$ is the probability of rolling an even number *or* a number less than 5. Now, if we try to compute this using Axiom 3, we get

$$P(E \cup F) = P(E) + P(F) = \frac{1}{2} + \frac{2}{3} = \frac{7}{6}$$

which violates Axiom 2! The difficulty here is that E and F are *not mutually exclusive* events; outcomes 2 and 4 are in both of them.

If we define a third event $G = \{1, 3\}$ (rolling an odd number less than 5), then E and G are mutually exclusive, and we can find the probability that one or the other will happen by using Axiom 3:

$$P(E \cup G) = P(E) + P(G) = \frac{1}{2} + \frac{1}{3} = \frac{5}{6}$$ □

Other Rules of Probability

These three simple axioms allow us to prove several useful rules about how probability works in any experiment. Before presenting them, however, it might be useful to review the interconnection among events, sets, and logic.

An event is a subset of the sample space of an experiment. When we say that an event E occurs, we mean that the actual outcome of the experiment lies within that particular subset E. To say an event E does not occur, then, means that the actual outcome lies outside of E (but still in the sample space, of course). That is, if an event E does not occur, then its complement E' must occur.

For two events E and F in a sample space, saying that one or the other of the events occurs means that the outcome of the experiment lies in $E \cup F$; saying that both of the events happen (at the same time) means that the outcome lies in $E \cap F$. Thus, knowing how to compute the probabilities of complements, unions, and intersections of events whose individual probabilities are known is a valuable tool for analyzing experiments. The rest of this section is devoted to developing and applying three basic probability rules for complement, union, and intersection of events.

The complement of an event E, written E', is the set of all outcomes in the sample space S but not in E. Since E and E' are mutually exclusive, we know, by Axiom 3, that

$$P(E) + P(E') = P(E \cup E')$$

But $E \cup E' = S$ by the definition of a complement, and Axiom 1 says that $P(S) = 1$. Thus,

$$P(E) + P(E') = 1$$

Subtracting $P(E)$ from both sides of this equation, we get the rule for complements:

(14.1)

> If E is any event in a sample space S, then
> $$P(E') = 1 - P(E)$$

Example 2 The probability of drawing a heart from a well-shuffled standard deck of cards is $\frac{1}{4}$ (because *hearts* is one of four 13-card suits of the deck). Thus, the probability of *not* drawing a heart is $1 - \frac{1}{4}$, or $\frac{3}{4}$. □

Example 3 If a pair of fair dice is thrown, there are 36 possible outcomes in the sample space. (Why?) Six of them add up to 7. (Which ones are they?) If E is the event of throwing a 7, then $P(E)$ is $\frac{6}{36}$, or $\frac{1}{6}$. Thus, $P(E')$, the probability of *not* throwing a 7, is $1 - \frac{1}{6} = \frac{5}{6}$. □

To compute the probability of the union of two events E and F, we need to generalize Axiom 3 to the case where E and F might have some outcomes in common. That is, we must consider the case where $E \cap F$ is not empty. Just adding the probabilities of E and F, ignoring the requirement that they be mutually exclusive, will give us incorrect information, as we saw in Example 1.

Another example of this situation is shown in Figure 14.4. If the sample space $S = \{0, 1, 2, \ldots, 9\}$ and two events are

$$E = \{1, 2, 3, 4, 5\} \qquad \text{and} \qquad F = \{4, 5, 6, 7, 8, 9\}$$

then $P(E) = \frac{5}{10}$ and $P(F) = \frac{6}{10}$. But $P(E) + P(F) = \frac{11}{10}$, which is greater than 1 and hence cannot be the probability of any event, by Axiom 2. Moreover, counting the outcomes in $E \cup F$ tells us that $P(E \cup F) = \frac{9}{10}$. In this case, the difference $(\frac{2}{10})$ between $P(E) + P(F)$ and $P(E \cup F)$ is accounted for by the fact that the two elements in $E \cap F$ were counted in computing $P(E)$, and again in computing $P(F)$. Thus, the sum $P(E) + P(F)$ counts these elements twice, whereas they only appear once in $E \cup F$.

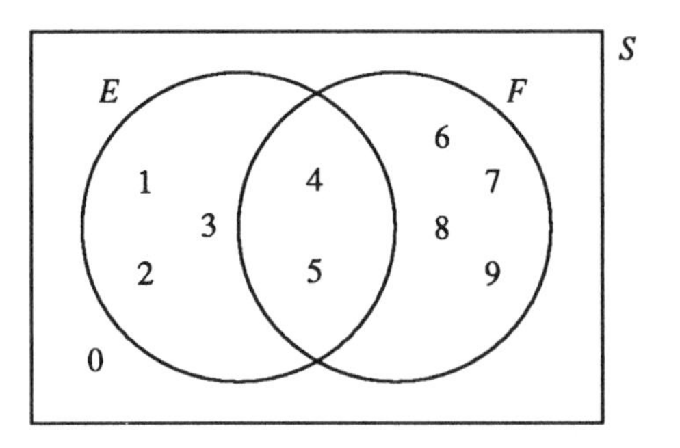

Figure 14.4 The union of two events.

This example typifies the general solution to the problem: Since the sum of $P(E)$ and $P(F)$ "counts" the common outcomes in $E \cap F$ twice, we must subtract the number of elements in $E \cap F$ from the total in order to get the correct number of elements in $E \cup F$:

$$n(E \cup F) = n(E) + n(F) - n(E \cap F)$$

Now, using this formula together with the definition of the probability of an event, we get

$$\begin{aligned} P(E \cup F) &= \frac{n(E \cup F)}{n(S)} \\ &= \frac{n(E) + n(F) - n(E \cap F)}{n(S)} \\ &= \frac{n(E)}{n(S)} + \frac{n(F)}{n(S)} - \frac{n(E \cap F)}{n(S)} \\ &= P(E) + P(F) - P(E \cap F) \end{aligned}$$

This becomes our second basic rule:

(14.2) If E and F are any two events in a sample space, then
$$P(E \cup F) = P(E) + P(F) - P(E \cap F)$$

Note that if E and F are mutually exclusive events, then $P(E \cap F) = P(\emptyset) = 0$. Thus, Rule 14.2 applies in this case, as well.

Example 4 Suppose a single die is thrown. Let E be the event of getting an even number and F the event of getting a number less than 4. Then $E = \{2, 4, 6\}$, $F = \{1, 2, 3\}$, and $E \cap F = \{2\}$. The set $E \cup F$ represents getting either an even number or a number less than 4; its probability can be computed from Rule 14.2 as follows:

$$\begin{aligned} P(E \cup F) &= P(E) + P(F) - P(E \cap F) \\ &= \frac{n(E)}{n(S)} + \frac{n(F)}{n(S)} - \frac{n(E \cap F)}{n(S)} \\ &= \frac{3}{6} + \frac{3}{6} - \frac{1}{6} \\ &= \frac{5}{6} \end{aligned}$$

□

Example 5 If a single card is drawn at random from a standard deck, the probability it will be a heart is $\frac{13}{52}$, the probability it will be a face card (king, queen, or jack) is $\frac{12}{52}$, and the probability it will be the king, queen, or jack of hearts is $\frac{3}{52}$. Using Rule 14.2, the probability that the card will be either a heart or a face card is

$$\frac{13}{52} + \frac{12}{52} - \frac{3}{52} = \frac{22}{52} = \frac{11}{26}$$ □

Example 6 A survey of results from a recent mathematics quiz showed that 30% of the students misspelled "commutative," 20% misspelled "associative," and 12% misspelled both words. The probability that a student chosen at random from that class misspelled at least one of the two words is

$$30\% + 20\% - 12\% = 38\%$$ □

Independence

Dealing with the intersection of two events is a bit more complicated than handling union or complement. If you know the probabilities of two events E and F *and* the probability of their union, then the probability of their intersection can be found using Rule 14.2. However, if only $P(E)$ and $P(F)$ are known, then there is no simple general formula for finding the probability of their intersection. One important special case, however, deserves some attention. This is the case when E and F are "independent."[1] Let us look first at an example.

Example 7 Suppose we are going to draw two cards from a throughly shuffled standard deck of playing cards and we are *not* going to put the first card back into the deck before drawing the second one. Let the event of getting a heart on the first draw be F and the event of getting a heart on the second draw be S. Since there are 13 hearts in a standard 52-card deck,

$$P(F) = \frac{13}{52} = \frac{1}{4}$$

Now, let us draw the first card but not look at it, and then ask: What is $P(S)$? The answer to this question depends on whether or not F actually happened. If it did, then there are only 12 hearts left among the remaining 51 cards, so

$$P(S) = \frac{12}{51} = \frac{4}{17}$$

[1] A rigorous discussion of independent events would require considering conditional probability and thereby take us beyond the scope of this chapter, so we shall content ourselves with an informal definition of independence.

If F didn't occur, then there are still 13 hearts left, so

$$P(S) = \frac{13}{51}$$

Thus, S is said to be *dependent* on F because its probability is affected by the occurrence or nonoccurrence of F.

By way of contrast, if we put the first card drawn back into the deck and shuffle throughly before drawing the second card, then the probability of S is the same, regardless of whether or not F occurred:

$$P(S) = \frac{13}{52} = \frac{1}{4}$$

In this situation, we should expect the probability that *both* cards drawn will be hearts is $\frac{1}{4} \cdot \frac{1}{4}$ because, if this experiment were repeated many times, the first card would be a heart in about $\frac{1}{4}$ of the cases and, in about $\frac{1}{4}$ of these successful first draws, the second draw would also be a heart. That is,

$$P(F \cap S) = P(F) \cdot P(S) = \frac{1}{4} \cdot \frac{1}{4} = \frac{1}{16}$$ □

Two events are said to be **independent** if the occurrence or nonoccurrence of one event has no effect on the likelihood of the other. In other words, if the value of $P(E)$ is the same whether or not we know that event F has occurred, then E and F are independent events. The rule for finding the probability of the intersection in this case is:

(14.3)

> If E and F are independent events, then
> $$P(E \cap F) = P(E) \cdot P(F)$$

Example 8 An experiment consists of flipping a coin three times. If E and F represent getting *heads* on the first toss and on the third toss, respectively, then E and F are independent events because the occurrence of E has no effect on the occurrence of F. $P(E)$ and $P(F)$ both equal $\frac{1}{2}$. To find the probability of getting *heads* on both the first and third tosses, we apply Rule 14.3:

$$P(E \cap F) = P(E) \cdot P(F) = \frac{1}{2} \cdot \frac{1}{2} = \frac{1}{4}$$ □

Example 9 The text excerpt shown in Figure 14.5 illustrates how Rule 14.3 is introduced at the sixth-grade level. Notice how the term "independent events" is explained there. □

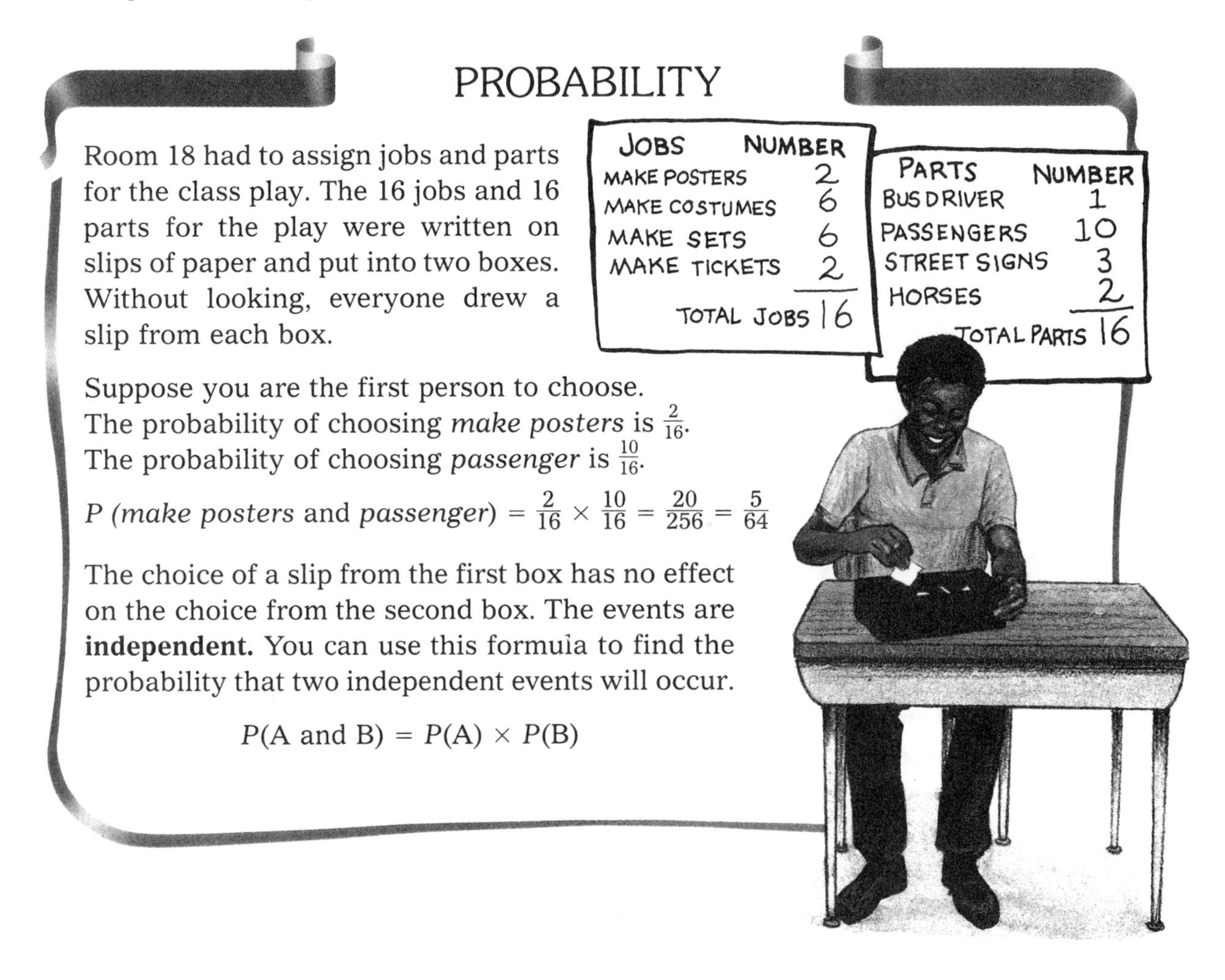

PROBABILITY

Room 18 had to assign jobs and parts for the class play. The 16 jobs and 16 parts for the play were written on slips of paper and put into two boxes. Without looking, everyone drew a slip from each box.

JOBS	NUMBER
MAKE POSTERS	2
MAKE COSTUMES	6
MAKE SETS	6
MAKE TICKETS	2
TOTAL JOBS	16

PARTS	NUMBER
BUS DRIVER	1
PASSENGERS	10
STREET SIGNS	3
HORSES	2
TOTAL PARTS	16

Suppose you are the first person to choose.
The probability of choosing *make posters* is $\frac{2}{16}$.
The probability of choosing *passenger* is $\frac{10}{16}$.

P (*make posters* and *passenger*) $= \frac{2}{16} \times \frac{10}{16} = \frac{20}{256} = \frac{5}{64}$

The choice of a slip from the first box has no effect on the choice from the second box. The events are **independent.** You can use this formula to find the probability that two independent events will occur.

$$P(\text{A and B}) = P(\text{A}) \times P(\text{B})$$

Figure 14.5 HOUGHTON MIFFLIN MATHEMATICS, Grade 6, p. 382

Example 10 Remember that Rule 14.3 only applies when the events are independent; in other cases it will give you incorrect information. For instance, if a single die is thrown, the event E of getting an even number has probability $\frac{1}{2}$ and the event F of getting an odd number also has probability $\frac{1}{2}$. Therefore,

$$P(E) \cdot P(F) = \frac{1}{2} \cdot \frac{1}{2} = \frac{1}{4}$$

However, this is *not* $P(E \cap F)$. The event $E \cap F$ is empty, so it has probability 0. Getting an even number and getting an odd number are not independent; if either event happens, the probability of the other drops from $\frac{1}{2}$ to 0. □

Exercises 14.3

For Exercises 1–8, let E and F be mutually exclusive events of a sample space S. Suppose that $P(E) = .45$ and $P(F) = .30$. Compute the probability in each case.

1. $P(E \cap F)$
2. $P(E \cup F)$
3. $P(E')$
4. $P(F')$
5. $P(E \cap E')$
6. $P(E \cup E')$
7. $P(E' \cap F)$
8. $P(E' \cup F)$

For Exercises 9–14, let E and F be independent events of a sample space S. Suppose that $P(E) = 25\%$ and $P(F) = 60\%$. Compute the probability in each case.

9. $P(E')$
10. $P(F')$
11. $P(E \cap F)$
12. $P(E \cup F)$
13. $P(E' \cup F')$
14. $P(E' \cap F')$

For Exercises 15–20, let E and F be events of a sample space S. Explain why the given situation is impossible.

15. $P(E) = .40$ and $P(E') = .75$.

16. $P(E) = .40$ and $P(E') = .50$.

17. E and F are mutually exclusive, $P(E) = .5$, and $P(F) = .7$.

18. $P(E) = P(F) = .6$ and $P(E \cup F) = .4$.

19. $P(E) = .35$, $P(F') = .70$, and E is a subset of F.

20. E and F are independent, $P(E) = .5$, E is a subset of F, and $P(F) = .6$

For Exercises 21–26, suppose that a kindergarten classroom has a box with 40 crayons in it. Five of the crayons are red, six are blue, and eight are green; the rest are assorted other colors. A child reaches in without looking and picks a single crayon. What is the probability that the child picks a crayon of the given kind?

21. A red one.

22. A blue one.

23. One that is not red.

24. One that is not green.

25. One that is neither red nor green.

26. Either a red one or a blue one.

For Exercises 27–36, suppose that a single card is drawn from a regular deck, and consider these events:

D: getting a diamond
F: getting a face card (Jack, Queen, King)
H: getting a heart

Find the probability of the given event.

27. D
28. F
29. H
30. $D \cap F$
31. $F \cap H$
32. $D \cap H$

33. getting either a heart or a diamond

34. getting either a diamond or a face card

35. getting a black card

36. getting a black card that is not a face card

For Exercises 37–42, let E and F be independent events of a sample space S containing 200 equally likely outcomes, and suppose that $P(E) = \frac{1}{5}$ and $P(F') = \frac{3}{4}$. Compute the required number in each case.

37. $n(E)$
38. $P(F)$
39. $n(F)$
40. $P(E \cap F)$
41. $P(E \cup F)$
42. $P(E' \cap F)$

14.4 Descriptive Statistics

The word *statistics* is used in a variety of senses and, in fact, is often misused in an attempt to lend credibility to otherwise doubtful opinions. This word really has two different meanings, one plural and one singular. In the plural sense it denotes collections of facts that can be stated numerically, such as mortality tables or census reports. In the singular, it is the science that deals with such facts, collecting and classifying them in such a way that general predictions may be based on them. This second meaning is the one that concerns us here.

Probability and statistics are closely related fields, in that they investigate the same basic situation from two different viewpoints. Both approaches deal with a large collection of things and a subset of that collection. In terms of the vocabulary we have just developed, probability begins with a *sample space*, a set whose elements are known, and evaluates the likelihood that some outcome of an *event*—a subset of the sample space—will occur in an experiment. Statistics begins with a **sample**—a known subset of a larger set that is mostly unknown—and by an analysis of that sample attempts to infer the composition of the larger set, which is called the **population**. Although the different words reflect the separate origins of their two fields, the terms *sample space* and *population* are analogues of each other, as are *event* and *sample*.

Once a sample has been obtained, the data must be organized and analyzed in such a way as to yield a concise description of the sample as a whole. That is the job of **descriptive statistics**, the subject of this section. The two key questions in this regard are:

> What single number can be used to characterize the sample?

and

> How are the elements of the sample spread about this single number?

The first question is asking for a way of indicating some sort of "center" for a sample, and any number used in this way is called a measure of the **central tendency** of the sample. A number that describes the way a sample is spread about some central value is said to measure the **dispersion** of the sample.

Measures of Central Tendency

Of the three common measures of central tendency we consider, the first is probably the most familiar; it is the process of finding an *average*.

Definition Let A be a sample of elements represented by the numbers $a_1, a_2, \ldots, a_n$. The **(arithmetic) mean** of A is the number

$$\frac{a_1 + a_2 + \ldots + a_n}{n}$$

The numbers representing the elements of a sample may not all be distinct. For instance, if the sample consists of examination results for a class of students, several students may have gotten the same score. It is sometimes convenient to deal with the *distinct* numbers of a sample and multiply each one by the number of times it occurs.

Definition The number of elements of a sample that are represented by a particular number k is called the **frequency** of k.

Note: Examples 1–5 are used repeatedly throughout this section; they are identified by the numbering given here.

Example 1 A class of ten students scored as follows on a test:

50, 60, 60, 65, 70, 70, 70, 75, 80, 80

The sum of these scores is 680. The mean score is 68, which is $\frac{680}{10}$. The frequency of 70 is 3; 60 and 80 have frequency 2; the other scores have frequency 1. □

Example 2 Another class of students took the same test and got scores of

30, 40, 50, 65, 70, 75, 75, 80, 95, 100

The sum of these scores is also 680, so the mean score is again 68. The frequency of 75 is 2; all the other scores have frequency 1. □

Example 3 A survey is taken of the net incomes of twenty-five families. The results, rounded to the nearest thousand, are:

$20,000 — 2 families; $22,000 — 6 families;
$24,000 — 2 families; $26,000 — 5 families;
$28,000 — 5 families; $30,000 — 3 families;
$90,000 — 2 families.

The frequency of each income is the number of families that have it, and the mean income may be found by computing as follows:

2	×	\$20,000	=	\$ 40,000
6	×	22,000	=	132,000
2	×	24,000	=	48,000
5	×	26,000	=	130,000
5	×	28,000	=	140,000
3	×	30,000	=	90,000
2	×	90,000	=	180,000
25				\$760,000

$$\text{Mean income} = \frac{\$760,000}{25} = \$30,400$$

□

Example 4 The same survey as explained in Example 3 was made in a small housing development, and it was discovered that 5 families had net incomes of \$20,000, 15 were at the \$30,000 level, and the remaining 5 made \$40,000. The mean income is \$30,000; the frequencies of the three different income levels are 5, 10, and 5, respectively. □

Example 5 Because the improvement of a certain stretch of road was being considered, the Highway Commission decided to determine the amount of traffic that used the road. They made a daily count of the number of cars using it during a 30-day test period. The results of the survey, rounded to the nearest hundred, were:

Number of days	Number of cars per day	Frequency times number per day
4	1100	4400
6	1600	9600
3	1900	5700
2	2000	4000
6	2300	13800
6	2500	15000
3	2900	8700
30		61200

$$\text{Mean, cars per day} = \frac{61,200}{30} = 2040$$

□

The mean of a sample is the most common measure of central tendency and it is often quite reliable, but its use sometimes obscures certain characteristics of the sample that relate to central tendency. For instance, consider Examples 1 and 2. Both classes contain ten students, and the mean test score of both classes is 68. Yet even a casual glance at the individual scores reveals a vast difference between the two classes. A similar situation exists

Definition Let A be a sample of elements represented by the numbers $a_1, a_2, \ldots, a_n$. The **(arithmetic) mean** of A is the number

$$\frac{a_1 + a_2 + \ldots + a_n}{n}$$

The numbers representing the elements of a sample may not all be distinct. For instance, if the sample consists of examination results for a class of students, several students may have gotten the same score. It is sometimes convenient to deal with the *distinct* numbers of a sample and multiply each one by the number of times it occurs.

Definition The number of elements of a sample that are represented by a particular number k is called the **frequency** of k.

Note: Examples 1–5 are used repeatedly throughout this section; they are identified by the numbering given here.

Example 1 A class of ten students scored as follows on a test:

50, 60, 60, 65, 70, 70, 70, 75, 80, 80

The sum of these scores is 680. The mean score is 68, which is $\frac{680}{10}$. The frequency of 70 is 3; 60 and 80 have frequency 2; the other scores have frequency 1. □

Example 2 Another class of students took the same test and got scores of

30, 40, 50, 65, 70, 75, 75, 80, 95, 100

The sum of these scores is also 680, so the mean score is again 68. The frequency of 75 is 2; all the other scores have frequency 1. □

Example 3 A survey is taken of the net incomes of twenty-five families. The results, rounded to the nearest thousand, are:

$20,000 — 2 families; $22,000 — 6 families;
$24,000 — 2 families; $26,000 — 5 families;
$28,000 — 5 families; $30,000 — 3 families;
$90,000 — 2 families.

The frequency of each income is the number of families that have it, and the mean income may be found by computing as follows:

2	×	$20,000	=	$ 40,000
6	×	22,000	=	132,000
2	×	24,000	=	48,000
5	×	26,000	=	130,000
5	×	28,000	=	140,000
3	×	30,000	=	90,000
2	×	90,000	=	180,000
25				$760,000

$$\text{Mean income} = \frac{\$760,000}{25} = \$30,400$$

□

Example 4 The same survey as explained in Example 3 was made in a small housing development, and it was discovered that 5 families had net incomes of $20,000, 15 were at the $30,000 level, and the remaining 5 made $40,000. The mean income is $30,000; the frequencies of the three different income levels are 5, 10, and 5, respectively. □

Example 5 Because the improvement of a certain stretch of road was being considered, the Highway Commission decided to determine the amount of traffic that used the road. They made a daily count of the number of cars using it during a 30-day test period. The results of the survey, rounded to the nearest hundred, were:

Number of days	Number of cars per day	Frequency times number per day
4	1100	4400
6	1600	9600
3	1900	5700
2	2000	4000
6	2300	13800
6	2500	15000
3	2900	8700
30		61200

$$\text{Mean, cars per day} = \frac{61,200}{30} = 2040$$

□

The mean of a sample is the most common measure of central tendency and it is often quite reliable, but its use sometimes obscures certain characteristics of the sample that relate to central tendency. For instance, consider Examples 1 and 2. Both classes contain ten students, and the mean test score of both classes is 68. Yet even a casual glance at the individual scores reveals a vast difference between the two classes. A similar situation exists in Examples 3 and 4; although the mean incomes of the groups differ very little, there are many differences between the central tendencies of the two samples.

Other measures of central tendency help to describe these differences. One such measure singles out the "middle number" of a sample arranged in order of size. This "middle number" is called the **median** of the sample. Of course, this number is unique only if there is an odd number of elements in the sample. A sample containing an even number of elements has two "middle numbers"; in this case, the mean of these two numbers is used.

Example 1 (*continued*) The scores are already arranged in order of size, and there are 10 elements in the sample. Thus, the median of the sample is the mean of the fifth and sixth scores in the order given. Both of these scores are 70, so the median is 70. □

Example 2 (*continued*) The fifth and sixth scores in this sample are 70 and 75, so the median is

$$\frac{70 + 75}{2} = 72.5$$

□

Example 3 (*continued*) If the 25 individual incomes are arranged in order of size, the thirteenth (or "middle") one is \$26,000. This is the median. □

Example 4 (*continued*) Again the median income is the thirteenth one in order of size — \$30,000. □

Example 5 (*continued*) Since there are 30 elements in this sample, the median is the mean of the fifteenth and sixteenth numbers in order of size. These numbers are 2000 and 2300, so the median number of cars per day is 2150. □

It is often helpful to know the most common numerical value in a sample. However, there may be several values that occur at least as many times as any other, so this measure may not yield a unique result.

Definition A **mode** of a sample A is an element of A whose frequency is greater than or equal to the frequency of every other element of A.

There may be more than one mode for a sample, as in Example 5 (which follows). Samples with two modes are said to be **bimodal**; samples with three modes are called **trimodal**. If no element of a sample is repeated, then the definition implies that every element is a mode. However, since this yields no useful information about the central tendency of the sample, it is common to speak of such a sample as having no mode.

Example 1 (*continued*) The frequency of 70 is greater than the frequency of any other score, so 70 is the (only) mode of this sample. □

Example 2 (*continued*) Similarly, 75 is the mode of this set of scores. □

Example 3 (*continued*) The mode of this income sample is $22,000 because its frequency, 6, is greater than any other frequency in the sample. □

Example 4 (*continued*) The mode is $30,000. Notice that, even though the mean incomes in Examples 3 and 4 differ by only $400, the medians differ by $4000, and the modes differ by $8000. □

Example 5 (*continued*) There are three values whose frequencies are greater than or equal to the frequency of any other element in the sample, so this sample has three modes — 1600, 2300, and 2500. Thus, this sample is trimodal. □

The primary value of the median and the mode as measures of central tendency is that neither one is greatly influenced by occasional extreme results. For instance, if we were to delete the two $80,000 incomes from Example 3, the mean would then be approximately $15,200, a change of $5200, but the median and the mode would remain the same. This characteristic of the median and the mode is especially useful when the available information indicates that a sample contains a few extreme results, but there is no way of finding their numerical value. For instance, suppose that an income survey had a final category of "more than $100,000" and this category had frequency 2 (or any relatively small positive number). In such a case it would be impossible to compute the mean, but the median and the mode could still be found.

Measures of Dispersion

The information supplied by the measures of central tendency usually does not describe a sample adequately. For instance, in the income sample of Example 4, the mean, median, and mode are all $30,000. If we chose another sample of 25 incomes, each one $30,000, again the mean, median, and mode all would be $30,000, but the samples themselves obviously are not the same. Another sample with the same measures of central tendency is one containing twenty-three $20,000 incomes, one $1000 income, and one $39,000 income. In order to distinguish among these samples or any others whose central tendencies are similar, we must consider their *dispersion.* In other words, we must find numerical methods to reflect the way elements of a sample are distributed around some central point. The simplest measurement of this kind gives the maximum difference between elements.

Definition The **range** of a sample is the difference between the largest and smallest numbers of the sample.

Example 1 (*continued*) The range of scores is $80 - 50 = 30$. □

Example 2 (*continued*) The range of scores is $100 - 30 = 70$. □

Example 3 (*continued*) The income range is $\$90,000 - \$20,000 = \$70,000$. □

Example 4 (*continued*) The income range is $\$40,000 - \$20,000 = \$20,000$. □

Example 5 (*continued*) The range of cars per day is $2900 - 1100 = 1800$. □

The range is not a very precise measure of the dispersion of a sample because it is the same for any two samples with the same extreme values, regardless of how many elements take on those values and where the rest of the numbers lie. For instance, the two samples with values

$$1, 1, 1, 5, 5, 5 \qquad \text{and} \qquad 1, 2, 3, 3, 4, 5$$

have the same range, even though most numbers of the latter sample are closer to the mean. We might try to describe this situation more precisely by considering the average of the individual differences between the mean and the elements of the sample. That is, if we define the **deviation** of an element a of a sample whose mean is m to be the difference $a-m$, then we can compute the average of all the individual deviations. Unfortunately, since the mean of a sample is actually a sort of "balance point" for the elements, the sum of the individual deviations is always zero! (You will be asked to verify this in an exercise.) Hence, this average-difference approach must be modified slightly to provide a meaningful description of the distribution.

The average of the deviations of a sample becomes useful if all deviations are considered as *distances* from the mean, without regard to the directions indicated by their signs. This is done by using absolute values.

Definition Let A be a sample consisting of the n numbers $a_1, a_2, \ldots, a_n$ and having mean m. The **mean deviation** of A is

$$\frac{|a_1 - m| + |a_2 - m| + \ldots + |a_n - m|}{n}$$

Example 1 (*continued*) Since the mean test score is 68, the mean deviation is

$$|50 - 68| + 2 \cdot |60 - 68| + |65 - 68| + 3 \cdot |70 - 68| + |75 - 68| + 2 \cdot |80 - 68|$$

divided by 10. That is,

$$\frac{18 + 2 \cdot 8 + 3 + 3 \cdot 2 + 7 + 2 \cdot 12}{10} = \frac{74}{10} = 7.4$$

□

Example 2 (*continued*) Similarly, the mean deviation of this sample is 17.4. □

Example 3 (*continued*)

Frequency	$\lvert a - m \rvert$		
2	$ 10400	=	$ 20,800
6	8400	=	50,400
2	6400	=	12,800
5	4400	=	22,000
5	2400	=	12,000
3	400	=	1,200
2	59600	=	119,200
25			$238,400

$$\text{Mean deviation} = \frac{\$238,400}{25} = \$9536$$ □

Example 4 (*continued*) The mean deviation of this income sample is

$$\frac{5 \cdot \$10,000 + 15 \cdot \$0 + 5 \cdot \$10,000}{25} = \$4000$$ □

Example 5 (*continued*) By a computation just like those above, the mean deviation is $13,800 \div 30$; that is, 460 cars per day. □

The mean deviation describes the dispersion of a sample simply and accurately, but the use of absolute value makes it unwieldy for algebraic manipulation because of the sign changes required. The "either-or" decision required by absolute value is especially annoying if the computation is being done with the aid of a calculator or computer. An alternative way of making the individual deviations positive is by squaring them. This leads to another measure of dispersion that is as descriptive as the mean deviation, but easier to handle mechanically.

Definition Let A be a sample consisting of the n numbers $a_1, a_2, \ldots, a_n$ and having mean m. The **variance** of A is

$$\frac{(a_1 - m)^2 + (a_2 - m)^2 + \ldots + (a_n - m)^2}{n}$$

The **standard deviation** of A is the (positive) square root of its variance.

Example 1 (*continued*) The variance of this sample of scores is

$$\frac{(-18)^2 + 2 \cdot (-8)^2 + (-3)^2 + 3 \cdot 2^2 + 7^2 + 2 \cdot 12^2}{10} = \frac{810}{10} = 81$$

The standard deviation is $\sqrt{81} = 9$. □

Example 2 (*continued*) For these scores a similar computation yields a variance of 456 and a standard deviation of about 21.4, giving us a clear measure of the difference between this sample of scores and that of Example 1. □

Example 3 (*continued*) Variance = \$318,080,000; standard deviation = \$17,835. □

Example 4 (*continued*) Variance = \$40,000,000; standard deviation = \$6325. Again the standard deviations clearly indicate the difference between the samples of this example and the previous one. □

Example 5 (*continued*) Variance = 288,400; standard deviation = 537 cars. □

These definitions of variance and standard deviation are used to describe the characteristics of a sample considered simply as data, without reference to any larger population that may contain it. However, in situations where a sample is used to predict the characteristics of a much larger population, it has been shown that dividing by n when computing the variance and standard deviation of the sample leads to a consistent underestimation of the standard deviation for the population as a whole. This problem is corrected if we divide by $n-1$, instead. Thus, the definitions of *variance* and *standard deviation* are altered in this way for the purposes of inferring the characteristics of a (large) population from the data contained in a (small) sample.

Example 1 (*continued*) If we were trying to infer the performance of a large group of students who took the test from this sample of ten scores, we would divide by 9 in computing the variance:

$$\frac{(-18)^2 + 2\cdot(-8)^2 + (-3)^2 + 3\cdot 2^2 + 7^2 + 2\cdot 12^2}{9} = \frac{810}{9} = 90$$

The standard deviation is $\sqrt{90} \approx 9.5$. □

Example 2 (*continued*) Similarly, the standard deviation of this set of scores would be $\sqrt{4560 \div 9} \approx 22.5$. □

Exercises 14.4

In these exercises, the variance and standard deviation of an n-element sample should be calculated by dividing by n.

For Exercises 1–6, find the mean, median, mode(s), range, mean deviation, variance, and standard deviation of the given sample. (Round all answers to one decimal place.)

1. Sample: 3, 1, 5, 10, 8, 1, 5, 2, 1

2. Sample: 13, 10, 2, −5, 7, 10, −2

3. Sample: 3.9, 8.05, 5.2, .11, 10.6, 2.0

4. Twelve students scored as follows on an exam:

   75, 60, 60, 72, 70, 86, 93, 54, 72, 50, 72, 85

5. The daily totals of people shopping at a particular market during a certain week are:

Monday: 700	Tuesday: 450
Wednesday: 630	Thursday: 520
Friday: 810	Saturday: 1050

6. A college hockey team achieved these scores during a 15-game season:

   2, 0, 3, 3, 1, 5, 4, 7, 0, 2, 3, 12, 6, 0, 1

7. (a) Compute the mean and all the individual deviations for the sample: 1, 3, 7, 4, 5, 4, 9, 1, 2

   (b) Add up all the deviations you found in part (a) to verify that their sum is 0.

   (c) Prove that the sum of all the individual deviations of any sample must be 0.

8. Construct three samples with the same range but with different standard deviations.

9. Construct two seven-element samples with a mean of 5 and a range of 10, but such that the standard deviation of the first sample is at least 1.5 more than the standard deviation of the second.

10. A restaurant makes a two-week survey of customers ordering roast beef so that it may better calculate its meat orders. The daily totals are:

    56, 72, 65, 73, 35, 96, 104,
    58, 68, 70, 62, 41, 82, 99

    Find the mean, range, and standard deviation of this sample.

11. A school system is recruiting teachers. As part of its publicity it has issued the following statement of teacher salaries it pays, rounded to thousands of dollars:

Number of teachers	Salary
2	$ 20,000
3	21,000
5	22,000
5	23,000
8	24,000
13	25,000
7	26,000
4	28,000
1	30,000
6	above $30,000

    Compute all measures of central tendency that are possible for this sample. If you were considering teaching in this school system and did not have access to this table, which single measure of central tendency would tell you the most about your expectations for the future, and why?

14.5 Generalization and Prediction

As one passes from the purely descriptive part of statistics to the uncertainties of making general predictions from a sample, the mathematical waters suddenly deepen. The background required to discuss predictive statistics rigorously is extensive, and any pretense of completeness in a treatment as brief as this would be far less than honest. We therefore limit ourselves to a general outline of the fundamentals of *predictive* or **inferential statistics,** the study of how the characteristics of a sample can be used to obtain a description of the population it represents.

It would be ideal if we could say that the distribution of any sample exactly reflects the distribution of the population from which it was drawn. Unfortunately, that is hardly ever true. We know from our study of probability that, even in the case of a known population (sample space), the samples drawn from it (the results of an experiment) are not always the same. The best we can do in such situations is to appeal to the Law of Large Numbers and say that the more an experiment is tried, the closer the ratio of successes to total trials will approximate the probability of success on a single trial. Applied to statistics, this law essentially says:

> The larger the sample taken, the closer the characteristics of the sample will approximate the characteristics of the population.

Of course, this statement may be made much more precise, but the mathematics required would take us far afield. A full, readable discussion of the Law of Large Numbers appears in Chapter XI of Warren Weaver's *Lady Luck* (Garden City, NY: Doubleday & Co., 1963).

The validity of the Law of Large Numbers depends to a great extent upon the condition that the sample be chosen in a *random* manner. That is, the elements must be selected without regard to any predetermined pattern, in such a way that each element of the population has an equal chance of being chosen. This type of sample is naturally called a **random sample**, and it has been shown that this is the only truly reliable method of sampling. *From this point on, every sample is assumed to be random.*

Estimates

Generalizing from a sample involves two questions: "What statements can be made about the characteristics of the population as a whole?" and "How reliable are these statements?" The simplest way to answer the first question is to use the summary statistics of the sample, the so-called **point estimates**, as an approximation for the corresponding summary statistics of the population as a whole, which are usually called population **parameters**. It is common to use the Greek letters μ (*mu*) and σ (*sigma*) for the population mean and standard deviation, respectively. Thus, if the mean and standard deviation of a random sample were $m = 68$ and $s = 2$, then point estimates for the population mean and standard deviation would be $\mu = 68$ and $\sigma = 2$.

It has been shown that such point estimates satisfy the Law of Large Numbers, provided that the standard deviation of an n-element sample is calculated by dividing by $n - 1$, instead of by n. In other words, as the sample size increases, the sample mean m and standard deviation s tend to approach the actual values of the population mean μ and standard deviation σ.

This is not the first time we have used point estimates, although we have not used the term before. In Section 10.7, when we approximated a true value N by a number $\overline{N}$, we were making a point estimate of N. That is, we were estimating an unknown quantity by a single number. Also, recall that when we calculated with approximate numbers and then rounded the answer to a certain number of significant digits, we constructed an interval estimate for the answer by first using a point estimate, a single value that was "close enough" to the exact value. In doing this, we needed to predict the maximum error in the calculation. In other words, we said that if $\overline{N}$ is a point estimate for N with a maximum error of e, then $[\overline{N} - e, \overline{N} + e]$ is an interval estimate for N; that is, $\overline{N} - e \leq N \leq \overline{N} + e$.

Confidence Intervals

In inferential statistics we do precisely the same thing. A sample statistic (point estimate) and an estimate of maximum error are used to construct an interval estimate of the corresponding population parameter. Thus, if we have a random sample with mean m and we want to find an interval estimate for the population mean μ, we need to estimate a maximum error e. Then $(m - e, m + e)$ becomes an interval estimate for μ:

$$m - e \leq \mu \leq m + e$$

Thus, the problem becomes: How do we estimate a maximum error e? This is really the reliability question posed earlier. In other words, how much confidence do we want to place in our interval estimate? This is a question of probability. We are seeking to quantify the likelihood that μ is in the interval we claim it is in. The specific number that gives this probability varies with the choice of e; the larger the "margin of error" e (and hence the larger the interval estimate), the greater the likelihood that μ is in that interval. The numerical value of this probability is called the **confidence level** or **confidence coefficient** of the interval estimate; the interval itself is called a **confidence interval**. Common confidence coefficients are .90, .95, .99, and .995, but any positive number ≤ 1 may be used if it is appropriate to the situation. Informally, what we are saying when we speak of a prediction having a confidence coefficient of .95, for example, is that the true value has a 95% chance of being within the predicted interval.

The value of e does not depend solely on the confidence coefficient; it also depends on the sample size n and the standard deviation d. Rather than go deeper into the theory of how e is determined, let us turn instead to the interpretation of the confidence intervals derived from e. Suppose, for instance, that the mean m of a particular random sample is 37, and a 95% confidence interval for the population mean μ is required. From the confidence coefficient .95, the sample size n, and the sample standard deviation d, it is possible to determine a maximum error, say $e = 2.6$. In this case, then, the 95% confidence interval for μ is $(37 - 2.6, 37 + 2.6)$, or $(34.4, 39.6)$. This means there is a 95% chance that this interval contains the mean of the population from which the sample came.

Notice that this process merely *approximates* the mean of the population. That is the best we can do under the circumstances of statistical inference. If the generalization process, with all its uncertainties, were forced to select a single exact number, such a choice would be highly unreliable. In fact, the more reliability we require of a prediction, the larger the confidence interval will have to be, and hence the less precise the prediction will be. Nevertheless, by adjusting the size of the sample used, it is usually possible to work out a convenient compromise between approximation and reliability so that the mean of a population can be known well enough to work with. This approach can also be used to approximate by confidence intervals the standard deviation and other population parameters.

Confidence intervals are especially useful for comparing the population of a particular sample to a known population. Suppose, for example, that the executives of a breakfast-food company want to know if the actual net weights of the corn-flakes boxes filled by their machinery conform (within reasonable limits) to the advertised net weight. They can settle the problem by taking a random sample of the filled boxes and weighing them. If they decide they must be 99% sure, then the confidence coefficient .99 can be used, along with the mean sample weight, the sample standard deviation, and the sample size, to produce a confidence interval for the population mean. If the desired uniform net weight lies within this interval, the company may be satisfied that its machinery is working well. If, however, the uniform weight lies outside the interval, the chances are 99 to 1 that this sample comes from a different population of weights than the one anticipated, and hence the machinery is not functioning properly.

This corn-flakes problem exemplifies **quality control**, an application of statistical theory to industrial production whereby acceptable product characteristics are checked by regular sampling. Other applications of statistics are almost too numerous to mention. The life-insurance industry is completely dependent on statistical predictions, as are advertising agencies and some divisions of commercial television and radio. Statistics is an essential tool in genetics, and the application of statistics to the movement of particles explains the physical phenomenon known as "Brownian motion." The social sciences and education depend on statistics for the proper formulation and interpretation of surveys and tests. These examples of applied statistics could be multiplied many times over.

The particular strength of statistics is its ability to deal quantitatively with uncertainty, thereby bridging the gap between theoretical results and practical applications. But this strength is also its weakness. As we have seen, many applications of statistical theory depend on *assumptions* about the population from which a sample is drawn, the *choice* of an appropriate confidence coefficient, and so forth. These assumptions and choices, if not made properly, make the statistical conclusions based on them invalid and thereby useless or misleading. Nevertheless, inferential statistics, carefully applied and properly understood, is an important tool for interacting with the world around us.

Statistics in the Psychology of Learning

We close this final section with an extended example of how statistics is applied to an important area of contemporary education—identifying students who have learning disabilities. Let us begin with a brief discussion of a classic educational testing device, the IQ test.

The IQ test is one of the most widely known measurement instruments in educational psychology. Almost all of us have had our IQ's measured at least once, and some of us may even recall the score. Yet relatively few people know what an IQ score really means (Do you?), except for some vague sense that it measures intelligence level in some way. In fact, the IQ test is a psychological testing tool whose meaning depends almost entirely on the concepts you have studied in this chapter.

"IQ" is an abbreviation for *Intelligence Quotient.* A quotient is, of course, a number formed from the division of one number by another. In this case the quotient is formed by dividing a person's mental age by his or her chronological age. It is standard practice to multiply the resulting fraction by 100 and then round after the decimal point, so that IQ scores appear as whole numbers. Thus,

$$\text{IQ} = \frac{\text{mental age}}{\text{chronological age}} \times 100$$

rounded to the nearest whole number. For example, a nine-year-old child with a mental age of 10 would have an IQ of 111 because

$$\frac{10}{9} \times 100 = 111.11\ldots$$

The obvious question is: What is *mental age*? This is where statistics enters the scene. In order to measure a child's mental age, educational researchers devise a test that can be administered to children over a broad span of ages. (The most common IQ tests in use today are the Wechsler Intelligence Scale for Children [WISC], spanning ages 6 through 15, and the Stanford-Binet Intelligence Scale, spanning all ages from preschool through high school.) To be standardized, these tests are administered to a large random sample of children of every age group, and the mean score in each age group is used to represent the mental age. Thus, a child whose individual test score matches the mean score for twelve-year-old children has a mental age of 12, regardless of that child's actual (chronological) age.

For instance, suppose the mean score for ten-year-old children on a particular intelligence test is 92. Then

- an eight-year-old child who scores 92 on that test has a mental age of 10 and, therefore, has an IQ of 125 [that is, $\frac{10}{8} \times 100$];
- a twelve-year-old child who scores 92 also has a mental age of 10 and thus has an IQ of 83 [obtained from $\frac{10}{12} \times 100$ by rounding];
- a ten-year-old child who scores 92 has a mental age of 10 and an IQ of 100 [$= \frac{10}{10} \times 100$].

In general, a person with an IQ of 100 is someone whose intelligence, or intellectual ability, tests out at the mean for his/her age group.

Notice that the computation process just described has actually *defined* mental age in a completely statistical way. That's fine, *if* mental age is understood solely in that way. Unfortunately, even the term itself suggests something more, as if mental age were some sort of fixed characteristic of individual minds that is discoverable by testing. It is quite clear, however, that an eight-year-old child with a mental age of 10 and a fifteen-year-old child with a mental age of 10 have very different kinds of minds.

Moreover, the IQ formula implies that an eight-year-old girl with an IQ of 125 is two years ahead in mental age [because $\frac{125}{100} \times 8 = 10$], but when that same child reaches age twelve (presumably with the same IQ), she must be three years ahead in mental age [because $\frac{125}{100} \times 12 = 15$]. According to the formula, on her 24th birthday this adult with an IQ of 125 would be fully 6 years ahead in mental age! The obvious extension of this idea leads to absurd conclusions, so the interpretation of variance from the norm is adjusted to mean different things at different ages. Such difficulties with the conceptual validity of "mental age" have led in recent years to IQ testing based on a more formal statistical foundation. Some of the major intelligence tests have now abandoned "mental age" in favor of scoring tables designed so that the (standardized) mean for each age group is 100 and the standard deviation is 15 or 16.

This little exercise in applied statistics can affect people's lives in many ways. Let us look at one important example. Since about 1970, many school systems around the country have begun to provide supportive educational services to a special class of handicapped children known as "learning disabled." These special services are now required (and partially funded) by the federal government. But children qualifying for these services are vaguely defined by federal regulations, which say only that such a child must have "a severe discrepancy between achievement and intellectual ability" in oral and/or written language or mathematics [*Federal Register*, 42 (240): 65083,

Dec. 29, 1977]. The question is: What is a "severe discrepancy"? The answer to this question has profound implications for parents and children, school boards and teachers, towns and taxpayers. Who qualifies for this help? How much service must be provided? How much will it cost? All these answers flow from the definition of "severe discrepancy," a definition the federal government does not provide.

Among the states attempting to provide legal definitions for themselves is Connecticut, whose answer is a striking illustration of the effect of statistics on public policy:

> [A] severe discrepancy is said to exist whenever the difference between academic achievement (AA) and intellectual functioning (IQ) (each expressed as a standard score) is equal to or greater than one-and-one-half standard deviations—that is,
>
> $$\text{IQ} - \text{AA} = 1\tfrac{1}{2}\ \text{SD}$$

[*Guidelines for Identification and Programming of Learning Disabilities.* Hartford, CT: Connecticut Department of Education, draft dated 11/27/81, p. 30.]

In other words, to qualify for special services in this area, children must satisfy a statistical criterion which rests, in turn, on statistically based test measurements. Their intellectual potential is measured by an IQ test and then their achievement is measured by other tests whose scores are standardized for purposes of comparison. (The Connecticut guidelines even provide a formula for converting other scoring scales to a mean of 100 and a standard deviation of 15.) Only if the achievement scores fall $1\frac{1}{2}$ standard deviations (22 points) below the IQ scores do the Connecticut guidelines declare these children to be learning disabled. Thus, the concept "learning disabled" in Connecticut is defined in a fundamentally statistical way.

The foregoing is but one illustration of how statistical concepts may be so thoroughly intermixed with ideas from education or other disciplines that there is no way to separate them. This is particularly true in the area of testing, a fundamental part of the teaching profession at any level. As you take your teacher-preparation courses, be on the lookout for concepts that depend on statistics. Recognizing them is a critical first step in truly understanding them and their implications for you and your future students.

Exercises 14.5

Exercises 1–8 refer to Example 5 of Section 14.4, concerning a survey of cars using a particular road, in which the sample mean $m = 2040$, the standard deviation $s = 537$, and the sample size $n = 30$.

1. Find a point estimate for the population mean μ.

2. Find a point estimate for the population standard deviation σ.

3. If it has been decided that the maximum error for a 95% confidence interval for the population mean is about 195 cars, what is the 95% confidence interval? (You will need the answer to Exercise 1.)

4. Give an informal interpretation of your answer to Exercise 3.

5. If it has been decided that the maximum error for a 99% confidence interval for the population mean is about 257 cars, what is the 99% confidence interval? (You will need the answer to Exercise 1.)

6. Give an informal interpretation of your answer to Exercise 5.

7. If you were to calculate a 90% confidence interval for this same sample, how would you expect it to compare with your answers to Exercises 3 and 5?

8. If you were to draw a sample of size 60 from the same population, how would you expect the 95% confidence interval for that sample to compare with your answer to Exercise 3? Why?

For Exercises 9–11, assume that, for a particular sample mean, the 99.5% confidence interval for the population mean was computed to be (102.4, 150.6).

9. In your own words, what does this information mean?

10. What is the maximum error for this approximation?

11. What is the sample mean for this approximation?

12. A probability-and-statistics test for seventh graders is shown in Figure 14.6. Which terms and concepts of this chapter appear explicitly in the test? Which appear implicitly? Identify as many of them as possible.

CHAPTER 14 TEST

You spin the spinner twice.
What is the probability? *(pages 400–409)*

1. *P*(1 and 5) 2. *P*(3 and 6) 3. *P*(4 and 5)

What are the range, the mean, the median, and the mode? *(pages 410–413)*

4. 7, 9, 5, 3, 6, 6
5. 42, 71, 38, 25, 65, 71
6. 105, 80, 97, 111, 101, 105, 94
7. 71.6, 83.4, 69.8, 57.4, 83.4

Quality control inspectors at the QT car tire factory test tires. The results are shown in the table. What is the probability? *(pages 414–415)*

8. A Monday tire is defective.
9. A Wednesday tire is not defective.
10. A Monday tire is not defective.

RESULTS	Mon	Wed
Defect	7	2
No Defect	43	48

Figure 14.6 HOUGHTON MIFFLIN MATHEMATICS, Grade 7, p. 424

Review Exercises for Chapter 14

For Exercises 1–10, indicate whether the given statement is *true* or *false*.

1. The question asked of probability is: Given a known sample of an unknown collection of objects, what can be said about the characteristics of the collection as a whole?
2. The theory of probability is valid only in situations in which the occurrences being studied are truly random.
3. If a fair coin is tossed repeatedly and comes up *heads* three times in a row, it is extremely unlikely that it will come up *heads* on the next toss.
4. For any events E and F, $P(E \cap F) = P(E) + P(F)$.
5. For any events E and F, $P(E \cup F) = P(E) + P(F)$.
6. *Sample space* is to *event* as *population* is to *sample*.
7. *Mean* and *median* are two different names for the same statistic.
8. The variance of a sample is the square of its standard deviation.
9. The more reliability we require of a prediction, the larger the confidence interval will be, and hence the less precise the prediction will be.
10. The maximum error e of a confidence-interval estimate depends solely on the confidence coefficient.

In Exercises 11–20, complete each statement by filling in the blank(s).

11. A situation or problem involving uncertain results is called an ____. The various possible results are called ____, and the set of all these possible ____ is called the ____.
12. As the number of repetitions of an experiment increases, the ____ of the number of occurrences of an event E to the total number of trials approaches ____.
13. If E and F are ____ events, then $P(E \cup F) = P(E) + P(F)$.
14. Two events are said to be ____ if the occurrence or nonoccurrence of either event has no effect on the likelihood of the other.
15. The arithmetic mean, the median, and the mode of a sample are all measures of ____
16. The range, the mean deviation, and the standard deviation of a sample are all measures of ____.
17. ____ statistics is the study of how the characteristics of a sample can be used to describe the population it represents.
18. Summary statistics about an entire population are called population ____.
19. When a sample mean is used to represent the mean of the entire population, the sample mean is called a ____ of the population mean.
20. The maximum error of a confidence interval depends on the ____, the ____, and the ____.

In Exercises 21–28, the experiment consists of throwing a pair of (fair) *regular octahedral* dice. (Each die has eight equally likely faces, numbered 1, 2, ... 8).

21. List all the equally likely outcomes for this experiment; that is, find a suitable sample space.
22. Does the set $\{2, 3, 4, \ldots, 16\}$ represent a suitable sample space for this experiment? Why or why not?
23. List all the outcomes in each of these events:
    E: The sum of the numbers thrown is 8.
    F: The sum of the numbers thrown is 9.
    G: The sum of the numbers thrown is a prime number.
    H: The sum of the numbers thrown is more than 12.
24. Find the probability of each event listed in Exercise 23.

25. List the elements of $G \cup H$ and of $G \cap H$.

26. Find $P(G \cup H)$ and $P(G \cap H)$.

27. List the elements of $E \cup G$ and of $E \cap G$.

28. Find $P(E \cup G)$ and $P(E \cap G)$.

For Exercises 29–34, let E and F be mutually exclusive events in a sample space S. Suppose that $P(E) = .48$ and $P(F) = .36$. Compute each probability.

29. $P(E \cup F)$

30. $P(E \cap F)$

31. $P(E')$

32. $P(F \cup F')$

33. $P(E' \cap F)$

34. $P(E \cup F')$

For Exercises 35–40, let E and F be independent events in a sample space S. Suppose that $P(E) = 35\%$ and $P(F) = 45\%$. Compute each probability.

35. $P(E \cup F)$

36. $P(E \cap F)$

37. $P(E')$

38. $P(E' \cup F')$

39. $P(E' \cap F)$

40. $P(E \cup F')$

For Exercises 41 and 42, calculate the mean, median, mode(s), range, mean deviation, variance, and standard deviation of the given sample. Round your answers to one decimal place. (The variance and the standard deviation should be calculated by dividing by n.)

41. 4, 3, 1, 6, 2, 5, 4, 1, 4, 7, 9

42. 4.2, 4.7, 7.5, 8.4, 4.2, 9.5, 7.5, 2.1

In Exercises 43–47, the lengths of a sample of 100 fish were recorded and the mean length was 15.4 inches. The standard deviation of the sample was 2.5 inches.

43. Find a point estimate for the population mean μ.

44. Find a point estimate for the population standard deviation σ.

45. If the maximum error for a 95% confidence interval for the population mean must be about 4.1 inches, what is the 95% confidence interval?

46. Give an informal interpretation of your answer to Exercise 45.

47. If you were to calculate a 99% confidence interval for this sample, how should it compare with your answer to Exercise 45?

Answers to Odd-numbered Exercises

Chapter 1

Section 1.2, page 13

1. (a) 12 (b) 4092 **3.** (a) 1, 6, 15, 20, 15, 6, 1; 1, 7, 21, 35, 35, 21, 7, 1; 1, 8, 28, 56, 70, 56, 28, 8, 1 (c) 2^{99} (e) 4851 (g) 4950 **5.** (a) 4940 (b) 4850 **7.** White **9.** $\frac{100!-1}{100!}$

Section 1.3, page 16

3. $10^6 = 1{,}000{,}000$ **5.** 1000 **Note**: The answers for 7–15 give one of many possible estimated values for each. **7.** 1200 **9.** 22 **11.** 5000 **13.** $\frac{1}{4}$ **15.** 11

Section 1.4, page 21

3. *generic*: a person or thing; *specific*: [it] serves as a model for one of a later period. **5.** *generic*: a rule or principle; *specific*: [It is] established or basic. **7.** *generic*: a truth, law, doctrine, or force; *specific*: [It is] fundamental or motivating; others are based [on it]. **9.** *generic*: a number of sheets of paper, parchment, etc.; *specific*: [They have] writing or printing on them [and are] fastened together along one edge, usually between protective covers. **11.** A triangle is **equilateral** if (and only if) all three of its sides are equal in length. **13.** A statement is **canonical** if it is authoritative or generally accepted. **15.** Something is **principal** if it is first in rank, authority, importance, degree, etc. **17.** Two persons or things are **complementary** if each makes up what is lacking in the other. **19.** Circular: "addition" and "adding" are forms of the same word. **21.** Not characteristic: "like" does not specify a process; also circular: "+" is not defined until "addition" is. **23.** Circular: "add" and "+" are used in the definition; also not characteristic: even if "add" and "+" have been previously defined, the phrase "when..." is too restrictive; there are many other times when we do addition.

Review Exercises, page 22

1. T **3.** F **5.** F **7.** T **9.** F **11.** See page 2. **17.** (a) 21, 28, 36 (c) 1, 4, 9, 16, 25; 10,000 **19.** 400 **21.** 55 **23.** 0.9 **25.** Circular: "defined" is a form of "definition." **27.** Not characteristic: it doesn't distinguish between horses and other four-legged animals.

Chapter 2

Section 2.1, page 31

1. (a) yes (b) The sun is not shining here. **3.** (a) yes (b) Somebody loves me. **5.** (a) no **7.** (a) no **9.** (a) no **11.** (a) yes (b) There is no number x such that $3 + x = 8$. **13.** (a) *existential* (b) No flowers are yellow. **15.** (a) *existential* (b) No unicorns are on the front lawn.

17. (a) *universal* (b) Some word is not important. **19.** (a) *particular* (b) Those trees are not green. **21.** (a) *universal* (b) Some man is an island. **23.** (a) *universal* (b) There is a number n such that $n + n \neq n \cdot n$. **25.** This is not a mathematics book. **27.** Some roses are not red. **29.** $8 + 3 \neq 11$ **31.** Some rabbits chase mice. **33.** There are no more than two trees in the field. **35.** Some rose has no thorns. Not all roses have thorns. **37.** Not everyone will pass the next exam. Someone will fail the next exam.

Section 2.2, page 37

1. This book is interesting and I am falling asleep. **3.** This book is not interesting and I am falling asleep. **5.** (Either) this book is not interesting or I am falling asleep. **7.** This book is not interesting and I am not falling asleep. **9.** It is false that this book is interesting and I am falling asleep.

11.

s	t	s and t
T	T	T
F	T	F

13.

s	c	s and c
T	F	F
F	F	F

15. 12: Any argument involving a disjunction with a contradiction can be simplified by eliminating the contradiction. 13: Any conjunction of a statement with a contradiction is a contradiction. 14: Any disjunction of a statement with a tautology is a tautology.

17.

p	q	$\sim(p$ or $\sim q)$	$(\sim p)$ and $(\sim q)$
T	T	F	F
T	F	F	F
F	T	T	T
F	F	F	F

19.

p	q	$\sim(p$ or $q)$	p and $\sim(p$ or $q)$
T	T	F	F
T	F	F	F
F	T	F	F
F	F	T	F

21.

p	q	r	p and $(q$ or $r)$	$(p$ and $q)$ or $(p$ and $r)$
T	T	T	T	T
T	T	F	T	T
T	F	T	T	T
T	F	F	F	F
F	T	T	F	F
F	T	F	F	F
F	F	T	F	F
F	F	F	F	F

23. (a) exclusive
(c) exclusive
(e) exclusive; inclusive
(g) inclusive

Section 2.3, page 45

1. *hypothesis*: There is snow on the ground. *conclusion*: This is winter. *converse*: If this is winter, then there is snow on the ground. **3.** *hypothesis*: Rabbits eat carrots. *conclusion*: Rabbits have good eyesight. *converse*: Rabbits must eat carrots if they have good eyesight. **5.** *hypothesis*: Something is a turkey. *conclusion*: Something is a bird. *converse*: All birds are turkeys. **7.** *hypothesis*: Something is an airplane. *conclusion*: Something does not fly like a bird. *converse*: Anything that does not fly like a bird is an airplane. **9.** *hypothesis*: The hypothesis [of a conditional] is false. *conclusion*: The conditional is true. *converse*: If a conditional is true, then its hypothesis is false. **11.** *inverse*: If there is no snow on the ground, then this is not winter. *contrapositive*: If this is not winter, then there is no snow on the ground. **13.** *inverse*: Rabbits must not have good eyesight if they don't eat carrots. *contrapositive*: Rabbits must not eat carrots if they don't have good eyesight. **15.** *inverse*:

Anything that is not a turkey is not a bird. *contrapositive*: Anything that is not a bird is not a turkey. **17.** *inverse*: Anything that is not an airplane flies like a bird. *contrapositive*: Anything that flies like a bird is not an airplane. **19.** *inverse*: A conditional is false if its hypothesis is true. *contrapositive*: If a conditional is false, then its hypothesis is true.

21.

p	q	$p \to (\sim q)$	$\sim(p \text{ and } q)$
T	T	F	F
T	F	T	T
F	T	T	T
F	F	T	T

23.

p	q	$\sim p$	$\sim(p \to q)$	$\sim p \text{ and } \sim(p \to q)$
T	T	F	F	F
T	F	F	T	F
F	T	T	F	F
F	F	T	F	F

25.

p	q	$\sim(p \text{ and } q)$	$p \text{ and } \sim(p \text{ and } q)$	$\sim(p \to q)$
T	T	F	F	F
T	F	T	T	T
F	T	T	F	F
F	F	T	F	F

27. Something is an elephant and it doesn't have a trunk. **29.** (26) Birds do not have wings or they can fly. (27) Something is not an elephant or it has a trunk. **31.** (1) A 30-60-90 triangle is a triangle with one right angle. (2) An equilateral triangle is a triangle with no right angle. (3) A square is not a triangle but it has a right angle. (4) A parallelogram is not a triangle and it may not have a right angle. The given statement is false. **33.** (1) 12. (2) Not possible. (3) Not possible. (4) 15. The given statement is true.

Section 2.4, page 50

1. It is Sunday. The sun is shining. (Simplification of *and*). **3.** I will go swimming. (Modus Ponens) **5.** No valid conclusion. **7.** No valid conclusion. **9.** Sue Ellen bought chocolate ice cream. (Law of the Excluded Middle.) **11.** Henry is not 23. (Modus Tollens) Henry is 21. (Law of the Excluded Middle.) **13.** No valid conclusion. **15.** n is not divisible by both 3 and 5. (Modus Ponens) n is not divisible by 3 or n is not divisible by 5. (DeMorgan's Laws) **17.** If n is divisible by 2 and 9, then it is divisible by 18. (Modus Ponens) **19.** By saying "No," Candidate X is declaring opposition to cutting taxes or to increasing the defense budget, (or possibly both). [DeMorgan's Laws] Since neither alternative is specified, she cannot make a valid voting decision.

Review Exercises, page 51

1. F **3.** T **5.** T **7.** F **9.** F **11.** Not a statement. **13.** False statement. Its negation is "Some spiders do not have polka-dot legs." **15.** True statement. Its negation is "For all numbers x, $x + 5 \neq 12$." **17.** Tautology. **19.** Tautology.

21.

p	q	r	$p \text{ or } (\sim q)$	$p \text{ or } (\sim q) \to r$
T	T	T	T	T
T	T	F	T	F
T	F	T	T	T
T	F	F	T	F
F	T	T	F	T
F	T	F	F	T
F	F	T	T	T
F	F	F	T	F

23. *converse*: Anything made of snow is an igloo. *inverse*: Anything that is not an igloo is not made of snow. *contrapositive*: Anything not made of snow is not an igloo. **25.** Philosophy is not fun and history is interesting. **27.** For some x, $x + 5 \neq 10$ and $2x + 1 \neq 7$. **29.** No valid conclusion. **31.** Zonks are kirps or zonks are laps. (Hypothetical syllogism) A traz is not a zonk. (Modus tollens) **33.** Sarah is 18. (Law of the Excluded Middle)

Chapter 3

Section 3.2, page 64

1. yes (for a particular day) **3.** no **5.** yes **7.** no **9.** (a) {Jan., Feb., Mar., Apr., Sep., Oct., Nov., Dec.} (b) All months of the year. **11.** (a) {0, 1, 2, ..., 20} (b) All whole numbers. **13.** $\{x \mid x$ is a weekday.$\}$ **15.** $\{x \mid x$ is a nonzero, even whole number.$\}$ **19.** F **21.** F **23.** F **25.** F **27.** F **29.** F **31.** F **33.** F **35.** F **37.** T **39.** T **41.** F **43.** {0, 6, 7, 8, 9} **45.** {a} **47.** Let R = {a}, S = {b}, T = {a, b}.

Section 3.3, page 73

1. {1, 2, 3, 4, 5, 6} **3.** {1, 2, 3, 5, 6} **5.** {3, 4, 5, 6} **7.** {(1,5), (2,5), (3,5), (1,6), (2,6), (3,6)} **9.** {0, 7, 8, 9} **11.** {0, 1, 2, 4, 5, 6, 7, 8, 9} **13.** {a, b, c, d, e, i, o} **15.** {f, g, h, ..., z} **17.** {g, i, j, ..., z} **19.** $\emptyset$ **21.** {a, b, c, d, e} **23.** $\emptyset$

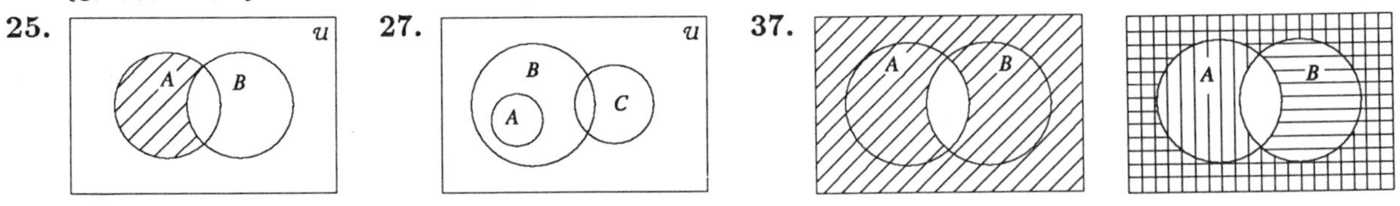

Section 3.4, page 84

5. yes **7.** no; not transitive **9.** $2\,\mathcal{R}\,1$; $3 \not\mathcal{R}\, 1$; not reflexive, not symmetric, not transitive **11.** $-2\,\mathcal{R}\,1$; $-2 \not\mathcal{R} -1$; not reflexive, symmetric, not transitive **13.** $2\,\mathcal{R}\,1$; $2 \not\mathcal{R}\, 2$; not reflexive, symmetric, not transitive **15.** $2\mathcal{R}\,1$; $2\not\mathcal{R}\,3$; reflexive, not symmetric, transitive **17.** $2\mathcal{R}\,3$; $2\not\mathcal{R}\,1$; not reflexive, symmetric, not transitive **19.** $\{\ldots,-14,-7,0,7,14,\ldots\}$, $\{\ldots,-13,-6,1,8,15,\ldots\}$, $\{\ldots,-12,-5,2,9,16,\ldots\}$, $\{\ldots,-11,-4,3,10,17,\ldots\}$, $\{\ldots,-10,-3,4,11,18,\ldots\}$, $\{\ldots,-9,-2,5,12,19,\ldots\}$, $\{\ldots,-8,-1,6,13,20,\ldots\}$ **21.** [Many other correct answers are possible for this exercise.] (a) {a, b, c}, {d, e, f, g}, {h, i, j, k, l}, {m, n, o, ..., z} (b) a is equivalent to b; a is not equivalent to d. (c) {a, b, c, r}, {d, e, f, ..., p, q}, {s, t, u}, {v, w}, {x, y, z} (d) no; no (e) x, y, z **23.** reflexive, transitive **25.** symmetric **27.** equivalence relation

Section 3.5, page 91

1. (each, e), (word, w), (in, i), (this, t), (sentence, s); function **3.** (t, e), (i, l), (l, r), (o, f), (t, s); not a function **5.** (2, of), (3, the), (3, set), (4, with), (5, every); not a function **7.** function; not a 1-1 correspondence **9.** not a function **11.** not a function **13.** function; not a 1-1 correspondence **15.** function; not a 1-1 correspondence **17.** (a) $\frac{4}{9}$ (b) -6 (c) 80 **19.** (a) 3 (b) 11 (c) 17 (d) 201 (e) 1 (f) -13 **29.** (a) 7 (b) 15 (c) 19 (d) 19

Review Exercises, page 92

1. F **3.** T **5.** T **7.** F **9.** F **11.** {8, 9, 10, 11} **13.** {Thursday, Friday, Saturday} **15.** {a, e, i, o, u, v, w, x, y, z} **17.** T **19.** T **21.** T **23.** T **25.** F **27.** F **29.** T **31.** T **33.** {30, 50, 60, 70} **35.** {10, 20, 40, 80, 90, 100} **37.** {30, 50, 70} **39.** {30, 50, 60} **41.** {50, 60, 70} **43.** See Exercise 37 of Section 3.3. **45.** equivalence relation **47.** reflexive, symmetric, not transitive **53.** impossible

Chapter 4

Section 4.1, page 102

11. {"one," "two," "three"} **13.** {"one," "two," "three," "four"} **15.** six **17.** nine **19.** six **21.** zero **23.** nine **25.** three **27.** no **29.** no **31.** yes **33.** T **35.** F **37.** F **39.** T **41.** F **43.** F **45.** F **47.** T **49.** F **51.** T **53.** F **55.** T **57.** (a) If $A = \{a, b\}$ and $B = \{c\}$, then $A \times B = \{(a, c), (b, c)\}$, but $B \times A = \{(c, a), (b, a)\}$. Match each pair in $A \times B$ with its "reverse" pair in $B \times A$. (b) The pairs of $B \times A$ are always just the pairs of $A \times B$ with the order reversed, so the matching described in (a) is always possible. (c) The cardinalities are equal.

Section 4.2, page 114

1. (a, b), (b, c), (c, d), (d, e), (e, f) is a 1-1 correspondence between A and $B \cup C$. **3.** (f, d) is a 1-1 correspondence between $A' \cap C$ and $C \cap B$. **5.** $A' \cap B = \emptyset$ because $B \subset A$. **7.** $B \cup \emptyset = B$. **9.** Let $A = \{a\}$, $B = \{b\}$. Then $1 + 1 = n(A \cup B) = n(\{a, b\}) = 2$ because $\{a, b\} \sim$ {"one," "two"}. **11.** Let $A = \{a, b, c, d, e, f, g\}$, $B = \{h, i, j, k\}$. Then $7 + 4 = n(A \cup B) = n(\{a, b, c, d, e, f, g, h, i, j, k\}) = 11$ by 1-1 correspondence with {"one," "two," ..., "eleven"}). **13.** Let $A = \{a, b, c, d, e\}$, $B = \emptyset$. Then $5 + 0 = n(A \cup \emptyset) = n(A) = 5$. **15.** Let $A = \{a, b\}$, $B = \{c, d, e\}$, $C = \{f, g, h, i\}$. Then $(2 + 3) + 4 = n[(A \cup B) \cup C] = n(\{a, b, c, d, e, f, g, h, i\}) = 9$ (by 1-1 correspondence with the counting set). **17.** Let $A = B = \emptyset$. Then $0 + 0 = n(A \cup B) = n(\emptyset) = 0$. **19.** $((11 + 19) + (13 + 17)) + 15 = (30 + 30) + 15 = 75$ **21.** $((28 + 12) + (5 + 15)) + (39 + 11) = (40+20)+50 = 110$. **23.** $3-1 = 2$ because $1+2 = 3$. **25.** $3-6$ is impossible in **W** because there is no whole number x such that $6+x = 3$. **27.** $6-3 = 3$ because $3+3 = 6$. **29.** $3-0 = 3$ because $0+3 = 3$. **31.** $0-0 = 0$ because $0+0 = 0$. **37.** commutative and associative **39.** commutative and associative **41.** not commutative; associative

Section 4.3, page 126

3. Let $4 = n(\{a, b, c, d\})$, $2 = n(\{a, b\})$. Then $4 \cdot 2 = n(\{a, b, c, d\} \times \{a, b\}) =$

$$\begin{array}{cccccccccc} n(\{ & (a, a), & (a, b), & (b, a), & (b, b), & (c, a), & (c, b), & (d, a), & (d, b) & \}) = \\ & \updownarrow & \updownarrow & \updownarrow & \updownarrow & \updownarrow & \updownarrow & \updownarrow & \updownarrow & \\ n(\{ & \text{"one,"} & \text{"two,"} & \text{"three,"} & \text{"four,"} & \text{"five,"} & \text{"six,"} & \text{"seven,"} & \text{"eight"} & \}) = 8 \end{array}$$

Also, $4 \cdot 2 = 2 + 2 + 2 + 2 = 8$. **5.** $2 \cdot 6 = n(\{a, b\} \times \{a, b, c, d, e, f\}) =$

$$\begin{array}{cccccccccc} n(\{ & (a, a), & (a, b), & \ldots & (a, f), & (b, a), & (b, b), & \ldots & (b, f) & \}) = \\ & \updownarrow & \updownarrow & & \updownarrow & \updownarrow & \updownarrow & & \updownarrow & \\ n(\{ & \text{"one,"} & \text{"two,"} & \ldots & \text{"six,"} & \text{"seven,"} & \text{"eight,"} & \ldots & \text{"twelve"} & \}) = 12 \end{array}$$

Also, $2 \cdot 6 = 6 + 6 = 12$. **7.** $1 \cdot 1 = n(\{a\} \times \{a\}) = n(\{(a, a)\}) = 1$. Repeated addition is not possible here. **9.** Let $A = \{a, b\}$, $B = \{c, d, e\}$, $C = \{f, g, h, i\}$. Then $(2 \cdot 3) \cdot 4 = n((A \times B) \times C) =$

$$\begin{array}{cccccccc} n(\{ & ((a, c), f), & ((a, c), g), & ((a, c), h), & ((a, c), i), & ((a, d), f), & \ldots & ((b, e), i) & \}) = \\ & \updownarrow & \updownarrow & \updownarrow & \updownarrow & \updownarrow & & \updownarrow & \\ n(\{ & \text{"one,"} & \text{"two,"} & \text{"three,"} & \text{"four,"} & \text{"five,"} & \ldots & \text{"twenty-four"} & \}) \end{array}$$

$= 24$. Also, $(2 \cdot 3) \cdot 4 = (3 + 3) \cdot 4 = 6 \cdot 4 = 4+4+4+4+4+4 = 24$. **11.** $(2 \cdot 5) \cdot (3 \cdot 11) = 10 \cdot 33 = 330$. **13.** $[(25 \cdot 8) \cdot 2] \cdot 13 = (200 \cdot 2) \cdot 13 = 5200$. **17.** $2 \div 6 = x$ only if $6 \cdot x = 2$. Clearly, $x \neq 0$, so x must be at least 1. But this implies that any representative set for $6 \cdot x$ will contain at least six ordered pairs. **19.** $10 \div 2 = 5$ because $2 \cdot 5 = 10$. **21.** Cannot be done in **W** because there is no whole number x such that $10 \cdot x = 2$. **23.** $4 \div 1 = 4$ because $1 \cdot 4 = 4$. **25.** $0 \div 7 = 0$ because $7 \cdot 0 = 0$. **27.** $2 + (3 \cdot 4) = 1 + 12 = 13$, but $(2+3) \cdot (2+4) = 5 \cdot 6 = 30$. **35.** $0 \div 0$ is not meaningful because there is not a *unique* number x such that $0 \cdot x = 0$. Any number could be used for x, implying that division would not be a function (because (0, 0) would have many different images). **37.** (b) Part (i): multiplication is commutative; part (ii): multiplication is associative. **39.** $(a - b) \cdot c = (a \cdot c) - (b \cdot c)$; always true **41.** $a \div (b + c) = (a \div b) + (a \div c)$; not always true

Section 4.4, page 132

1. $10 \cdot 12$ **3.** 23^{12} **5.** 21^{10} **7.** 5^{16} **9.** 3^5 **11.** 1 **13.** 2^6 **15.** $5 \cdot 2^5$ **17.** 1 **19.** F **21.** F **23.** T **25.** F **27.** F **29.** F **31.** T **33.** T **35.** T **37.** F **39.** F **41.** T **43.** T **45.** $3(a + b)$ **47.** 3^{a+b} **49.** 3^{2n} **51.** can't **53.** $3^{2(n-1)}$ **55.** 1 **59.** $3^{20} = (3^2)^{10} = 9^{10}$, so a rough estimate is $10^{10} = 10{,}000{,}000{,}000$. Better: $3^{20} = (3^4)^5 = 81^5$, approximately $80^5 = 3{,}276{,}800{,}000$. Exact value: $3{,}486{,}784{,}401$.

Section 4.5, page 141

1. $2 \not< 2$; $2 < 3$, but $3 \not< 2$. **3.** Let $7 = n(\{a, b, c, d, e, f, g\})$ and $4 = n(\{p, q, r, s\})$. Then $\{p, q, r, s\} \sim \{a, b, c, d\} \subset \{a, b, c, d, e, f, g\}$. **5.** $0 = n(\emptyset)$, $5 = n(\{a, b, c, d, e\})$, and $\emptyset \subset \{a, b, c, d, e\}$. **7.** "$4 \leq 9$" means "$4 < 9$ or $4 = 9$," and $4 < 9$ can be shown by using representative sets (as in Exercise 3). **9.** $5 < 7$ can be shown by using representative sets (as in Exercise 3), implying $7 > 5$ *or* $7 = 5$. **11.** $8 = 8$ implies $8 = 8$ *or* $8 > 8$. **13.** For any nonzero whole number a, we know $0 + a = a$, implying $0 < a$ by (4.17). **15.** true **17.** true **19.** true **21.** true

Review Exercises, page 141

1. T **3.** T **5.** F **7.** T **9.** T **11.** {a, b, c} and {e, f, g} **13.** impossible **15.** {a, b, c, d, e, f} **17.** $\emptyset$ **19.** $A = \{a, b\}$ and $B = \{a, b, c, d, e\}$ **21.** Let $A = \{a, b, c\}$ and $B = \{d, e, f, g\}$. Then $3 + 4 = n(A \cup B) = n(\{a, b, c, d, e, f, g\}) = 7$. **23.** $7 - 5 = 2$ because $5 + 2 = 7$. **25.** $0 \div 3 = 0$ because $3 \cdot 0 = 0$; however, $3 \div 0 = x$ implies $0 \cdot x = 3$ for some whole number x, which is impossible. **27.** false **29.** true **31.** true **33.** true **35.** true **37.** $(3 \cdot 2)^{10} = 6^{10}$ **39.** 4^5 **41.** $4^0 \cdot 4^2 = 4^{0+2} = 4^2$, so $4^0 \cdot 16 = 16$, implying 4^0 is the multiplicative identity, 1.

Chapter 5

Section 5.1, page 144

1. 36, $1 + 35$, $2 + 34$, $12 \cdot 3$, $4 \cdot 9$, 6^2, $40 - 4$, $50 - 14$, etc. **3.** Incorrect; if a and b are *names* for the same number, then they are numerals, not numbers.

Section 5.2, page 155

1. 32 **3.** 43 **5.** 153 **7.** 1946 **9.** 𓁨 𓂭𓂭 𓆼𓆼𓆼 𓍢𓍢𓍢 𓎆𓎆 |||| **11.** 𓍢 𓎆𓎆𓎆𓎆𓎆𓎆𓎆 |||

13. 𒌋 𒌋𒁹𒁹𒁹 **15.** 𒁹 𒁹𒁹𒁹𒁹 𒌋𒌋𒌋𒁹𒁹𒁹𒁹𒁹𒁹𒁹𒁹𒁹 **17.** 𝋭
𝋠 **19.** 𝋡
𝋮
𝋩

21. CXXIV **23.** $\overline{\text{MMCCCLXXVI}}\text{CDLI}$ **25.** $2 \cdot 10^2 + 3 \cdot 10^1 + 8 \cdot 10^0$ **27.** $2 \cdot 10^6 + 0 \cdot 10^5 + 0 \cdot 10^4 + 0 \cdot 10^3 + 3 \cdot 10^2 + 5 \cdot 10^1 + 1 \cdot 10^0$ **29.** $4 \cdot 10^5 + 7 \cdot 10^4 + 2 \cdot 10^3 + 3 \cdot 10^2 + 9 \cdot 10^1 + 8 \cdot 10^0$ **31.** 56,842 **33.** 40,375 **35.** 427,300

Section 5.3, page 164

1. 342, 343, 344, 345, 350, 351, 352, 353, 354, 355, 400, 401, ..., 423 **3.** 101, 110, 111, 1000, ..., 10001 [*Note*: For Exercises 5–13, expanded forms are written in base-ten numerals.] **5.** $3 \cdot 5^2 + 2 \cdot 5 + 1$; 86 **7.** $3 \cdot 8^2 + 2 \cdot 8 + 1$; 209 **9.** $10 \cdot 12^3 + 11 \cdot 12^2 + 8 \cdot 12 + 3$; 18,963 **11.** $1 \cdot 2^5 + 0 \cdot 2^4 + 0 \cdot 2^3 + 1 \cdot 2^2 + 1 \cdot 2 + 1$; 39 **13.** $3 \cdot 6^3 + 5 \cdot 6^2 + 0 \cdot 6 + 2$; 830 **15.** 100110_{two}; 123_{five}; 102_{six}; 32_{twelve} **17.** 1101011000_{two}; 11411_{five}; 3544_{six}; $5E4_{twelve}$ **19.** sixteen; $FF_{sixteen} = 255$ **21.** 1100011_{two}; 143_{eight}; $63_{sixteen}$ **23.** 10000000_{two}; 200_{eight}; $80_{sixteen}$ **25.** 10011010010_{two}; 2322_{eight}; $4D2_{sixteen}$ **27.** approximately 64,000 bytes (512,000 bits); 65,536 bytes (524,288 bits) **29.** approximately 1,024,000 bytes (8,192,000 bits); 1,048,576 bytes (8,388,608 bits) **31.** 01001100, 01000101, 01010111, 01001001, 01010011, 00100000, 01000011, 01000001, 01010010, 01010010, 01001111, 01001100, 01001100 **33.** 01010100, 01001000, 01010010, 01001111, 01010101, 01000111, 01001000, 00100000, 01010100, 01001000, 01000101, 00100000, 01001100, 01001111, 01001111, 01001011, 01001001, 01001110, 01000111, 00100000, 01000111, 01001100, 01000001, 01010011, 01010011 **35.** TWEEDLEDUM **37.** OFF WITH HIS HEAD! **39.** WONDERLAND **41.** TWAS BRILLIG, AND...

Review Exercises, page 166

1. T **3.** F **5.** T **7.** F **9.** F **11.** 101,212 **13.** 108,734 **15.** 227 **17.** 3454 **19.** 𓍢𓍢𓍢 𓎆𓎆𓎆𓎆𓎆 |||||| **21.** 𒁹 𒌋𒌋𒌋𒁹𒁹𒁹𒁹𒁹𒁹𒁹𒁹𒁹 **23.** 𝋠 **25.** CCXCV **27.** 407 **29.** 1035 **31.** 834 **33.** 43103_{five} **35.** 10725_{eight} **37.** 100010_{two}; 42_{eight}; $22_{sixteen}$

Chapter 6

Section 6.2, page 174

1. Expanded form; commutativity and associativity of +; basic addition fact; distributivity of · over +; basic addition fact; standard form **3.** Expanded form; commutativity and associativity of +; basic addition fact; expanded form; commutativity and associativity of +; $[(1 + 2) + 3] \cdot 10 + 4$; $6 \cdot 10 + 4$; standard form.

5. One estimate is 50.
$(1 \cdot 10 + 7) + (3 \cdot 10 + 2)$ Expanded form
$(1 \cdot 10 + 3 \cdot 10) + (7 + 2)$ Assoc. & comm. of +
$(1 \cdot 10 + 3 \cdot 10) + 9$ Basic addition fact
$(1 + 3) \cdot 10 + 9$ Distributivity of · over +
49 Standard form

7. One estimate is 90.
$(6 \cdot 10 + 8) + (2 \cdot 10 + 5)$ Expanded form
$(6 \cdot 10 + 2 \cdot 10) + (8 + 5)$ Assoc. & comm. of +
$(6 \cdot 10 + 2 \cdot 10) + 13$ Basic addition fact
$(6 \cdot 10 + 2 \cdot 10) + (1 \cdot 10 + 3)$ Expanded form
$[(1 \cdot 10 + 6 \cdot 10) + 2 \cdot 10] + 3$ Assoc. & comm. of +
$[(1 + 6) + 2] \cdot 10 + 3$ Distrib. of · over +
$9 \cdot 10 + 3$ Basic addition facts
93 Standard form

9. One estimate is 800.

$3 \cdot 100 + 2 \cdot 10 + 4$

$\underline{5 \cdot 100 + 1 \cdot 10 + 3}$

$8 \cdot 100 + 3 \cdot 10 + 7$

837

11. One estimate is 2200.

$1 \cdot 1000 + 2 \cdot 100 + 3 \cdot 10 + 4$

$\underline{\qquad\qquad 9 \cdot 100 + 8 \cdot 10 + 7}$

$1 \cdot 1000 + 11 \cdot 100 + 11 \cdot 10 + 11$

$1 \cdot 1000 + 10 \cdot 100 + 1 \cdot 100 + 10 \cdot 10 + 1 \cdot 10 + 1 \cdot 10 + 1$

$1 \cdot 1000 + 1 \cdot 1000 + 1 \cdot 100 + 1 \cdot 100 + 1 \cdot 10 + 1 \cdot 10 + 1$

$2 \cdot 1000 + 2 \cdot 100 + 2 \cdot 10 + 1$ 2221

13. Same as Exercise 7 analysis, except for the fifth and sixth steps:

$[(6 \cdot 10 + 2 \cdot 10) + 1 \cdot 10] + 3$ Assoc. of +

$[(6 + 2) + 1] \cdot 10 + 3$ Distrib. of $\cdot$ over +

15. 4120_{five} **17.** 13331_{five}

19. Basic addition, base six:

+	0	1	2	3	4	5
0	0	1	2	3	4	5
1	1	2	3	4	5	10
2	2	3	4	5	10	11
3	3	4	5	10	11	12
4	4	5	10	11	12	13
5	5	10	11	12	13	14

21. 110151_{six} **23.** 12122_{six}

25. 125_{six}; 114_{seven}; 103_{eight}; 82_{nine}; 81_{ten}; 80_{eleven}; $7E_{twelve}$

Section 6.3, page 182

1. Expanded form; Sum-Difference Property; basic subtraction fact; distributivity of $\cdot$ over $-$; basic subtraction fact; standard form **3.** $(8 \cdot 10 + 4) - (2 \cdot 10 + 9)$; basic addition fact; distributivity of $\cdot$ over +; associativity of +; standard form; $(7 \cdot 10 - 2 \cdot 10) + (14 - 9)$; $(7 \cdot 10 - 2 \cdot 10) + 5$; distributivity of $\cdot$ over $-$; basic subtraction fact; 55

5. One estimate is 400.

$(7 \cdot 100 + 9 \cdot 10 + 5) - (4 \cdot 100 + 3 \cdot 10 + 1)$ Expanded form

$(7 \cdot 100 - 4 \cdot 100) + (9 \cdot 10 - 3 \cdot 10) + (5 - 1)$ Sum-Diff. Prop.

$(7 - 4) \cdot 100 + (9 - 3) \cdot 10 + (5 - 1)$ Distributivity of $\cdot$ over $-$

$3 \cdot 100 + 6 \cdot 10 + 4$ Basic subtraction facts

364 Standard form

7. One estimate is 70.

$(2 \cdot 100 + 3 \cdot 10 + 4) - (1 \cdot 100 + 5 \cdot 10 + 9)$ Expanded form

$[(1 + 1) \cdot 100 + (2 + 1) \cdot 10 + 4] - (1 \cdot 100 + 5 \cdot 10 + 9)$
Basic addition facts

$[(1 \cdot 100 + 1 \cdot 100) + (2 \cdot 10 + 1 \cdot 10) + 4] - (1 \cdot 100 + 5 \cdot 10 + 9)$
Distributivity of $\cdot$ over +

$[1 \cdot 100 + (1 \cdot 100 + 2 \cdot 10) + (1 \cdot 10 + 4)] - (1 \cdot 100 + 5 \cdot 10 + 9)$
Associativity of +

$[1 \cdot 100 + (1 \cdot 10 + 2) \cdot 10 + (1 \cdot 10 + 4)] - (1 \cdot 100 + 5 \cdot 10 + 9)$
Distributivity of $\cdot$ over +

$(1 \cdot 100 + 12 \cdot 10 + 14) - (1 \cdot 100 + 5 \cdot 10 + 9)$ Standard form

$(1 \cdot 100 - 1 \cdot 100) + (12 \cdot 10 - 5 \cdot 10) + (14 - 9)$ Sum-Diff. Prop.

$(1 - 1) \cdot 100 + (12 - 5) \cdot 10 + (14 - 9)$ Distributivity of $\cdot$ over $-$

$0 \cdot 100 + 7 \cdot 10 + 5$ Basic subtraction facts

75 Standard form

9. One estimate is 20.

$7 \cdot 10 + 3$

$\underline{-5 \cdot 10 + 5}$

$6 \cdot 10 + 1 \cdot 10 + 3$

$\underline{5 \cdot 10 \qquad + 5}$

$6 \cdot 10 + 13$

$\underline{5 \cdot 10 + \ 5}$

$1 \cdot 10 + \ 8$

18

11. One estimate is 400.

$8 \cdot 100 + 0 \cdot 10 + 2$

$\underline{- 3 \cdot 100 + 5 \cdot 10 + 7}$

$7 \cdot 100 + 1 \cdot 100 + 0 \cdot 10 + 2$

$\underline{3 \cdot 100 \qquad + 5 \cdot 10 + 7}$

$7 \cdot 100 + 10 \cdot 10 + 2$

$\underline{3 \cdot 100 + \ 5 \cdot 10 + 7}$

$7 \cdot 100 + 9 \cdot 10 + 1 \cdot 10 + 2$

$\underline{3 \cdot 100 + 5 \cdot 10 \qquad + 7}$

$7 \cdot 100 + 9 \cdot 10 + 12$

$\underline{3 \cdot 100 + 5 \cdot 10 + \ 7}$

$4 \cdot 100 + 4 \cdot 10 + \ 5$

445

13. 14_{five} **15.** 4333_{five}

17. $42_{five} - 23_{five}$
$(4 \cdot 5 + 2) - (2 \cdot 5 + 3)$ Expanded form
$[(3+1) \cdot 5 + 2] - (2 \cdot 5 + 3)$ Basic addition fact
$[(3 \cdot 5 + 1 \cdot 5) + 2] - (2 \cdot 5 + 3)$ Distrib. of · over +
$[3 \cdot 5 + (1 \cdot 5 + 2)] - (2 \cdot 5 + 3)$ Associativity of +
$[3 \cdot 5 + 12_{five}] - (2 \cdot 5 + 3)$ Standard form
$(3 \cdot 5 - 2 \cdot 5) + (12_{five} - 3)$ Sum-Difference Property
$(3 - 2) \cdot 5 + (12_{five} - 3)$ Distrib. of · over −
$1 \cdot 5 + 4$ Basic subtraction facts
14_{five} Standard form

19. 12043_{six}

21. 3043_{six}; 3054_{seven}; 3065_{eight}; 3076_{nine}; 3087_{ten}; 3098_{eleven}; $30T9_{twelve}$

Section 6.4, page 188

1. 36 **3.** Expanded form; $(6 \cdot 10^2) \cdot 7 + (5 \cdot 10) \cdot 7 + 8 \cdot 7$; associativity and commutativity of ·; $42 \cdot 10^2 + 35 \cdot 10 + 56$; Powers-of-ten Product Property; Addition Algorithm **5.** $658 \cdot (3 \cdot 10 + 7)$; Distributivity of · over +; n-by-1 Multiplication Algorithm; 19,740 + 4606; Addition Algorithm

7. One estimate is 2000.
$(2 \cdot 10^2 + 5 \cdot 10 + 7) \cdot 8$
Expanded form
$(2 \cdot 10^2) \cdot 8 + (5 \cdot 10) \cdot 8 + 7 \cdot 8$
Distrib. of · over +
$(2 \cdot 8) \cdot 10^2 + (5 \cdot 8) \cdot 10 + 7 \cdot 8$
Assoc. & comm. of ·
$16 \cdot 10^2 + 40 \cdot 10 + 56$
Powers-of-ten Product Prop.
2056 Addition Algorithm

9. One estimate is 3500.
$98 \cdot (3 \cdot 10 + 5)$
Expanded form
$98 \cdot (3 \cdot 10) + 98 \cdot 5$
Distrib. of · over +
$(98 \cdot 3) \cdot 10 + 98 \cdot 5$
Associativity of +
$294 \cdot 10 + 490$
Powers-of-ten Product Prop.
3430 Addition Algorithm

11. One estimate is 360,000.
$368 \cdot (8 \cdot 10^2 + 7 \cdot 10 + 2)$
Expanded form
$368 \cdot (8 \cdot 10^2) + 368 \cdot (7 \cdot 10) + 368 \cdot 2$
Distributivity of · over +
$(368 \cdot 8) \cdot 10^2 + (368 \cdot 7) \cdot 10) + 368 \cdot 2$
Associativity of ·
$2944 \cdot 10^2 + 2576 \cdot 10 + 736$
n-by-1 Mult. Algorithm
294,400 + 25,760 + 736
Powers-of-ten Product Prop.
320,896 Addition Algorithm

13. One estimate is $8 \cdot 10^8$.
(Reasons as in Exercise 11 analysis.)
$35{,}789 \cdot (2 \cdot 10^4 + 3 \cdot 10^3 + 6 \cdot 10^2 + 7 \cdot 10 + 5)$
$35{,}789 \cdot (2 \cdot 10^4) + 35{,}789 \cdot (3 \cdot 10^3) + 35{,}789 \cdot (6 \cdot 10^2) + 35{,}789 \cdot (7 \cdot 10) + 35{,}789 \cdot 5$
$(35{,}789 \cdot 2) \cdot 10^4 + (35{,}789 \cdot 3) \cdot 10^3 + (35{,}789 \cdot 6) \cdot 10^2 + (35{,}789 \cdot 7) \cdot 10 + 35{,}789 \cdot 5$
$71{,}578 \cdot 10^4 + 107{,}367 \cdot 10^3 + 214{,}734 \cdot 10^2 + 250{,}523 \cdot 10 + 178{,}945$
$715{,}780{,}000 + 107{,}367{,}000 + 21{,}473{,}400 + 2{,}505{,}230 + 178{,}945$
847,304,575

15. 231_{five}

17. 4444_{five}

19. (Reasons as in Exercise 11.)
$132_{five} \cdot (2 \cdot 10_{five} + 4)$
$132_{five} \cdot (2 \cdot 10_{five}) + 132_{five} \cdot 4$
$(132_{five} \cdot 2) \cdot 10_{five} + 132_{five} \cdot 4$
$314_{five} \cdot 10_{five} + 1133_{five}$
$3140_{five} + 1133_{five}$
4323_{five}

21. 232_{six} **23.** 12204_{six}

25. 4212_{six}; 3636_{seven}; 3504_{eight}; 3362_{nine}; 3240_{ten}; 3119_{eleven}; $2EE8_{twelve}$

Section 6.5, page 198

1. $52 = 6 \cdot 8 + 4$ **3.** $5698 = 22 \cdot 251 + 176$ **5.** $8658 = 149 \cdot x + 16$ **7.** $2379 = z \cdot x + y$ **9.** $17 \div 7 = 2\ R\ 3$ **11.** $3729 \div 68 = 54\ R\ 57$ **13.** $804 \div 31 = 25\ R\ 29$ **15.** $1036 \div 37 = 28$; $1036 \div 28 = 37$ **17.** $26{,}071 \div 128 = 203\ R\ 8$; $26{,}071 \div 203 = 128\ R\ 87$ **19.** $587 \div 87 = 6\ R\ 65$ **21.** none **23.** II: Subtract 2000, then 500, then 70, then 2 42s. III: Partial quotients: 2000, 500, 70,

2 IV: 2573 R 9 **25.** II: Subtract 2000, then 50 17s. III: Partial quotients: 2000, 000, 50, 0 IV: 2050 R 4

27.

108,075	34,854
− 42	− 17
108,033	34,837
− 42	− 17
107,991	34,820
− 42	− 17
107,949	34,803
⋮	⋮
51	21
− 42	− 17
9	4

29. Insert some zeros in the display to make its format like that of Algorithm III. **31.** $101{,}313_{five}$ R 3 **33.** $12{,}530_{six}$ R 2 **35.** 31_{six} R 14_{six}; 33_{seven} R 31_{seven}; 37_{eight} R 2; 41_{nine} R 27_{nine}; 45_{ten} R 10_{ten}; 49_{eleven} R 14_{eleven}; 52_{twelve} R 8 **37.** $10{,}101_{two}$ R 100_{two}

Review Exercises, page 199

1. F **3.** F **5.** F **7.** T **9.** F

11. One estimate is 60.
$(3 \cdot 10 + 7) + (2 \cdot 10 + 6)$ Expanded form
$(3 \cdot 10 + 2 \cdot 10) + (7 + 6)$ Assoc. & comm. of +
$(3 \cdot 10 + 2 \cdot 10) + 13$ Basic addition fact
$(3 \cdot 10 + 2 \cdot 10) + (1 \cdot 10 + 3)$ Expanded form
$[(1 \cdot 10 + 3 \cdot 10) + 2 \cdot 10] + 3$ Assoc. & comm. of +
$[(1 + 3) + 2] \cdot 10 + 3$ Distrib. of · over +
$6 \cdot 10 + 3$ Basic addition fact
63 Standard form

13. One estimate is 3200.
$76 \cdot (4 \cdot 10 + 3)$ Expanded form
$76 \cdot (4 \cdot 10) + 76 \cdot 3$ Distrib. of · over +
$(76 \cdot 4) \cdot 10 + 76 \cdot 3$ Assoc. of ·
$304 \cdot 10 + 228$ n-by-1 Mult. Algorithm
$3040 + 228$ Power-of-ten Prod. Prop.
3268 Addition Algorithm

15. I: $55{,}173 - 67 = 55{,}106; \ldots; 99 - 67 = 32$ II: Subtract 800, then 20, then 3 67s. III: Partial quotients: 800, 20, 3 IV: $55{,}173 \div 67 = 823\,R\,32$ **17.** 233_{five} **19.** $112_{five}\,R\,22_{five}$ **21.** 244_{six} **23.** $112_{six}\,R\,22_{six}$

Chapter 7

Section 7.1, page 204

1. is a multiple of, is divisible by **3.** is a factor of, is a divisor of, divides **5.** is a multiple of, is divisible by **7.** is a factor of, is a divisor of, divides **9.** none **11.** is a factor of, is a divisor of, divides **13.** All the phrases apply. **15.** Not true; is a factor of, is a divisor of, divides. Could be true; is a multiple of, is divisible by. **17.** is a factor of, is a divisor of, divides; is a multiple of, is divisible by **19.** 12, 24, 36 **21.** infinitely many **23.** 2, 3, 5 **25.** 1, 3, 5, 15 **27.** 1, 2, 4, 8, 16, 32 **29.** 1, 2, 5, 7, 10, 14, 35, 70 **31.** {5, 10, 15, 20, ...} **33.** {18, 36, 54, 72, ...}

Section 7.2, page 210

1. 2 **3.** 2, 4, 8 **5.** 2, 4, 8 **7.** 3 **9.** 3 **11.** 3, 9 **13.** 5, 10 **15.** 5, 10, 100, 1000 **17.** 5, 10, 100 **19.** no **21.** yes **23.** yes **25.** yes **27.** no **29.** yes **31.** 2, 4, 5, 7, 8, 10, 11 **33.** 2, 3, 4, 5, 6, 8, 10, 11 **35.** 2, 3, 4, 5, 6, 7, 8, 9, 10, 11 **37.** A number is divisible by 16 if and only if the number formed by the last four digits is divisible by 16. (True) **39.** T **41.** T **43.** F

Section 7.3, page 218

1. $3 \cdot 19$ **3.** $2^4 \cdot 3^2$ **5.** $7 \cdot 17 \cdot 19$ **7.** $2^2 \cdot 3^3 \cdot 5^2 \cdot 7$ **9.** prime **11.** $7 \cdot 73$ **13.** (a) 31 (b) $31^2 = 961$ **15.** 67 **17.** (a) odd; odd; odd (b) no (c) $p - 1$, p, and $p + 1$ are consecutive numbers, so one of them must be a multiple of 3. But p is a prime. (d) If $p-1$ is a multiple of 3, then so is $p+2$. If $p+1$ is a multiple of 3, then so is $p+4$. **19.** $3+3$ **21.** $5+5$ **23.** $5+11$ **25.** $3+17$ **27.** $11+89$ **29.** (a) prime (b) prime (c) $11\cdot 19$ (d) prime **31.** If there were a greatest prime P, then by Exercise 30, the product $2\cdot 3\cdot 5\cdot \ldots P-1$ either is a prime or is a composite whose prime factorization contains a prime greater than P. In either case, we have a contradiction.

Section 7.4, page 226

1. (a) 1, 2, 3, 4, 6, 8, 12, 24 (b) 1, 2, 3, 4, 6, 9, 12, 18, 36 (c) 1, 2, 3, 4, 6, 12 (d) 12 **3.** (a) 12, 24, 36, 48, 60, 72, 84, 96, 108, 120 (b) 15, 30, 45, 60, 75, 90, 105, 120, 135, 150 (c) 60, 120, 180, 240, 300 (d) 60 **5.** 3, 4 **7.** 3, 3 **9.** 6, 12 **11.** 6; 1620; no **13.** 1; 210; yes **15.** 1; 311,850; yes **17.** 6; 3240 **19.** 60; 415,800 **21.** 2; 336 **23.** 3^{21}, $3^{23}\cdot 5^7\cdot 7^5\cdot 29^3\cdot 31^3$ **25.** a **27.** 1 **29.** a **31.** 1

Section 7.5, page 236

1. 7 **3.** 8 **5.** 11 **7.** 0 **9.** 1 **11.** 1 **13.** 5 **15.** 0 **17.** 7, 19, 31, 43, 55 **19.** 2, 5, 8, 11, 14 **21.** 6, 13, 20, 27, 34 **23.** $\mathbf{W}_3$ is $\underline{0}$, $\underline{1}$, $\underline{2}$; $\underline{0}=\{0, 3, 6, 9, \ldots\}$, $\underline{1}=\{1, 4, 7, 10, \ldots\}$, $\underline{2}=\{2, 5, 8, 11, \ldots\}$. $\mathbf{W}_6$ is $\underline{0}$, $\underline{1}$, $\underline{2}$, $\underline{3}$, $\underline{4}$, $\underline{5}$; $\underline{0}=\{0, 6, 12, \ldots\}$, $\underline{1}=\{1, 7, 13, \ldots\}$, $\underline{2}=\{2, 8, 14, \ldots\}$, $\underline{3}=\{3, 9, 15, \ldots\}$, $\underline{4}=\{4, 10, 16, \ldots\}$, $\underline{5}=\{5, 11, 17, \ldots\}$ **25.** $\underline{3}$, $\underline{2}$, $\underline{0}$, $\underline{8}$ **27.** $\underline{0}$, $\underline{2}$, $\underline{1}$, $\underline{1}$ **29.** $(\underline{5}\odot\underline{8})\odot\underline{10}=\underline{4}\odot\underline{10}=\underline{4}$ and $\underline{5}\odot(\underline{8}\odot\underline{10})=\underline{5}\odot\underline{8}=\underline{4}$

Review Exercises, page 237

1. F **3.** F **5.** F **7.** F **9.** F **11.** is a multiple of, is divisible by **13.** is a factor of, is a divisor of, divides **15.** is a factor of, is a divisor of, divides **17.** composite **19.** prime **21.** composite **23.** 15, 30, 45, 60, 75 **25.** 560 **27.** 25,410 **29.** 5040 **31.** 91 and 119 are not primes; $1547=7\cdot 13\cdot 17$. Sarah's two factorizations are considered the same because the only difference between them is the order in which the primes are written. **33.** 2 **35.** 0 **37.** 3 **39.** $\underline{6}$, $\underline{5}$, $\underline{2}$, $\underline{9}$

Chapter 8

Section 8.1, page 243

1. $1-2$, $3-5$, $9-18$, $13-25$, $6-100$ do not; $2-1$, $5-3$, $18-9$, $25-13$, $100-6$ do **3.** $8-1$, $9-2$, $10-3$, $11-4$, $12-5$ **5.** $1-1$, $2-2$, $3-3$, $4-4$, $5-5$ **7.** 5 **9.** The temperature is 0° and then drops 5°. A submarine is at sea level, then submerges 5 fathoms.

Section 8.2, page 249

1. (a) $10-3$, $11-5$, $4-10$, $19-13$, $8-0$, $0-6$, $16-0$, $16-10$, $7-11$, $46-52$ (b) $(11,5)$, $(19,13)$, $(16,10)$ (c) $(4,10)$, $(0,6)$, $(46,52)$ **3.** 16 **5.** 6 **7.** 11 **9.** 6 **11.** 8 **13.** 14 **15.** 0 **17.** 4 **19.** 2 **21.** 6 **23.** T **25.** T **27.** F **29.** F **31.** T **33.** F **35.** $(10,1)$, $(11,2)$, $(12,3)$, $(13,4)$, $(14,5)$ **37.** $(4,0)$, $(5,1)$, $(6,2)$, $(7,3)$, $(8,4)$ **39.** $(0,0)$, $(1,1)$, $(2,2)$, $(3,3)$, $(4,4)$ **41.** $(0,25)$, $(1,26)$, $(2,27)$, $(3,28)$, $(4,29)$ **43.** $[0,7]$ **45.** $[0,0]$ **47.** $[0,9]$ **49.** $[25,0]$ **51.** Key idea: Add $b+d$ to both sides. **53.** *Reflexive*: $(a,b)\sim(a,b)$ for any whole numbers a and b because $a+b=b+a$ by commutativity of $+$ on $\mathbf{W}$. *Symmetric*: If $(a,b)\sim(c,d)$, then $(c,d)\sim(a,b)$ because $a+d=b+c$ implies $c+b=d+a$, by commutativity of $+$ (and symmetry of $=$) on $\mathbf{W}$. *Transitive*: If $(a,b)\sim(c,d)$ and $(c,d)\sim(e,f)$, then $a+d=b+c$ and $c+f=d+e$. Adding these two equations, we get $a+d+c+f=b+c+d+e$. Subtracting c and d from both sides, we get $a+f=b+e$, implying $(a,b)\sim(e,f)$, as required.

Section 8.3, page 257

1. $[9,7] = {}^{+}2$ **3.** $[6,6] = 0$ **5.** $[8,0] = {}^{+}8$ **7.** $[3,5] = {}^{-}2$ **9.** $[0,4] = {}^{-}4$ **11.** $[1,4]$ **13.** $[6,6]$ **15.** ${}^{+}5 = [5,0]$ **17.** If ${}^{+}5 \oplus x = 0$, then ${}^{-}5 \oplus ({}^{+}5 \oplus x) = {}^{-}5 \oplus 0$, so $({}^{-}5 \oplus {}^{+}5) \oplus x = {}^{-}5$, implying $x = {}^{-}5$. **19.** $[8,5] = {}^{+}3$ **21.** $[15,21] = {}^{-}6$ **23.** $[7,3] = {}^{+}4$ **25.** $[0,10] = {}^{-}10$ **27.** $[2,0] = {}^{+}2$ **29.** $[2,2] = 0$ **31.** $[0,4] = {}^{-}4$ **37.** Fundamental Subtraction Property; definition of additive inverse; definition of negative integer; definition of addition; definition of negative integer; standard form. **39.** Since $p \oplus -p = -p \oplus p = 0$ by definition of $-p$, then p must be the additive inverse of $-p$; that is, $-(-p) = p$.

Section 8.4, page 265

1. $[43,23]$ **3.** $[43,23]$ **5.** $[28,0]$ **7.** $[0,28]$ **9.** $[0,0]$ **11.** $0 \odot {}^{+}7 = [0,0] \odot [7,0] = [0,0] = 0$ **13.** Let $p = [a,b]$ and $q = [c,d]$. Then $p \odot -q = [a,b] \odot [d,c] = [ad+bc,\ ac+bd] = [bc+ad,\ bd+ac] = [b,a] \odot [c,d] = -p \odot q$. The other part is done similarly. **15.** If $[2a,2b] = [1,0]$, then $2a + 0 = 2b + 1$, implying that some even number equals an odd number; contradiction. **19.** Both are true. **21.** (a) T (b) F (c) T (d) F **23.** $\underline{1}$ is its own multiplicative inverse, as is $\underline{5}$. **25.** $-\underline{0} = \underline{0}$, $-\underline{1} = \underline{2}$, $-\underline{2} = \underline{1}$ **27.** $\underline{2}$, $\underline{1}$, $\underline{1}$, $\underline{0}$ **29.** (a) Main idea: $3x$ is always congruent to a multiple of 3 (mod 12), so it cannot also be congruent to 1 (mod 12). (b) Each of $\underline{1}$, $\underline{5}$, $\underline{7}$, and $\underline{11}$ is its own multiplicative inverse; there are no others. (c) Mimic (a). (d) Each of $\underline{1}$, $\underline{4}$, $\underline{11}$, and $\underline{14}$ is its own multiplicative inverse; $\underline{2}$ and $\underline{8}$ are inverses of each other, as are $\underline{7}$ and $\underline{13}$.

Section 8.5, page 274

1. ${}^{+}3 \oplus {}^{+}5 = {}^{+}8$ **3.** ${}^{-}5 \oplus {}^{+}4 = {}^{-}1$ **5.** $0 \oplus {}^{+}4 = {}^{+}4$ **7.** ${}^{-}6 \oplus {}^{+}14 = {}^{+}8$ **9.** Since ${}^{-}8 ⧀ {}^{-}2$ and ${}^{-}2 ⧀ {}^{+}5$, we have ${}^{-}8 ⧀ {}^{+}5$. **11.** ${}^{+}1 ⧀ {}^{+}1$ is false; ${}^{-}5 ⧀ {}^{+}3$, but ${}^{+}3 \not⧀ {}^{-}5$. **13.** ${}^{+}7$ **15.** ${}^{+}13$ **17.** ${}^{+}5$ **19.** ${}^{+}36$ **21.** ${}^{+}13$ **23.** ${}^{+}36$ **25.** ${}^{-}13$ **27.** ${}^{-}5$ **29.** ${}^{+}13$ **31.** ${}^{+}10$ **33.** ${}^{+}5$ **35.** T **37.** T **39.** T **41.** T **43.** F **45.** F **47.** F **49.** T **51.** (a) If ${}^{+}a$ is any positive integer, then $0 \oplus {}^{+}a = {}^{+}a$, so $0 ⧀ {}^{+}a$ by the definition of ⧀. (b) If ${}^{-}a$ is a negative integer, then $-({}^{-}a) = {}^{+}a$ is a positive integer such that ${}^{-}a \oplus {}^{+}a = 0$, so ${}^{-}a ⧀ 0$ by the definition of ⧀.

Section 8.6, page 279

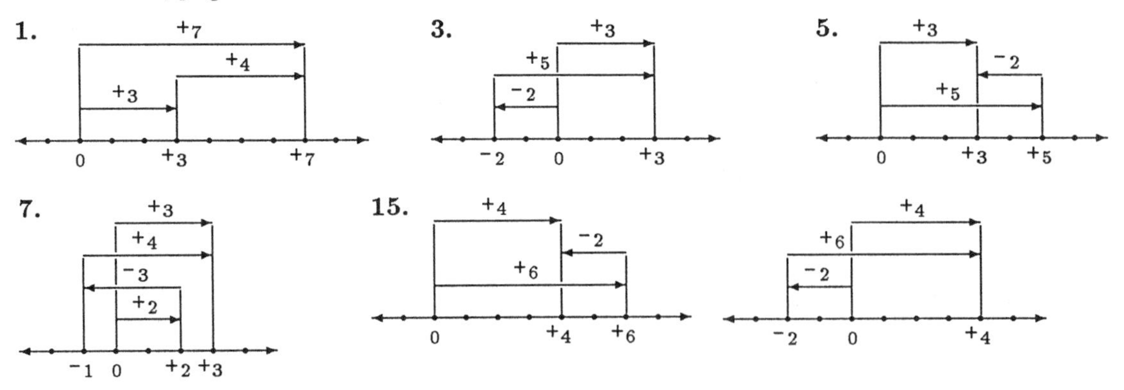

9. Gain: ${}^{+}17 + {}^{-}9 = {}^{+}8$ **11.** Increase: ${}^{-}7 + {}^{+}15 + {}^{+}20 = {}^{+}28$ **13.** Decrease: ${}^{+}15 + {}^{-}20 = {}^{-}5$ **17.** The middle-level vector goes from ${}^{+}3$ to ${}^{+}1$ and represents ${}^{-}2$. **19.** The middle-level vector goes from ${}^{+}3$ to ${}^{+}8$ and represents ${}^{+}5$.

21. **23.** **25.**

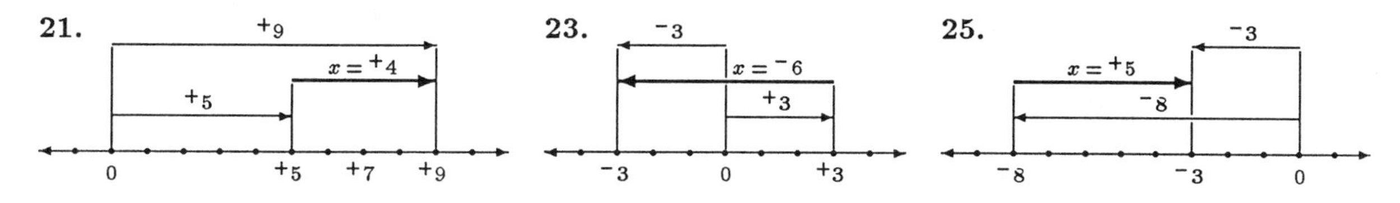

Section 8.7, page 289

1. $^-15$, II **3.** $^+8$, III(*b*) **5.** $^-18$, III(*c*) **7.** $^-44$, II **9.** 0, III(*a*) **11.** $^-60$; $^-59$, III(*c*) **13.** 0; 0, III(*a*) **15.** $^-160$; $^-163$, II **17.** $^-180$; $^-187$, III(*c*) **19.** $^+1$ **21.** $^+24$ **23.** $^+32$ **25.** $^+8$ **27.** $^-26$ **29.** $^-100$; $^-103$ **31.** $^+100$; $^+106$ **33.** $^-105$; $^-107$ **35.** $^+285$; $^+283$

37. **39.** **41.**

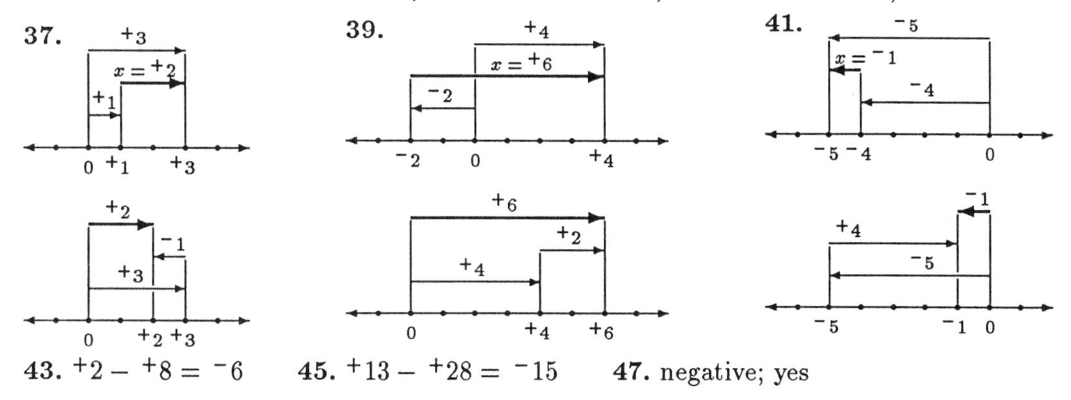

43. $^+2 - {}^+8 = {}^-6$ **45.** $^+13 - {}^+28 = {}^-15$ **47.** negative; yes

Section 8.8, page 295

1. $^+40$, I **3.** $^+32$, III **5.** $^+231$, III **7.** $^-672$, II **9.** $^+2340$, III **11.** $^+2 + {}^+2 + {}^+2 + {}^+2 + {}^+2 = {}^+10$ **13.** impossible **15.** impossible

17.

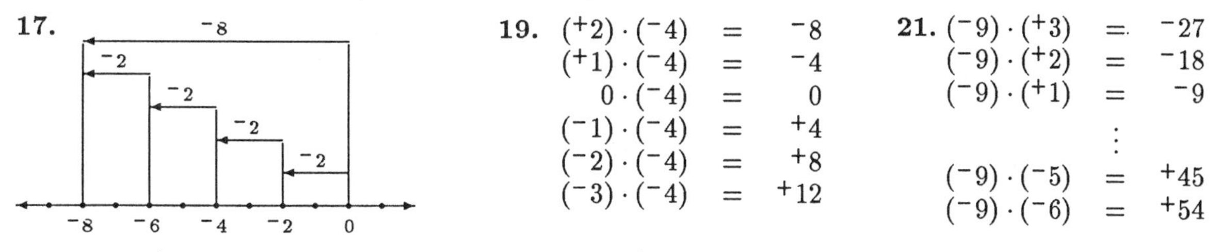

19.
$$\begin{aligned}(^+2)\cdot(^-4) &= {}^-8\\ (^+1)\cdot(^-4) &= {}^-4\\ 0\cdot(^-4) &= 0\\ (^-1)\cdot(^-4) &= {}^+4\\ (^-2)\cdot(^-4) &= {}^+8\\ (^-3)\cdot(^-4) &= {}^+12\end{aligned}$$

21.
$$\begin{aligned}(^-9)\cdot(^+3) &= {}^-27\\ (^-9)\cdot(^+2) &= {}^-18\\ (^-9)\cdot(^+1) &= {}^-9\\ &\vdots\\ (^-9)\cdot(^-5) &= {}^+45\\ (^-9)\cdot(^-6) &= {}^+54\end{aligned}$$

23. $(^-5)\cdot(^+2) = {}^-10$ and $((^-5)\cdot(^-2)) + ((^-5)\cdot(^+2)) = {}^-5\cdot((^-2)+(^+2)) = {}^-5\cdot 0 = 0$ **25.** If a player takes 2 steps forward at every turn, then 3 turns from now she will be 12 steps ahead of her present position. **27.** 3 turns ago, the player in Exercise 25 was 12 steps behind her present position. **29.** If a videotape of a pool filling at the rate of 4 mm per minute is run forward for 3 minutes, you will see the water rise 12 mm. **31.** If the videotape in Exercise 29 is run in reverse for 3 minutes, you will see the water drop 12 mm. **33.** If a football team gains 4 yards on every down, then 3 downs from now they will be 12 yards ahead of the original line of scrimmage. **35.** Three downs ago, the football team in Exercise 33 was 12 yards further back. **41.** $^+2$; $(^+6)\cdot(^+2) = {}^+12$ **43.** $^-4$; $(^-6)\cdot(^-4) = {}^+24$ **45.** undefined; there is no x such that $0\cdot x = {}^-3$ **47.** $^-3$; $(^-16)\cdot(^-3) = {}^+48$ **49.** $^-45$; $(^+5)\cdot(^-45) = {}^-225$; **51.** $^+3$; $(^-27)\cdot(^+3) = {}^-81$ **53.** $^-3$; $(^-15)\cdot(^-3) = {}^+45$ **55.** $^-13$; $(^-4)\cdot(^-13) = {}^+52$

Review Exercises, page 296

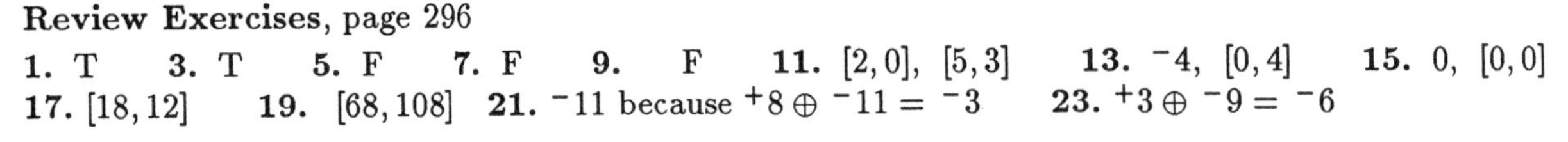

1. T **3.** T **5.** F **7.** F **9.** F **11.** $[2,0]$, $[5,3]$ **13.** $^-4$, $[0,4]$ **15.** 0, $[0,0]$ **17.** $[18,12]$ **19.** $[68,108]$ **21.** $^-11$ because $^+8 \oplus {}^-11 = {}^-3$ **23.** $^+3 \oplus {}^-9 = {}^-6$

25. **27.** **29.**

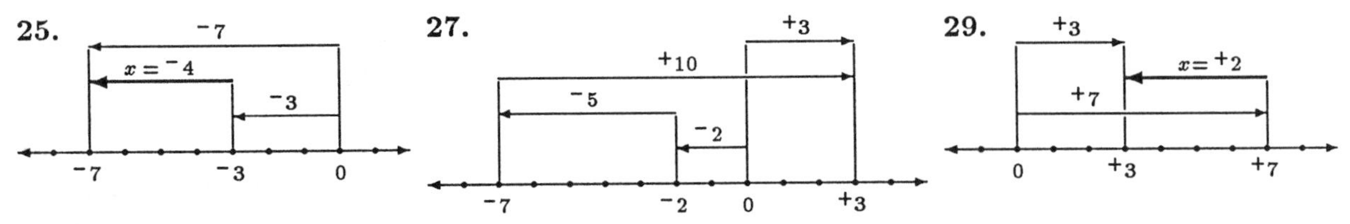

31.

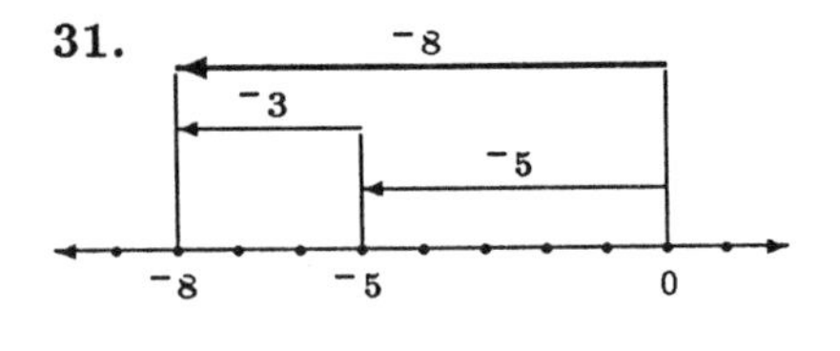

33. If the temperature drops 3° and then drops 7° more, it has dropped 10° altogether. **35.** $(^+2)\cdot(^-4) = {}^-8$ and $(^-2)\cdot(^-4) + (^+2)\cdot(^-4) = (^-2 + {}^+2)\cdot(^-4) = 0\cdot(^-4) = 0$. Thus, $(^-2)\cdot(^-4) = -(^-8) = {}^+8$. **37.** $^+6$, III(b) **39.** $^+10$, III(b) **41.** $^-17$, III(c) **43.** 0, III(a) **45.** $^+18$ **47.** $^-7$ **49.** $^-13$ **51.** $^-23$

Chapter 9

Section 9.1, page 303

3. $^-5 \div 1$, $5 \div {}^-1$, $25 \div {}^-5$, $^-100 \div 20$, $35 \div {}^-7$ **5.** $1 \div 1$, $^-1 \div {}^-1$, $2 \div 2$, $^-3 \div {}^-3$, $100 \div 100$ **7.** $1 \div {}^-1$, $^-1 \div 1$, $^-2 \div 2$, $3 \div {}^-3$, $100 \div {}^-100$ **9.** 6; yes **11.** If you divide 7 pizzas equally among 3 (big) people, how much pizza does each person get? If you walked at a constant rate of 7 miles in 3 hours, how many miles did you walk in 1 hour?

Section 9.2, page 312

1. (a) $5 \div 3 = ?$, $^-7 \div {}^-1 = ?$, etc. (b) $(14, {}^-2)$, $(^-70, 10)$ (c) $(12,20)$, $(^-21, {}^-35)$, $(30,50)$, $(15,25)$ **3.** 6 **5.** 0 **7.** 16 **9.** $^-5$ **11.** 16 **13.** 25 **15.** 3 **17.** $^-12$ **19.** F **21.** T **23.** F **25.** T **27.** F **29.** F **31.** $(7,5)$, $(^-7,{}^-5)$, $(14,10)$, $(^-14,{}^-10)$, $(21,15)$ **33.** $(6,1)$, $(^-6,{}^-1)$, $(12,2)$, $(^-12,{}^-2)$, $(18,3)$ **35.** $(8,3)$, $(^-8,{}^-3)$, $(16,6)$, $(^-16,{}^-6)$, $(24,9)$ **37.** $(20,4)$, $(^-20,{}^-4)$, $(40,8)$, $(^-40,{}^-8)$, $(60,12)$ **39.** $(3,1)$, $(^-3,{}^-1)$, $(6,2)$, $(^-6,{}^-2)$, $(9,3)$ **41.** $(1,1)$, $(^-1,{}^-1)$, $(2,2)$, $(^-2,{}^-2)$, $(3,3)$ **43.** $[4,5]$ **45.** $[35,4]$ **47.** $\frac{5}{8}$ **49.** $\frac{77}{1}$ or 77 **51.** $\frac{26}{9}$ **53.** $\frac{^-11}{12}$ **55.** Mimic the answer to Exercise 53 of Section 8.2, substituting multiplication for addition.

Section 9.3, page 321

1. $[3,8]$ **3.** $[^-16,15]$ **5.** $\frac{5}{6}$ **7.** $\frac{20}{17}$ **9.** $\frac{^-25}{28}$ **11.** $[5,3]$ **13.** $\frac{7}{5}$ **15.** $\frac{1}{3}$ **17.** Multiply both sides of the equation by $\frac{5}{3}$. **19.** Multiply both sides of the equation by $\frac{5}{3}$. **21.** $[21,10]$ **23.** $[5,8]$ **25.** $\frac{8}{15}$ **27.** $\frac{^-13}{24}$ **29.** 15 **31.** $\frac{2}{7}$ **37.** If $r = [x,y]$ and $s = [u,v]$, then $(r \ominus s)\odot(s \ominus r) = ([x,y]\odot[v,u])\odot([u,v]\odot[y,x]) = [xv,yu]\odot[uy,vx] = [xvuy,yuvx] = 1$. **39.** The multiplicative inverse of a rational number greater than 1 is a positive rational less than 1; the multiplicative inverse of a positive rational less than 1 is a rational number greater than 1.

Section 9.4, page 327

1. $[5,2]$ **3.** $[7,12]$ **5.** $\frac{47}{30}$ **7.** $\frac{^-3}{14}$ **9.** 5 **11.** $[^-3,4]$ **13.** $[1,6]$ **15.** $\frac{^-9}{4}$ **17.** $[1,3]$

19. $[5,8]$ **21.** $\frac{19}{55}$ **23.** $\frac{41}{30}$ **25.** $\frac{13}{8}$ **27.** (b) $(r \oplus s) \oplus (-r \oplus -s) = (r \oplus -r) \oplus (s \oplus -s) = 0$, so $-r \oplus -s$ is the additive inverse of $r \oplus s$.

Section 9.5, page 334

1. $2{\cdot}5 < 3{\cdot}4$ **3.** $\frac{6}{7} \oplus \frac{2}{7} = \frac{8}{7}$ **5.** $5{\cdot}5 < 6{\cdot}6$ **7.** $^-9{\cdot}3 < 4{\cdot}2$ **9.** $1 = \frac{1}{1}$ and $11{\cdot}1 > 7{\cdot}1$ **15.** $\frac{7}{8}$ is larger. **17.** $\frac{9}{11}$ is larger. **19.** $\frac{1}{100}$ is larger. **21.** $\frac{1}{2}$ **23.** $\frac{55}{18}$ **25.** $\frac{8}{7}$ **27.** F **29.** T **31.** F **33.** F **35.** F **37.** T **39.** T **41.** $\frac{19}{72}, \frac{20}{72}, \frac{21}{72}, \frac{22}{72}, \frac{23}{72}$ **43.** $\frac{193}{96}, \frac{194}{96}, \frac{195}{96}, \frac{196}{96}, \frac{197}{96}$ **45.** $\frac{2}{13} \olessthan \frac{2}{9} \olessthan \frac{1}{4} \olessthan \frac{3}{11} \olessthan \frac{2}{7}$ **47.** If $r \olessthan s$, then there is some $d \ogreaterthan 0$ such that $r \oplus d = s$. Use $d \odiv 2$ to get the desired result.

Section 9.6, page 340

1. $\frac{1}{64}$ **3.** 625 **5.** $^-625$ **7.** 1 **9.** $\frac{1}{49}$ **11.** $\frac{^-1}{8}$ **13.** $\frac{16}{81}$ **15.** 144 **17.** $\frac{1}{256}$ **19.** $3^2 = 9$, $3^1 = 3$, $3^0 = 1$, $3^{^-1} = \frac{1}{3}$, $3^{^-2} = \frac{1}{9}$; Decreasing the exponent by 1 has the effect of dividing by the base each time. Also, $3^{^-2} \cdot 3^2 = 3^{^-2+2} = 3^0 = 1$, so $3^{^-2} = \frac{1}{3^2}$. **21.** $3^{^-5} = \frac{1}{243}$ **23.** $(^-2)^{^-2} = \frac{1}{4}$ **25.** $3^{^-2} = \frac{1}{9}$ **27.** $5^1 = 5$ **29.** $2^{^-2} = \frac{1}{4}$ **31.** $\frac{5}{7}$ **33.** $\frac{^-63}{8}$ **35.** 573,000 **37.** .00000004723 **39.** 5.67

Review Exercises, page 341

1. T **3.** F **5.** T **7.** F **9.** F **11.** $[a,b] \sim [c,d]$ if and only if $ad = bc$. **13.** For every $a \in \mathbf{Q}$, $a \cdot 1 = 1 \cdot a = a$. **15.** For all $a, b \in \mathbf{Q}$ with $b \neq 0$, $a \div b = a \odot \frac{1}{b}$. **17.** $[13,30]$ **19.** $[^-10,21]$ **21.**

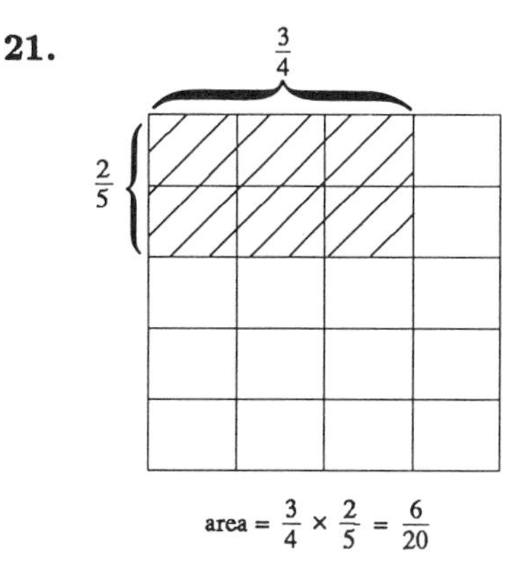

23. $\frac{^-3}{8}$ **25.** See Exercise 19 of Section 9.6. **27.** $\frac{1}{40}$ **29.** $\frac{^-9}{8}$ **31.** .000043276 **33.** $-.000000789$

Chapter 10

Section 10.2, page 350

1.

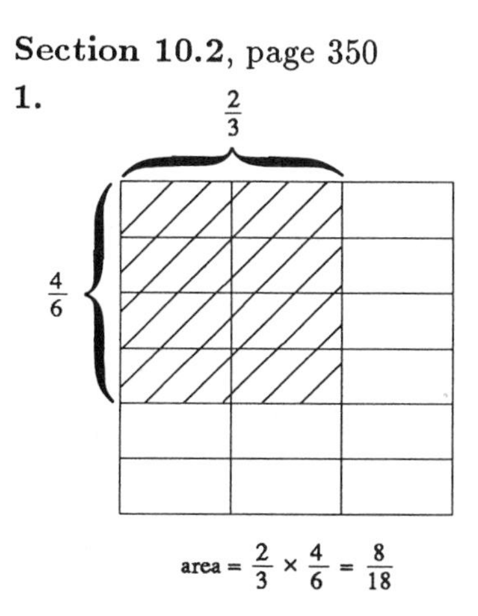

3. $\frac{7}{8}$ **5.** $\frac{25}{3}$ **7.** $\frac{14}{9}$ **9.** $4 \cdot \frac{a}{b} = \frac{a}{b} + \frac{a}{b} + \frac{a}{b} + \frac{a}{b} = \frac{4a}{b} = \frac{4}{1} \cdot \frac{a}{b}$, so it is reasonable to say that $4 = \frac{4}{1}$. **11.** $\frac{9}{12} = \frac{3\cdot3}{3\cdot4} = \frac{3}{4} \cdot \frac{3}{3} = \frac{3}{4} \cdot 1 = \frac{3}{4}$ **13.** $\frac{7}{8} = \frac{7}{8} \cdot 1 = \frac{7}{8} \cdot \frac{12}{12} = \frac{84}{96}$ **15.** $\frac{180}{77} \cdot \frac{143}{1350} = \frac{180\cdot143}{77\cdot1350} = \frac{2\cdot2\cdot3\cdot3\cdot5\cdot11\cdot13}{2\cdot3\cdot3\cdot3\cdot5\cdot5\cdot7\cdot11} = \frac{2\cdot13}{3\cdot5\cdot7} \cdot \frac{2}{2} \cdot \frac{3}{3} \cdot \frac{5}{5} \cdot \frac{11}{11} = \frac{26}{105} \cdot 1 \cdot 1 \cdot 1$ **17.** gcf is 60; $\frac{7}{8}$ **19.** gcf is 26; $\frac{4}{11}$ **21.** gcf is 1; $\frac{207}{209}$ **23.** (a) Each piece would be a 1-by-1-inch square. You would need a ruler marked in inches. (b) Each piece would be a 5-by-1-inch rectangle. You would need a ruler marked in inches. (c) Each piece would be a pie-shaped wedge with a central angle of 72°. You would need a protractor and a straightedge.

Section 10.3, page 358

1. **3.**

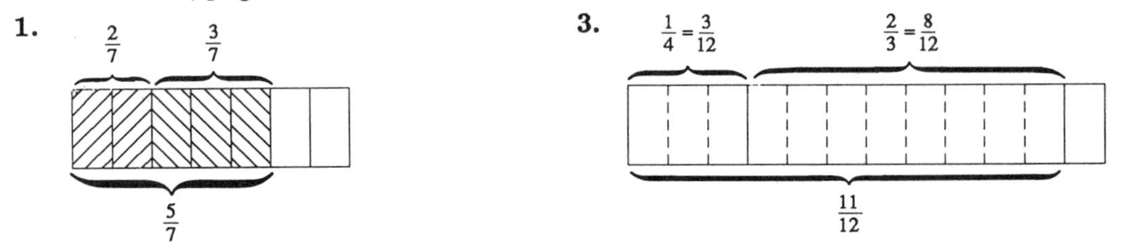

5. lcd $= 70$; $\frac{27}{70}$ **7.** lcd $= 120$; $\frac{193}{120}$ **9.** For $\frac{a}{b}+\frac{c}{b}$, lcd $= b$ and the sum is $\frac{a+c}{b}$. By formula, $\frac{ab+bc}{b^2} = \frac{b(a+c)}{b^2} = \frac{a+c}{b}$ **11.** $\left(\frac{a\cdot \text{lcm}(b,d)}{b} + \frac{c\cdot \text{lcm}(b,d)}{d}\right) \div \text{lcm}(b,d)$ **13.** $5\cdot\left(3+\frac{2}{5}\right) = 5\cdot 3 + 5\cdot\frac{2}{5} = 17$ **15.** lcd $= 70$; $\frac{3}{70}$ **17.** lcd $= 120$; $\frac{17}{120}$ **19.** $\frac{5}{4}$ **21.** $\frac{105}{88}$ **23.** $a\cdot\frac{1}{a} = \frac{a}{1}\cdot\frac{1}{a} = \frac{a}{a} = 1$; $\frac{a}{b}\cdot\frac{b}{a} = \frac{ab}{ba} = 1$. **25.** $\frac{16}{35}, \frac{^-44}{35}, \frac{^-12}{35}, \frac{^-7}{15}$ **27.** $\frac{^-109}{66}, \frac{1}{66}, \frac{15}{22}, \frac{54}{55}$ **29.** $\frac{^-215}{132}, \frac{71}{132}, \frac{13}{22}, \frac{72}{143}$ **31.** $\frac{^-3}{5}, \frac{3}{^-5}, -\frac{^-3}{^-5}$ **33.** $6\frac{1}{3} - 1\frac{2}{3} = \left(6+\frac{1}{3}\right) - \left(1+\frac{2}{3}\right) = \left((5+1)+\frac{1}{3}\right) - \left(1+\frac{2}{3}\right) = 5 + \left(1+\frac{1}{3}\right) - \left(1+\frac{2}{3}\right) = \left(5+\frac{4}{3}\right) - \left(1+\frac{2}{3}\right) = (5-1) + \left(\frac{4}{3}-\frac{2}{3}\right) = 4 + \frac{2}{3} = 4\frac{2}{3}$

Section 10.4, page 366

1. $3\cdot 10^2 + 5\cdot 10^1 + 7\cdot 10^0 + 9\cdot 10^{-1} = 300 + 50 + 7 + \frac{9}{10}$ **3.** $3\cdot 10^0 + 5\cdot 10^{-1} + 7\cdot 10^{-2} + 9\cdot 10^{-3} = 3 + \frac{5}{10} + \frac{7}{100} + \frac{9}{1000}$ **5.** $2\cdot 10^0 + 7\cdot 10^{-1} + 0\cdot 10^{-2} + 8\cdot 10^{-3} = 2 + \frac{7}{10} + \frac{8}{1000}$ **7.** $1\cdot 10^2 + 3\cdot 10^1 + 7\cdot 10^0 + 5\cdot 10^{-1} + 0\cdot 10^{-2} + 8\cdot 10^{-3} + 9\cdot 10^{-4} = 100 + 30 + 7 + \frac{5}{10} + \frac{8}{1000} + \frac{9}{10{,}000}$ **9.** $1\cdot 10^3 + 0\cdot 10^2 + 0\cdot 10^1 + 7\cdot 10^0 + 0\cdot 10^{-1} + 0\cdot 10^{-2} + 9\cdot 10^{-3} + 0\cdot 10^{-4} + 8\cdot 10^{-5} = 1000 + 7 + \frac{9}{1000} + \frac{8}{100{,}000}$ **11.** .2373 **13.** 50,300.00206 **15.** 5050.505 **17.** 9000.009 **19.** $\frac{34}{100}$ **21.** $\frac{2578}{100}$ **23.** $\frac{23{,}521}{100}$ **25.** $\frac{5{,}000{,}009}{10{,}000}$ **27.** $.34 = \frac{3}{10} + \frac{4}{100} = \frac{30}{100} + \frac{4}{100} = \frac{34}{100}$ **29.** $235.21 = 235 + \frac{2}{10} + \frac{1}{100} = \frac{23{,}500}{100} + \frac{20}{100} + \frac{1}{100} = \frac{23{,}521}{100}$

31.

```
   3917
 - 3600        90
    317
  - 280         7
     37
 = 370/10
 - 360/10     9/10
   10/10
 = 100/100
 - 80/100     2/100
   20/100
 = 200/1000
 - 200/1000   5/1000
      0
```

```
        5/1000
        2/100
        9/10
        7
        90
  40 | 3917
     - 3600
        317
      - 280
       370/10
     - 360/10
       100/100
      - 80/100
       200/1000
     - 200/1000
          0
```

```
        97.925
  40 | 3917.000
       360
        317
        280
         37 0
         36 0
          1 00
            80
           200
           200
             0
```

33. .1875 **35.** $4.\overline{36}$ **37.** 1.4 **39.** $.\overline{013}$

Section 10.5, page 372

1. 1.879; $\frac{13}{10} + \frac{579}{1000} = \frac{1300}{1000} + \frac{579}{1000} = \frac{1879}{1000}$ **3.** 34.0894 **5.** 23.73 **7.** 234.9146 **9.** 39.52605 **11.** .0004824 **13.** .30625 **15.** 3.788 **17.** 2.375 **19.** $\frac{9}{14}$

Section 10.6, page 376

1. .26 **3.** .572 **5.** 2.54 **7.** .0032 **9.** .00005 **11.** .005 **13.** $\frac{47}{100}$ **15.** $\frac{3}{2}$ **17.** $\frac{1}{200}$ **19.** $\frac{19}{500}$ **21.** $\frac{49}{300}$ **23.** $\frac{1}{10{,}000}$ **25.** 59% **27.** 2500% **29.** 300% **31.** $83\frac{1}{3}$% **33.** 3450% **35.** 75% **37.** 12.24 **39.** 568 **41.** 4.82 **43.** .75% **45.** 720 **47.** 27.5 **49.** 7.5% **51.** $p =$ \$38.10, $b =$ \$254, $r =$ 15%; \$215.90 **53.** $p =$ \$.05, $b =$ \$1.25, $r =$ 4% **55.** $p =$ 1394, $b =$ 16,400, $r =$ 8.5% **57.** $p =$ \$9240, $b =$ \$33,000, $r =$ 28% **59.** $p =$ \$5.92, $b =$ \$78.95, $r =$ 7.5%

Section 10.7, page 385

1. 7.6821×10^2 **3.** 7.6821×10^0 **5.** 9.854×10^6 **7.** 3.21×10^{-7} **9.** 12,370 **11.** .07332 **13.** .000004975 **15.** .000000578 **17.** .2, $\approx$ .4% **19.** 20, $\approx$ .5% **21.** 3 **23.** 6 **25.** 4 **27.** 44,000 $= 4.4 \times 10^4$; 43,500 $= 4.35 \times 10^4$; 43,500 $= 4.350 \times 10^4$ **29.** 4.7 $= 4.7 \times 10^0$; 4.68 $= 4.68 \times 10^0$; 4.679 $= 4.679 \times 10^0$ **31.** 2,400,000 $= 2.4 \times 10^6$; 2,380,000 $= 2.38 \times 10^6$; 2,379,000 $= 2.379 \times 10^6$ **33.** 2200 $= 2.2 \times 10^3$; 2200 $= 2.20 \times 10^3$; 2200 $= 2.200 \times 10^3$ **35.** .000054 $= 5.4 \times 10^{-5}$; .0000544 $= 5.44 \times \delta 10^{-5}$; .00005440 $= 5.440 \times 10^{-5}$ **37.** 50; (62,950, 63,050) **39.** .005; (3.135, 3.145) **41.** .005; (3.135, 3.145) **43.** 1360; (1355, 1365) **45.** .0025; (.00245, .00255) **47.** .002468; (.0024675, .0024685) **49.** 2.24×10^{-5} **51.** 5.799×10^3

Review Exercises, page 386

1. F **3.** T **5.** T **7.** T **9.** F **11.** $\frac{4}{5} = \frac{4}{5} \cdot 1 = \frac{4}{5} \cdot \frac{7}{7} = \frac{28}{35}$ **13.** lcd = 252; $\frac{59}{252}$ **15.** lcd = 105; $\frac{23}{105}$ **17.** $\frac{^-14}{15}$ **19.** $\frac{2}{25}$ **21.** $\frac{124}{10}$ **23.** $\frac{52}{1000}$ **25.** $\frac{3275}{100}$ **27.** 41.684 **29.** −4.452 **31.** 30.551 **33.** 43.863 **35.** .375 **37.** $.\overline{857142}$ **39.** \$450 **41.** 24 **43.** $\approx$ 13.6% **45.** $\frac{21}{20}$ **47.** $\frac{32{,}000}{7589}$ **49.** 2.41×10^2

Chapter 11

Section 11.1, page 390

1. Not characteristic. **3.** Not characteristic; the specific part is not specific enough. **5.** Not characteristic; the generic and specific parts are not clearly described. **7.** The definition of *circle* is not characteristic; the generic and specific parts are not clearly described. If you accept the first definition, however, then *center* is well-defined.

Section 11.2, page 395

3. Postulates 1–4 are as they appear at the beginning of this section. Postulate 5 says: "If **l** is a committee and P is a club member who does not belong to **l**, then there is exactly one committee to which P belongs that has no members in common with **l**." 7; 7; 3 **5.** There must be at least 3 *points*, 3 *lines*, and one *plane*. One picture is a triangle, with the vertices, sides, and triangle itself representing the *points*, *lines*, and *plane*, respectively.

Section 11.3, page 405

1. There are 5 lines; the names and points can vary: $\overleftrightarrow{AF}$, G; $\overleftrightarrow{AG}$, F; $\overleftrightarrow{BE}$, A; $\overleftrightarrow{GE}$, A; $\overleftrightarrow{GF}$, D

3. P R Q t

5. R P Q

7. By Postulate P-2, there are infinitely many points on **l**. Every point of **l** is on a distinct line containing Point A (by P-3). Therefore, there are infinitely many lines through A. **9.** W, T, U, V **11.** $\emptyset$ **13.** WVS, VUR, URQ, WPQ, QVS, RPW, SUT, RVW, SWT, PVU, RVT, QSU, QSW, RPT, RPV, SUW **15.** $\overleftrightarrow{PS}$, $\overleftrightarrow{RU}$; skew **17.** $\overleftrightarrow{PS}$ and $\overleftrightarrow{SV}$ **19.** no **21.** By Postulate T-2, the line **l** contains at least two points, say A and B. Let C be the given point not on the line. Then A, B, and C are noncollinear, so (by T-3) there is exactly one plane containing A, B, and C. By Postulate P-5, the line **l** is also in this plane. **23.** yes **25.** yes **27.** no **29.** no **31.** no **33.** T **35.** T **37.** T **39.** T **41.** T **43.** T

Section 11.4, page 415

1. 2 **3.** ray **5.** T is between R and S. **7.** on the same side of **l**. **9.** on opposite sides of **l**. **11.** (a) Y, J, R (c) $\overleftrightarrow{YZ}$ and $\overleftrightarrow{JK}$ (1) $\overleftrightarrow{GH}$ (3) R (5) $\overrightarrow{PA}$, $\overrightarrow{PQ}$, and $\overrightarrow{PF}$ (7) $\overline{HG}$ (9) $\overline{CD}$, $\overline{CS}$, and $\overline{SR}$ (11) $\overline{AP}$, $\overline{AQ}$ (13) Q **13.** $\overline{SD}$ **15.** A, B, D, H **17.** H, I **19.** C **21.** F, C **23.** E **25.** 6 **27.** From P-12: a half-plane and its boundary; two lines in a plane and all the points between the two lines. From P-13: a half-space and its boundary; two planes and all the points in space between the two planes. **29.** Each great circle of a sphere separates the sphere into two disjoint subsets, called *hemispheres*. The great circle, which is called the *boundary* of each hemisphere, is not a subset of either one. Yes.

Section 11.5, page 426

1. $\angle XPQ$, $\angle MPX$, $\angle MPT$, $\angle TPY$ **3.** $\angle XPQ$, $\angle TPY$ **5.** $\angle QPQ$, $\angle XPX$ **7.** M, T, Y **9.** $\angle XPQ$, $\angle MPT$ **11.** a, d, e, and h are simple closed curves; b, f, and g are not because they intersect themselves; c is not closed; you can't trace i without picking up the pencil.

13.

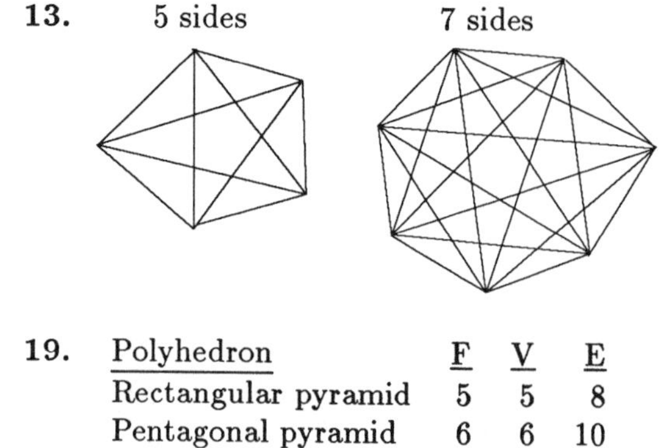

15. In a convex polygon, all the points of each diagonal (except the endpoints) are inside the polygon. **17.** Tetrahedron – triangular pyramid; pentahedra – rectangular pyramid, triangular prism, figure in (a); hexahedra – pentagonal pyramid, rectangular prism, figures in (2) and (4). The figure in (2) appears to be regular.

19.

Polyhedron	F	V	E
Rectangular pyramid	5	5	8
Pentagonal pyramid	6	6	10
Triangular prism	5	6	9
Rectangular prism	6	8	12

Review Exercises, page 430

1. F **3.** F **5.** F **7.** T **9.** F

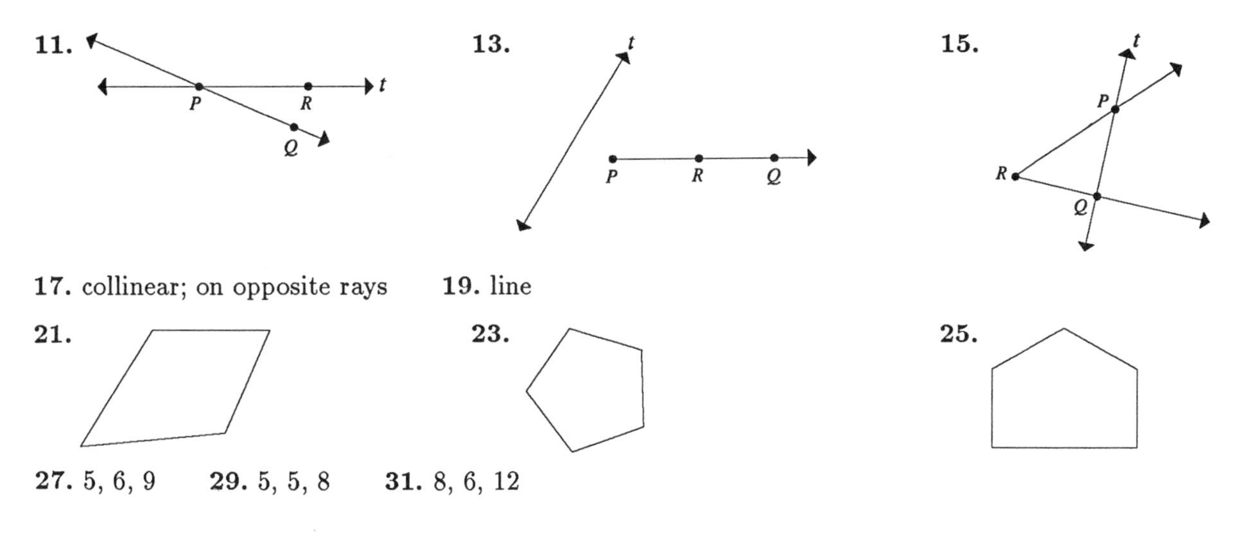

17. collinear; on opposite rays **19.** line

27. 5, 6, 9 **29.** 5, 5, 8 **31.** 8, 6, 12

Chapter 12

Section 12.1, page 439

1. 2 **3.** 6 **5.** 3 **7.** 2 **9.** $\overline{BC}$, $\overline{CD}$ **11.** $\overline{AC}$ **13.** the half-line on the A-side of point D. **15.** $\{x \mid -1 \leq x \leq 5\}$ **17.** $\{x \mid x \leq 2\}$ **19.** $\{x \mid 2 < x < 5\}$

Section 12.2, page 443

1. scalene **3.** isosceles **5.** 35 units **7.** 360 feet **9.** measurement **11.** line segment **13.** measurement **15.** 18.8 **17.** 13.6 **19.** .342 **21.** 21.4 **23.** minor arc **25.** semicircle **27.** major arc **29.** $\overline{XY}$, $\overline{AB}$ **31.** $\overline{XY}$, $\overline{AB}$, $\overline{BC}$, $\overline{BY}$, and $\overline{YC}$ **33.** (a) $3.\overline{142857}$ (b) The answer can only be accurate to at most 3 significant figures, which is the same as using 3.14. If you use 3.14159, the accuracy is at most 6 significant figures.

Section 12.3, page 455

1. m, 12, 20, 8, 28 **3.** $\angle POA$, $\angle AOP$ **5.** $\angle AOQ$, $\angle ROQ$ **7.** (These measures are estimates.) $\angle A$: acute, 45°, $\frac{\pi}{4}$ rad; $\angle B$: obtuse, 140°, $\frac{7\pi}{9}$ rad; $\angle C$: acute, 75°, $\frac{5\pi}{12}$ rad; $\angle D$: obtuse, 160°, $\frac{8\pi}{9}$ rad; $\angle E$: right, 90°, $\frac{\pi}{2}$ rad **9.** $\frac{\pi}{4}$ rad **11.** $\frac{5\pi}{6}$ rad **13.** 180° **15.** 108° **17.** 155° **19.** 55° **21.** $\pi - 3$ rad **23.** 68° **25.** $\frac{\pi}{4}$ rad **27.** $\frac{\pi}{2} - .5$ rad **29.** 2.5 rad **31.** 1 rad **33.** 3 **35.** $\approx$ 2.0 **37.** $\frac{18}{\pi}$ **39.** $\approx$ 1.73 **41.** $\frac{\pi^2}{2}$ **43.** 5.2 **45.** 24.77 **47.** no **49.** yes **51.** yes **53.** yes **55.** yes **57.** no

Section 12.4, page 466

1. 5 m, 2 dm, 7 cm, and 6 mm **3.** 3 dm, 6 cm, and 4 mm **5.** 3 g, 7 dg, 6 cg, and 8 mg **7.** 7.35 m **9.** 4.002 m **11.** 62.5 **13.** 4120 **15.** 75.4 **17.** 3.251 **19.** 214,600 **21.** 23,500 **23.** .0127 **25.** 34,670 **27.** 657,200 **29.** .0000324 **31.** 23,400 **33.** .00005723 **35.** .000025895 **37.** 5.7×10^7 **39.** 3.67 **41.** 2.698×10^{13} **43.** 3 g **45.** 2.58×10^5 kg **47.** 46 g **49.** 2000 cm^3; 2 l **51.** .0785 cm^3; .0785 ml **53.** 2.17 cm^3; 2.17 ml **55.** approximately 38°, 100°, 311 **57.** approximately 50°, 120°, 323 **59.** approximately 2°, 35°, 275

Section 12.5, page 477

1. 148°; they are supplementary. **3.** 58°; they are complementary. **5.** $\overline{VU}$; $\overline{VW}$; $\overline{WU}$ **7.** $\angle QPR$ **9.** (a) x°; congruent angles have equal measure. (b) $180° - x°$; definition of supplementary angles (c) $180° - x°$; definition of supplementary angles (d) They have equal measure. **11.** (b) ASA (c) no **13.** no **15.** Given: $\triangle ABC$ with $\angle A \cong \angle B$. Then $\angle B \cong \angle A$ by the symmetric property, and $\overline{AB} \cong \overline{BA}$ by the reflexive property. Then $\triangle ABC \cong \triangle BAC$ by ASA, and therefore $\overline{AC} \cong \overline{BC}$. **17.** No; consider a rhombus. **19.** If two lines intersect to form congruent adjacent angles, then (by T-7) these angles are supplementary. Since they are congruent, they have equal measure. Therefore, the measure of each must be 90°; that is, they are right angles, so the lines are perpendicular.

Section 12.6, page 483

1. T-17 **3.** T-23 **5.** T-19 **7.** Suppose that **l** and **m** are two lines cut by a transversal **t**, and that $\angle 1$ and $\angle 2$ are congruent corresponding angles with $\angle 1$ an exterior angle. Let $\angle 3$ be the angle vertical to $\angle 1$, so that $\angle 3 \cong \angle 1$. By transitivity, $\angle 3 \cong \angle 2$. Since $\angle 3$ and $\angle 2$ are alternate interior angles, the lines are parallel, by T-17. **9.** Let $\angle 1$ and $\angle 2$ be a pair of corresponding angles with $\angle 1$ an exterior angle. Let Let $\angle 3$ be the angle vertical to $\angle 1$, so that $\angle 3 \cong \angle 1$. Since $\angle 2$ and $\angle 3$ are alternate interior angles, $\angle 3 \cong \angle 2$ by T-21. Then $\angle 1 \cong \angle 2$ by transitivity. **11.** If two lines are cut by a transversal so that a pair of alternate exterior angles are congruent, then the lines are parallel. *Proof*: Suppose that **l** and **m** are two lines cut by a transversal **t**, and that $\angle 1$ and $\angle 2$ are congruent alternate exterior angles. Let $\angle 3$ be the angle vertical to $\angle 1$, so that $\angle 3 \cong \angle 1$. By transitivity, $\angle 3 \cong \angle 2$. Since $\angle 3$ and $\angle 2$ are corresponding angles, the lines are parallel by T-18. (*Note*: The converse of this statement is another correct answer to the question.) **13.** . . . two interior angles on the same side of the transversal are not supplementary, then the lines are not parallel (that is, they intersect).

Section 12.7, page 489

1. E, L, R **3.** E **5.** E, L, R **7.** E, L **9.** E, R **11.** E, R **13.** E, L **15.** R, L **17.** E, L **19.** Suppose that $\angle 1$ and $\angle 2$ are alternate interior angles. Let $\angle 3$ be vertical to $\angle 1$, so that $\angle 3 \cong \angle 1$. Then $\angle 2$ and $\angle 3$ are corresponding angles and, by T-22, $\angle 2 \cong \angle 3$. By transitivity, $\angle 1 \cong \angle 2$. **21.** By Exercises 18 and 19, T-21 is equivalent to T-22. By Exercise 20, T-22 is equivalent to T-23. Then T-21 is equivalent to T-23 by transitivity. (Equivalence of theorems is an equivalence relation.)

Review Exercises, page 490

1. F **3.** F **5.** F **7.** F **9.** F **11.** $\overline{CD}$ **13.** 10 **15.** A, F **17.** $\overrightarrow{BA}$ **19.** isosceles **21.** $\frac{\pi}{180}$ **23.** kilo-, milli- **25.** vertical **27.** greater than **29.** 0°, 100° **31.** meter **33.** 5250 **35.** 472,000 **37.** .00000895 **39.** 19.5 **41.** 8.61 **43.** 20° **45.** $\angle SNE$ **47.** $\angle QNE$, $\angle QNU$ **49.** 120° **51.** 47.8

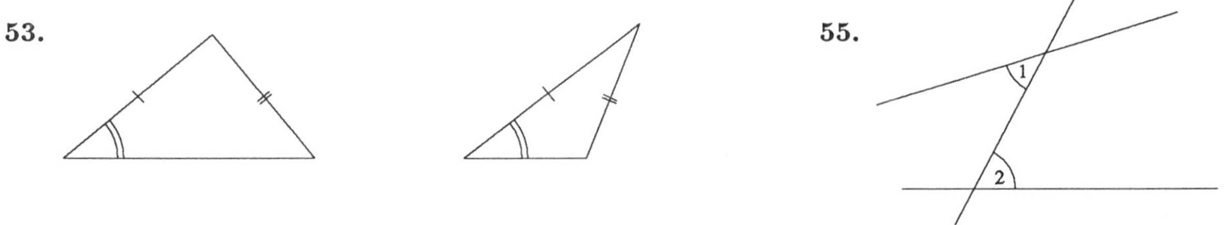

Chapter 13

Section 13.1, page 496

1. The prime factors of n^2 are exactly those of n. (Why?) Thus, if 2 divides n^2, it must also divide n, so n is even. **3.** Substitute p for 2 in the answer for Exercise 1. **5.** If $5\sqrt{2}$ were rational, then $\frac{1}{5} \cdot 5\sqrt{2} = \sqrt{2}$ would also be rational (by closure of multiplication in **Q**), contradicting a known result. **7.** Consider $n^{-1} \cdot n\sqrt{2}$. **9.** Let h be the length of the hypotenuse (in centimeters). By the Pythagorean Theorem, $h^2 = 5$. Assume $h = \frac{a}{b}$ in lowest terms, then show algebraically that 5 must be a factor of both a and b; contradiction.

Section 13.2, page 503

1. $\frac{10}{32}$; 0, 1, 0, 1, 0 **3.** $\frac{19}{32}$; 1, 0, 0, 1, 1 **5.** $\frac{5}{32}$; 0, 0, 1, 0, 1 **7.** $\frac{409}{512}$; 1, 1, 0, 0, 1, 1, 0, 0, 1 **9.** $\frac{455}{512}$; 1, 1, 1, 0, 0, 0, 1, 1, 1 **11.** $\frac{27}{16}$; $\frac{886}{512}$; 1., 1, 0, 1, 1, 1, 0, 1, 1, 0 **13.** $\frac{10}{8}$; $\frac{161}{128}$; 1., 0, 1, 0, 0, 0, 0, 1

Section 13.3, page 508

1. 1.41421; $\left[\frac{141{,}421}{100{,}000}, \frac{141{,}422}{100{,}000}\right]$ **3.** $\frac{2645}{1000} = 2.645$ **5.** $\frac{3142}{1000} = 3.142$ **7.** $\frac{1259}{1000} = 1.259$ **9.** (a) The decimal-fraction points are a dense subset of the number line (13.4), and every power of 10 is a multiple of 10. (b) Every multiple of 5 is a multiple of 10; use part (a).

Section 13.4, page 513

1. $\frac{1}{1000}$; yes **3.** $\frac{0}{1000}$; yes **5.** $\frac{1}{10}\left(= \frac{100}{1000}\right)$; no **7.** $\frac{0}{1000}$; no — The step from 9.1 to 9.2 is too big. **9.** $\frac{2}{1000}$; yes **13.** $\frac{1}{2}$: 0, $\frac{5}{10}$, ...; $\frac{2}{3}$: 0, $\frac{6}{10}$, ... **15.** $\frac{3}{4}$: 0, $\frac{7}{10}$, $\frac{75}{100}$, $\frac{750}{1000}$, $\frac{7500}{10{,}000}$, ...; $\frac{743}{990}$: 0, $\frac{7}{10}$, $\frac{75}{100}$, $\frac{750}{1000}$, $\frac{7505}{10{,}000}$, ... **17.** (a) 2, 2.5, 2.53, 2.533, 2.5333, ...; 6, 6.5, 6.53, 6.533, 6.5333, ... (b) They differ first at s_0. (c) [3, 5] **19.** (a) 1, 1.8, 1.84, 1.849, 1.8499, ...; 1, 1.8, 1.85, 1.850, 1.8500, ... (b) They differ first at s_2. (c) There is no such interval. **21.** 1.999 ... **23.** 6.458999 ...

Section 13.5, page 522

1. max. = 4, actual = 1 **3.** max. = 36, actual = 3 **5.** max. = 6, actual = 6 **7.** max. = 13, actual = 0 **9.** $.\overline{1}$ **11.** $2.1\overline{6}$ **13.** $.\overline{45}$ **15.** $1.\overline{428571}$ **17.** $3.\overline{027}$ **19.** $\frac{1}{45}$ **21.** 1 **23.** $\frac{5}{33}$ **25.** $\frac{10259}{165}$ **27.** $\frac{40}{99}$ **29.** $\frac{101}{999}$ **31.** $.\overline{3}$ **33.** $1.\overline{1}$ **35.** .3030030003 ... **37.** $.\overline{0141}$, $.0\overline{141}$, .141, $.\overline{141}$, $.1\overline{41}$, 1.41, $1.4\overline{1}$, 1.414, $1.\overline{41}$, $\sqrt{2}$

Section 13.6, page 528

1. $(0,2)$ **3.** $(2,1)$ **5.** $(\sqrt{3},0)$ **7.** $(2,\sqrt{2})$ **9.** $(3,8)$ **11.** $(\frac{19}{3},2)$ **13.** $(7,4)$ **15.** $(5,2) \oplus (0,0) = (5+0, 2+0) = (5,2)$ **17.** $(a,b) \oplus (0,0) = (a+0, b+0) = (a,b)$ **19.** $(a,b) \odot (0,0) = (a \cdot 0, b \cdot 0) = (0,0)$ **25.** (a) $(-2,-3)$ (b) $(3,-2)$ **27.** $(a,0) \oplus [(b,0) \odot (0,1)] = (a,0) \oplus (b \cdot 0 - 0 \cdot 1, b \cdot 1 + 0 \cdot 0) = (a,0) \oplus (0,b) = (a,b)$

Section 13.7, page 532

1. $4(5) - 20 = 20 - 20 = 0$ **3.** $8(\frac{1}{4})^2 - 6(\frac{1}{4}) + 1 = \frac{8}{16} - \frac{6}{4} + 1 = 0$ **5.** $(-7i)^2 + 2i(-7i) + 35 = -49 + 14 + 35 = 0$ **7.** **I** **9.** **Q** **11.** **C** **13.** **R** **15.** $x^2 + 3x + 2 = 0$ **17.** $x^2 + x - 1 = 0$

Review Exercises, page 532

1. F **3.** T **5.** F **7.** T **9.** F **11.** $\frac{3}{8}$ **13.** $\frac{71}{32}$ **15.** $\frac{875}{1000} = .875$ **17.** $.\overline{4}$ **19.** 2.5 $= 2.4\overline{9}$ **21.** $3.\overline{142857}$ **23.** $\frac{14}{9}$ **25.** $\frac{1}{33}$ **27.** $\frac{235}{999}$ **31.** $8.\overline{4}$ **33.** $.\overline{5}$ **35.** $(-6,0)$ **37.** $(3,\sqrt{5})$ **39.** $(4,0)$ **41.** $(-2i)^2+4 = 4\cdot(-1)+4 = 0$

Chapter 14

Section 14.2, page 543

1. $\{hhh,\ hht,\ hth,\ thh,\ htt,\ tht,\ tth,\ ttt\}$ **3.** $P(E) = \frac{1}{2}$, $P(F) = \frac{1}{2}$, $P(G) = \frac{1}{4}$ **5.** $\{hhh,\ hht,\ hth,\ tht,\ tth,\ ttt\}$; $\frac{3}{4}$ **7.** $\{hhh,\ hth,\ tht,\ ttt,\ htt\}$; $\frac{5}{8}$ **9.** $E = \{2h, 4h, 6h, 2t, 4t, 6t\}$; $F = \{5h, 5t\}$; $G = \{1h, 2h, 3h, 4h, 5h, 6h\}$, $H = \{1t, 2t, 3t\}$ **11.** $\{2h, 4h, 6h\}$; $\frac{1}{4}$ **13.** There are 36 of them—all ordered pairs of the numbers 1, 2, 3, 4, 5, 6. **15.** $E = \{(1,1)\}$; $F = \{(1,6), (2,5), (3,4), (4,3), (5,2), (6,1)\}$; $G = \{(1,1), (1,3), (1,5), (2,2), (2,4), (2,6), (3,1), (3,3), (3,5), (4,2), (4,4), (4,6), (5,1), (5,3), (5,5), (6,2), (6,4), (6,6)\}$; $H = \{(1,2), (1,3), (2,2), (3,1), (2,1)\}$ **17.** None; $\{(1,1), (1,6), (6,1), (2,5), (5,2), (3,4), (4,3)\}$ **19.** $G \cap H = \{(1,1), (1,3), (2,2), (3,1)\}$; $G \cup H = \{(1,1), (1,2), (1,3), (1,5), (2,1), (2,2), (2,4), (2,6), (3,1), (3,3), (3,5), (4,2), (4,4), (4,6), (5,1), (5,3), (5,5), (6,2), (6,4), (6,6)\}$ **21.** There are 64 of them—all ordered triples of the numbers 1, 2, 3, 4. **23.** $E = \emptyset$; $F = \{(1,2,4), (2,1,4), (1,4,2), (2,4,1), (4,1,2), (4,2,1), (1,3,3), (3,1,3), (3,3,1), (2,2,3), (2,3,2), (3,2,2)\}$; $G = \{(1,2,3), (2,1,3), (1,3,2), (2,3,1), (3,1,2), (3,2,1), (1,4,3), (4,1,3), (1,3,4), (4,3,1), (3,1,4), (3,4,1), (1,2,1), (1,1,2), (2,1,1), (1,4,1), (1,1,4), (4,1,1), (3,2,3), (3,3,2), (2,3,3), (3,4,3), (3,3,4), (4,3,3), (2,2,4), (2,4,2), (4,2,2), (2,2,2), (2,4,4), (4,2,4), (4,4,2), (4,4,4)\}$; $H = \{(1,1,1), (1,1,2), (1,2,1), (2,1,1)\}$ **25.** $E \cap F = \emptyset$; $E \cup F = F$ **27.** $G \cap H = \{(1,1,2), (1,2,1), (2,1,1)\}$; $G \cup H = G \cup \{(1,1,1)\}$ **29.** $\frac{1}{13}$ **31.** $\frac{1}{52}$ **33.** $P(E) = \frac{1}{7}$; $P(F) = \frac{2}{7}$; $P(G) = \frac{2}{21}$

Section 14.3, page 553

1. 0 **3.** .55 **5.** 0 **7.** .30 **9.** 75% **11.** 15% **13.** 85% **15.** They do not add up to 1. **17.** Then $P(E \cup F) > 1$. **19.** $P(F) = .30$, but $E \subseteq F$ implies $P(E) \le P(F)$. **21.** $\frac{5}{40} = \frac{1}{8}$ **23.** $1 - \frac{1}{8} = \frac{7}{8}$ **25.** $\frac{27}{40}$ **27.** $\frac{13}{52} = \frac{1}{4}$ **29.** $\frac{13}{52} = \frac{1}{4}$ **31.** $\frac{3}{52}$ **33.** $\frac{26}{52} = \frac{1}{2}$ **35.** $\frac{26}{52} = \frac{1}{2}$ **37.** 40 **39.** 50 **41.** $\frac{2}{5}$

Section 14.4, page 561

1. 4; 3; 1; 9; 2.7; 9.6; 3.1 **3.** 5.0; 4.6; each number (no useful mode); 10.5; 3.0; 12.5; 3.5 **5.** 693.3; 665; each number (no useful mode); 600; 160; 39,022.2; 197.5 **7.** (a) $m = 4$ (b) $-3-1+3+0+1+0+5-3-2=0$ (c) Denote the elements of the sample by $a_1, \ldots, a_n$ and the mean by m. Then the sum of the deviations is $(a_1 - m) + \ldots + (a_n - m) = (a_1 + \ldots + a_n) - n \cdot m = (a_1 + \ldots + a_n) - n \cdot \left(\frac{a_1 + \ldots + a_n}{n}\right)$ (by the definition of mean), and this equals 0. **9.** 0, 0, 0, 5, 10, 10, 10 and 0, 5, 5, 5, 5, 5, 10 work. **11.** The mean cannot be found; median = \$25,000; mode = \$25,000. The median is probably the most revealing of the three because it indicates a minimum pay for half the teachers in the system. (Other defensible answers are possible.)

Section 14.5. page 568

1. 2040 **3.** $(1845, 2235)$ **5.** $(1783, 2297)$ **7.** The maximum error would be smaller, making the confidence interval smaller than both of these. **9.** If many samples were randomly drawn from this population, 99.5% of them would fall within the given interval. **11.** 126.5

Review Exercises, page 570

1. F **3.** F **5.** F **7.** F **9.** T **11.** experiment; outcomes; outcomes; sample space **13.** mutually exclusive **15.** central tendency **17.** inferential **19.** point estimate **21.** There are 64 of them—all possible ordered pairs of the numbers 1, 2, ..., 8. **23.** $E = \{(1,7), (2,6), (3,5), (4,4), (5,3), (6,2), (7,1)\}$; $F = \{(1,8), (2,7), (3,6), (4,5), (5,4), (6,3), (7,2), (8,1)\}$; $G = \{(1,1), (1,2), (2,1), (1,4), (2,3), (3,2), (4,1), (1,6), (2,5), (3,4), (4,3), (5,2), (6,1), (3,8), (4,7), (5,6), (6,5), (7,4), (8,3), (5,8), (6,7), (7,6), (8,5)\}$; $H = \{(5,8), (6,7), (7,6), (8,5), (6,8), (7,7), (8,6), (7,8), (8,7), (8,8)\}$ **25.** $G \cup H = \{(1,1), (1,2), (2,1), (1,4), (2,3), (3,2), (4,1), (1,6), (2,5), (3,4), (4,3), (5,2), (6,1), (3,8), (4,7), (5,6), (6,5), (7,4), (8,3), (5,8), (6,7), (7,6), (8,5), (6,8), (7,7), (8,6), (7,8), (8,7), (8,8)\}$; $G \cap H = \{(5,8), (6,7), (7,6), (8,5)\}$ **27.** $E \cup G = G$; $E \cap G = \emptyset$ **29.** .84 **31.** .52 **33.** .36 **35.** 64.25% **37.** 65% **39.** 29.25% **41.** 4.2; 4; 4; 8; 1.9; 5.6; 2.4 **43.** 15.4 **45.** (11.3, 19.5) **47.** It would be larger.

Index